D0787004

The
Functional
and
Evolutionary
Biology
of
Primates

The Functional and Evolutionary Biology of Primates

Edited by

Russell Tuttle

ALDINE · ATHERTON
Chicago/New York

THE EDITOR

Russell Tuttle is Associate Professor of Anthropology and Evolutionary Biology at the University of Chicago. He organized the 1970 Burg-Wartenstein Symposium on "Functional and Evolutionary Biology of Primates: Methods of Study and Recent Advances," on which this book is based. Professor Tuttle received his Ph.D. from the University of California at Berkeley in 1965, and specializes in research in the history and theory of human evolution, primate behavior, and comparative functional morphology. His contributions to the literature have appeared in the *American Journal of Physical Anthropology, Science, Science Journal,* and other publications.

First published 1972 by
Aldine · Atherton, Inc.
529 South Wabash Avenue
Chicago, Illinois 60605

ISBN 0–202–02011–8
Library of Congress Catalog Number 72–169508

Printed in the United States of America

to
THEODORE D. McCOWN

Introduction

Burg Wartenstein Symposium No. 48 on *Functional and Evolutionary Biology of Primates: Methods of Study and Recent Advances* convened July 18–26, 1970, at the European Conference Center of the Wenner-Gren Foundation for Anthropological Research near Gloggnitz, Austria. Twenty-one scientists participated in six daily sessions, and Professor Adolph H. Schultz honored us with an evening lecture on the history, progress, and prospects of primatological research.

Invitations to the conference contained notice of the topics that I wished to be developed by each participant. The invitations also listed the following "statement of general purpose":

> The major objective of the conference is to bring several perspectives to bear on a variety of problems in evolutionary primatology. We hope to emphasize experimental design, quantification, and the strategies of the comparative method and intensive longitudinal studies in morphology and behavior. Since the most convincing proofs of the validity of a particular technique or approach are the quality of results obtained through its application and the credibility of inferences that may be based on such studies, participants are asked to emphasize the results and implications of their research for basic theoretical problems.

Subsequently, I wrote to each participant, suggesting more specifically the nature of the expositions that they might attempt on the problems selected for discussion at the conference.

Five broad groups of basic subject matter on functional and evolutionary primatology were represented at the symposium: paleoprimatology; cranial morphology; comparative neurobiology and endocasts; postcranial morphology; and aspects of behavior and ecology. In order to treat these five subject areas with an acceptable level of thoroughness, we were regrettably

induced to omit detailed representation of certain other important disciplines, such as biomolecular primatology and prehistory which have been
discussed at length in several previous conferences.

Paleoprimatology provides the only direct evidence for detailing the
phylogenetic histories of primate lineages. The chronology of evolutionary
events is wholly dependent upon accurately dated fossils. But because of the
fragmentary nature of most fossils, postdeposition perturbations at fossil
sites, limited methods of recovery and analysis, the complexity and diversity
of datable minerals and their associations with fossils, and many other com-
plicating factors, paleoprimatologists are somewhat limited on the empirical
data that they can extract from available fossils. Thus, like other comparative evolutionary biologists, they must depend heavily upon inference in
constructing models and hypotheses on the course of evolution and the life
habits of principals in the primate career. In order to establish broader bases
of inference and more complete profiles on hypothetical form and functions
in extinct species, paleoprimatologists often consult other branches of comparative evolutionary biology in addition to searching constantly for more
fossils.

The difficulties that beset the evolutionary primatologist who attempts to
infer the mechanical and genetical factors that may be causally related to
features of bones and teeth in extant primates are increased many-fold
should he turn his attention to extinct species, upon which most kinds of
experiments and behavioral observations cannot be conducted. Further, not
only the incompleteness of certain fossils but also their sparse representation
in available collections may limit the scope of information that may be recovered from the fossils themselves and the nature of quantitative comparisons between them and other taxa.

Among evolutionary primatological sciences, investigations on cranial
morphology occupy a preeminent position for the documentation of taxonomic affinities among living and fossil species. Except for certain aspects
of the masticatory apparatus, orbital dimensions, and perhaps some neurocranial dimensions, many features of the skull evidence negligible detectable
functional correlates. It is therefore commonly believed among evolutionary
biologists that the postcranial skeleton may reflect more fully the functional
and ecological particularities of a species than its skull can.

Whereas selected features of primate skulls may be expressed quantitatively and persist in fossilized specimens, certain soft structures of the head,
in particular the facial muscles, are generally difficult to measure in living
forms and leave little impression on underlying bony features. Yet it is
these perishable structures which are employed by taxonomists to distinguish
colobine from cercopithecine monkeys.

One of the greatest sets of challenges that face comparative primatologists
and anthropologists is the precise description and evolutionary explanation
of particular morphological features in carefully chosen series of primate

brains. The evolution of the neurological bases of human speech, tool-making capacities, and other symbolic behaviors also remain poorly eluci-dated today.

Two basic kinds of evidence may be brought to bear on problems of brain evolution in primate lineages, *viz.,* results of comparative studies on brains of extant species and comparative studies on fossil and extant primate endocasts. Both approaches entail a great amount of painstaking effort in order to accumulate representative comparative and evolutionary series. Comparative neurobiologists and paleoneurobiologists must possess a broad knowledge of neurophysiology, naturalistic behavior, and ecology in living primates in order to explain the results of their studies in functional and evolutionary contexts.

Whereas comparative neurobiologists studying extant primates may ex-amine both external and internal neural features, the paleoneurobiologist is restricted to available surface features of naturally occurring fossil endocasts and endocasts prepared in the laboratory from fossil crania. There are no known instances of brains themselves being fossilized such that paleohis-tological studies could be conducted on them. The occurrence of such fossil specimens is not to be anticipated due to the paucity of supporting connec-tive tissues and concordant rapid postmortem degeneration of the brain.

Studies on the postcranial morphology of primates and especially investi-gations on primate locomotive systems have long been of central interest to evolutionary anthropologists and other biologists. During the past two decades, a profusion of experimental and analytical techniques have been developed and refined which not only permit but also necessitate their several employment in functional and evolutionary studies on primate loco-motion and other motile behaviors.

Studies on the morphology of extant and fossil primates, in view of biomechanical principles known to be operant in living systems, may gen-erate credible hypotheses on the functions of selected parts of an individual and perhaps on certain aspects of its life habits. But functional models based on morphology alone are greatly limited in scope and often remain equivocal until they are confirmed, revised, or rejected on the basis of naturalistic behavioral observations, laboratory behavioral studies, or both.

The increased number and intensity of behavioral studies on nonhuman primates, though unfortunately rarely focused on posture and locomotion, are providing somewhat firmer bases for functional interpretations of mor-phological features than were possible during the more strictly anecdotal era of primate field studies.

Direct observations of behavior and ecology are impossible for extinct organisms, so we must carefully delimit the nature of variations in the be-havior and ecology of living primates and attempt to determine the extent to which uniformitarian arguments about these features might be employed to formulate refined models on the life modes of each fossil species.

It is imperative that the ecologies of all species be thoroughly surveyed with regard to seasonal and geographic variations, and that longitudinal studies be conducted on the population dynamics, aging, ontogeny of loco-motive behaviors and other behavior patterns, responses to environmental perturbations, and many other features, sometimes dramatic but more often subtle, in free-ranging primates. Such documentation is especially important if we intend to utilize knowledge of living primates to interpret evolutionary events.

Further, in many instances, including most aspects of social behavior, we are virtually entirely dependent upon living primates to provide bases of evolutionary inference and a comparative background for understanding human behavior.

Among evolutionary primatologists it is perhaps the behavioralists who face the greatest set of challenges regarding the selection of most meaningful constituents in a total suite of features and the means of recording, quanti-fying, and transforming data so that proper inferences and intergroup com-parisons may be made. To a considerable extent, the process of deposition and fossilization selects the bits that the paleoprimatologist may study. The comparative primate morphologist has far greater freedom of choice than the paleoprimatologist in the materials that he may elect to study and the methods that he may employ on the same or different species. For instance, certain procedures which destroy materials may be conducted by compara-tive primate morphologists since their specimens commonly are not unique. Both the paleoprimatologist and the primate morphologist may study avail-able specimens at leisurely rates and return repeatedly to the same or very similar contexts for further examination. They can have tactile contact with most materials.

By contrast, behavioralists, especially those working in free-ranging con-texts, must contend with a series of passing events, sometimes characterized by imperceptible subtlety and speed, which can only be revisited if captured on film and sound recordings or by fortuitous replications of the same event under nearly identical circumstances. Tactile perceptions by the observer of behavioral bits are generally impossible, but he may utilize auditory and olfactory senses, neither of which are employed commonly by paleoprima-tologists and, to a lesser extent, by comparative morphologists during their particular investigations.

I have arranged the 19 papers in this volume in a manner that would group the major types of subject matter together, since many readers may have restricted special interests in fossil materials, cranial morphology, com-parative neurobiology, primate locomotive systems, or behavior and ecology.

This is not the order in which the papers were presented at the confer-ence. At Burg Wartenstein, I juxtaposed papers that exhibited contrasting perspectives and approaches in order to provoke interdisciplinary discus-sions. Certain readers of this volume also may wish, for example, to first peruse the comparative morphological papers that describe methods for elucidating functional problems, and then consult the papers on paleoprimatology to see what materials are in fact available to provide directly phylogenetic perspectives.

Readers who have special interests in multivariate statistical applications to biological problems and mathematical modeling may wish to read papers by Howells (Chapter 5), Oxnard (Chapter 14), and Cohen (Chapter 19) as a contiguous set in order to contrast the employment of such methods on different materials.

In the three papers constituting Part I, Frederick S. Szalay, Elwyn L. Simons and David R. Pilbeam, and Phillip V. Tobias survey the available evidence and prospects for additional recoveries of Paleogene primates, Neogene primates, and early hominids, respectively.

In Chapter 1, Szalay updates knowledge on the earliest known primates, discusses their morphological particularities, and suggests possible niches occupied by certain species of each family. Szalay stresses that careful taxonomic assessments of fossils are imperatively prerequisite to the construction of phylogenies and to placing the results of functional analyses on fossils in an evolutionary perspective. He also discusses the knotty problem of identifying the earliest primates among the total suite of Cretaceous and Paleocene mammals. In particular, he discusses middle ear morphology in prosimian primates and demonstrates that past evolutionary inferences based on this complex region have been oversimplified and misleading.

Finally, Szalay presents a critique of the Napier-Walker hypothesis on the purported vertical clinging and leaping adaptations of all Eocene primates. Szalay particularly criticizes their employment of the paleontological literature in the construction of a vertical clinging and leaping model.

In Chapter 2, Simons and Pilbeam state a convincing case for reopening and excavating "old" sites in order to reclaim more complete remains and greater numbers of specimens of known fossil taxa instead of virtually ignoring these localities while continuing to search for new superlatives elsewhere. Further systematic research is requisite at many "old" localities in order to properly date specimens that were previously recovered from them at times when paleontological techniques were less sophisticated than they are today.

Simons and Pilbeam discuss certain morphological features and concordant functional and evolutionary inferences on the products of Simons' diligent, well organized paleontological studies in the Fayum of Egypt and the Siwalik Hills region of the Indian subcontinent. They also provide an updated survey on our knowledge of available Neogene primates. Finally, they briefly review perspectives on the relationships of Paleogene primates of the Eastern and Western hemispheres and suggest certain experimental studies that might elucidate problems of special interest to paleoprimatologists.

In Chapter 3, Tobias reviews the status of ongoing research on early hominids of eastern and southern Africa and catalogues available specimens, many of which have been discovered quite recently due to the persistent efforts of the Leakey family in Kenya, Tanzania, and Ethiopia and two international multidisciplinary research teams directed by F. C. Howell, Y. Coppens, and the late C. Arambourg in the Omo River Valley, Ethiopia. Recent advances in geochronological information also are discussed by Tobias. He pinpoints anatomical regions that are poorly, substantially, and intermediately represented in the fossil record of early Hominidae and thereby directs the attention of comparative evolutionary biologists, who may not be well acquainted with these specimens, to anatomical regions upon which they might contribute bases of inference by conducting detailed studies on counterpart features in extant primates.

Tobias asks that specialists cooperate in multidisciplinary studies on early hominids both in the field and subsequently in the laboratory. He mentions a number of ongoing or recently completed studies on diverse aspects of australopithecine functional and evolutionary biology.

Tobias submits a critique on the theory recently expounded by B. G. Campbell that the "robust" and "gracile" australopithecines are simply males and females, respectively, of one species instead of representing different hominid species. This leads him to a discussion on the status of research on sexual dimorphism in extant hominoid primates. He closes with a list of unanswered questions about australopithecine functional and evolutionary biology that remain to perplex and otherwise challenge the scientific community.

In Part II, Matt Cartmill (Chapter 4) and W. W. Howells (Chapter 5) reevaluate several problems on cranial morphology of long-standing interest to anthropologists. They not only apply sophisticated mathematical techniques to extensive series of specimens, thereby elucidating patterns of intra- and interpopulation variation, but also employ a broad set of perspectives that enable them to transcend their techniques and render their data into models that should be of considerable interest to evolutionary anthropologists and other biological scientists.

Cartmill critically reviews the Smith-Jones hypothesis that many features of primate cranial morphology are adaptations to special vicissitudes of arboreal habitation. He applies the comparative strategy of searching for convergent morphological trends in counterpart features of the skull in nonprimate arboreal mammals.

On the basis of his comprehensive biometrical studies on orbital convergence and rostral reduction and the synthesis of these morphological data with knowledge obtained from a thorough survey of the natural historical literature on numerous species of small mammals, Cartmill concludes that instead of the arboreal habitat *per se* having selected for certain primate trends, it was selection for eye-hand coordination, as part of an adaptive complex for the manual capture of nocturnal insect prey, that led to the emergence of tarsian and anthropoid cranial trends during the Eocene period. Cartmill rejects the notion of universality in Napier and Walker's vertical clinging and leaping model of Eocene primates.

Further, by contrast with Elliot Smith, F. Wood Jones, Hermann Klaatsch, and other anatomists, Cartmill concludes that the predecessors of arboreal Eocene primates were probably forest-floor insectivoran predators. He suggests that the Paleocene progenitors of Eocene primates are undiscovered or unrecognized among fossil forms. Cartmill removes the microsyopoids (plesiadapids, picrodontids, microsyopids) and perhaps also the paromonyids from the Primates and relegates them to the Insectivora.

Howells employs two multivariate statistical methods—factor analysis and multiple discriminant functions—in order to determine major sources of intrapopulation variation among cranial measurements in man and the chief distinctions between human populations in cranial morphology, respectively. Further, the population differences observed in "functions" appear to be compounds of "factors" found within the study populations. Thus in Howells' study the patterns of variation within populations correspond with those between populations. Such a result is not necessarily expected. It is therefore particularly challenging for evolutionary anthropologists to determine possible causation underlying this association.

Howells provides a lucid review of the strategies, prospective applications, and limitations of each technique, supplying abundant examples from his own extensive studies on the skulls of historical and prehistorical human populations.

The comparability of results that Howells obtained in separate discriminant function analyses of males and females of available human populations greatly enhances the credibility of his methods and the inferences and hypotheses that may be based upon them. Further, skulls drawn from the same or similar populations were correctly allocated to these populations in a high proportion of cases. Thus Howells' application of discriminant function analysis holds considerable promise for indicating the "racial" affinities of prehistoric skulls of *Homo sapiens* and other fossil men.

When Howells applied discriminant function scoring to three Neanderthal skulls, they either fell at an extreme limit or completely outside of the male range of certain functions of *Homo sapiens,* and further specified those regions of the skull where the major differences were situated. This points the way to future experimental, functional, and genetical studies on the adaptive and geographical group differences between Neanderthal man and modern man.

In Part III, Heinz Stephan (Chapter 6), presents an overview of recent efforts to express relative brain weight and to quantify certain internal structures of the brain in a broad comparative series of primates and insectivores; Leonard Radinsky (Chapter 7) proffers some functional and evolutionary inferences based on his studies of primate endocasts; and Ralph Holloway (Chapter 8) presents results and inferences from his exciting new studies on early hominid endocasts, and reviews the status of neurobiological information that might be employed to construct models of hominid brain evolution and behavior.

Stephan and his associates have developed a method for expressing a relationship between brain weight and body weight in primates and insectivores on a double logarithmic scale in a manner that may correct for certain allometric factors and thus permit quantitative assessments of "encephalization" (e.g. brain size increase above the level evidenced in "basal" insectivores) in primates and other mammals. This approach is more promising for evolutionary biological studies than previous estimates of relative brain size based on methods that do not attempt to correct for allometry. Stephan also presents estimates on relative development of neocortex and subcortical structures in primates, again using counterpart neural structures in "basal" insectivores as principal referents.

Stephan concludes that "encephalization" is not more pronounced in the Primates than in other orders of living mammals and that only Recent man surpasses all other animals in this feature. He notes that brains of fossil men indicate stages of encephalization between those of *Homo sapiens* and extant nonhuman primates.

While enlargement of the brain in an ascending scale of primates is generally accompanied by increased size and complexity of the neocortex and most internal structures, Stephan notes that certain other structures, such as the olfactory apparatus, evidence reduction and reversions in the same series.

Radinsky has prepared an extensive series of endocasts from extant primates and other mammals in order to provide a broad base for comparisons with fossil forms. He describes a simple rapid procedure for producing latex endocasts from recent crania and properly cleaned fossil crania.

Radinsky reports that endocasts of small-bodied mammals provide a highly accurate replication of surface features on their brains but that certain details may be severally indistinct on endocasts of some large-brained mammals, including *Megaladapis,* individual monkeys, pongids, and hominids. He subsequently discusses the types of information that may be derived from endocasts, providing frequent examples of functional and evolutionary interpretations of features in available fossil forms and the collateral evidence and perspectives requisite to such inferences. Radinsky's contribution indicates that certain aspects of the chapter entitled "evidence from the brain" in LeGros Clark's widely used textbook, *The Antecedents of Man,* may require reevaluation and revision.

Holloway strategically approaches the study of brain evolution in the Hominoidea with broad perspectives and firsthand experience in neurobiology, endocast studies, and evolutionary anthropology. He presents a new comprehensive survey of available australopithecine endocasts, demonstrating that in many instances previous values on their volumes were overestimated.

Holloway stresses the imperative need to consider the nature and mechanisms of neural reorganization that must have accompanied expansion of the brain during hominid evolution. Determination of endocranial volumes is a modest, but nonetheless important, initial step toward elucidating this problem. He points out the existent paucity of thorough comparative neuro-biological studies on extant hominoids that might provide solid foundations for evolutionary inferences and functional interpretations of neural development in fossil hominids.

Holloway also urges the refinement of concepts on behavioral parameters and the specific neuroendocrinological mechanisms upon which they are based. Finally, he proffers a three-stage working model of early hominid evolution based on available morphological and cultural evidence.

The seven papers in Part IV represent a wide sampling among possible approaches and perspectives that have been employed to elucidate functional and evolutionary problems on the postcranial morphology of primates.

In Chapter 9, Owen J. Lewis demonstrates that a great deal of qualitative information, that is profoundly important for inferring functional and evolutionary characteristics of fossil forms, may be garnered from available specimens if the researcher approaches these materials with a sophisticated repertoire of descriptive and comparative morphological experience.

Lewis describes certain morphological features in the wrist bones of *Dryopithecus (Proconsul) africanus* that evidence its close affinites with extant Pongidae. On the basis of the unmistakable hominoid configuration of its carpus, Lewis infers that *D. africanus* was adapted for suspensory arm-swinging behaviors. Thus, he provides an answer to part of the long debated riddle on the chronology of "brachiation" in the Hominoidea, *viz.,* that by middle Miocene times a key adaptive complex for arm-swinging had developed in the pongid wrist.

Lewis also suggests that since *Homo sapiens* shares this unique hominoid organization of the carpus with *D. africanus* and extant Pongidae, the hominid career must have included a notable stage of suspensory arm-swinging locomotion.

In Chapter 10, Friderun Ankel describes various aspects of vertebral and costal morphology in primates that may provide a base for identifying the higher taxonomic affinities of isolated specimens. But, she cautiously concludes, precise functional interpretations of these anatomical regions (and especially attempts to relate them to specific locomotor behaviors in selected species) meet with less success than do similar studies on the limb bones of primates (see Lewis, Oxnard).

Ankel has developed an ingenious biometrical technique for estimating the size of tails in primates from dimensions of their sacral canals. Utilizing an extensive set of comparative data derived from extant primates, she persuasively demonstrates that *Pliopithecus vindobonensis* probably possessed a tail within the size range of long-tailed monkeys. This information, combined with Lewis' preliminary observations on the monkey-like carpus of *Pliopithecus* (Chapter 9), require reassessments of the role that *Plio-*

pithecus may have played in hominoid evolution, particularly in the hylobatid career.

In Chapter 11, Donald Wilson strategically applies the comparative
method in a study of tail reduction in *Macaca,* a genus that contains species
with a wide range of tail sizes but which members are so closely related that
they have produced hybrid offspring in the laboratory and, with lesser frequency, in free-ranging contexts. Wilson's study is both intensively and comprehensively executed within the limits of materials available and serves well
as a model for future studies of its kind. It evidences the wealth of information that can be derived from a complex anatomical region through careful
dissection, mensuration, and decription of properly prepared specimens.

In Chapter 12, I present an overview of quantitative studies on muscles,
certain limitations on the functional inferences that may be based upon
them, and examples of possible functional explanations that may be advanced for the relative development of selected cheiridial muscles in a
large comparative series of catarrhine primates.

In Chapter 13, John V. Basmajian reviews certain biomechanical principles of bone-joint-muscle relationships, exemplified chiefly by electromyographic studies on human posture and locomotion.

He demonstrates the inaccuracy of many statements in the biomedical
and anthropological literature on the purported inefficiency of human bipedal
posture. He suggests, *per contra,* that once achieved, the human pattern of
orthograde bipedalism is one of the most economical and durable postural
habituses among mammals.

Electromyography may prove to be as robust a research technique in
studies on comparative primate morphology and evolutionary biology as it
is in clinical medicine. During the next five years, Basmajian and I, and our
associates at the Yerkes Regional Primate Research Center and the Rehabili-
tation Research and Training Center of Emory University expect to test
experimentally certain hypotheses on the mechanisms of knuckle-walking,
arm-swinging, bipedalism, object manipulation, and other motile behaviors
that are of vital concern to evolutionary anthropologists who wish to construct refined models of hominoid phylogeny.

In Chapters 14 and 15, Charles Oxnard surveys several mathematical
and physical techniques that also promise to generate hypotheses and help
elucidate certain questions on the functional and evolutionary biology of
primate locomotor systems. Oxnard reviews results of several different and
independent multivariate statistical studies on an extensive series of primate
shoulder girdles, and compares these with similar studies on the shoulders
of nonprimate mammals. Data from these studies converge to evidence a
consistent picture of interpopulation and intertaxonal variation in primate
shoulders that may be associated readily with certain basic functionally
important features and locomotive modes. If results of similar ongoing studies on the pelvic girdle of primates evidence as clear a distribution of inter-

population variation as those on the shoulder girdle, this particular combination of mathematical techniques will constitute a notable advance in primate studies and would merit serious consideration in the training of students of primate locomotion and functional postcranial morphology.

Oxnard reviews certain experimental biomechanical techniques that may be used to further refine and test hypotheses based on results of biometrical studies. He also surveys several mathematical and experimental methods for extracting functional and evolutionary inferences from fossils and other enigmatic specimens.

The four papers in Part V constitute a small sample of possible approaches and perspectives in primate behavioral and ecological studies that provide insights into the functional and evolutionary biology of selected species.

In Chapter 16, Benjamin Beck and I discuss the particular strategies and concordant limitations of conducting stationary observations and detailed studies on vital segments of a species' habitat, in addition to the commonly employed technique of pursuing primate subjects as they engage in their daily activity cycles. These two approaches—stationary and mobile —are not mutually exclusive, since observers may elect to concentrate strategically on waterholes, sleeping sites, arboreal pathways, and special food sources for only part of the total study period in each "season." Or probably with greater promise of thorough results, researchers may work in teams, some of which monitor (through observation by individual team members or by camera) behavior at stationary points while other persons concurrently move with the subjects.

The wealth of information on subsistence behaviors and interspecies relationships that may be obtained during relatively brief but persistent presence of observers at waterholes is demonstrated by Beck's and my study on Ceylonese gray langurs and associated fauna. It is remarkable that with the exception of studies by Beck, Cartmill, and me, and our associates, no major studies of free-ranging primates have included intensive samplings of waterhole ecology. We suggest that investigators engaging in future long-term studies on cercopithecoid terrestrial behaviors and ecology should contemplate employment of the waterhole strategy.

In Chapters 17 and 18, Donald Stone Sade and Irwin S. Bernstein respectively demonstrate the types of information that may be derived from longitudinal studies on relatively free-ranging and caged primate groups. Sade convincingly demonstrates the manner in which longitudinal studies may lead to conclusions totally different from those based on episodic and other short-term investigations on dominance hierarchies and other aspects of primate social organization. The longitudinal studies of rhesus monkeys on Cayo Santiago Island are elucidating the subtle dynamics and mechanisms of mother-son mating inhibitions, female dominance ordering, male intragroup roles, and other features in rhesus sociobiology that are notably influenced by geneological relationships. Sade's studies clearly exhibit not only the merits and imperative need for longitudinal vectors in primate sociobiological research, but also a mode whereby interobserver inconsistencies and vagaries may be minimized to the extent that longitudinal studies can be conducted by successive observers.

Finally, Sade discusses the advantages and limitations of sociobiological studies on primate colonies such as those of Cayo Santiago Island. His candor on this subject greatly augments the value of inferences that may be based on results of ongoing studies at the newly reorganized Caribbean Primate Research Center, and it might bear replication by students of primate sociobiology elsewhere.

Bernstein critically examines concepts of "natural habitat" and discusses the different kinds of situations, ranging from virtually undisturbed tropical forest to laboratory cages, in which primates may be studied. He stresses the need to choose carefully the proper context in which to conduct each behavioral study. In research reports, the possible influences of the study site or cage on results should be discussed thoroughly.

Bernstein notes the particular advantages of studies on captive groups and suggests that most behavioral projects would profit by including both laboratory and free-ranging constituents. He suggests that intertaxonal comparisons of primate social groups under similar captive conditions may generate hypotheses on intergroup differences. Generalizations based on such studies may be tested by further controlled laboratory experiments and field observations. He provides examples of longitudinal and comparative studies on captive groups of cercopithecine monkeys at the Yerkes Regional Primate Research Center Field Station.

During the past 15 years, several primate behavioralists have devoted considerable effort to develop methods for quantifying behavioral events, often with time as principal referent, and to constructing predictable models of population dynamics, interindividual relations, and other behavioral parameters.

In Chapter 19, Joel E. Cohen discusses the prospective utility and modes of formulating nonmathematical and mathematical models on primate behavior and other aspects of primate evolutionary biology. He provides several examples of mathematical models and statistical renderings of behavioral data derived from available studies on free-ranging primates. Cohen is one of those rare mathematical biologists who, when in need of certain types of data, will suffer the vicissitudes of field work and the absence of wide expanses of blackboard, in order to collect it. Other mathematically inclined scientists could profitably follow his suit.

I elected not to include edited transcripts of discussions that occurred at the conference table. Instead, I asked the participants to incorporate principal constructive comments in revised papers for the volume.

The symposiasts did not attempt collectively to proffer resolutions on one or more particular major theoretical problems in evolutionary biology. Instead, each participant's contribution is meant to convey the status of progress in a carefully prescribed area of special interest and to point the way toward further elucidation of the functional biology and phylogeny of primates through the application of relatively new techniques, or by the employment of available methods more intensively and with more comprehensive comparative series.

Most participants agreed that evolutionary biologists should not become so engrossed with the fascinating intrinsic properties of particular techniques and experiments that they lose sight of empirical data *per se* or of broader theoretical problems. Perhaps the best way for us to proceed toward culminant solutions to fundamental problems in evolutionary biology is through the careful step-wise resolution of selected bits, with special emphasis on underlying mechanisms, whether these be principally biomechanical, neuroendocrinological, sociobiological, ecological, or some combination of these and other factors.

Progress toward handling and synthesizing greater amounts of information may be accelerated greatly not only by modern gadgetry, such as electronic computers, but also through more intensive systematic collaborative multidisciplinary efforts among morphologists, paleoprimatologists, anthropologists, bioengineers, earth scientists, biomolecular scientists, behavioralists, and ecologists.

Equally imperative for the future progress of evolutionary primatology and anthropology is that some individuals in each subdiscipline keep abreast of new knowledge in collateral fields so that informed syntheses may keep pace with accruals of empirical data and with quantitative and functional analyses of particular data sets.

Recurrently at the conference it was noted by scientist *x* that often detailed information in subdiscipline *y* is not available for synthesis with his own data. In particular, morphologists claimed that behavioralists do not collect the kinds of detailed information requisite to proffer functional and evolutionary explanations of morphological data. Most primate behavioralists and ecologists concentrate on particular sociobiological features of their subjects and merely sketch broad ecological patterns of locomotion, feeding, and other subsistence activities. Some behavioralists asserted that the morphologist himself should venture to the field and collect such data as he may require. Other participants offered to help implement studies of special interest to morphologists if the latter could pursuade them to the high priority of such projects for elucidating the adaptive complex of a species.

Participants from all subareas of evolutionary primatology generally recognized the need for much greater quantities of properly rendered baseline data on body weight, brain size, sexual dimorphism, demography, growth, dental eruption and wear, asymmetry, aging, mortality rates, and other features in extant and fossil species which may serve as referents for elucidating many types of problems on primate evolution. If the resources and facilities of the national and international primate research centers are properly employed along with modern equipment for information storage and retrieval, great progress could be made toward supplying such critical information within the next two decades.

Finally, there is an urgent need for the scientific community to apply existing knowledge of ecology and game management intensively in order to preserve selected habitats in which primates are still living, and to exert considerable effort to study primate behavior and population dynamics in protected areas and in regions where land is being cleared or the habitat otherwise perturbed by the activities of man.

This volume is not intended to serve as a textbook or overview treatment of problems in functional and evolutionary primatology. But I suggest that instructors in introductory courses may profit by consulting many papers in this book as they prepare their lectures on research trends in comparative primate studies. Instructors of advanced courses in evolutionary anthropology, evolutionary biology, comparative morphology, and primate behavior also should find much food for thought and discussion with students in succeeding chapters. Although it is clearly impossible for a preponderant part of the benefits derived by the participants themselves to be transmitted to colleagues and other interested persons who did not attend the conference, I hope that these papers may serve as a base for original creative discussions in student seminars and other academic contexts.

On behalf of the participants, I should like to thank Mrs. Lita Osmundsen, Director of Research of the Wenner-Gren Foundation for Anthropological Research, for her encouragement and continued support of this project. The work that we accomplished during this scientific adventure is in large part attributable to the enriched environment of Burg Wartenstein, which was so graciously supervised and hostessed by Mrs. Osmundsen.

I should like also to thank Dr. and Mrs. Karl Frey, Patty Cassell, Arlene Sheiken, Judy Webb, and Peter Murray for their sustaining, patient efforts to maximize the time and scholarly resources available to us for pursuing scientific questions. Such good company is not soon forgotten.

Finally, I acknowledge with deep appreciation the special assistance of John Basmajian, Ben Beck, Bill Howells, and Lita Osmundsen in preparing this essay and Mrs. Susan Toibin's assistance with the index.

Contents

Introduction   vii
   Russell Tuttle

I. PALEOPRIMATOLOGY   1

   1. Paleobiology of the Earliest Primates   3
      Frederick S. Szalay

   2. Hominoid Paleoprimatology   36
      Elwyn L. Simons and *David R. Pilbeam*

   3. Progress and Problems in the Study of Early Man in
      Sub-Saharan Africa   63
      Phillip V. Tobias

II. CRANIAL MORPHOLOGY   95

   4. Arboreal Adaptations and the Origin of the Order Primates   97
      Matt Cartmill

   5. Analysis of Patterns of Variation in Crania of Recent Man   123
      W. W. Howells

III. COMPARATIVE NEUROBIOLOGY AND ENDOCASTS   153

   6. Evolution of Primate Brains: A Comparative Anatomical
      Investigation   155
      Heinz Stephan

   7. Endocasts and Studies of Primate Brain Evolution   175
      Leonard Radinsky

   8. Australopithecine Endocasts, Brain Evolution in the
      Hominoidea, and a Model of Hominid Evolution   185
      Ralph L. Holloway

IV. POSTCRANIAL MORPHOLOGY   205

     9. Evolution of the Hominoid Wrist   207
         Owen J. Lewis

    10. Vertebral Morphology of Fossil and Extant Primates   223
         Friderun Ankel

    11. Tail Reduction in *Macaca*   241
         Donald R. Wilson

    12. Relative Mass of Cheiridial Muscles in Catarrhine Primates   262
         Russell Tuttle

    13. Biomechanics of Human Posture and Locomotion:
         Perspectives from Electromyography   292
         John V. Basmajian

    14. Functional Morphology of Primates: Some Mathematical
         and Physical Methods   305
         Charles E. Oxnard

    15. The Use of Optical Data Analysis in Functional
         Morphology: Investigation of Vertebral
         Trabecular Patterns   337
         Charles E. Oxnard

 V. ASPECTS OF BEHAVIOR AND ECOLOGY   349

    16. The Behavior of Gray Langurs at a Ceylonese Waterhole   351
         Benjamin B. Beck and *Russell Tuttle*

    17. A Longitudinal Study of Social Behavior of
         Rhesus Monkeys   378
         Donald Stone Sade

    18. The Organization of Primate Societies:
         Longitudinal Studies of Captive Groups   399
         Irwin S. Bernstein

    19. Aping Monkeys with Mathematics   415
         Joel E. Cohen

References   437

Index   469

Plates—between pages 234 and 235

I

Paleoprimatology

1

FREDERICK S. SZALAY
HUNTER COLLEGE, CITY UNIVERSITY OF NEW YORK
AND THE AMERICAN MUSEUM OF
NATURAL HISTORY

Paleobiology of the Earliest Primates

Primates of the late Cretaceous and Paleocene are gradually becoming better known through the reexamination of old specimens and the collection of new ones. A continuous evaluation of the Paleocene primate radiation, although so far necessarily restricted to North America and Europe, makes it apparent that known Paleocene species, despite the presence of some specializations, including enlarged incisors coupled with tooth reduction and correlated minor changes in the snout, represent a level of evolutionary development more primitive than that displayed by the better known and in some respects more primitive later primates.

In addition to an account of the Paleocene families, this paper will focus on aspects of evolutionary trends and functional inferences. But first I will discuss briefly the approach that paleontologists take prior to the actual publication of scientific data and its multifaceted interpretations and applications. The various activities of paleontological research are closely interdependent, yet time and human limitations make its practice multiphased and time consuming. The discovery and collection of the fossils is preceded by planning, geological reconnaissance, and resolution of problems particular to the particular locality. Because modern quarrying methods generally involve patient work for crews of varying sizes at any one locality for

I wish to thank Paul O. McGrew, Malcolm C. McKenna, Bryan Patterson, and Donald E. Russell for loan of specimens and other favors. Critical reading of the manuscript by Malcolm C. McKenna and Russell Tuttle is greatly appreciated. Discussions with James S. Mellett on diet and species size have been most enlightening. Figures 1–2, 1–4, and 1–5 were drawn by Daria Dykyj, Plate 1–10 by Anita J. Cleary, figure 1–1 by Herbert Goldman, and figures 1–6, 1–7, and 1–8 by Biruta Ackerbergs.

I am especially grateful to Daria Dykyj and Miriam Siroky for technical assistance in the preparation of the manuscript.

The paper was written based on research supported by NSF Grants GB–7418 and GB–20085.

3

several field seasons, a relatively complete description of any taxon must await the collection of an adequate sample. The study of the fossils focuses on several closely related objectives. The faunistic and biostratigraphic appraisals of all species at a locality or in a faunal level are amalgams of separate studies on the sundry species from many perspectives. In studies on fossils of any one species, probably the most important and certainly the most fundamental problem is the correct delineation of the taxon. This alpha taxonomy, the cornerstone of all ecological and functional considerations as well as supraspecific taxonomy, must solve not only problems of geographical and temporal variation in related samples from other localities of known stratigraphic relationships, but also problems of variation in the study sample itself.

The greatest problem facing students of fossil primates is the recognition and correct allocation of sometimes abundant and potentially very important postcranial elements. Only quarries or fields which have been thoroughly collected may be expected to produce associations with high probability. The correct allocation of primate postcranials to "dental taxa" from the Paleocene is not very likely unless the student knows all specimens of a collected sample and is familiar with taxa of other higher categories. This problem is pernicious at the beginning of the Age of Mammals because of the relative absence of morphological divergence among the species at that time.

It is becoming apparent, even to the most conservative descriptive paleontologists, that even a rudimentary understanding of the functional anatomy of a species has profound effects on the phylogenetic evaluations of its morphology. Mammal teeth have long been yielding phylogenetic dividends when asked biological questions about the form and function of their isolated components, of the whole tooth, or of whole dentitions. A constant evaluation of heritage components, as well as functional aspects, in any biological feature is necessary when concentrating on the taxonomy, phyletic affinities, or mechanical and behavioral functions of one or more species. Whereas many taxonomic studies suffer from an almost complete neglect of function, equally sterile "taxonomic" undertakings which concentrate purely on functional attributes of biological features also appear from time to time. Some attempts to classify *Oreopithecus,* for example, consist of anatomical comparisons concerning the degree of locomotor adaptations reflected in the individual bones of the Tuscan catarrhine, rather than a systematic evaluation of the genus with the necessary weighting of characters clearly indicative of its relationships. Unless the geometry of phylogeny (that is, the degree of genetic relationship between the various species or genera) is understood in a time sequential framework, functional studies by themselves are unlikely to yield answers to taxonomic questions. Without a careful and continuous search for and scrutiny of the fossil record, this precarious but most important distinction between "heritage" and "habitus" features of organisms is not likely.

One might briefly point to the relationship of paleontological studies to more specialized undertakings of fossil specimens. The most advanced available techniques and tools which yield reasonable dividends for time and effort invested should be utilized to study biological attributes of living and extinct species to elicit functional answers. Only then will it be possible to evaluate to the fullest extent fossil taxa ecologically in the context of their respective faunas.

The most sophisticated analyses of individual bones by specialists well versed in engineering and mathematical techniques are based not only on planned discovery of fossils but also on the presumably correct allocation of unassociated postcranials to a particular taxon by the practicing paleontologist. Before a postcranial element can be allocated to the Primates, for example—and then, or even before, to a particular species—a great deal of primarily paleontological expertise is employed, with most variables difficult to quantify. The number and kinds of species in the fauna, and their abundance and relative size, weigh heavily. But even more important is the recognition of the balance of characters on a given part of the bones (based on the knowledge of associated skeletal remains of known early Tertiary mammals). This recognition of characteristic bony elements in one fauna may then serve as a hypothesis that can be tested with faunas of similar age but which contain a different combination of taxa.

Ordinal Characters and the Origins of Primates

Practicing paleontologists cannot be overly concerned about the "first" species of primates, lagomorphs (i.e., rabbits, hares, and pikas), odd- or even-toed ungulates, or other higher categories. This "first" species, even if the paleontological records were so complete as to yield the ancestral species, would have only a primarily taxonomic meaning. Nevertheless, because zoological communication is nearly impossible without taxonomy, it is important to delineate origins of higher categories, preferably so that they might coincide with the first appearance of various ordinal features.

In spite of the relatively consistent balance of characters in the cheek teeth of the earliest primates (Szalay 1968a, 1969a) it is apparent that many Cretaceous or even Paleocene primates are not readily recognized on dental evidence alone. On the basis of recent discoveries, it has become evident that some features of the hard anatomy accessible to paleontological inquiry may be used as taxonomic criteria to delineate the earliest primates from their insectivoran ancestors. The accumulating evidence from the fossil record strongly supports the hypothesis (based on dental evidence) that all undoubted primates possessed a middle ear enclosed by a petrosal derived auditory bulla in addition to either a ring-like or tubular ectotympanic for the support of the tympanic membrane. All living primates have a petrosal bulla in combination with either a tympanic bone forming a ring inside the middle ear or a lateral extension of the bulla, the auditory tube.

(For a recent treatment of the tupaiid auditory bulla and its phylogenetic significance see Van Valen 1965).

McDowell (1958) and McKenna (1966) suggested that the primate bulla originated in a leptictid ancestor as a result of a cartilaginous entotympanic being ontogenetically incorporated into a descending wing of the petrosal. This is a plausible hypothesis and it may well become substantiated by the fossil record. There is some indirect evidence, however, that some early palaeoryctids, which evolved into apatemyids and early erinaceotans (Szalay 1968c), probably had a composite bulla consisting partly of the petrosal. It is probably also significant that astute students of Paleocene primates and insectivorans still occasionally mistake erinaceoid dentitions for those of paromomyid primates. Early erinaceotans, such as the Cretaceous *Batodon* and the Paleocene *Leptacodon,* show a deceptively paramomyid-like balance of characters in their molars, although not in the antemolar dentition. Primates are among the earliest mammals to surround the middle ear with an ossified bulla, and this structure is most like the erinaceotan one in sharing a petrosal component. Considering the evidence from the dentition, and the still very incomplete but suggestive data of the bulla construction, primate origins from a "palaeoryctoid"-erinaceotan stock of insectivorans appears more plausible to me than derivation from *bona fide* leptictids. By the combined criteria of teeth and ear regions (favoring the basicranial evidence), neither the Eocene Microsyopidae nor the recent Tupaiidae can be viewed as primates. Both of these families appear to have leptictid ties, although the tupaiids may prove to have evolved independently from some stocks of palaeoryctid insectivorans.

Given the petrosal-ectotympanic combination as an ordinal character of the Primates, it is difficult to judge which of the several common combinations of these two components in the auditory bulla is the earliest. Because of the mistaken homologies of tupaiid ear regions and because of a tacit acceptance of most lemuroid characters as primitive, a ring-like ectotympanic enclosed in the bulla has become a common textbook example of the primitive primate condition. As new early primate cranial evidence comes to light, however, it necessitates a reassessment of the primitive primate bulla condition. Although a detailed examination of this problem will be pursued elsewhere in a study of the new cranial evidence of *Phenacolemur,* some pertinent comments follow.

The basicranial region of the Eocene adapids, which are the earliest known lemuroids, is difficult to assess in terms of what is primitive or advanced compared to homologous structures of other primates. Although changes in the ear region from the Eocene level of organization to that of living species of lemurids and indriids have been relatively small, this rate of evolution should not be used to extrapolate from the adapid condition the earliest, Cretaceous representatives of the order. In general, the basicranial anatomy inside of the bulla (and probably the inferred arterial and venous circulation) is probably slightly more primitive in *Phenacolemur* than in any known lemuroid or *Plesiadapis*. Although the only known skull of the genus is early Eocene, this taxon makes its appearance in the late Paleocene. *Phenacolemur* has a broad tympanic (Plate 1–1) forming an external auditory tube, similar to the condition seen in *Plesiadapis,* tarsiids, and microchoerids. Although it is far from certain that a broad ectotympanic outside the bulla is the primitive primate condition, the lorisoid auditory bulla might be viewed as more primitive than the lemuroid condition, and intermediate between the Cretaceous-Paleocene level of organization, represented by *Phenacolemur* and *Plesiadapis,* and that seen in lemuroids. Furthermore, it is not unlikely that the earliest primates had a bulla construction more similar to that of the lorisoids than to that of known lemuroids.

The suggestion above is merely a tentative working hypothesis, and in spite of the very primitive condition of the middle ear proper of *Phenacolemur,* its ectotympanic may represent an early primate specialization both in that form and in *Plesiadapis*. The presence of a ring-like ectotympanic in some quasi-primitive placentals, such as some tenrecoids, tupaiids, and leptictids, does not necessarily indicate that this condition was primitive among the earliest insectivorans. The total combined evidence as to the nature of the ectotympanic in earliest primates suggest to me a C-shaped, slightly plate-like ectotympanic, rather than a ring-like one exemplified by the tupaiids. What must also caution the apparently logical but not necessarily true inference that the ring-like ectotympanic inside of the bulla was primitive in the earliest primates is the knowledge of ancient insectivorans (*sensu lato*) which probably possessed composite (therefore presumably advanced) ossified bullae with an ectotympanic outside of the bulla proper as seen in marsupials in general, and in numerous primates in particular. As noted above, indirect evidence suggests primate origins from insectivorans with composite bulla construction. Yet for proof of the primitive condition of auditory bulla construction in the earliest primates[1] the earliest fossils of the order are necessary. Convergences of the condition of the

1. R. D. Martin (1968) in a paper dealing with the definition of the Order Primates touched on the problem of primate origins. Although he spent some time critically examining previous views on the derivation of primates from various groups of insectivorans, Martin's lack of firsthand familiarity with the Cretaceous-early Tertiary fossil record bearing on insectivoran phylogeny and primate origins renders his discussion a somewhat sterile exercise in logic. He suggests that in a new ordinal diagnosis of the primates, the loss of the entotympanic and its replacement by the petrosal should be included. The petrosal bulla is undoubtedly a feature diagnostic of primates; but the loss of the entotympanic and its replacement by a petrosal wing (although suggested as probable by McDowell 1958, and McKenna 1966) is not an established fact of phylogeny. In addition to the auditory bulla, another osteological character proposed by Martin as diagnostic of primates is the "os planum" in the orbit, and an inferred but paleontologically testable feature, the retention of opposable pollex and hallux. This latter, presumably primate, feature if held over from insectivoran ancestors should not be considered as a diagnostic primate character. The earliest known primate skulls, such as those of *Plesiadapis,* the Eocene adapids, and *Necrolemur,* do not show an ethmoid component, the "os planum," in the orbital region.

bulla structure and relative position of the ectotympanics to that of tupaiids, and some living insectivorans, and the relatively primitive lemuroids might be easily mistaken for the primitive condition in the Primates.

Known Paleocene Radiation of the Primates: A Systematic Account

The earliest records of primates from the late Cretaceous and Paleocene of Africa and Southeast Asia have not been discovered. These mammals, along with other elements of the faunas, may not be collected in the foreseeable future. Fortunately, the combined Paleocene record of both North America and Europe provide more than a glimpse of the extent of the first major primate radiation and allow estimates on the levels of organization in feeding mechanisms and locomotor systems in some very early primates. In my judgment, all so far known and described undoubted primates available from Paleocene sediments of Europe and North America may be classified in one of four families. These families, the Paromomyidae, Picrodontidae, Plesiadapidae, and Carpolestidae, represent a natural superfamily, the Plesiadapoidae,[2] which radiated from taxa that might be classified as paromomyids. Although I have previously placed the Picrodontidae in a monotypic superfamily of its own (Szalay 1968b), I now prefer to classify this family as one of the families in the Plesiadapoidea.

THE PAROMOMYIDAE

Paromomyids are the most ancient of the known families of primates. They were predominantly tiny to small mammals, only a few species of *Phenacolemur* attaining sizes slightly above a common rat. Aside from the specializations of the enlarged incisors and the reduced number of incisors and premolars, Paleocene representatives of the family are probably the most primitive forms of the order as far as the lower jaw morphology, cheek tooth structure, and the organization of the basicranium are concerned.

The original concept of the Paromomyidae is based on Simpson's (1940)

---

2. Van Valen's (1969) concept of the Microsyopoidea is essentially the Plesiadapoidea of this paper without the Microsyopidae. I believe that the Microsyopoidea should be restricted to the probably leptictid derived microsyopids (Szalay 1969b). Van Valen continues the use of a Prosimii-Anthropoidea subordinal arrangement in his formal classification. This schema has been the most important concept concerning the classification of primates, and this underlying theoretical framework has channeled the most widely accepted classifications and evolutionary research in all aspects of the order. As a biological grade concept, it has been helpful in sorting out the spotty and uneven fossil record and certain information from living species, and it still might be justified for heuristic purposes. But as the fossil record improves, it appears that the adaptive radiation of the order was diverse and many-faceted, but most of all that known phylogenetic relationships in primates do not support a dichotomous "prosimii-anthropoidea" taxonomic arrangement.

Paromomyinae of the Anaptomorphidae. He grouped together *Paromomys, Palaechthon, Plesiolestes,* and *Palenochtha,* and considered *Trogolemur* "or" the Necrolemurinae (the latter was part of his 1940 concept of the Anaptomorphidae) as possibly derivable from this subfamily. In 1955 in a new family, the Phenacolemuridae, Simpson included *Phenacolemur, Paromomys,* and *Palaechthon,* but he inexplicably omitted the horizontally closely allied genera *Palenochtha* and *Plesiolestes.* Simpson disregarded the fact that he previously had erected the Paromomyinae, and his discussion clearly indicated the Phenacolemuridae to be a near clade, or at least that the three genera included by him represented three successive levels of organization, with *Phenacolemur* representing the most advanced condition. McKenna (1960, p. 70) summarized the various objections to Simpson's concept of the Phenacolemuridae; McKenna's views are now widely accepted. Van Valen and Sloan (1965) followed McKenna's logic and divided the Paromomyidae (which they raised to the family rank level) into two subfamilies: the Paromomyinae for *Paromomys* and *Phenacolemur,* and the Purgatorinae for *Purgatorius, Plesiolestes, Palaechthon, Palenochtha* and *McKennatherium.* I have indicated (Szalay 1969a, p. 313) that *"Mckennatherium libitum"* Van Valen 1965, is a junior synonym of the erinaceotan insectivoran *Leptacodon ladae* Simpson 1935. My arrangement of the Paromomyidae (Szalay 1968a, 1969a) is not very different from that of Van Valen and Sloan (1965), and my phylogenetic conclusions are essentially the same as the interpretations of these authors. The tribe Palaechthonini is virtually the same as their Purgatoriinae, except for *Purgatorius* and *"Mckennatherium,"* and the Paromomyini is identical to their Paromomyinae. My continued reluctance to group *Purgatorius* with *Palaechthon, Palenochtha, Plesiolestes,* and *Navajovius* stems from our ignorance of the anterior dentition of *Purgatorius.* I have no doubts about the paromomyid affinities of *Purgatorius,* and if the genus proves to have the full eutherian dental formula, one could then accept the Purgatoriinae Van Valen and Sloan 1965 (including only the genus *Purgatorius*) in addition to the Paromomyinae Simpson 1940, as a second subfamily of the Paromomyidae Simpson 1940. Like Van Valen and Sloan, I believe that *Paromomys* should be separated from the remaining paromomyids (within the confines of the family) because *Paromomys* is clearly the ancestor of the polytypic, very successful *Phenacolemur.* By contrast, *Plesiolestes, Palaechthon, Palenochtha* (Plate 1–2), and *Navajovius* represent a relatively close assemblage, separated from each other by relatively minor differences. In summary, the differences between Van Valen and Sloan's (1965) Paromomyinae and Purgatorinae, and my Paromomyini and Palaechthonini are mainly those of taxonomic expression of an essentially agreed on phylogeny for most of the genera involved. Recent doubts expressed by some experts on fossil primates as to the primate ties or even generic distinctness of *Purga-*

torius are based mainly on lack of erudition concerning Mesozoic and Paleocene Mammalia rather than substantiated, objective criticism.

Purgatorius is known from the latest Cretaceous sediments of Montana, but it is best represented from the early Paleocene Purgatory Hill locality of that state (Van Valen and Sloan 1965). The Palaechthonini (*Palaechthon, Palenochtha, Plesiolestes,* and *Navajovius*) is predominantly Paleocene, as far as known. Because the late Paleocene *Berruvius* (D. E. Russell 1964) from Europe appears to be indistinguishable on the generic level from *Navajovius,* the latter genus is then known from late Paleocene sediments of both North America and Europe.

A probable species of *Navajovius, N.? mckennai* (Szalay 1969a) is present from the early Eocene of New Mexico. Of the Paromomyini, *Paromomys* (Plate 1–3) occurs exclusively in the middle Paleocene of North America, whereas *Phenacolemur* is known from the late Paleocene to medial Eocene (Robinson 1967) of North America, and it has been recently reported from early Eocene sediments in France (D. E. Russell et al. 1967).

Dental formulae of the taxa comprising the Paromomyidae have been consistently misinterpreted. I have previously dealt with this subject in some detail (Szalay 1968a, pp. 37–40) and therefore will present only a summary of the formulae of various paromomyid genera. My recent studies on virtually all available specimens of Paleocene primates resulted in conclusions of relatively high degree of probability concerning the numbers of teeth and their homologies of the species listed below.

Palaechthon spp.	$I_{1(?),2(?)}^{1(?),2(?)}$; C_1^1; $P\ _{2,3,4}^{2,3,4}$; $M_{1,2,3}^{1,2,3}$
Palenochtha minor:	$I_{1(?),2(?)}^{1(?),2(?)}$; C_1^1; $P\ _{3,4}^{?,2,3,4}$; $M_{1,2,3}^{1,2,3}$
Plesiolestes spp.	$I_{1(?),2(?)}^{1(?),2(?)}$; C_1^1; $P\ _{2,3,4}^{2,3,4}$; $M_{1,2,3}^{1,2,3}$
Navajovius kohlhaasae:	$I_{1(?),2(?)}^{?}$; C_1^1; $P\ _{?,3,4}^{2,3,4}$; $M_{1,2,3}^{1,2,3}$
Paromomys maturus:	$I_{1(?),2(?)}^{1(?),2(?)}$; C_1^1; $P\ _{2,3,4}^{2,3,4}$; $M_{1,2,3}^{1,2,3}$
Phenacolemur spp.:	$I\ _{1(?)}^{?}$; C_0^1; $P\ _4^{2,3,4}$; $M_{1,2,3}^{1,2,3}$

Among paromomyids, only species of *Phenacolemur* have highly specialized mandibular morphology. Lower jaws of *Palaechthon, Palenochtha, Plesiolestes,* and *Navajovius* display a somewhat shortened and deepened mandible, and a tall and large coronoid process. Some well preserved specimens of the first two genera have long mandibular angles. *Phenacolemur* possesses a relatively short mandibular angle, being, thus, morphologically the most advanced inferred herbivore in the Paromomyidae.

Striking dental specializations in the family are confined again to the species of *Phenacolemur*. In addition to the extreme reduction of the lower antemolar dentition, its molars are reconstructed, compared to the more conservative members of the family. In general, the teeth flatten out and the major cusps become partly incorporated into the crests leading to them (Plate 1-4). The conules lose their distinctness and the hypocones spread out, so in effect the most advanced species of the genus had square upper molars. In addition, the extreme enlargement of the hypocone on M^3 compared to the size of the hypocone on M^1 and M^2 is one of the hallmarks of the genus. It is correlated with the equally unusual enlargement of the M_3 talonid.

The enlarged incisor of *Phenacolemur* is relatively the longest among plesiadapoids. Rather than being short and robust as in plesiadapids, it tends to be long, gliriform, but never with the faceted apex of rodent incisors, nor with the attribute of permanent growth that characterizes the latter.

Navajovius is unusual in its unique emphasis on the premolar dentition—the trenchant premolars increase in size anteriorly from the two-rooted third premolar, and this tendency might have culminated in an exceptionally large trenchant canine, the type seen in *Lemur*.

There are two skulls known of paromomyid primates. One of them is the skull of *Phenacolemur,* that of *P. jepseni* from the early Eocene of New Mexico first reported by Simpson (1955): it is under study by me, and soon will be published. The other cranium is a crushed specimen of *Palaechthon* sp., part of the sample of primates from the Torrejonian of New Mexico that is presently being described by Dr. Robert W. Wilson and myself. This latter specimen is the earliest known skull of a primate.

THE PICRODONTIDAE

The ordinal affinities of the dentally very advanced, tiny picrodontids of the North American Paleocene have long been a vexing problem for students of early Tertiary mammals. The highly altered molar morphology of picrodontids caused earlier workers, such as Matthew and Simpson, to suspect rather vague relationships either with insectivorans, bats, or primates. The steadily accumulating evidence from new specimens and new studies, however, indicates with a sufficiently high degree of probability that picrodontids are paromomyid primate derivatives (Szalay 1968b). Picrodontids are known only from two species of two genera, *Picrodus* and *Zanycteris, Picrodus silberlingi* occurring both in medial and late Paleocene sediments, and *Z. paleocenus* in late Paleocene deposits only. Subsequent to review of these primates, additional specimens of *Picrodus* have been collected by a Hunter College party under my direction, and additional research has been conducted on known specimens of both genera. Some of these studies are incorporated into the new reconstructions of the mandible of *Picrodus* and the palate of *Zanycteris,* presented here for the first time (Figures 1–1 and 1–2).

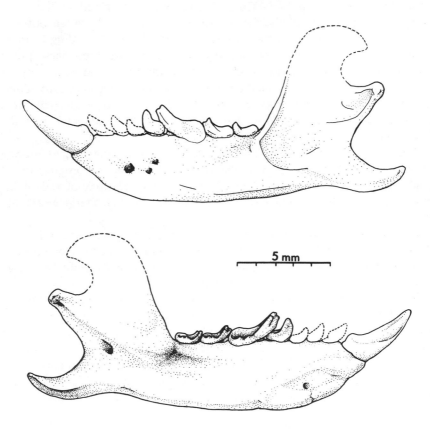

Figure 1–1. *Picrodus silberlingi,* medial and late Paleocene. Lateral and medial views of the left mandible, bearing two incisors, canine, and P_3–M_3. Broken lines indicate parts of morphology for which specimens are lacking. Reconstruction is based on specimens from Gidley Quarry, Swain Quarry, and the Saddle locality of Bison Basin.

The major changes during the evolution of the picrodontids from a more primitive tritubercular ancestry are almost completely restricted to the cheek teeth among known anatomical features.

As in paromomyids such as *Palaechthon, Palenochtha,* or *Paromomys,* the mandible of *Picrodus* is shortened and moderately deep (Figure 1–1). The masseteric fossa is extensive, and although a complete coronoid process

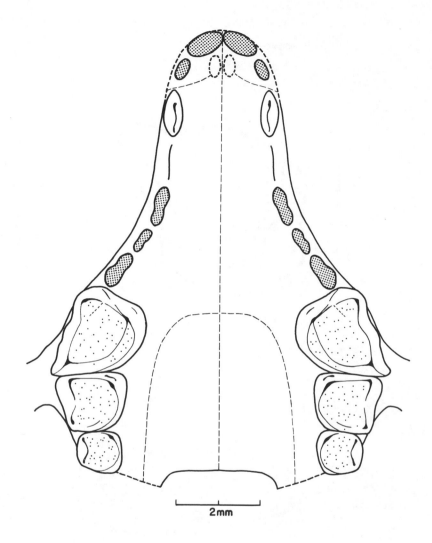

Figure 1–2. *Zanycteris plaeocenus,* late Paleocene. Reconstructed ventral view of palate, based on the type skull, showing two incisors, the canine, alveoli for P²–P⁴ and M¹⁻³.

is not known, it was probably large, indicating that both the ectental and orthal components of the masticatory stroke were important. Like in paromomyids, the angle of the mandible in *Picrodus* is long and relatively narrow, primitive in its proportions, and very similar to those of primitive early Tertiary Insectivora. The articular condyle is small and rounded, not suggestive of any particular specialization restricting masticating actions of

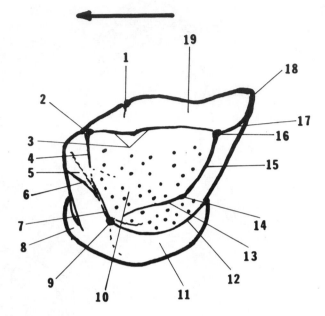

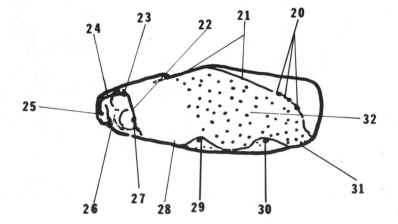

Figure 1–3. Suggested homologies of the structures
on picrodontid molars.

1. Small cuspule, probably equivalent to a small mesostyle. 2. Paracone.
3. Centrocrista (=postparacrista and premetacrista combined). 4. Crest on lingual slope of paracone probably homologous to similar, less-pronounced crest
of paromomyids. Possibly (there is no evidence) the lingual part of the crest

the mandible. If anything, this structure evidences a considerable freedom of movement at the jaw articulation.

The only known cranium of a picrodontid is the type specimen of *Zanycteris paleocenus,* a partial, crushed skull consisting mainly of the palatal portion. The muzzle was relatively narrow compared to the expanded posterior half of the palate. The root of the zygoma is above the first and second molars, and the relative dimensions of this structure do not suggest any particular specialization. The skull reveals a great wealth of information, yet at the same time poses many tantalizing unanswered questions. The reconstruction presented here (Figure 1–2) is an improvement over the one in Szalay (1968b, fig. 22). Deduced from the occlusal relationships of the anterior teeth, there were two incisors anterior to the large, buccolingually constricted canine; and posterior to the long diastema behind the canine, there were three double-rooted premolars, probably the second, third and fourth. Judged from the alveoli, these teeth were elongated, and P^4, usually the most molar-like of the upper premolars, had no lingual alveolus and hence presumably no protocone, or at least not a prominent one.

In general, in both genera there is a great reduction in the height of the cusps of the upper molars and the conules are eliminated. The extensive enlargement of the metacone on the first upper molar results in the disprotionate emphasis on the posterobuccal components of the ancestral upper molar, resulting in the distorted, unusual appearance of the picrodontid M^1.

In the lower molars the trigonids are drastically reduced, whereas the talonid, particularly on the first molar, is greatly enlarged. The hypoconid is obliterated, and as a result of the posterobuccal extension of the talonid, the entoconid and hypoconulid are anterolingually displaced. To appreciate the homologies of the components of picrodontid molars, Figure 1–3 should be consulted.

The enamel of the basined parts of both upper and lower molars is unusually papillated, thus greatly increasing the total available surface for occlusal contact. It appears that the unorthodox flattening and spreading of the upper molars and the elongation of the buccal and posterior crests of the lower molars is the result of selection for crushing and shredding plant materials which were not very abrasive. The low crowned teeth of picro-

---

is homologous with part of the postparaconule crista. 5. Paracingulum (?), a shelf originally formed by the paraconule at the base of the paracone. 6. Preparaconule crista. 7. Preprotocrista. 8. Precingulum. 9. Protocone. 10. Trigon basin. 11. Hypocone. 12. Postcingulum. 13. Postprotocrista. 14. Metaconule (vestigial). 15. Lower part of this crest probably premataconule crista. 16. Metacone. 17. Postmetacrista. 18. Metastyle. 19. Stylar shelf. 20. Posterobuccal cuspules; one or more may be homologues of the hypoconid. 21. Cristid obliqua. 22. Protocristid. 23. Protoconid. 24. Paracristid. 25. Paraconid. 26. Area homologous to the obliterated trigonid basin. 27. Metaconid. 28. Talonid notch. 29. Entoconid. 30. Hypoconulid. 31. Area probably homologous to the notch between the hypoconid and hypoconulid of more primitive eutherians. 32. Talonid basin.

dontids could not have withstood abrasive attrition for even a very short life-span of mammals their size. Their peculiar, phyllostomatid bat-like dental adaptation might have been the result of selection for a juicy fruit or nectar feeding diet.

Flattening and elongating of the molars is probably significantly coupled with the transversely strongly constricted premolars. The combined effect of the transversely narrow canine and premolars might serve as a very effective slicing device against the lower premolars, canine and enlarged incisor.

The enlarged lower anterior incisor is not noticeably different from those of the more primitive paromomyines. Its apical half (i.e. the edge of the elongated tooth) is narrower than the basal part, so in effect this enlarged tooth is knife-like with a long edge on it. The homologies of the three anterior lower teeth—the two incisors and the small, premolariform canine —are judged to be the same as those of known paromomyids other than *Phenacolemur,* carpolestids, and *Pronothodectes.*

THE PLESIADAPIDAE

This family, subdivided into the Plesiadapinae and Saxonellinae, is un-doubtedly one of the most successful Paleocene families of primates, both in the number of known species as well as in terms of numerical abundance of individual specimens. *Plesiadapis* remains are relatively common even in some early Eocene localities in North America.

There are only four described valid genera within the Plesiadapinae. These are *Pronothodectes, Plesiadapis, Platychoerops,* and *Chiromyoides.* A fifth genus, from early Paleocene sediments of Purgatory Hill, is presently being described by Van Valen (personal communication).

Previous suggestions (McKenna 1961, Wood 1962) that plesiadapids seem to be in the ancestry of the Rodentia appear to be based on some con-vergent resemblances between *Plesiadapis* and early rodents. There is no meaningful resemblance between them in details of the dentition, cranial morphology, or the basicranium. All known rodents form the auditory bulla by the ectotympanic bone, whereas *Plesiadapis* has a petrosal bulla with a large ectotympanic tube. This is not a peculiarity of *Plesiadapis,* since the hitherto undescribed ear region of *Phenacolemur* shows the same bulla construction. Paromomyids and plesiadapids diverged no later than the early Paleocene, hence the auditory bulla construction of the common early Paleocene ancestor was probably the same as those of *Plesiadapis* and *Phenacolemur.* It is unlikely, therefore, that the auditory bulla in rodents (the earliest forms have no ossified bulla) evolved from the petrosal bulla of primates.

Unfortunately, it is still often stated that plesiadapids are "lemuroid" primates. This was very emphatically stressed by Simpson (1935, pp. 23–30) who placed great weight on dental similarities, which was summarily reiter-ated by Simons (1963, p. 80). Evidence from the petrosals of the earliest known notharctine (presently under study), *Pelycodus* from early Eocene sediments, clearly shows, however, that the early Eocene adapid middle ear construction was already characteristically lemuroid, very unlike the middle ear of *Phenacolemur* or *Plesiadapis*.

The dental formulae of plesiadapids are as follows.

Pronothodectes: $\qquad I_{1(?),2(?)}^{1(?),2(?)} \; ; \; C_1^1 \; ; \; P_{2,3,4}^{2,3,4} \; ; \; M_{1,2,3}^{1,2,3}$

Plesiadapis

(probably also *Platychoerops*): $\qquad I_{1(?)}^{1(?),2(?)} \; ; \; C_0^{0-1} \; ; \; P_{(2),3,4}^{2,3,4} \; ; \; M_{1,2,3}^{1,2,3}$

Chiromyoides: $\qquad I_{1(?)}^{1(?),2(?)} \; ; \; C_0^? \; ; \; P_{(2),3,4}^{2,3,4} \; ; \; M_{1,2,3}^{1,2,3}$

Saxonella: $\qquad I_{1(?)}^{1(?),?} \; ; \; C_0^? \; ; \; P_{3,4}^{2,3,4} \; ; \; M_{1,2,3}^{1,2,3}$

Species of *Pronothodectes* (two or three valid species) were roughly in the size range of the living *Lepilemur leucopus*. Although the deep-jawed, medial Paleocene *Pronothodectes* (Plate 1–5) retain an additional small incisor and a small canine, its enlarged procumbent anterior incisor is already robust and has very massive roots, indicating the establishment of a *Plesiadapis*-like habitus at least as far as incisor structure and function are concerned (Plate 1–6). Another character of the upper molars which diagnoses plesiadapids, the alignment of the protocone in a less mesial position than in the dentally more primitive paromomyids and other early Tertiary groups of primates, is already seen on *Pronothodectes*.

Species of plesiadapids ranged from forms which were like *Lepilemur leucopus* to *Lemur catta* in size and larger, but none—with the exception of *Saxonella*—were as tiny as some of the paromomyids, picrodontids, or carpolestids. *Plesiadapis tricuspidens,* the only species represented by an adaquate skull, was in the size range or larger than *Lemur catta*. It had a skull about 10 cm. in length.

There is general agreement that *Pronothodectes* probably gave rise to *Plesiadapis*. Whether the genus is derived from a single, ancestral species or derived from separate species of *Pronothodectes* cannot be determined without a thorough revision of the family.

Within *Plesiadapis* there are two accepted subgenera, *Plesiadapis* and *Menatotherium*.[3]

A new reconstruction of the relatively highly specialized *Plesiadapis tricuspidens* is presented here. Previous figures of the almost complete, but crushed skull (CR 125), those of Simons (1964b) and D. E. Russell

3. Since Russell (1967) established the identity of the enigmatic *Menatotherium* Pitton 1940 as *Plesiadapis* near *P. walbackensis, Menatotherium* has come to replace *Ancepsoides* as a subgenus of *Plesiadapis*.

(1964) have shown the teeth in incorrect occlusal relationship, and consequently the entire muzzle was forced dorsally on the neurocranium in an unnatural position. These authors showed the tips of the incisors in occlusion when the cheek tooth rows were locked. The more probably occlusal relationships are shown in Figure 1–4. Other differences of the reconstruction

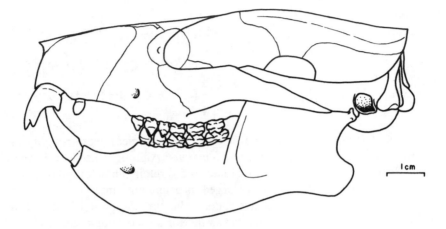

Figure 1–4. *Plesiadapis tricuspidens,* late Paleocene, new reconstruction of lateral view of the skull and mandible.

in Figure 1–4 from previous ones (which will be discussed elsewhere) lie in the relatively lower neurocranium, less posteriorly projecting occipital condyles, less inflated petrosal bulla, and the less posteriorly extended dorsal wings of the premaxilla, in addition to the considerably shorter diastema on the lower jaw which might have belonged to the known skull (CR 125).

Postcranial elements of *Plesiadapis* are known from a species from North America (Simpson 1935) and another species from Europe (Russell 1964, Simons 1964b). Based on the published figures, it appears that the intermembral index was only slightly below 100. It may be that once postcranial elements of the Paleocene primate radiation become more common, *Plesiadapis* might become recognized as a relatively more aberrant form than the majority of early Tertiary primates. The peculiar morphology of the multiple hypotympanic sinuses of the middle ear in *Plesiadapis* certainly hints to this. The several species of this "dental" genus might have had highly diverse locomotor adaptations.

Chiromyoides was derived either from *Pronothodectes,* or from a primitive late medial Paleocene *Plesiadapis,* such as *P. walbeckensis* (see D. E. Russell 1964).

I estimate the skull length of *Chiromyoides campanicus,* based on the partial mandible of the type, to be between 7 and 8 cms. This genus, known only by the type species, is a very rare component of the late Paleocene European record.

The short-faced *Chiromyoides* (Plate 1–7) appears to have a heavily buttressed symphysis of particular importance. The symphysis, which was probably fused at least during part of the ontogeny, possesses two distinct tori which may be referred to as superior and inferior transverse tori. Undoubtedly these structures were an adaptive response to buttress the mandible against stresses in a transverse direction during very rigorous ectental chewing. The degree of mobility of the symphysis in life of this plesiadapid is difficult to judge. One can infer, however, that selective premium may have been placed on an increasingly immobile, strong symphysis.

Because of the brevity of the mandible, the distance between the incisor and P_3 is very short. Although *Chiromyoides* is dentally a relatively primitive plesiadapid, the great depth of the mandible and the robust incisor of this superficially daubentonioid-like form indicates a mode of life considerably more specialized than that of other plesiadapids.

Differences of the early Eocene European (and possibly North American) *Platychoerops* from species of *Plesiadapis* are stated by Donald E. Russell et al. (1967) to lie principally in the possession by *Platychoerops richardsoni* of two instead of three cusped upper incisors, lack of a paraconule on P^4, well developed molar mesostyles, and in general, having rugose crenulated enamel which sometimes produces *de novo* cuspules on the teeth. Judged from the palate of *P. richardsoni,* the length of the skull was over 10 cm.

The recently described very small *Saxonella* (D. E. Russell 1964, pp. 128–132) from late medial Paleocene fissure fills near Walbeck, Germany, represents a new group of early primates (Plate 1–8). Although the closest affinity of *Saxonella crepaturae* is with primitive plesiadapids (and not with carpolestids, as correctly recognized by Van Valen 1969) such as *Pronothodectes,* an excellently preserved left mandible clearly shows adaptations different from those of other plesiadapids.

On the basis of premolar morphology (which shows adaptations possibly different from carpolestid premolars), incisor structure, and a *Pronothodectes*-like upper molar morphology, *Saxonella* is best considered as a plesiadapid (Van Valen 1969) rather than a carpolestid (D. E. Russell 1964). Russell's original subfamily designation, however, is fully justified.

The type mandible[4] of *Saxonella* is one of the best preserved Paleocene primate lower jaws (Figure 1–5). The coronoid process, the angle, and the articular condyle are intact. The horizontal ramus is short and deep and the coronoid process is high and relatively wide anteroposteriorly. The angle is

---

4. Unfortunately, my recent attempts to locate this specimen at Halle, D.D.R., have failed, and so the only record of this fossil is the photographs in D. E. Russell 1964 (pl. 8, figs. 4 and 5).

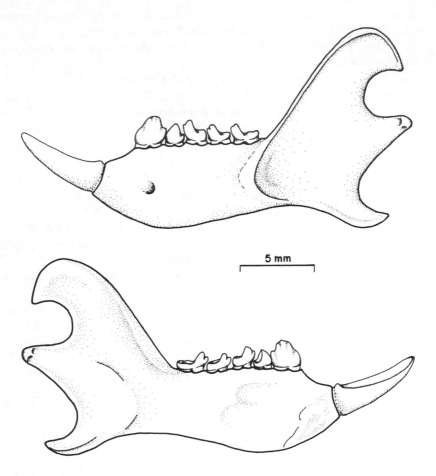

Figure 1–5. *Saxonella crepaturae,* lateral view of left mandible, based mainly on the type WA/351, late medial Paleocene. Lateral view above and medial view below.

somewhat reduced in length and it is not particularly deep. It appears, judging from the deep masseteric fossa and the trenchant P_3, that the orthal and ectental (side to side) components of the bite were more important than the propalinal one. One observation relates to the fully plesiadapid-like, mitten-shaped upper incisors (Plate 1–9). Unlike plesiadapine incisors, those of *Saxonella* are relatively wider transversely, and the prong-like cusps are more separated from each other on a horizontal plane. This may reflect a more efficient grasping, anchoring, husking ability of those incisors than the known homologous plesiadapid teeth may have been capable of. Yet extensive inferred functional comparison between *Saxonella* and plesiadapines is not very meaningful considering the size difference between *Saxonella* and other members of the family.

THE CARPOLESTIDAE

Carpolestids are known from the period between the medial Paleocene and early Eocene in North America. The diagnostic characters of the carpolestids lie in the construction and progressive specialization of the fourth lower premolar, the third and fourth upper premolars, and the first lower molar. Comparison of the molars of the primitive carpolestid *Elphidotarsius* with those of the plesiadapid *Pronothodectes* reveal that these genera share a

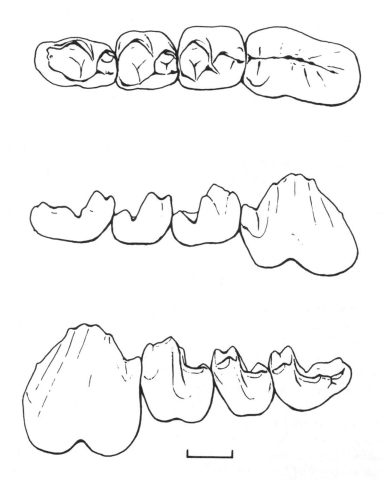

Figure 1–6. *Carpodaptes aulacodon*, AMNH No. 17367, left P_4–M_3, occlusal (above), lingual (middle), and buccal (below) views. Scale represents one millimeter.

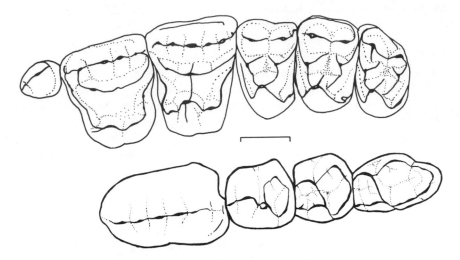

Figure 1–7. *Carpodaptes hazalae*, AMNH No. 33980, left P²–M³ and P₄–M₃ (lower dentition is reversed), late Paleocene, occlusal views. Scale represents one millimeter.

balance of characters different from those found in other Paleocene primates. There is little doubt that the shared similarities are indicative of special ties between the two families instead of convergence. Of the two closely related families, the plesiadapids tended to be relatively larger forms. The known species of carpolestids ranged from mouse- to rat-sized animals; the following species have been described:

1. *Elphidotarsius florencae* Gidley 1923 (Torrajonian, Montana).
2. *Carpodaptes aulacodon* Matthew and Granger 1921 (Tiffanian, Colorado).
3. *Carpodaptes hazelae* Simpson 1936 (Tiffanian, Montana).
4. *Carpodaptes hobackensis* Dorr 1952 (Tiffanian, Wyoming).
5. *Carpolestes nigridens* Simpson 1928 (Tiffanian, Montana).
6. *Carpolestes aquilae* Simpson 1929 (Tiffanian, Montana).
7. *Carpolestes dubius* Jepsen 1930 (Tiffanian, Wyoming).
8. *Carpolestes cygneus* L. Russell 1967 (Tiffanian, Alberta).

The only published specimen[5] of *Elphidotarsius* is a mandible fragment, containing P₄–M₃, the type of *E. florencae* Gidley 1923. The medial Paleocene form is a small carpolestid with a relatively shallower mandible than the deep-jawed late Paleocene species of the family. *E. florencae* has a

5. For several decades, excellent specimens of *Elphidotarsius* and other carpolestids from northwestern Wyoming have been collected by G. L. Jepsen. These undescribed specimens number close to one hundred and include cranial elements.

moderately enlarged blade-like P_4 with four cusps and a cuspate remnant of the talonid. Unlike younger, more advanced carpolestids, the M_1 trigonid in *E. florencae* maintains the primitive arrangement of the early primate trigonid. By contrast, in species of *Carpodaptes* and *Carpolestes* the trigonid cusps of the first molar align themselves mesiodistally to form a suplementary cutting edge distal and continuous to that of the P_4 blade. This latter trend is not supposed to be as pronounced in *Carpodaptes* as in the slightly more advanced species making up the genus *Carpolestes*.

Differences between the late Paleocene carpolestid samples are minor, and not as great as differences between any of these samples and the genotype of *Elphidotarsius*. A preliminary study of carpolestids now in progress reveals that all other known, described, taxa previously allocated either to *Carpodaptes* or *Carpolestes* could be accommodated within *Carpodaptes,*[6] the generic name with priority. This procedure would place the genera of Carpolestidae in a meaningful systematic balance with other polytypic Paleocene primates, such as *Paromomys, Phenacolemur,* or *Plesiadapis.*

When Simpson (1928, pp. 7–10) described *Carpolestes nigridens,* the genotype of *Carpolestes,* he gave no adequate generic diagnosis, primarily because it could not be given. Until the range of variation in the number of cuspules on P_4 is studied in an adequate sample, this will remain a precarious generic feature. The same applies to the supposed distinction of the more talonid-like posterior cusp on the P_4 in *Carpodaptes* compared to that of *Carpolestes.* The degree to which specializations in the third and fourth upper premolars are diagnostic species characters among various late Paleocene samples of carpolestids is unknown. Thus, features of the upper premolars also must await judgment as to their usefulness as generic characters. I can find no features of generic magnitude in the lower molars that might separate the two late Paleocene genera. As Simpson (1928, p. 8) states, the "M_{1-2} are almost identical in structure" on the genoholotypes of *Carpodaptes* and *Carpolestes.*

The carpolestids are badly in need of a monographic revision that would encompass all the known but hitherto undescribed specimens of the family.

I believe that the dental formulae of known carpolestids have been misinterpreted. Jepsen (1930) gave the tentative lower dental formula for *Carpolestes dubius* as ?1, 0, 4, 3; whereas Simpson (1935, 1937) considered the dental formula of *Carpolestes* to be

$$\frac{?, ?, 3, 3}{1, 1, 3, 3} .$$

This latter interpretation was accepted and followed by Dorr (1952).

My interpretation is slightly different from that of Simpson. I consider

---

6. Including proposed species of *Carpolestes* in *Carpodaptes* would not alter the validity of the family-group name, Carpolestidae (see Art. 40 in the Int. Code of Zool. Nom., 1961).

the first three teeth of all known carpolestids to be the enlarged first (?) incisor, the second (?) incisor, and the small, reduced canine. I believe premolar reduction proceeded from front to back, thus the third antemolar tooth of *Elphidotarsius* is P_2, whereas the second antemolar tooth of late Paleocene carpolestids is P_3. The dental formula of late Paleocene carpolestids may be written then as:

$$I^{1(?),2(?)}_{1(?),2(?)};\ C^?_1;\ P^{2,3,4}_{3,4};\ M^{1,2,3}_{1,2,3}.$$

Although speculation about the homologies of the carpolestid upper dentition need to be confirmed from new specimens, I strongly suspect that the homologies of the upper dentition were probably identical to those of the medial Paleocene paromomyids. Thus, the carpolestids probably possessed two upper incisors, one canine, three premolars, and three molars.

One of the best preserved, described carpolestid specimens is a fragmentary mandible, the holotype of *C. hobackensis* Dorr 1952. In addition to the presence of a nearly complete enlarged incisor, its alveoli and natural casts of roots clearly show three teeth between the enlarged incisor and the fourth premolar. These teeth are interpreted to have been a single-rooted P_3, a small canine, and a small incisor. There is no reason to suppose that the lower tooth homologies were different in any of the other published *Carpodaptes* or *Carpolestes*. New illustrations and a reconstruction of *C. hobackensis* are given in Figure 1–8.

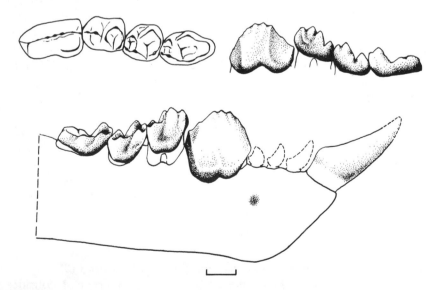

Figure 1–8. *Carpodaptes hobackensis*, UMMP No. 27233, occlusal view (upper left) and lingual view (upper right) of P_4–M_3; below is lateral view of slightly restored mandible with $I_{1(?)}$, $I_{2(?)}$, C, and P_3–M_3. Scale represents one millimeter.

I have recently identified an upper incisor (Figure 1–9) of the Bear Creek *Carpolestes aquilae*. This is the first known published carpolestid upper incisor. By comparison to those of species of *Plesiadapis,* it reveals a considerable degree of similarity, presumably due to common inheritance probably in early medial Paleocene times. This carpolestid incisor has two distinct cusps on the mesial border of the crown but it lacks the distal cuspule which characterizes most known plesiadapids, including *Saxonella.*

It appears that the carpolestids represent taxa adapted to an unknown vegetable diet of high fiber content. The detailed convergence (at least of the lower premolars) between distantly related mammals such as multituberculates of the Mesozoic and early Cenozoic, caenolestoid marsupials of the early and later Tertiary (abderitine caenolestids, polydolopids), and the late Tertiary phalengeroid marsupials (phalengerids and macropodids), is probably due to adaptations to similar vegetable diet (Simpson 1933).

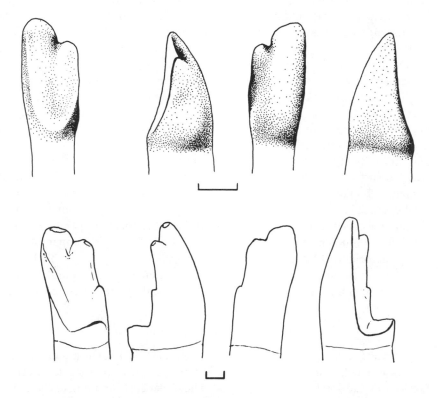

Figure 1–9. Enlarged upper incisors of *Carpolestes nigridens* (?) (AMNH No. 22152) and *Plesiadapis gidleyi* (AMNH No. 17171) from the late Paleocene. The views from left to right are occlusal (and distal), buccal, mesial, and lingual. Scales represent one millimeter.

Unfortunately, although there are several extant species of marsupials with plagiaulacoid-type of premolars, no detailed study of the morphology, ecology, and diet of these forms has been undertaken. Merely as a hopefully research provoking hypothesis, the generalization may be ventured that the plagiaulacoid premolars, at least in small- to medium-sized mammals, evolved in response to a diet which contained a large percentage of tough fibers. Perhaps this food was roots, bark, or fruits with a tough fibrous coat. This inference is largely unsubstantiated, and it is hoped that detailed comparative work on relevant extant species will be carried out in the future.

The inference that the carpolestid premolar-molar dentition functioned differently from that of other known therians with a plagiaulacoid dentition may be derived from the combined upper and lower cheek tooth morphology and the wear facets on the teeth. In living mammals such as the macropodid *Hypsiprymnodon* or the phalangerid *Burramys,* the plagiaulacoid lower premolars occlude against trenchant upper premolars, and this arrangement is true of most fossil caenolestoids. Unlike other plagiaulacoid therians, however, carpolestids utilized their cheek teeth in a different manner. The cutting edge of the lower dentition is supplied primarily by the blade-like, hypsodont P_4, whose cutting edge is supplemented by the peculiar trigonid elongation of the first molar in more advanced carpolestids. It appears that once lateral excursion of the mandible has occurred (the preparatory stroke of Crompton and Hiiemae 1970), the apical edge of P_4 contacted the upper cheek teeth commencing medial excursion of the active mandible (power and/or shearing stroke of Crompton and Hiiemae 1970). As P_4 moved across the occlusal surface of P^3 and P^4, moving slightly upward, mesially and lingually, the cutting edge of its blade worked against the file-like, studded surface of the third and fourth upper premolars. This arrangement explains why the carpolestid blade is so much higher crowned than the upper premolars. The relatively narrow occlusal surface of P_4 (compared with the large occlusal surface of the combined third and fourth upper premolars) is worn at a much more rapid rate, but the speed of wear is compensated by the great height of its crown. Thus the hypsodont carpolestid blade is a compromise solution for a device that is both trenchant and long-lasting in occlusion with the flat upper premolars of unusually large surface area.

Evolutionary Trends and Function

There are several important aspects of the earliest known major primate radiation which will be of increasing importance in evaluating the various families as the fossil record improves. The Paleocene and Eocene families undoubtedly differed in their geographical distribution, in the number of species evolved, and presumably in the population density of the various species. In spite of the fact that large quarry samples of sundry species now make it possible to ask questions of the fossil record related to the population biology, no one has tackled problems of this nature for early primates. Studies on the population biology of Paleocene and Eocene primates will surely supply additional criteria for illuminating the biology of these species in addition to the taxonomic and functional criteria utilized in this report.

FEEDING ADAPTATIONS

Before I attempt to characterize some of the main evolutionary trends in the feeding mechanisms of Paleocene primates, I would like to discuss briefly a frequently neglected facet in interpretations of feeding adaptations. It is important for the paleobiologist making inferences from fossils and from the inferred functional evidence to consider the absolute body size of members of a radiation, to the extent that this can be determined from the usual representatives of a species, its teeth and jaws.

It appears (but, of course, it cannot be proven) that the various Paleocene families of primates fed predominantly on plant materials. However, consideration of the absolute and relative sizes of different species in conjunction with their dental morphology may allow more detailed interpretations of individual species. Absolute size of an animal, particularly at the extremes of the size range, affects its dietary requirements and preferences. Thus, large herbivores such as gorillas or kangaroos, process great bulks of vegetable matter with low concentrations of nutrients. This is possible because their small surface-to-volume ratio allows for relatively slow energy assimilation with relatively little loss of energy through the radiation of heat. By contrast, a tiny nectar-feeding bat or a shrew, each with very large surface-to-volume ratios, need highly concentrated nutrients—sugars for the bat and animal protein for the shrew—in order to maintain the necessary amount of energy assimilation concomitant with continuous high heat loss. Thus, to regard the small, insectivorous and carnivorous *Microcebus murinus,* for example, as retaining the alleged primitive feeding preferences in primates (contra Crook and Gartlan 1966) would be a mistake. It might be more meaningful, however, to consider this tiny member of the overwhelmingly herbivorous lemur radiation to be a specialized feeder on animal proteins, due to the requirements of its small size.

Thus postulating feeding on fruit pulp, seeds of fruits, or nectar for the tiny paromomyid derived Paleocene picrodontids is warranted on the basis of their size and the convergent similarity of their cheek teeth with nectar and fruit pulp feeding bats. The high concentrations of quick energy found in nectar would be essential for the high metabolic rates in such a tiny herbivore.

The Paleocene radiation of primates in North America and Europe is unique in regard to the particular development of incisor enlargement coupled with (a) premolarization of the canine or its complete elimination, (b) occasional specialization of the third or fourth lower premolar, and (c) the loss of one or two incisors and one, two, or three premolars. Incisor and premolar reduction in any one species was usually less extensive on the palate than on the mandible.

It appears that the great variety of molar patterns in the Paleocene species, derived from ancestors similar to the more primitive paromomyids such as the palaechthoninines, were not adapted for the mastication of fibrous materials of animal bodies such as chitin or muscle. Instead, primitive primate molar adaptations can be interpreted as responses to a herbivorous-frugivorous shift in feeding from a primarily insectivorous (i.e., proteinivorous) ancestry sometime in the Cretaceous (Szalay 1968a, 1969a). Subsequent specializations in the antemolar dentition of the Paleocene descendants of the earliest primates may be related then to diversified specialization of herbivores and frugivores.

Occasionally, individual dental remains can be excellent in depiciting dietary preferences of a mammal (e.g. the fish-catching teeth of a delphinid odontocete whale, or the grinders of an elephantid). Yet they can be deceptive or, at best, noncommittal in many ancient or even geologically young species of primates. Although the molar structure that characterized the earliest primate (a hypothetical form at present) was presumably adapted for more efficient mastication of a diet more herbivorous than that of its insectivoran ancestor, the recurrence of this particular crown pattern with minor modifications tells us very little about the diet of its numerous Paleocene descendants. Similarly, the molar morphology of *Tarsius* would tell us little about the fierce predatory habits of this taxon were we to recover it as a fossil. Thus, although molar characters might be highly informative regarding the initial adaptations of a supraspecific category, later species of the radiation can retain characters as heritage features and merely use them as building blocks in differently adapted feeding mechanisms. The total dentition, however, when considered as a functional unit, particularly in conjunction with the jaw and the skull, can be more informative for fossils. Were *Tarsius* a fossil, its vertically implanted dagger-like incisors, canines, and premolars might indicate adaptations for prey catching. *Microchoerus* and *Necrolemur,* alleged tarsiids, in spite of the adaptive similarity of the skull to the living *Tarsius,* show no apparent piercing adaptations in their antemolar teeth. As a general guideline, then, the total dentition of Paleocene primates, along with the size of the animals, can be informative of their dietary adaptations.

It is a well-known fact that living primates have solidly fused mandibular symphyses, yet the earliest members of the order, like all primitive mammals —with the possible exception of *Chiromyoides*—had a mobile joint at the symphysis. In all later lineages, however, including genera of the Madagascan radiation, the symphysis fuses early in ontogeny. It appears that among the primates, symphyseal fusion is an adaptive response to withstand great, horizontally directed stresses during chewing of plant materials, as opposed to the primitive placental condition where the stresses of orthal shearing occlusion were primarily vertically directed.

Enlargement of a median pair of incisors above and below, the lowers forming a spoon-like device and the uppers a multicusped, broad anchor, are clearly features related to food procurement and manipulation. Reduction of the primitively piercing canines and the nipping incisors in favor of the two pairs of specialized incisors in paromomyids, picrodontids, plesiadapids, carpolestids, and somewhat similar specializations, although not homologous, in the primarily Eocene anaptomorphids and microchoerids indicate an ubiquitous invasion of numerous early primates into a broad herbivore-frugivore adaptive zone. The particular emphasis on the incisors and de-emphasis of the canines would indicate a general lack of predatory habit, at least among the plesiadapoids.

In three Paleocene suprageneric categories there was independent specialization of one of the posterior premolars. In the paromomyid *Phenacolemur,* in different species of the genus (e.g. *P. pagei),* the fourth lower premolar becomes a highly premolariform, tall, and robust tooth, larger and taller in some species than the first molar. Its function is best considered to be a seed and nut cracker. The Paleocene carpolestids evolved dental specializations that are unique among primates and functionally unique among all other plagiaulacoid mammals such as multituberculates and several groups of marsupials (see above, under Carpolestidae). In contrast to *Phenacolemur,* carpolestids did not develop incisors to the degree seen in the former genus.

Saxonella shows a unique adaptation among Paleocene primates in the combination of a single enlarged incisor and a slightly plagiaulacoid P_3. Unlike the M_1 trigonid of carpolestids which is transversely constricted to supplement the cutting edge of P_4, both the P_4 and M_1 of this tiny plesiadapid are relatively unmodified.

Only the picrodontids show drastically unusual molar specializations among the Paleocene primates, probably in adaptation for a fruit pulp and fruit nectar diet.

Although the horizontal ramus is shortened and deepened among paromomyids, picrodontids, plesiadapids, and carpolestids, the ascending ramus and the mandibular angle remained largely unmodified among available species of these families. As in most small, primitive Paleocene mammals, the articular condyle is not far above the level of the tooth row and the angle of the mandible is long and tapers to a point. Whether the anteroposteriorly elongated thick dorsal edge of the coronoid process is a primitive trait among early primates or merely an independent specialization in different groups is difficult to evaluate. Although this dorsal segment of the coronoid process is short in adapids, a probable advanced condition, an anteroposteriorly long dorsal segment of the coronoid process is present in known fragments of early leptictids and small early condylarths, both of which have a full eutherian dentition and occlusal pattern not drastically different from adapids with a full dentition.

It may be inferred from the relatively large coronoid process and the thickened bone (due to great stressing) running from the articular condyle to the muscle scar on the antero-medial side of the coronoid process (for temporalis attachment) that paromomyids, picrodontids, plesiadapids, and carpolestids had unusually large temporalis muscles for herbivores. The explanation for the inferred large temporalis can be sought partly in the retention of the primitively important insectivoran temporalis mass where stressing at the canines was mainly resisted by that muscle. In the plesiadapoid families which eliminated canine function but greatly increased specialized incisor activities, the large temporalis is retained to resist stressing of the enlarged incisors. This is an unusual musculoskeletal arrangement for herbivores which are more often characterized by the importance of the masseter and internal pterygoid musculature. Although the temporalis continued to play the important masticatory function of the insectivoran ancestors in Paleocene primates which reduced the canines and enlarged the incisors, almost all lineages of post-Paleocene primates placed greater emphasis on masseter and pterygoid functions, the usual specialization among herbivorous mammals.

The known articular condyles of Paleocene primates are more or less bilaterally symmetrical. Both medial and lateral halves are about of equal proportions and dimensions. It is difficult to assess the primitive placental condition (see Szalay 1969c), or even the pre-primate condition of the condyle. This highly informative part of the mandible is undoubtedly correlated with chewing habits, and these in turn with diet of the species. Diet and subsequent masticatory adaptations are without question of great importance because in Paleocene insectivorans, many of which are in the size range of paromomyid primates, one finds relatively small, rounded articular condyles as well as robust, transversely very wide ones (unpublished information from the late Paleocene Bison Basin fauna). Whenever known (in *Plesiadapis* and *Phenacolemur*), the mandibular condyle is not restricted anteriorly by a preglenoid process on the mandibular fossa, although a very peculiar condition of the mandibular fossa is present in *Phenacolemur*.

Table 1–1 depicts levels of specializations in the loss of lower teeth among Paleocene primates. Although an obvious continuum exists from the primitive expression of incisors, canines, premolars, and molars to the hyperemphasis of certain teeth or to their complete loss, the table merely sets up heuristic boundaries to indicate degrees of importance of the antemolar dentition in the different families. In the medial Paleocene, most primates with known dental formulae lost one incisor and the first premolar and deemphasized canine function by relegating this tooth to be a premolariform, small tooth behind one moderately and one greatly enlarged incisor. These forms were *Palaechthon, Plesiolestes, Paromomys,* and perhaps also *Elphidotarsius*. Other primarily medial Paleocene forms such as *Palenochtha* and *Picrodus* have lost two anterior premolars, a single incisor, and reduced the size of the canine. While *Pronothodectes* and carpolestids lost only one incisor and two premolars, *Plesiadapis, Chiromyoides, Platychoerops,* and *Saxonella* functioned with one greatly enlarged incisor and the third and fourth premolars. In primitive *Plesiadapis,* such as *P. gidleyi* of North America, a peg-like P_2 persisted. As early as the late Paleocene, *Phenacolemur* evolved a combination of one enlarged incisor and only the fourth premolar in place in addition to the molars. This highly successful taxon lacks six pairs of its lower teeth and persists into the early Tertiary longer than any other Paleocene genus of therian mammals. From the late Paleocene to the late Eocene, a period of 16 to 18 million years, *Phenacolemur* is part of the North American mammal fauna, whereas *Plesiadapis,* the second most specialized genus among the Paleocene forms in terms of tooth reduction, persisted into early Eocene times on both sides of the North Atlantic.

Table 1–1. Levels of specialization in the loss of lower teeth in the Plesiadapoidea

	Families of Primarily Paleocene Primates			
Teeth Lost	*Paromomyidae*	*Picrodontidae*	*Plesiadapidae*	*Carpolestidae*
Two incisors, canine, and three premolars	*Phenacolemur*			
Two incisors, canine, and two premolars			*Chiromyoides* *Platychoerops* *Plesiadapis* *Saxonella*	
One incisor, and two premolars	*Palenochtha* *?Navajovius*	*Picrodus*	*Pronothodectes*	*Carpolestes* *Carpodeptes*
One incisor, and one premolar	*Paromomys* *Plesiolestes* *Palaechthon*			*?Elphidotarsius*
One incisor		(hypothetical ancestor)		

In conjunction with incisor emphasis and antemolar tooth reduction in the Paleocene radiation the hypertrophy of M_3 must be noted. No known Paleocene primate lost its M_3, but instead there is a clear, possibly independent, trend in all four families to increase the surface area of the last molar by elongating the talonid.

It appears that the survival of the last plesiadapoids is correlated with feeding specializations involving a combination of strong incisivation and unreduced molar function of the masticatory system. Curiously, extreme cheek tooth specialists, such as the picrodontids, and *Carpodaptes* to some extent, do not appear to have survived in known Eocene faunas. Whether the demise of the Paleocene species was due solely to competition from rodents and microsyopids, as well as from the invading anaptomorphid and adapid primates is difficult, if not impossible to judge at the present.[7]

The most specialized cheek teeth of Paleocene primates are those of picrodontids and carpolestids. With these extreme dental specialists in mind there appears to be no particular correlation between molar specialization and the extent of tooth reduction.

The most extreme changes in mandibular morphology and inferred cranial modifications do not appear to be correlated with extremes of dental evolution. Thus, although *Chiromyoides* retains a relatively primitive molar morphology, it had evolved an extremely short and deep mandible with a greatly enlarged angle, a symphysis specialized to resist great stresses, and a muzzle reduced almost to the anterior border of the zygomatic arch. *Picrodus,* by contrast, although showing extreme specializations in molar development, retained a mandible not very different from the primitive paromomyid stock.

LOCOMOTOR ADAPTATIONS

Napier and Walker (1967a) recently described a newly recognized locomo-for category[8] among primates—vertical clinging and leaping. This particular locomotion was characterized to consist of leaping almost exclusively by the use of the hindlimb and in clinging vertically in the rest posture (Napier and Walker 1967a, 1967b; Napier 1967). The authors explicitly state in several instances (with particular emphasis in Napier and Walker 1967b, p. 69) that this specialized category represents the primitive condition among the primates. They further note that quadrupedalism "undoubtedly . . . bridges the gap between vertical clinging and leaping and brachiation" (Napier and Walker 1967a, p. 214).

Among the morphological characters listed by the authors, two can be seriously questioned as having any relevance to vertical clinging and leaping or, for that matter, to a vertical posture in general. The centrally placed foramen magnum is correlated with relatively large brains in absolutely small and tiny mammals and to relatively very large brains in larger mammals. A shortened facial region can be correlated with selection for stereoscopy. Both of these points are clearly made by Biegert (1963); further

7. Further speculations as to competitive relationships involving rodents, microsyopids, *Phenacolemur, Plesiadapis* and the Eocene primates of North America is not meaningful until locomotor adaptations and habitats of these taxa are better understood.

8. Separation of the locomotor pattern of any one species into locomotor categories might be considered artificial or of no scientific value. No doubt, a similar objection might be raised against supraspecific taxonomic categories or descriptive phenomena of evolutionary biology other than those dealing with species biology. Clearly, the heuristic advantages of delimiting parallel, convergent, or divergent evolutionary pathways in locomotor specializations and concomitant structural and behavioral modifications amply justify the usage of locomotor categories.

restatement of his arguments is unnecessary. Although these two features are of no value in assessing posture or locomotion, they have been extensively used to do just that.

In addition, citations by Napier and Walker of osteological characters allegedly known from the fossil record are based on inaccurate, outdated literature. The evidence cited to imply that *Tetonius* was a vertical clinger and leaper (which it might or might not have been) is the centrally placed foramen magnum and the widely divergent mandibular rami. Both of these alleged characters of *Tetonius,* particularly the position of the foramen magnum, are incorrectly inferred from the fragmentary evidence. In the only known skull of this genus, the entire occipital region is missing. Divergent rami were also cited as evidence for vertical clinging and leaping in *Aeolopithecus, Parapithecus, Amphipithecus, Microchoerus,* and *Necrolemur.*

The much disputed Microsyopidae apparently heavily influenced Napier and Walker's primitive primate locomotor pattern. The elongated calcaneum and navicular is not known in *Microsyops* (contra Napier and Walker 1967a, p. 213), and with the exception of a fragmentary tibia (Szalay 1969a, p. 303) there are no recognized postcranial remains which can be allocated to any of the numerous early and middle Eocene species of the family. The original reference to an elongated calcaneum (and inferred elongated navicular) in *Microsyops* was made by Wortman (1903). Simpson (1940) allocated this specimen to *Hemiacodon,* a medial Eocene omomyid, a fact which was never widely publicized in the secondary literature. Since the discovery of the ear region of *Phenacolemur,* the total available evidence strongly suggests that microsyopids should be considered as leptictid derived Insectivora rather than having close ties with any Paleocene or Eocene primates.

The fossils used by Napier and Walker to argue for the ordinally primitive nature of vertical clinging and leaping are all post-Paleocene, and most are medial Eocene in age. Some of these possibly correctly identified vertical clingers and leapers[9] such as *Smilodectes, Hemiacodon, Necrolemur,* and *Microchoerus* are at least twenty million years younger than the origins of the order in late Cretaceous times. Most, if not all, of the osteological characters enumerated by Napier and Walker could have easily evolved convergently.

Napier (1967, pp. 334–335) makes subtle and I believe incorrect use of the concept of "truncal uprightness" to support the all-pervasive primitive nature of vertical clinging and leaping among primates. Although truncal uprightness is commonly practiced among numerous species of rodents, the

9. The osteological characters that might be invariably associated with vertical clinging and leaping have not been deduced as yet, if such clear-cut features exist at all.

implied necessary vertical clinging and leaping correlate does not follow in that order. In spite of the fact that vertical clinging and leaping had become a recurrent locomotor pattern among many groups of primates, this does not bear on the systematically precise meaning of the term "primitive" as far as locomotion is concerned for the order Primates. As a similar example, few scientists dispute the hypothesis that rigid requirements of an arboreal existence resulted in binocular adaptations of greater or lesser degree in many lineages, yet the paleontological record unequivocally shows that Paleocene and even some Eocene forms were not stereoscopically adapted.

It is of some value to compare the brachial, crural, and intermembral indices of two of the best known early Tertiary postcranial skeletons, those of the late Paleocene *Plesiadapis* (see Simpson 1935, and composite figure in Simons 1964b, from two species of the genus) and the medial Eocene *Notharctus* (*Notharctus tenebrosus*). The indices are 93 and 94 for the pectoral limb, 93 and 84 for the pelvic limb, and 81 and 64 for the intermembral proportions in the two genera, respectively. Perusal of the brachial indices of genera categorized as vertical clingers and leapers in Napier and Napier (1967, pp. 393–395) will show that the brachial indices of *Plesiadapis* and *Notharctus* are well below those of living taxa of that locomotor category. Although an intermembral index of 64 for *Notharctus tenebrosus* is clearly within the range of living vertical clingers and leapers, *Plesiadapis* with an index of 81 cannot be considered a vertical clinger and leaper. The similarity between the brachial indices of *Plesiadapis* and *Notharctus* is striking, and it might represent the primitive size relationship between the radius and humerus of the earliest primates. An intermembral index between 70–80 might have been primitive, and the respective indices of 81 and 64 might represent specializations primarily achieved by altering length proportions of the hindlegs.

The presence of claws rather than nails in early primates, as seen in *Plesiadapis,* is probably one of the most decisive lines of indirect evidence against vertical clinging and leaping on a Paleocene level of organization. Although claws are undoubtedly helpful in claw-climbing and walking on trees, the necessary sudden and bold jumps which characterize extant vertical clingers and leapers would be seriously impeded by claws catching on bark and upsetting the animal's precarious balance.

In conclusion, it appears to me that living species such as *Tarsius,* galagines, and various lemuroids attained their vertical clinging and leaping specializations independently. If species of the earliest primate radiation which were ancestral to later radiations had intermembral indices between 75–80, and crural and brachial indices in that range or higher, and were able climbers and jumpers, then they possessed all the necessary preadaptations for some lineages to become vertical clingers and leapers whenever selection favored that specialized locomotor pattern. *Plesiadapis* is an advanced form in some respects, compared to some post-Paleocene members of the order, but its known postcranial anatomy cannot be interpreted to reflect a secondarily quadrupedal mode of locomotion from an ancestry adapted to vertical clinging and leaping.

2

ELWYN L. SIMONS AND DAVID R. PILBEAM
YALE UNIVERSITY

Hominoid Paleoprimatology

In a document of this length, it is almost impossible to deal adequately with all the various aspects of paleoprimatology of the Old World hominoids. Probably the most important single thing that should be said about future study of the fossil primates of this area is that much more effort should be directed toward recovery of new and better material of them. What we have is by no means adequate to interpret their evolution. Surface prospecting for fossils seems a deceptively simple procedure; there are many self-trained fossil collectors. In spite of the fact that many areas in Europe, Eurasia, and North, South, and East Africa have been collected for higher primate fossils for more than forty years, in only a small fraction of these has a thorough search been made for concentrations of small fossil vertebrates ("microfaunas") and for quarryable sites. In consequence, many regions which might once have been considered exhausted, worked out, or now subject to the law of diminishing returns, are not really in such condition and deserve further field exploration. For instance, although it is unlikely that new species of fossil apes will be found in the Kenya Miocene, there can be little doubt that long-term quarry excavation will yield more complete cranial and postcranial material. Such intense investigation involving many hundreds of man-hours in collecting has so far only been carried out in one place and time: by the Leakeys at Olduvai Gorge, Tanzania.

Areas for Future Excavation

The principal limitation to collection of new fossil primates, particularly in the Old World, is not so much that they are difficult to find but that so few scholars are actually prepared to search for them. Recovery of fossil primates is principally dependent on the choice of areas which were ecologically suitable for these primarily forest-dwelling animals. Scientists should also concentrate on localities which yield remains of fossils in sufficient abundance that the probability of recovering primates as part of the total fauna is high. European sites have been much studied, but this is far from meaning that that particular area is exhausted. The most dramatic example of the results of new work at old sites is the discovery in the late 1950's of the nearly complete skull and numerous other associated remains of the archaic prosimian *Plesiadapis* by Russell's expeditions near Cernay-lès-Reims, in northern France. These Paleocene sites, near the villages of Cernay and Berru, were first discovered in the 1870's and worked by Lemoine in the late nineteenth century. Lemoine's sites were of known location, but for a whole lifetime no further serious excavations in the region were carried out. It remained for Russell to prove, by the success of his excavations, which produced a whole suite of vertebrate fossils, how profitable further digging there could be. Remains included three Paleocene primates, *Plesiadapis, Cheiromyoides,* and a new form, *Berruvius* (see Russell 1964). Other European early Tertiary sites remain available as potential objectives for future excavation. Two of these are the deposits, latest Eocene in age, at Hordwell, Hampshire, on the southern coast of England, and early Oligocene sites on the Isle of Wight, both of which have produced significant fossil finds, including primates. Within the limits of London there is an early Eocene site at Abbey Wood in Kent, where adapid prosimians have been found. Another area is the region of the Eocene Quercy phosphorites in south central France. These phosphatic deposits were extensively quarried in the nineteenth century for economic purposes, and are often said to be exhausted. We believe that this is not the case. In the course of quarrying for phosphates, many fossil mammals were discovered including a large number of highly significant primate skulls and jaws. A definite possibility of locating further phosphorite fissures exists and should be explored.

The mid-Tertiary deposits in the vicinity of Saint Gaudens, in France, should also be further excavated with the objective of locating additional material of the type species of *Dryopithecus, D. fontani.* The cliff face from which the three original jaws of *Dryopithecus,* several isolated teeth, and one humerus were removed, could be quarried more extensively for a relatively low cost. Numerous finds of fossil ape specimens of genus *Dryopithecus* have recently been located by Crusafont in northern Spain. These indicate the potential of this region of southern Europe for contributions to our knowledge of higher primate evolution. None of the *Dryopithecus* specimens which Crusafont's group has found in the last five or ten years is very complete. The best of them is a left maxillary fragment with C through M^2 (see Crusafont 1965). The principal sites, Can Ponsich and Can Llobateres are, nevertheless, in sediments which have yielded skulls and associated skeletons of other mammals. Adequate financing of his projects there should eventually produce skulls or associated postcranial remains of *Dryopithecus.* Such new finds could be of the utmost importance for the understanding of hominoid evolution.

Turning to a consideration of North African sites, the significance of the fossil primates recovered from the Fayum Oligocene deposits of Egypt is well known. Further work there should be carried out whenever political conditions make this feasible. The Eocene and Oligocene continental and coastal deposits of the Fayum trace laterally into Libya and there is a high probability that Oligocene vertebrates will be discovered in some abundance in central or northern Libya if sufficient field exploration is carried out. In southern Libya in the Jebel Coquin region, field groups working first with Arambourg and more recently with Savage of the University of Bristol have already established that Eocene mammalian faunas deposited in continental sediments are present there. As these become better known it may be possible to determine how long ago primates arrived in Africa. It would be advisable to see that a variety of trained collectors reach this region because, here, as in general, different training and abilities at locating sites increase the effectiveness of exploratory programs.

There are indications that early Tertiary sediments which might yield continental vertebrates exist in Nigeria and there may be deposits of middle Tertiary age in Saudi Arabia, Oman, and possibly in southwestern Sudan, as well as the better known sites in Uganda and Kenya. The westernmost Congo has already produced a Miocene vertebrate fauna described by Hooijer (1963). Primates are known in great abundance from Miocene, Pliocene, and Pleistocene sites in Ethiopia, Uganda, Kenya, and Tanzania. Every effort should be made by scientists and funding agencies to support and promote further expeditions in these areas. In spite of the magnificent collections assembled by the Leakeys and other British-based groups, the region is still largely untapped. The more complete associations of skeletal bones or relatively complete parts of skulls, which could certainly be recovered by extensive quarrying in the Rusinga Island sites, for instance, are yet to be found because so much of the field work to date has been restricted to surface prospecting alone, and not to long-term quarrying of the sites richest in fossil remains. Vast areas east of Lake Rudolph have never been even surface prospected. Because of continued weathering, even the classic East African Miocene sites around the Kavirondo Gulf of Lake Victoria should receive constant attention.

All recent field work in Ethiopia, Tanzania, and South Africa indicates that those areas which have in the past yielded remains of *Australopithecus* are likely to produce further finds of great significance. Saudi Arabia, Iraq, and Iran represent paleontologically relatively underdeveloped regions in which known exposures of middle Tertiary continental rocks occur. Further, more intensive exploration in these countries is likely to add to our knowledge of Cenozoic mammalian faunas, with contained primates as well.

The Siwalik hills of North India and West Pakistan have been a traditional hunting ground for fossil vertebrates for more than a hundred years. From these areas scores of fossil primates have been recovered, many of which have been central objects in consideration of the origin and evolution of Higher Primates and man. The recently concluded Yale field programs in North India produced large additions to our knowledge of the mammalian microfaunas of the Nagri zone there, as well as several new primate finds of significance. Nevertheless, it seems unlikely that rich quarry sites containing associated skeletons will be found, even in the most productive regions around Haritalyangar, India. It may prove preferable to concentrate future Siwalik excavations in West Pakistan, where the exposures of such sediments are both more extensive and more devoid of forest and grass cover than those in India. However, those expeditions which have collected in West Pakistan so far have failed to find much in the way of microfaunal elements. This could be due to employment of less intensive collecting techniques than are now possible. This was the case at Haritalyangar, India. Before the Yale expeditions of 1968 and 1969, only about a dozen small mammal fossils had ever been found out of thousands of specimens collected there. Our group recovered hundreds of small rodents, insectivores, and primates with no difficulty.

Indonesia remains an important area for paleoprimatological research. Several new *Homo erectus* fossils have been found in recent years. Now that potassium-argon dating by Curtis (personal communication) of the University of California at Berkeley has indicated that a tuff in the Djetis beds of Java is 1.9 million years old, the entire question of the relationships of *"Meganthropus"* to African *Australopithecus* is reopened. The probability is increased that Tobias and von Koeingswald (1964) were right in suggesting that *"Meganthropus" paleojavanicus* more nearly resembled African gracile australopithecines than it does Indonesian *Homo erectus* from the considerably younger Trinil beds. Since *"Meganthropus"* and so-called *Australopithecus "habilis"* differ dentally from other *Australopithecus* in showing slightly narrower lower premolars, and in known parts differ little from each other, if they are truly of about the same age they could well prove to be conspecific members of *Australopithecus*. Thus the "handy man" may have to become the "old java man."

Paleoecological Setting and Dating of North African and Asian Fossil Primates

Tertiary fossil primates from North Africa come from the earliest Pliocene of Oran in Algeria (Arambourg 1959) and from three sites in Egypt. The Egyptian fossil-bearing regions which have yielded primates are a Pliocene site at Wadi Natrun in northern Egypt (see Stromer 1913, 1920); a Miocene fossil-bearing level at Wadi Moghara, north of the Qattara Depression and south of El Alamein in northern Egypt (see Fourtau 1918, and Simons 1969b); and the well-known Oligocene bone fields north of Birket Qarun in the Fayum Depression, southwest of Cairo (see Beadnell 1905, Andrews

1906, Schlosser 1911, and Simons and Wood 1968). Of all these areas, only the last has been indirectly dated by geochemical methods.

Two samples of lava-rock lying at the top of Fayum section have been run for potassium-argon dates. No sediments below this in the Fayum are chemically datable. The determinations of Armstrong (at Yale) and Curtis (at Berkeley) substantiate each other, giving a date of approximately 25 million years for the basalt. Vondra's stratigraphic studies in the Fayum have shown that the top of the Oligocene Jebel Qatrani formation has been truncated by erosion and in some areas the entire formation weathered away before the extrusion of the lava flow which caps the section. It would seem reasonable to add one or two million years at least for the time period during which this erosional phase took place. In addition, the youngest fossil-yielding horizons in the Fayum lie stratigraphically about 250 feet below the basalt (see Simons 1967b, p. 30). The time which it took to deposit these sediments must also be allowed for. In consequence, it seems probable that an age of 28 or 29 million years is the best estimate for the youngest Fayum fossils from the upper fossil wood zone. This is the level of Yale Quarry I which has yielded the majority of primates known from the Fayum.

Faunal correlation was also employed as a method of dating the age of the Fayum mammalian fauna by Osborn (1907, 1908) while describing remains of Fayum mammals in the first part of this century. It was first thought that the best correlation was with the Eocene faunas of Europe (see Beadnell 1905) but subsequently Osborn came to the conclusion that an early Oligocene correlation was more likely. Faunal correlation between North Africa and Europe at that time remains somewhat uncertain for there are very few forms in common. However, the Carnivora which appear to be in common are species of two genera originally described from the European Eocene: *Apterodon* and *Pterodon*. The Fayum species of these two genera are considerably larger and more advanced than the European type species of each. In consequence, a somewhat younger date seems possible. Jaws of an undescribed proviverrid carnivore in the recent Yale collections from the Fayum appear to resemble most closely certain European Eocene species (see Van Valen 1966). Such resemblance to European species is also the case for the mandible from the Fayum which Osborn (1908) named *Metasinopa*. In addition, the anthracotheres of the Fayum, although otherwise similar, appear to be larger and more advanced than those from the late Eocene of Burma. Consequently, they may well be of early Oligocene age. If the exotic Fayum mammal *Ptolemaia* is actually related to or is a surviving member of the pantolestid Insectivora, as seems possible, its survival into the African early Oligocene would be paralleled by the survival of this group in the form of *Chadronia* into the earliest Oligocene of North America. There are some similarities between the two genera. In any case, there is little evidence from the Fayum mammalian fauna to indicate that the classic mammal horizons in the lower fossil wood zone of the Jebel el Qatrani formation are any younger than earliest Oligocene. Potassium-argon dates from earliest Oligocene faunal sites in North America and Europe would indicate that this time period ranges from about 35 to 30 million years in age, which is in harmony with the projected dating of the upper part of the Fayum section.

Another indirect confirmation that all the Fayum sites are much older than those of the early Middle Miocene of East Africa comes from the fact that the skull and dentition of *Aegyptopithecus* is distinctly different, at the generic level, at least, from any East African Miocene dryopithecine and is clearly more primitive. Andrews (1970) has recently reviewed all the smaller Kenya Miocene apes with the possibility that some represented species of *Aegyptopithecus*.

Wadi Natrun has yielded a number of fossil monkeys, which Stromer (1913, 1920) placed in *Aulaxinus* (now considered a junior synonym of *Macaca*) and in a then new genus he named *Libypithecus*. The ecological setting for these monkeys is fairly well indicated in the associated fauna by the presence of an otter, *Lutra libyca,* seacow, and seal (*Pristiphoca* aff. *occitana*). The latter indicates the presence of standing water, and the seal suggests that the region, now an oasis, was then an area of the sea. Gar Maluk, the spot where the fossils occur, is approximately sixty miles from the present southern margin of the Mediterranean sea. Arambourg (1945) assigned the *Hippopotamus* present at Natrun to a species he had described from the Omo beds, Ethiopia, *H. protamphibius*. His assignment might suggest a late Pliocene (Astian) correlation for the fauna rather than the middle Pliocene age given by most workers (see Blankenhorn 1921, Stromer 1920, and Tobien 1936). Besides otter, seal, and hippo, this fauna contains a varied assemblage including an antelope species, a camel, a mastodon, a hare, a wolf, a pig, *Sivachoerus, Libytherium,* and a sabre tooth tiger (*Machaerodus*). The presence of these animals suggests that the area around Pliocene Lake Natrun was forested, but how densely would be uncertain. The macaque could have been terrestrially adapted, and *Libypithecus* also appears closest craniologically to the late Pliocene French species *Dolichopithecus ruscinensis* (Simons 1970). Robust limb bones are associated with the latter, suggesting a macaque-like terrestrial adaptation. Consequently, neither primate is of the sort which had to reach the Wadi Natrun area through continuous forest cover.

The Wadi Moghara fauna is harder to assess, being at once older, smaller, and with a higher fraction of members of unknown ecologic preference. It is usually correlated with the Burdigalian (earliest Miocene provincial age of Europe) on grounds that it comes from deposits which were thought to underlie sediments containing a marine fauna said to be of this particular age. Fourtau's correlation was, however, made long ago and should be rechecked. One of the most distinctive animals at Moghara is the large anthracothere *Masritherium,* an animal much larger than any related form from the East African Miocene deposits. However, Patterson (personal communication) believes that specimens he has recovered from the Turkana grit in Northern Kenya and which have been K/Ar dated to approximately 16.5 million years are more like *Masritherium* than any other anthracothere. If there is a relationship, the smaller Turkana specimens may be older than those from Moghara, which weighs against an early Miocene correlation for the latter fauna. The primates from Wadi Moghara were long considered to be apes (*Prohylobates* and *?Dryopithecus*) but Simons (1969b) indicated that these specimens are in all probability assignable to one species of monkey. The ecological setting is again that of a wet coastal plain, with marine as well as terrestrial elements. Fossils from Moghara include sharks, rays, sawfish, giant catfish, crocodiles, false gavials, and dolphins. Three species of anthracotheres suggest riverine conditions. These artiodactyls are often riparian, as is exclusively their modern derivative, the hippopotamus. *Prohylobates* is slightly larger than the East African Miocene monkey *Victoriapithecus* recently described by von Koenigswald (1969). It seems to be somewhat more colobine-like than cercopithecine. Since there is increasing evidence that several extinct colobine species were terrestrial or semi-terrestrial, *Prohylobates,* even if colobine, need not have been exclusively a forest form.

As far as the paleoecologic situation in the Fayum at the time of occurrence of the earliest apes is concerned, we are on safer ground. The fauna is much larger and a great deal more has been published on its ecologic interpretation (see Beadnell 1905, Osborn 1907, 1910, Schlosser 1911, and Simons and Wood 1968). In summary, the picture is one of major watercourses with channels 300 to 500 feet broad, fringed by tropical forest trees with average crown heights well over the 30 to 40 meter lengths of many of the silicified trunks preserved in these sediments. The survival, almost intact, of exceedingly fragile fossils such as milk dentitions of rodents in jaws barely more than a centimeter long, yet buried in sediments as coarse as gravels, suggests that many Fayum fossils were not transported significant distances before burial. This point needs to be made in correction of the impression given earlier (see Beadnell 1905) that the pertified logs are of trees that were swept downstream from far to the south of their present-day locations. It appears much more probable that the Fayum faunal and floral assemblage can be considered as a single community which existed near the site of deposition. This paleoecological reconstruction indicates gallery forest along the watercourses. Other than the logs, the only common plant fossils are seed pods which resemble those of species of three closely related plant genera occurring in Africa today. The latter are plants which grow in standing water. Of the associated fauna, the abundance of crocodiles, false gavials, and anthracotheres implies warm, wet conditions. The associated herbivores (primitive proboscideans, large and small hyraxes, arsinoetheres, the anthracothere *Brachyodus*) have dentitions of forest-browsing type and consequently there is little evidence that there were open savannahs in this region of the North African Oligocene coastal plain.

The primate postcranial bones recovered there also agree with this interpretation. Ulnae referred to as *Apidium* are concave anteriorly like arboreal primates, not retroflexed as in ground-living quadrupeds. A proximal metatarsal of the hallux assigned to *Aegyptopithecus* is virtually indistinguishable from the same bone in *Dryopithecus africanus,* except for its smaller size. (see Napier and Davis 1959).

New Additions to Knowledge of North African and Indian Fossil Primates

FAYUM HOMINOIDS

The only known hominoids of Oligocene age have been recovered from deposits on the north side of the Fayum depression, approximately 100 km. southwest of Cairo, U.A.R. (Simons 1965). Four primate species, assigned each to its own genus, were collected and described in the first two decades of this century: *Propliopithecus haeckeli, Moeripithecus markgrafi, Parapithecus fraasi, Apidium phiomense* (Osborn 1908, Schlosser 1911). Since 1961, Simons has organized and led yearly expeditions to the Fayum; these expeditions have collected a considerable fauna including many primate specimens, some of which represent new genera and species.

The primates have come from the younger of the Fayum deposits, the Jebel el Qatrani Formation (Simons 1965), and have probable ages of around 28 or 29 million years.

Aegyptopithecus zeuxis. In 1965 Simons described a new dryopithecine from the upper levels of the Jebel el Qatrani formation, *A. zeuxis.* At present it is known from several mandibular specimens, an almost complete cranium and upper dentition, various isolated teeth, and a small number of postcranial bones. These remains have yet to be described in detail, but a few preliminary remarks can be made.

A. zeuxis was approximately the size of *Cercopithecus aethiops.* The dentition of *A. zeuxis* resembles that of later dryopithecines, particularly those from earlier Miocene sites in Uganda and Kenya, in such features as relative proportions of anterior and posterior teeth, molar size progression, presence of molar cingula, and certain features of occlusal morphology. Other aspects of the dentition are more primitive than *Dryopithecus* species. Molar cusps are more puffy and more inflated than in *Limnopithecus* and the lingual molar cusps are relatively much larger. The cranium which in *A. zeuxis* is long-snouted with a relatively small brain-case is also primitive. The endocranial mold, according to Radinsky (Chapter 7) shows that *A. zeuxis* had reached a "monkey" level of brain organization in occipital lobe development. Almost complete postorbital closure had been attained. Moderate sagittal and nuchal crests were present, and canine size was apparently quite strongly dimorphic.

The postcranial skeleton is known from a phalange of the manus, a first metatarsal, and a (possible) caudal. These bones suggest that *A. zeuxis* was an arboreal quadruped like most *Cercopithecus* species, or species of such ceboid genera as *Cebus*.

The most recently discovered mandible of *Aegyptopithecus* preserves the canine socket and P_3 through M_3 (see Plate 2–1). This mandibular dentition is more complete than those previously found and confirms large lower canine size and premolar heteromorphy for *Aegyptopithecus zeuxis*, as was already suspected from the much damaged American Museum half-mandible and from the palatal dentition of the skull at Yale. This particular jaw is slightly damaged by chemical erosion of crown enamel of the teeth, but the salient features of the inferior dental mechanism can easily be made out. This species shows one of the most characteristic features of catarrhines: the front ridge of the P_3 was clearly utilized as a hone against the back side of the upper canine which we know to have been large and long (Simons 1965). This particular dental mechanism is seen fully developed only in catarrhine primates. Living members of this infraorder almost universally exhibit the interlocking canines, and premolar hone (or if, as in *Homo* they do not, they are presumed to descend from ancestral forms which did). Relatively large, interlocking canines with premolars reduced to only two pairs is thus a significant dental characteristic of catarrhines. Platyrrhines also show this hone-complex but it is often less developed. P_2^2 is present and P_2 does not always show the hone function of catarrhine P_3.

In our opinion *Aegyptopithecus* is closer to primitive *Dryopithecus* than to *Pliopithecus* or *Limnopithecus* and almost certainly is in or near the ancestry of all subsequent dryopithecines and the hominids. Like primitive *Dryopithecus,* the tooth-rows in this genus diverge markedly to the rear, which is not the case for *Limnopithecus*. The very early possession in *Aegyptopithecus* of long, interlocking canines with a lower premolar-hone settles the question as to whether or not the ancestors of man ever had enlarged canines.

Propliopithecus haeckeli. The type and only specimen was described in 1911 by Schlosser (see Figure 2–1). Its exact stratigraphic provenance is unknown, but it probably comes from approximately the middle of the Jebel el Qatrani formation (Simons 1965). The type consists of partial mandibular rami and fragmentary symphysis; right canine, and cheek teeth of both sides are preserved. The Yale expeditions have recovered some isolated teeth from the middle part of the Jebel el Qatrani formation which resemble very closely those of the type specimen; these are as yet undescribed (Simons 1965). Fragmentary specimens from the upper part of these beds, contemporaneous with *Aegyptopithecus zeuxis* and *Aeolopithecus chirobates,*

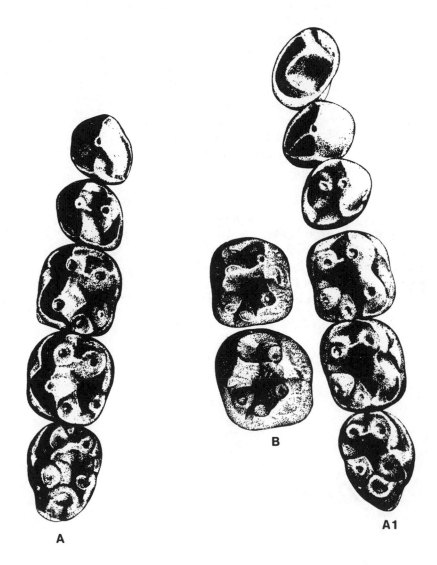

Figure 2–1. Comparison of *Propliopithecus haekeli* (Type), A, A1, from the Fayum Oligocene of Egypt with M_1, M_2 of *P. markgrafi*, B, from the same area. (From Kälin, 1961.)

may possibly belong in *Propliopithecus* or may be referable to *A. zeuxis*. In any case, they clearly do not belong to the species *P. haeckeli*.

This brings us to the question of the relationship between *P. haeckeli* and *A. zeuxis*. Cheek-tooth morphology is essentially similar in the two forms, although there are differences in relative tooth size (Simons 1965).

For example, the molars of *P. haeckeli* are subequal in length, whereas those of *A. zeuxis* increase in length markedly posteriorly. There is little or no upper canine wear on the non-sectorial P_3 in *P. haeckeli,* but a good deal of wear on the lower canines. This was probably not the case in *A. zeuxis,* where P_3 is elongated and hone-like. In addition *P. haeckeli* is considerably smaller than *A. zeuxis,* and is almost certainly geologically older. The two species may well represent samples drawn at different times from one evolving lineage (in which case some of the real differences between the two would have to be attributed to sexual dimorphism). Alternatively, two lineages could be involved. For the moment, the former hypothesis seems preferable, in which case *P. haeckeli* can be regarded as the oldest known dryopithecine.

"Moeripithecus" markgrafi. This taxon is based on a mandibular fragment containing first and second molars (Schlosser 1911, Simons 1965). The "species" is probably sampled from the same lineage as *P. haeckeli* and *A. zeuxis,* and seems in many ways intermediate between the two, though a little more similar to *P. haeckeli.* This genus is probably best regarded as a synonym of *Propliopithecus* (see Simons 1970, and Figure 2–1).

Aeolopithecus chirobates. This species was described by Simons in

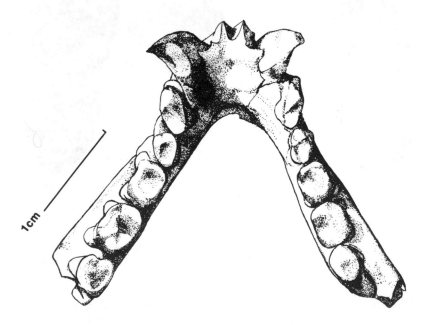

Figure 2–2. Occlusal view of *Aeolopithecus chirobates* from the Oligocene of the Fayum, Egypt. (Drawing by Mary F. Simons.)

1965, and, like *A. zeuxis,* comes from the upper levels of the Jebel el Qatrani formation, with an approximate age of 28 million years (see Figure 2–2). The type and only specimen is an incomplete mandible with all teeth present except incisors. Occlusal morphology is indistinct because of erosion. Simons (1965) has tentatively suggested a relationship with the Hylobatidae, although it must be emphasized that more material is needed (not only from the Oligocene) before his suggestion can become more than a possibility.

Oligopithecus savagei. This, the oldest known Fayum primate, was described by Simons in 1962, and comes from the lower levels of the Jebel el Qatrani formation; it is almost certainly more than 30 million years old (Simons 1965). Once again, the type is the only specimen known, consisting of a left mandibular fragment with C (see Plate 2–2). through M_2. Only two premolars are present, as in all the Fayum hominoids (assuming that *Parapithecus* and *Apidium,* which have three premolars, are not hominoids). The cheek teeth have an intriguing morphology, reminiscent in many ways of omomyine prosimians (Simons 1962a).

The relationships of *Oligopithecus savagei* and other hominoids, even in the Fayum, are obscure, but resemblances to *Aegyptopithecus* suggest that the Hominoidea might well be derived (as a number of students have suggested) from Eocene omomyids.

Conclusions. At least two, and possibly three species of Fayum hominoids, can be classified in the Dryopithecinae (*A. zeuxis, P. haeckeli, P. markgrafi*) and are the earliest known members of the subfamily. Meager postcranial material suggests that *A. zeuxis* was an arboreal quadrupedal form, in all probability adapted to living in dense tropical rain forest.

Aeolopithecus chirobates and *Oligopithecus savagei* are known only from single more or less complete mandibular fragments. Little can be said about their phylogenetic relationships until they are better known.

AFRICAN MIOCENE DRYOPITHECINAE

Following the period of hominoid evolution documented in the Fayum, we know something of the nature of fossil apes from the early middle Miocene collections made in Uganda and Kenya, East Africa during the past thirty to forty years. In spite of various claims, none of these fossil apes from the Kavarondo Gulf area of Lake Victoria in Kenya or from Mount Elgon, Napak, or Moroto in Uganda show any hominid features in their dental mechanism. No specimens show significant reduction in size of the canines. There is no evidence in these early apes of foreshortening of the jaw and tooth row, reduction in size of the front premolars, or thickening of enamel and flattening of molar crowns. The first specimen to show these hominid features dates back only to the end of the Miocene at Fort Ternan, Kenya. Similar finds have been made in North India and West Pakistan, at a time when there was a high degree of faunal community between the two areas.

These early *Dryopithecus* species, which occur about 8 to 10 million years after the Fayum fossil apes lived, come from earlier Miocene deposits in Kenya and Uganda. *Dryopithecus* species have been recovered from sites on Rusinga Island, Koru, Songhor, Napak, Moroto, Losodok, and possibly Maboko Island (Pilbeam 1969a). The first five groups of sites have been dated absolutely and range in age from a little under 20 million years to around 18 million (Bishop, Miller, and Fitch 1969). Strictly speaking, these sites are latest Early and Earliest Middle Miocene, in a European sense, rather than Early Miocene as has been generally stated (Pilbeam 1969a).

As far as can be deducted from geological and faunal studies, the sites fall into at least two ecological groups. Those at Napak, Moroto, Koru, and Songhor apparently sample the thickly forested slopes of active volcanoes. The sites at Rusinga sample a wider variety of habitats, including lowland forest, open woodland, and even swamp (Pilbeam 1969a).

Dryopithecus africanus. One of the best known of all dryopithecines is *D. africanus,* a vervet-sized species known from a number of sites in Kenya representing both "upland" and "lowland" habitats (Clark and Leakey 1951). Remains of the skull, dentition, forelimb, and parts of the foot are known (Pilbeam 1969a).

The dentition resembles that of *A. zeuxis* in certain features (presence of cingula), although there are a number of differences between the two species; for example, incisors are relatively a little larger in the Miocene form, and cheek tooth occlusal surface morphology is more reminiscent of living Pongidae, particularly *Pan troglodytes* (Pilbeam 1969a). In fact, the loss of cingula and further hypertrophy of incisors would leave *D. africanus* with a dentition very similar to that of the chimpanzee.

The cranium (Plate 2–3) resembles in a number of ways that of a small pygmy chimpanzee, although supraorbital tori are very small and the facial skeleton is less prognathous (Pilbeam 1969a). These differences are presumably due to the smaller size of *D. africanus,* and the relatively smaller anterior dentition. The mandible lacks a projecting inferior transverse torus, but in other features is essentially similar to that of *Pan paniscus.*

Postcranially, *D. africanus* is a very interesting form indeed. The forelimb has been studied in detail by Napier and Davis (1959), and more recently by Walker (personal communication) and Lewis (Chapter 9). It is clearly not that of a cercopithecoid-like quadruped, nor is it identical to that of any living hominoid. However, certain features do indicate that *D. africanus* was capable of suspensory posturing and arm-swinging. Both elbow and wrist joints suggest this, and indeed in many ways resemble specifically *Pan troglodytes.* The foot remains resemble those of arboreal quadrupeds, although Walker (personal communication) has suggested that there are certain features indicating ground living behaviors. Probably *D. africanus* was a quadrupedal arm-swinger, perhaps equally at home in the trees and on the ground. Walker (personal communication) has suggested that *D. africanus* was adapting to knuckle-walking behaviors. If this is the case, then the lineage leading to the modern chimpanzee has experienced an increase in overall body size, minor changes in dental morphology, differential incisor hypertrophy, various cranial changes, and the final development of knuckle-walking adaptations. Chimpanzees have become relatively large-bodied, semi-terrestrial, predominantly frugivorous primates, possibly as a result of the radiation of the very successful omnivorous/frugivorous arboreal cercopithecines.

Dryopithecus major. *D. major* is the largest of the earlier Miocene dryopithecines (Clark and Leakey 1951, Pilbeam 1969a). Remains come almost exclusively from "upland" sites in Kenya and Uganda, deposits indicating a Miocene habitat which was densely forested (Pilbeam 1969a). The species was apparently markedly sexually dimorphic in body and canine size (unlike *P. troglodytes,* and like *G. gorilla*); males were probably as big as large chimpanzees or small female gorillas, while the females were considerably smaller (see Plate 2–4).

What little is known of the facial skeleton and dentition indicates a form which, although possessing a number of primitive features, can be regarded as ancestral to or close to the ancestry of the living gorilla. For example, unworn cheek teeth show resemblances to other African *Dryopithecus* species but also possess certain morphological features which are characteristic of *G. gorilla* (Pilbeam 1969a).

The postcranial skeleton is known only to a limited extent. Parts of the forelimb, as yet undescribed, are said to be similar to homologous parts of the knuckle-walking African apes (Walker, personal communication). Walker and Rose (1968) have described a relatively complete lumbar vertebra of *D. major.* It is clearly that of a hominoid, and indicates that *D. major* had a short lumbar region. (This suggests that the thorax was shallow and broad, as in hominoids, rather than deep and narrow like cercopithecoids.) Although this vertebra shows resemblances to *Pan, Homo,* and *Gorilla,* certain features are characteristic of *Gorilla.* The femoral fragments assignable to *D. major* do not resemble Pongidae particularly closely (Walker, personal communication), but suggest a more agile and active form. A talus from Songhor, which probably represents *D. major* (Pilbeam 1969b), has recently been studied by Day and Wood (1969). Their study involved a canonical analysis, based on seven supposedly functionally significant angles and indices, of tali from *Pan, Gorilla,* and *Homo.* The *D. major* talus resembled the knuckle-walkers much more closely than it did bipedal man. A more recent unpublished analysis, with the addition of *Papio,* brings out resemblances to the ground-living quadrupedal baboons as well as to knuckle-walking apes (Day, personal communication).

D. major, like *D. africanus,* has no precise living locomotor equivalent.

The evidence suggests that it was at least as terrestrial as the chimpanzee, and may well have been capable of knuckle-walking behaviors, although it was probably a more agile creature than either African pongid.

It seems reasonable at present to consider *D. major* a probable gorilla ancestor (Pilbeam 1969a). Unless *D. major,* rather than *D. africanus,* is also ancestral to the chimpanzee, this would imply that proto-chimpanzees and proto-gorillas had diverged 20 million years ago. It is not impossible that the common ancestor of the African Pongidae was a knuckle-walker, at least incipiently. Since the early Miocene the two lineages have followed dissimilar adaptive pathways: one toward an arboreal/terrestrial predominantly frugivorous habitus, while the other has become a very large-bodied terrestrial herbivorous forager.

D. nyanzae. The third species of Miocene African dryopithecines is intermediate in size between *D. africanus* and *D. major* (Clark and Leakey 1951). It has been found in some abundance from sites on Rusinga Island, and is poorly represented elsewhere (Pilbeam 1969b). Such pattern of recovery suggests an ecological niche different from that of *D. major.*

In a number of facial and dental characters, *D. nyanzae* resembles *D. major* more closely than it does *D. africanus.* In certain features (for example, tooth proportions) *D. nyanzae* is more primitive than *D. major.* It is possible that *D. nyanzae* represents the little-changed descendant of the common ancestor of both *D. major* and *D. nyanzae;* possibly speciation occurred as *D. major* adapted to a more herbivorous niche on thickly forested volcanic slopes.

Other African Dryopithecine Species. At least one other species of dryopithecine is represented in Miocene sediments. This has been described recently by Leakey (1967) as *"Kenyapithecus africanus,"* and is based on a small number of mostly fragmentary and eroded maxillary and mandibular specimens (see Plate 2–5). Leakey has stated his belief that *"K. africanus"* is the oldest known hominid, and is ancestral to the late Miocene hominid *Ramapithecus.* Pilbeam (1968, 1969b) and Simons (1967a) have discussed this material in some detail and have concluded that the material assigned to *"K. africanus"* is probably not sampled from a single species; some specimens which do appear to represent a new species are similar to other *Dryopithecus* species, particularly to later Miocene and Early Pliocene *D. fontani* from Europe and *D. sivalensis* from India and Pakistan (Pilbeam 1968), similarities which were originally pointed out by LeGros Clark and Leakey in 1951. If indeed a distinct species does exist, it seems most likely that it is sampled from an African lineage (broadly) ancestral to the Eurasian dryopithecines. At present, the material is probably too fragmentary to permit a differential diagnosis or to warrant a new name.

Hominoid postcranial remains have been recovered from Miocene deposits on Maboko Island (Clark and Leakey 1951). The fossiliferous sediments have yielded Early and Late Miocene faunal elements, and it is possible that the primate material is Late Miocene in age. The specimens, all incomplete, are a clavicle, a humerus, and a femur, apparently associated. Little can be said about the clavicle; the limb bones are of interest, for the humerus was probably longer than the femur, a condition found only in pongids and certain ground-living cercopithecoids. The humerus resembles that of terrestrial quadrupedal monkeys in that the shaft is convex anteriorly and the deltoid insertion is very well marked. If these are the remains of a hominoid species, a unique set of postcranial features is once again indicated.

Curiously, the humerus of *Dryopithecus* from St. Gaudens, France (Pilbeam and Simons 1971), although about the same size as that from Maboko, is much more like that of *Pan paniscus* than is the latter.

Conclusions. The Early Miocene African *Dryopithecus* species are the oldest known of that genus. At present, *D. africanus, D. nyanzae,* and *D. major* are grouped together in a subgenus, (*Proconsul*) (Simons and Pilbeam 1965). Many of the characters they share appear to be primitive; it might be necessary in the future to revise the taxonomy of this group, but the present classification seems adequate at the moment. All three species of (*Proconsul*) can be derived from *Aegyptopithecus zeuxis,* or a similar form. *D. major* and *D. africanus* are possible ancestors respectively of the gorilla and the chimpanzee. The Miocene species are larger than *A. zeuxis,* and are probably more advanced postcranially, as well as cranially and dentally. At least one species, *D. major,* may have been capable of knuckle-walking behaviors. There is no evidence to suggest that any early *Dryopithecus* species was a gibbon-like brachiator, nor that they were cercopithecoid-like quadrupeds.

At least one other African *Dryopithecus* species is represented in Early and Late Miocene deposits. One of these Early Miocene taxa could be related to later Miocene Eurasian dryopithecines.

EUROPEAN DRYOPITHECINES

Dryopithecines occur in European deposits no earlier than Middle Miocene (approximate age, 16 million years). They have been recovered in France, Spain, Germany, Austria, Hungary, and Czechoslovakia (Simons and Pilbeam, 1965). Middle and Late Miocene European sediments contain faunas indicating a forested habitat, and warm, moist climatic conditions. At the beginning of the Pliocene (around 10 to 12 million years), the climate became drier, a little cooler, and more seasonal, resulting in a change to more open country conditions. Almost all the dryopithecines are associated with forest and woodland faunas (Simons and Pilbeam 1965).

The type species of *Dryopithecus, D. fontani,* is known from mandibular and maxillary remains recovered originally in France. Further, mainly dental material which may be assignable to this species has come from Germany, Austria, and Czechoslovakia, while possible *D. fontani* postcranials have been recovered in France, Austria, and Germany.

Dentally, *D. fontani* is somewhat less primitive (for example, in cingula size) than (*Proconsul*) species—as might be expected from its mainly younger geological age—but is nevertheless similar enough to the African species for all of them to be placed in a single genus (Pilbeam 1969a, Simons and Pilbeam 1965). Postcranially, *D. fontani* resembles the African dryopithecines more than any other group, extinct or extant. A generalized "quadrupedal" form, adapted for at least some suspensory posturing and locomotion, is once again indicated. Steininger (1967) has recently described a right lower third molar of *Dryopithecus* from the Vienna basin Miocene. In size, structure, and development of the external cingulum it is most markedly reminiscent of *D. (P.) nyanzae* from Kenya.

At least one dryopithecine other than *D. fontani* is represented in European sediments. An extensive series of pongids has been collected in the Vallés Panedés, north of Barcelona, Spain. Only a few of these have been described (Simons and Pilbeam 1965). At least one new species of dryopithecine, *D. laietanus,* is represented, although this form may well warrant a separate generic designation. *D. laietanus* is the smallest of the Eurasian dryopithecines, and while clearly pongid, shows a number of similarities to *Ramapithecus* species. Unfortunately, little more can be said about it until a fuller description is published.

ASIAN DRYOPITHECINES

The bulk of the Asian dryopithecines has been recovered from the Siwalik Hills in India and the Salt Range in Pakistan (Simons and Pilbeam 1965). Extensive primate-bearing sediments in these areas span the time range from Middle Miocene to Pleistocene. Dryopithecines are known from deposits yielding Kamlial faunas (Middle Miocene, approximate age 14 to 16 million years), Chinji faunas (Late Miocene, 12 to 14 million), Nagri faunas (Early Pliocene, 9 to 12 million), and possibly Dhok Pathan faunas (Middle Pliocene, 6 to 9 million years). Faunal and sedimentological evidence relevant to paleoecology indicates that Kamlial and Chinji faunal elements lived in warm, moist tropical rain-forested areas of low relief (Tattersall 1969). The Dhok Pathan represents an open country habitat with cooler, drier, more seasonal climate, while the Nagri stage is climatically intermediate between Chinji and Dhok Pathan.

Dryopithecines are very poorly represented in Kamlial formations and only questionably in Dhok Pathan faunas, the majority of the specimens coming from the Chinji and Nagri. Thus the Asian species have a very similar temporal distribution to European *Dryopithecus;* both are younger than the great majority of African dryopithecines, and presumably represent Middle Miocene migrants from Africa into Eurasia. Chinji and Nagri time spans at least 4 million years, possibly more. Unfortunately, until recently collectors have neglected to include in published reports on fossils such useful details as exact stratigraphic position, and this has made an understanding of these dryopithecines rather difficult.

Almost all the pongids from India and Pakistan have been classified in two species, *D. indicus* and *D. sivalensis* (Simons and Pilbeam 1965). These have been placed in a subgenus, (*Sivapithecus*), separate from both (*Proconsul*) and (*Dryopithecus*). Actually the Eurasian species are extremely similar to one another, and there remains little reason to maintain (*Sivapithecus*) as a separate subgeneric nomen.

Although many fossil ape species have been described from India and Pakistan and although they come in a range of sizes, those from Haritalyangar in India are not diversified. The ape fossils from that area all appear to represent the larger Indian species *Dryopithecus indicus,* although several names have been proposed. Two new discoveries concerning specimens of this species deserve discussion with particular reference to the idea that these could be males of a species, and *Ramapithecus* specimens females of the same animal. The *D. indicus* finds from Haritalyangar, North India all indicate a much larger animal than *Ramapithecus* with teeth on the average at least as much larger than those of *Ramapithecus* as gorilla teeth are larger than those of chimpanzees.

Most specimens of *D. indicus* were fragmentary. In the fall of 1968 David Pilbeam discovered that a cast of a left symphyseal region with canine from Haritalyangar belonging to the Geological Survey of India (D. 189) contacted perfectly with a mandibular ramus of *D. indicus* (Y.P.M. 13828) described in 1938 by Gregory, Hellman, and Lewis (see Plate 2–6). The symphysis had apparently been found much earlier—before 1915—and curiously, red paint marking the contact exists on both specimens, a fact confirmed by us in Calcutta two years ago. Apparently a local collector sold parts of one specimen separately to paleontologists visiting the region at different times. Although brought to scientific attention nearly a quarter of a century apart, the two finds comprise a mandibular ramus which runs from the mid-line to behind M_3, making it the most complete known Indian *Dryopithecus* mandible. It is distinctly larger than a chimpanzee and the comparatively large canine suggests that it is male. Von Koenigswald (1949) described a supposed new genus *Indopithecus giganteus,* the material of which consisted of a large M_3 from near Alipur in the Salt Range, West Pakistan and a large right P_3 from Haritalyangar. Hooijer (1951) rightly questioned the association of the two finds (occurring so far apart geographically) in one species and challenged the likelihood that together they justified a new generic name. His position proved prophetic because, if one measures the anteroposterior length of P_3 roots in the combined specimen of *D. indicus* (G.S.I. D. 189 + Y.P.M. 13821), the measurement is the same as the length of the right P_3 of supposed *Indopithecus,* showing (a) that the tooth of "*Indopithecus*" is not unusually large for *Dryopithecus,* and (b) that single teeth from far-flung sites are to be associated in species only with the greatest caution. But the story does not end there because what evidence there is as to the site of recovery of the large right P_3 indicates the same spot near Haritalyangar as that of the left mandibular remains.

Size and wear are compatible with their belonging to one individual. Even more surprisingly, Prasad later collected a jaw fragment of the right side with M_{2-3}, the teeth of which correspond exactly in size and degree of wear to those of the mandibular fragment described in 1938 by Gregory, Hellman, and Lewis. Locality data also indicates recovery at the same site. In consequence, it is most likely that four different fragments of one mandible were collected at different times (see Simons and Pilbeam 1971a). Most importantly in relation to *Ramapithecus,* this individual with large canine and very large P_3 (so large that von Koenigswald considered it of a different genus) occurs at Haritalyangar. This newer combined specimen shows no hominid feature. As mentioned before, it seems to be a male ape, much larger than any *Ramapithecus.* There are other finds from the region which can fairly be considered to represent the female of *D. indicus* at Haritalyangar. It is much larger in size and different in character from anything indicated for *Ramapithecus.* There is thus evidence at Haritalyangar for at least four different sizes of hominoids—two larger, two smaller. The best probability remains that these represent sexually dimorphic forms of two different species, one pongid, the other hominid, which were unlike each other in both size and structure. Naturally, further collections from the area should help to show more clearly how many size groups occurred there.

DRYOPITHECINE EVOLUTION

The earliest known dryopithecines occur in Oligocene deposits in Egypt. Limited evidence suggests that they were small, arboreal, quadrupedal forms. The Early Miocene primate faunas of East Africa are dominated by dryopithecines, comprising at least four species. These varied in size (*Cercocebus*-size to that of a small female gorilla); they were probably generalized vegetarians—although *D. major* was probably more herbivorous than the other species—and mainly arboreal creatures (*D. major* being the most terrestrial of the species). In locomotor terms, *D. africanus* and *D. major* are best described as quadrupeds (*D. major* may have been a knuckle-walker), although they were clearly more like atelines and alouattines, and even pongids, than like cercopithecoids, in being capable of suspensory posturing and locomotion. If these two (*Proconsul*) species are indeed ancestral to the living African apes, there is no evidence to support the view that the pongids have a gibbon-like phase in their phylogeny.

There is some evidence to suggest that one of the African Early Miocene dryopithecines is ancestral to the Eurasian forms *D. fontani* and *D. sivalensis. D. sivalensis* may, in turn, be ancestral to the orangutan.

PLIOPITHECINAE

Pliopithecines are known from Early and Late Miocene deposits in Kenya and Uganda (genus, *Limnopithecus*), and from Middle Miocene through Early Pliocene sediments in Europe (genus *Pliopithecus*) (Clark and Leakey 1951, Clark and Thomas 1951, Hürzeler 1954). One probable pliopithecine tooth has recently been recovered from the Nagri deposits (Early Pliocene) of India.

Two species of *Limnopithecus* are known from East Africa, *L. legetet* and *L. macinnesi*. Andrews (personal communication) has recently suggested that these two species should not be placed in the same genus since the characters they share are, in his opinion, primitive retentions. However, at present it is probably best to adhere to the older scheme. The sediments from which they come are between 23 million and 18 million years old (Bishop, Miller, and Fitch 1969), and contain faunas indicating tropical rain-forest habitats of various types. Younger *Limnopithecus* specimens have been recovered in Late Miocene beds at Fort Ternan in Kenya (Leakey 1968), and dated at 14 million years. These specimens are said to resemble European *Pliopithecus* more closely than do earlier *Limnopithecus*. This is not surprising, since the European species are also Later Miocene (and Early Pliocene) in age. It is quite possible that one of the Early Miocene African pliopithecines is ancestral to *Pliopithecus* species; migration from Africa probably occurred during the Middle Miocene. *Pliopithecus* species are associated in the main with forest faunas, although some of the later forms may have lived in more open country.

The principal resemblances between pliopithecines and living hylobatids are found in the skull and dentition (Zapfe 1958). To what extent these similarities are due to parallelism is at present unknown; detailed functional analyses of jaws and teeth in the living forms are needed to determine this. For the moment, however, we assume that the resemblances indicate some kind of ancestor/descendant relationship between pliopithecines and hylobatids.

There are a number of differences between the living and extinct groups. Pliopithecine species were dimorphic in body size and canine size; hylobatids are not; forelimbs and hindlimbs were of subequal length in *Pliopithecus* and *Limnopithecus,* indicating some kind of quadrupedal locomotion. Living hylobatids have extremely elongated forelimbs and are brachiators *par excellence*. At least one species of *Pliopithecus* apparently was tailed (Ankel, 1965, Chapter 10); none of the living hominoids have tails.

These morphological differences have led some workers to propose that pliopithecines and hylobatids are not related (Koenigswald 1968). However, this need not necessarily be the case if pliopithecine morphology is interpreted functionally. Thus it is evident from a detailed study of *Limnopithecus* and *Pliopithecus* postcranials that species of these genera were more similar to New World quadrupeds such as *Lagothrix* and *Ateles* than to Old World cercopithecoids such as *Macaca* or *Colobus*. Evidence from the scapula, clavicle, and humerus suggests that pliopithecines were capable of arm-swinging, even though they were basically quadrupedal. Certain features of the hindlimbs suggest that they were somewhat less well adapted to leaping than the cercopithecoids.

It can be postulated that hylobatid evolution involved an increasing reliance on frugivorous feeding in the terminal branches of the canopy which required suspensory posturing (as in hylobatids and *Ateles*). Selection acted to increase the efficiency of suspensory behaviors, resulting in forelimb elongation and associated changes in the shoulder girdle, thorax, and vertebral column. According to Ellefson (1968), brachiation is a relatively inefficient method of locomotion; accordingly, the development in living hylobatids of territoriality, monoganous reproductive groups, and males and females of similar body size and canine size, might be related in some way to the evolution of the unique hylobatid feeding/locomotor adaptations.

Unfortunately, fossil hylobatids from crucial areas (Asia) and time periods (Pliocene) are unknown (with the exception of one possible tooth from India), so at present this hypothesis cannot be tested. However, it does seem probable that the fossil record of hylobatids extends back at least to the Early Miocene (23 million years). If Simons (1965) is correct in assuming hylobatid affinities for the Fayum primate *Aeolopithecus,* then the origins of the family can be traced at least to the Oligocene (some 30 million years).

HOMINIDAE

Hominids can be differentiated from pongids morphologically in terms of three character complexes: the dentition and face; the postcranial skeleton, particularly pelvic girdle and hindlimbs; and the brain (Pilbeam 1968). The earliest specimens which can plausibly be interpreted as hominids are known only from dental and gnathic remains; they have been classified in the genus *Ramapithecus* (Simons 1961b, 1964a, 1968, Pilbeam 1968, 1969a).

R. punjabicus is represented by a dozen or so maxillae, mandibles, and isolated teeth from India and Pakistan. The remains have been recovered from deposits of Chinji and Nagri age, and hence have approximate ages of between 14 and 10 million years. Associated faunas and sedimentological evidence suggests a tropical rain-forest habitat, changing to rather more open country in late Nagri times (Tattersall 1969).

By far the richest site for earliest hominid fossils is the region around Haritalyangar, India, about 200 miles directly north of New Delhi, India. Unfortunately, the fossil sites exposed there do not contain rocks which can be dated geochemically. Faunal correlation suggests that the fauna recovered from spots distributed throughout nearly 2,000 feet of sediments of the region appears to cross the temporal boundary between Miocene and Pliocene times, the greater part of it apparently being about 12 ± million years old. The cuesta or "temple" scarp and the lower scarp near Dangar are the principal sites of discovery of hominoid fossils at Haritalyangar. Almost a dozen specimens of *R. punjabicus* have come from these two localities. As discussed previously, one of the questions raised about the dental and facial anatomy of *Ramapithecus* (which does show canine size reduction, facial foreshortening, broad flat molars, and thick mandibular rami) is whether or not the specimens assigned to this primate genus might be nothing more than females of ape species which then lived in the region. Mention is often made of the comparatively small canines of females among gorillas or pygmy chimpanzees. It has been pointed out elsewhere (Simons 1969a) that the proportions of the front of the face and the anterior dentition relative to jaw and cheek tooth size in *Ramapithecus* from Haritalyangar are relatively more gracile or diminutive than in female pygmy chimpanzees, yet some of the molars from Haritalyangar which are included in *Ramapithecus* are larger than any that can be found in the common chimpanzee *P. troglodytes.* Clearly, fossil evidence exists which shows that front and back tooth sizes differed markedly from those of the chimpanzee.

Recently, we have been restudying a mandibular specimen discussed by Pilgrim (1927) as probably referable to *?Paleopithecus,* which also came from Haritalyangar (see Plate 2–7). He initially recognized that this jaw ramus with P_4–M_3 was not compatible with a placement in *Dryopithecus,* and observed that these teeth were much broader than any specimen of the latter genus known to him. In our preliminary revision of *Dryopithecus* (Simons and Pilbeam 1965) we placed this specimen in *D. sivalensis* but not without reservations, as we had not then studied carefully the original. Working from casts and Pilgrim's illustration, it seemed to have too small an M_1 and too narrow a mandibular horizontal ramus to be *Ramapithecus.* This molar crown anatomy of M_{2-3} seemed suggestive of such an assignment which was nevertheless not done because we did not have then any direct observational data.

After we studied the original in Calcutta in March 1969, it was clear that both the internal and external faces of the mandibular horizontal ramus are spalled off so that the specimen in its present state does not reflect the original corpus thickness under the molars. However, external to M_{2-3} the alveolar border is in a horizontal plane at the place where the external laminae of bone are broken off, implying that the jaw was much thicker before breakage. In the same manner, much of the inner half of M_1 was broken off in this specimen so that in casts and unclear drawings the tooth looked narrower than it actually was in life. The final piece of evidence that this is a *Ramapithecus* lower dentition—the most complete known—is that there is high interstitital wear, a sharply decreasing wear-gradient posteriorly, and very broad flat teeth with thick enamel. These teeth are remarkably like posterior lowers of *Australopithecus* and are functionally identical in that abrasive wear on the thick enamel caps of the molars wears the tooth to an essentially flat surface before the enamel is significantly perforated. Except for its smaller size, the M_3 in G.S.I.D.199 is extraordinarily like a mirror image of the right M_3 of the largest Omo mandible of *Australopithecus*

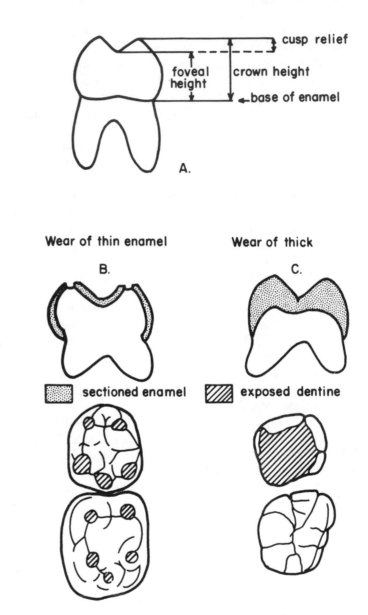

Figure 2–3. Diagram showing (A) various height measurements to be taken on hominoid teeth; (B) wear of thin enamel in pongids; and (C) thick enamel and differential wear in hominids.

which is worn to a similar stage. Since the Omo mandible is undoubtedly a hominid, the likeness with D.199 further strengthens such placement for D.199 and *Ramapithecus* as well. In both, the cusps have been worn down almost to the level of the valleys which separate them and in both the only cusp which has been barely perforated into the enamel is the protoconid which is the largest single cusp and which in both (viewed from above) bulges out laterally beyond the general ovoid outline of the tooth.

Unworn teeth of *Australopithecus* and *Ramapithecus* may have fairly high cusps. It is difficult without breaking to determine the enamel thickness, particularly since X-rays of these teeth seldom resolve the enamel-dentine junction. Nevertheless, if molar cusps can wear almost flat without perforation into the dentine, then clearly the cusp relief (see Figure 2–3) of these teeth must be built almost entirely of enamel. In consequence, in both, crown enamel has to be thick for this type of wear to obtain. This also provides a ready way to distinguish between hominoids with thick and thin enamel, assuming that cusp relief in unworn species is approximately the same (see Figure 2–3). It seems that the anatomy of this mandible forms a good example of how one goes about using the comparative method in relating primates from different periods of primate evolution. First, there must be multiple characters of similarity which mutually reinforce the probability of a relationship between two or more given fossils; but the functional uses of dental and other anatomical structures must be understood in assessing relationships. Moreover, the probability of independent acquisition of similar functional systems must be weighed before direct relatedness rather than parallelism or convergence can be advocated. One chimpanzee may show advanced interstitial wear with age, another a small canine, a third broad molars, a fourth markedly decreasing molar wear posteriorly, a fifth a thick mandibular ramus under the molars—all characters of similarity to species of *Australopithecus*. This neither makes the chimpanzee a hominid nor proves (because of the occurrence of such structures separately in a pongid) that *Ramapithecus* is not. G.S.I.D.199 combines all these features of resemblance to *Australopithecus* together in one individual as a multiple character complex. No ape exhibits such a combination.

Ramapithecus is also represented at Fort Ternan, Kenya, in Late Miocene deposits dated absolutely at 14 million years (Leakey 1962). Remains of at most two individuals are known. The East African form has been classified as *R. punjabicus* but it might be better to place it in a separate species, *R. wickeri* (previously "*Kenyapithecus wickeri*"). Most, if not all, *R. punjabicus* are probably younger than *R. wickeri,* and there is some evidence indicating differences in canine and premolar morphology and relative incisor size between *R. wickeri* and *R. punjabicus.*

In summary, *Ramapithecus punjabicus* had relatively flat-crowned, broad cheek teeth which became packed closely together during life and which showed marked differential wear (thus M_1, for example, was heavily worn before M_2 became worn; this pattern is characteristic of forms like later hominids, *Gigantopithecus,* and *Theropithecus,* which have a heavy dental load). Incisors were small relative to cheek tooth size and more vertically implanted than in pongids. The face was flatter and deeper, and mandibular corpus more robust and better buttressed at the symphysis than in an ape of equivalent size like *Pan paniscus.*

The differences indicate a more herbivorous diet than that of the *Pan* species, for example—one which, if Jolly's (1970) arguments are correct, might well have resembled that of the gelada baboon in consisting of large amounts of small, tough items such as seeds. If this hypothesis is correct, then *R. punjabicus* can be seen as a ground feeder at the forest fringe and on the grassy margins of seasonally flooding lakes within the forest.

Jolly's hypothesis goes a long way toward explaining hominid facial morphology and the relative proportions of incisors and cheek teeth. It does not, however, account for the fact that hominid canines are small in both sexes, and also incisiform. *R. punjabicus* had small canines, and there is some evidence to suggest that crown morphology was incisor-like. It has been suggested that hominid canine reduction was correlated with weapon use (Washburn 1960). Yet there is no evidence to indicate that *R. punjabicus* was a toolmaker or tool-user, at least no more so than *Pan troglodytes,* a pongid the males of which have elongated canines. Neither does the hypothesis account for changes in canine morphology. An alternative explanation is that canines became reduced in hominids as the main masticatory movements of the cheek teeth involved ever-larger components of transverse shearing and rotatory chewing. This accounts for canine reduction, while morphological changes probably indicate that the anterior dentition (hominid canines and incisors function as a single cutting unit) became adapted to powerful slicing. Possibly this was in response to an increasing amount of meat in the diet.

R. wickeri is probably more primitive than *R. punjabicus* in having somewhat larger incisors, more caniniform (though small) canines, and more sectorial anterior lower premolars. Possibly, *R. wickeri* represents the very earliest detectable stages of the hominid divergence from Miocene Pongidae, in which case the date of the hominid/pongid divergence might be set at some 15 to 20 million years.

GIGANTOPITHECUS

Two recent papers have discussed the aberrant primate genus *Gigantopithecus* in some detail (Pilbeam 1970, Simons and Ettel 1970). The younger species of the genus, *G. blacki,* comes from Early Pleistocene deposits in southern China, with a probable age of around 1 million years. Associated faunas suggest a nonforested habitat. *G. blacki* is known only from jaws and teeth. It was clearly a very large species, probably bigger than the mountain gorilla, and almost certainly therefore a ground-living form. Recently, an earlier species of *Gigantopithecus, G. bilaspurensis,* has been recovered in the Middle Pliocene (Dhok Pathan) deposits in India (Simons and Chopra 1969); these have an approximate age of between 6 and 9 million years. *G. bilaspurensis* is somewhat smaller than its probable descendant *G. blacki,* and more primitive in certain features. *G. bilaspurensis* can be plausibly derived from Late Miocene and Early Pliocene *D. indicus,* a presumed forest form.

Gigantopithecus shares a number of features with early hominids: a short, deep face, relatively large cheek teeth and small canines and incisors, and massive mandibles. These characters suggest an open-country herbivorous adaptation (as in geladas, and perhaps hominids). Although *Gigantopithecus* canines are small, they are not hominid-like incisiform teeth but robust teeth with broad occlusal surfaces.

Presumably, *Gigantopithecus* represents an Asian lineage which became adapted to ground feeding in open country much as the hominids did (possibly) further to the west. The lineage seems to have become extinct at the end of the Early Pleistocene.

Interrelationship between New and Old World Primates

This subject has recently been discussed in another paper (Simons 1969c) and consequently we will only summarize the state of our knowledge. Apparently about the last time in Tertiary history when there could have been a significant amount of migration among primates between North America and Eurasia was during the Early Eocene, the time of the Wasatchian and Sparnacian provincial ages. To quote from Simons (1969c): "During this time, many genera and perhaps even species of mammals (including primates) were common to Europe and North America. Such genera include *Peratherium, Entomolestes, Paleosinopa, Esthonyx, Phenacodus, Hyracotherium, Coryphodon, Paramys, Reithroparamys, Microparamys, Pelycodus, Plesiadapis* and *Phenacolemur.*" Recent geotectonic evidence indicates that the spreading of the North Atlantic was late and that in all probability in Wasatchian times eastern North America and Europe were closely approximated and both were further south. This may explain the presence of species of the same primate genera in both continents at that time. It also means that whatever the route of migration was, there was one; and that therefore when considering the forebears of Old World primate groups at that time, North American taxa cannot be discounted. Since the Wasatch, as far as we know, only *Homo sapiens* among the primates has crossed between the Eastern and Western hemispheres.

*Suggestions on the Kinds of Studies on Living Primates that
Could Provide Broader Inferential Bases for the
Study of Past Primates*

X-ray cinematography and electromyography provide new experimental
approaches to the study of chewing mechanisms of living primates as well
as delineating the range and kinds of movements at the joints, important
in understanding the functional mechanics of the jaws and limb bones of
fossil forms. Anatomists could clarify the function and purpose of such
structures as the tubercle on the posterior aspect of the neck of the femur,
absent in modern hominoids but present in modern lemurs and monkeys,
and which was widely distributed among extinct hominoids such as *Dryopithecus* and probably also *Oreopithecus* and *Australopithecus*. Another
structure which either may be present or entirely missing in both living and
fossil primates is the entepicondylar foramen of the distal end of the
humerus. What factors in the mechanics of the elbow joint and structure of
the forelimb control its presence or absence? Examples of this kind can be
multiplied. In general, however, the questions paleontologists would ask
anatomists are determined by the frequency of occurrence of postcranial
material. For instance, proximal and distal ends of humeri and femora are
fairly common, as are proximal ends of ulnae and tibiae; radii and fibulae
are seldom found. While tali and calcanea are often found intact, wrist bones
almost never are. In consequence, anatomical studies of the function of the
tarsus can be related to fossil finds with much greater frequency than can
those of the carpus.

Understanding the adaptive purposes of the dental mechanisms of primates both living and fossil can be enhanced if those studying field behavior
would also record in the greatest detail possible not only the kinds of food
eaten by various primate species but also their relative amounts and consistencies. Whenever possible it could be noted whether the food was processed
by the cheek teeth or by the front teeth and lips. Use of the teeth for purposes other than feeding should, if it occurs, be recorded.

Similarly, field observers could note as precisely as possible the different
proportions of time spent in the various categories of locomotion. For some
species such details are known; but the more species we have in which dental
and osteological structure can be related to feeding and locomotor function,
the easier it will be for paleontologists to draw analogies from recent forms
to primates of the past.

Finally, more sophisticated studies of paleoenvironments and modern
primate habitats should allow a synthesis of primate paleontological and
neontological data to produce a more coherent picture of higher primate
evolution.

3

PHILLIP V. TOBIAS
UNIVERSITY OF THE WITWATERSRAND

Progress and Problems in the Study of Early Man in Sub-Saharan Africa

This is a crucial moment in the history of paleoanthropology. An unprecedented spate of discoveries has flowed from systematic searches and excavations in a number of major sites in sub-Saharan Africa. The rate of new finds, which has always outdistanced the rate of description, analysis, and publication, is now threatening to leave it standing by decades rather than years; multivariate analysis and other statistical methods have provided powerful aids to the study of the fossils; the application of new methods of dating has shown that we are dealing with populations spread over at least 4 million years in time, rather than the more modest span formerly inferred.

We are also gaining new insights into primate variability and primate behavior, both of which are essential aids to the interpretation of fossils and of the erstwhile functioning, evolving populations that they represent.

Therefore, when so much meaningful research is planned or underway, and so little actually accomplished, this chapter will not pretend to solve any problems. Rather, it will try to indicate the extent of our problem-solving materials, the fossils themselves, their wealth and their simultaneous poverty, their potentialities for functional and evolutionary analyses, and

For the purposes of this chapter I shall interpret "Man" in the title as meaning a member of the Hominidae. By "early" I shall understand "dateable to the Upper Pliocene, Lower Pleistocene, and early Middle Pleistocene," as commonly understood by students of the Cenozoic and more specifically of the Quaternary Period.

I wish to thank Miss C. J. Orkin, Mrs. E. Hibbett, Miss J. Walker, Miss D. Rosenblatt, Miss C. Holdsworth, Mr. A. R. Hughes, Mr. C. S. Block. For helpful discussions, personal communications, and other kinds of assistance, I am indebted to Dr. and Mrs. L. S. B. Leakey, Mr. Richard Leakey, Prof. G. Sperber, Dr. F. Clark Howell, Mr. John Wallace, Dr. Yves Coppens, Dr. C. K. Brain, Mr. Brett Hendey, Mr. R. J. Clarke. My gratitude is extended to the Wenner-Gren Foundation for Anthropological Research, the L. S. B. Leakey Foundation, the Research Committee of the University of the Witwatersrand, the Council for Scientific and Industrial Research (S. Afr.), and the Bernard Price Institute for Palaeontological Research.

their limitations. A number of questions will be posed which are, or should be, engaging our attention, both in the study of the fossils themselves and in the collection of comparative data from living primates.

Geographical Distribution

Hominid fossils of the period under review have been recovered from 14 sub-Saharan African sites. Five of these sites are in the Republic of South Africa and nine in East Africa (Tanzania, Kenya, and Ethiopia) (Figure 3–1).

Geographically, the 14 sites lie between 5° 24′ North and 27° 32′ South; all are situated on the Great Central Plateau of Africa, between 25° and 36° East longitude. More precisely, the sites which have yielded australopithecine remains fall into two clusters, an *Equatorial* group in East Africa (E.A.) and a *Subtropical* group in South Africa (S.A.).

The S.A. dolomitic limestone deposits which have yielded australopithe-

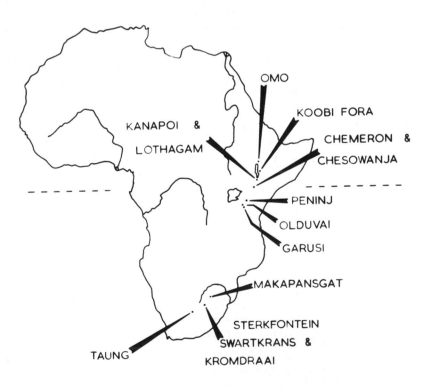

Figure 3–1. Map of Africa showing the geographical distribution of the 14 African fossil sites reviewed in this chapter.

cine remains are confined to a fairly small area, some 200 miles in N-S and some 300 miles in E-W extent. More exactly, they extend from 24° 12′S to 27° 32′S, and from 24° 45′E (at the most westerly) to 28° 57′E. The five sites in this area—Taung, Sterkfontein, Makapansgat, Kromdraai and Swartkrans—thus all lie just south of the Tropic of Capricorn. For convenience, they may be designated Subtropical.

In E.A., during the last decade, one new site after another was opened up, forming a veritable chain of nine sites up the Eastern Rift Valley, from Garusi and Olduvai about 3° south of the Equator, to the Lake Rudolf sites up to 5° north of the Equator. They may be called the Equatorial sites.

There is an enormous gap between the two groups of sites. One or more geographically intermediate deposits between the S.A. sites and those of Tanzania would be most valuable. Such an intermediate site, if one of appropriate age could be located, would make possible more meaningful faunal comparisons between the S.A. sites (for which we have only faunal and sedimentological dates) and the E.A. sites (for which potassium-argon and fission-track dates are available).

Search in the intervening zones has so far proved fruitless. The fossiliferous breccias and sediments of Rhodesia (e.g. Chelmer), Zambia (e.g. Twin Rivers and Kalambo Falls), and Southern Tanzania (e.g. Isimila) have all proven too young in age—either late Middle Pleistocene or Upper Pleistocene—to be relevant for this discussion. Research in the Lake Malawi Rift revealed that its oldest sedimentary unit, informally designated the Chiwondo, is of appropriate age: the vertebrate fauna covers the period from the Plio-Pleistocene junction to the Middle Pleistocene. Despite intensive searching, no signs of occupation by hominids have come to light: indeed it is not until the Middle Pleistocene that any cultural remains (Acheulian) betray the presence of man (Clark et al. 1966, Clark, Haynes and Mawby 1971, Tobias 1971a).

There remains no single australopithecine or other early hominid site between Makapansgat in the Transvaal (24° 12′S) and Garusi in Tanzania (about 3° 12′S). A wide geographical gulf separates the populations and the stamping grounds of the Equatorial australopithecines from those of the Subtropical australopithecines. It would of course be hazardous to draw any ecological inferences from this negative evidence.

A northernmost outlier of the Rift Valley sites is 'Ubeidiya in Israel, which is probably Middle Pleistocene in age and has yielded a few fragments attributable to *Homo* species (Tobias 1966). Between 'Ubeidiya in the north and Lake Malawi in the south, there are doubtless a number of additional Quaternary deposits awaiting discovery and development. All such deposits south of 3°S would help to close the geographical gap between the E.A. and S.A. sites, thereby facilitating relative dating.

In S.A., outside the tiny area enclosing the five australopithecine sites, searches have been made from the University of Cape Town and from the

South African Museum, at Langebaanweg, northwest of the Upper Pleistocene site of Hopefield (Elandsfontein) where the Saldanha calvaria was
found. From the fauna of Langebaanweg, some 60 mammalian species have
been identified, as well as the remains of sharks, rays, a skate, bony fishes,
frogs, lizards, snakes, tortoises, and birds. The fauna comprises both
marine and terrestrial forms in association, a unique occurrence in sub-
Saharan Africa (Hendey 1969). According to Hendey, the fauna includes
a number of genera which would indicate a Pliocene date if they were recorded in Eurasian contexts. Included are an agriotheriine bear (which is
the only bear recorded in Africa south of the Sahara), a giant otter (*Enhydriodon*), a primitive hyaenid, and a *Mammuthus* species. On palaeontological grounds, the deposit looks like a very early Pleistocene one; or even a
late Pliocene one, according to a recent unpublished reassessment which
Mr. Hendey has kindly permitted me to quote. No primates have yet been
found at Langebaanweg.

Other South African sites exist in the dolomitic limestone belt, such as
Bolt's Farm and Gladysvale not far from Sterkfontein; none of these has
yet yielded hominoid remains, although many mammalian species have
been identified from them.

Ongoing Excavations

The continued exploitation of known australopithecine-bearing sites in
Africa is at present being set forward at two sites in South Africa, one in
Tanzania, one area in Kenya, and one cluster of sites in Ethiopia. In S.A.,
C. K. Brain (1967a) is excavating at Swartkrans whilst, a mile away across
the valley, a new large-scale excavation has been under way at Sterkfontein
for the past three and a half years.

Swartkrans. Brain's (1967a) excavation is aimed at further elucidating
the geomorphology of the cave and its deposits; enquiring into the agency
or agencies which may have been responsible for the accumulation of the
bones, in this, the world's richest ancient hominid-bearing site; and collecting, developing, and identifying further fossils, including those of hominids.
This present program is now nearly complete. Meantime, the site, which
came upon the market following the death of the owner, has been pur-
chased by the University of the Witwatersrand. The latter university already
owns the adjoining site of Sterkfontein, through the generosity of the
Stegmann family. Thus, the world's two richest sites are in safe hands,
which ensures the freedom of scientific exploitation of the sites in perpetuity.

Sterkfontein. The new Witwatersrand University excavation of Sterkfontein is aimed at throwing light on the extent and layout of the cave; its
geomorphology and stratigraphy; the possibility of "absolute" dating on the
site; and at the systematic excavation of selected parts of the total breccia
body, to recover further hominid skeletal remains, associated nonhominid

fauna, materials which may be of value for dating, cultural remains, as well as samples of deposit for climatological and ecological analysis. After four years of continuous work, we consider the whole project at Sterkfontein may take 20 years. A detailed account of progress to date has been published (Tobias and Hughes 1969).

Makapansgat. Little new field work has been carried out at Makapansgat for several years, save for B. Maguire's collection and analysis of further stone objects from within the excavation area and the systematic emptying and elucidation of the sinks and swallows in the exposed upper surface of the undisturbed breccia. We are planning to resume an active program of field studies there in the near future.

Olduvai. Dr. M. D. Leakey returned to the field at Olduvai in 1968 and again more recently. The latest phase of field operations has swelled the total number of individuals represented by hominid fossils from Olduvai to at least 36. The new finds include an important gracile cranium, perhaps of *H. habilis* (Old. hom. 24 from DK I East). M. D. Leakey's archaeological analysis of Beds I and II and R. Hay's geologic analysis of Olduvai are soon to be published as two further volumes in the *Olduvai Gorge* series; Volume V on hominid remains is in preparation.

East Rudolf. Further searches and systematic work are being carried out east of Lake Rudolf by R. E. Leakey. Exciting finds have already fallen into his hands, including a new and well-preserved cranium of *A. boisei*, making East Rudolf the fifth richest in Africa in cranial remains. The area east of Lake Rudolf has been dated to about 2.4 million years *and over*. Mr. Leakey is still seeking a site with a well-stratified deposit suitable for systematic excavation.

Omo. The French and American teams have continued their operations in the basin of the Omo River, which flows into the northern end of Lake Rudolf, while the Kenyan team (under R. E. Leakey) has moved its operations to East Rudolf. In the few years since systematic work began at Omo, sufficient fossils have come to light to put Omo fourth on the list of 14 African sites, with 110 hominid fossil items on an incomplete count. Perhaps the most striking aspect of the Omo finds is that they take the fossil hominid record back over 1 million years earlier than the base of Olduvai Bed I and even ½ million years before the East Rudolf deposits. Indeed, the oldest of the Rudolf deposits so far dated—over 5 million years at Lothagam, southwest of the Lake—should almost certainly be regarded as falling in the Pliocene.

The other important inference from Omo, and from the rest of the Lake Rudolf sites, is that for the first time it now seems that robust and gracile hominids were sympatric and synchronic throughout the period spanned between the times of the B tuff in the main Omo Beds Succession and Upper Bed II at Olduvai (that is, from some time after 3.75 to about 1.00 million years B.P.). The new datum now requires supercession of the sequence which still obtained at the time of the Supper Conference of the Wenner-Gren Foundation in April 1968; at that stage, it seemed that in both S.A. and E.A., the earlier hominids were of the gracile kind, while only later deposits contained robust australopithecines, as well as more advanced hominids (Tobias 1968a).

The coexistence of robust and gracile hominids over such a long time poses anew the interesting question of their interrelationships (see pp. 86–91).

Available Fossils from 14 African Sites

In the main part of this paper, the nature of the available hominid assemblages is examined. To effect this, a detailed census has been made of all hominid fossils from the 14 sites. The census includes all published specimens from these 14 sites and a number of unpublished ones known to the author, from Swartkrans, Sterkfontein, Makapansgat, Kromdraai, Olduvai, Chesowanja, East Rudolf, and Omo. The list is believed to be complete, save for some isolated and unlisted teeth discovered at Omo by the French team and possibly some items from East Rudolf and Olduvai. The full list, which could not be given here for reasons of space, includes the individual catalogue numbers of virtually all items. Irrespective of the taxa and number of individuals involved, the fossils have been classified anatomically in the first instance, on the following scheme:

Cranial remains   calvaria
                                 face
                                 maxilla(e) and palate
                                 mandible
                                 endocranial cast

Dental remains   deciduous
                                     maxiliary—individual teeth
                                     mandibular—individual teeth
                                 permanent
                                     maxillary—individual teeth
                                     mandibular—individual teeth

Postcranial remains   upper limb girdle—individual bones
                                 upper limb—individual bones
                                 lower limb girdle—individual bones
                                 lower limb—individual bones
                                 axial skeleton—individual bones (postcranial)

Where a single fossil (e.g. a fairly complete cranium) embraces more than one area, its calvaria, its facial parts, its maxillae including palate, and its included teeth have each been itemized separately in the appropriate list. The reason is that so much material is now available that specialized studies of functional and structural complexes are being carried out in our labora-

*Table 3–1. Anatomical classification of early hominid fossils
from 14 African sites*

	South Africa	East Africa	Total
Cranial parts	175 (17.1%)	59 (15.2%)	234 (16.6%)
Teeth	769 (75.2%)	269 (69.3%)	1038 (73.6%)
Postcranial bones	78 (7.6%)	60 (15.5%)	138 (9.8%)
Grand Total	1022 (99.9%)	388 (100.0%)	1410 (100.0%)

tories. It is therefore most convenient to be able to specify, say, the number of calvariae or of deciduous cheek teeth we now possess.

Table 3–1 gives the totals of specimens from the five S.A. sites as a group and from the nine E.A. sites as a group (Figure 3–2).

Of the overall sample of 1,410 specimens, 1,038 or 73.6 percent are teeth, 234 or 16.6 percent are cranial parts, and 138 or 9.8 percent are postcranial bones.

From both S.A. and E.A., teeth form the lion's share of items, comprising 75.2 percent of the S.A. and 69.3 percent of the E.A. samples. Not all the teeth enumerated are measurable, although a large majority of the 1,038 are. The criterion for enumerating teeth in the census was whether they provide useful morphological information, metrical or nonmetrical. With this yardstick, a great number of additional teeth which are crushed, heavily worn, or fragmentary have been excluded from the count.

The E.A. assemblage is relatively though not absolutely richer in postcranial remains (60 specimens or 15.5 percent of the total E.A. sample) than is the S.A. one (78 or 7.6 percent); while in cranial parts, the S.A. sample (175 or 17.1 percent) is both absolutely and relatively richer than the E.A. sample (59 or 15.2 percent).

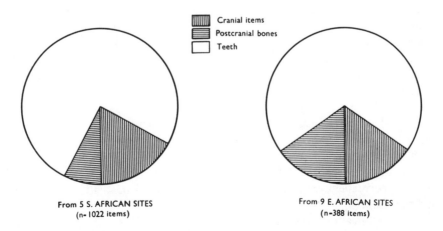

Figure 3–2. Relative proportions of early African fossil hominid specimens in three anatomic categories.

Table 3–2. Cranial parts represented among early hominid fossils
from 14 African sites

	Calvaria	Face	Maxilla(e) and Palate	Mandible	Endocast	Total
S.A. Hominids	34	15	62	55	9	175
E.A. Hominids	18	7	11	19	4	59
Grand Total	52	22	73	74	13	234

Table 3–2 gives a detailed breakdown of the cranial remains. In this classification, "Face" refers to the upper portion of the face, from brow-ridges to floor of nose. If only the upper jaw, and/or palate, and/or maxillary arcade is present, such a specimen is not classified under "Face" but under "Maxilla(e) and Palate."

It emerges that upper jaws are relatively and absolutely rather scarce among E.A. finds, while mandibles are somewhat more numerous, forming about the same proportion of the E.A. cranial finds (32.2 percent) as of the S.A. cranial samples (31.4 percent).

Of the *deciduous dentition* (Table 3–3), far more mandibular than maxillary teeth have been recovered. Over 90 percent of all deciduous teeth so far available come from the five S.A. sites. Anterior teeth remain rare; in this category, only a single c̄ has so far been found in E.A. (Olduvai), while the S.A. sample of maxillary anterior teeth is dominated by the complete deciduous dentition of the Taung child, which accounts for six of the nine specimens recorded. The 24 upper and 46 lower deciduous molars comprise more respectable samples, although the samples cited may include male and female teeth, and at least two species of *Australopithecus* and one or two species of *Homo* may be represented in the combined samples. When broken down taxonomically and, possibly, by sex, it is clear that the individual subsamples available (even if we could correctly identify all isolated teeth, which is virtually impossible) are far from adequate.

In contrast, the permanent teeth (Table 3–3) are more or less equally divided between uppers and lowers (469 and 425 respectively). In the S.A. sample, uppers (375) somewhat exceed lowers (299), while the opposite is true of the E.A. sample (94 and 126 respectively). As always, anterior teeth are less common than cheek teeth, the best represented anterior teeth being lower canines (39) and upper canines (36). Of all the permanent teeth thus far available for this analysis, 72.2 percent are from S.A. and 27.8 from E.A. The overwhelming majority (584 out of 674 or 86.6 percent) of the S.A. permanent teeth have come from the two richest sites, Swartkrans and Sterkfontein.

The numerical discrepancy between the S.A. and E.A. samples is far less for postcranial bones than for other anatomical categories (Table 3–4), through the large number of hand- and foot-bones from Olduvai. In fact,

Table 3–3. *Numbers of individual teeth represented among fossils*
from 14 African sites

	S.A. Hominids	E.A. Hominids	Total
	A. Deciduous Teeth		
di^1	4	—	4
di^2	2	—	2
dc	3	—	3
dm^1	6	2	8
dm^2	13	3	16
Maxillary Total	28	5	33
di$_1$	5	—	5
di$_2$	8	—	8
dc̄	12	1	13
dm$_1$	20	—	20
dm$_2$	22	4	26
Mandibular Total	67	5	72
Grand Total of Deciduous Teeth	95	10	105
	B. Permanent Teeth		
I^1	13	6	19
I^2	16	5	21
C	29	7	36
P^3	55	14	69
P^4	58	16	74
M^1	81	19	100
M^2	63	16	79
M^3	60	11	71
Maxillary Total	375	94	469
I$_1$	16	9	25
I$_2$	15	11	26
C̄	25	14	39
P$_3$	39	14	53
P$_4$	44	20	64
M$_1$	61	22	83
M$_2$	52	20	72
M$_3$	47	16	63
Mandibular Total	299	126	425
Grand Total of Permanent Teeth	674	220 +39 Omo teeth* 259	933

*These 39 teeth discovered by French scientists at Omo had not been individually
identified in the published works available at the time this chapter was written.

Table 3–4. Numbers of hominid postcranial bones represented from 14 African sites

	Upper Limb-girdle	Upper Limb	Lower Limb-girdle	Lower Limb	Axial Post-cranial	Total
S.A. Hominids	3 1 scapula 2 cl. vicles	19 6 humeri 5 radii 1 ulna 1 capitate 3 metacarpals 3 phalanges	9 5 ossa coxae 3 ilia 1 ischium	13 6 proximal femur 2 distal femur 1 talus 4 phalanges	34 20 presacral vertebrae 1 sacrum 4 ribs 9 rib fragments	78
E.A. Hominids	2 2 clavicles	35 4 humeri 1 radius 2 ulnae 3 carpals 6 metacarpals 19 phalanges	1 1 os coxae	22 6 femora 2 tibiae 1 fibula 7 tarsals 5 metacarpals 1 phalanx	0	60
Grand Total	5	54	10	35	34	138

the appendicular skeleton (except girdle-bones) is appreciably better represented from E.A. than from S.A., although two individuals—Olduvai hominids 7 and 8—contribute no fewer than 40 out of 60 E.A. bones. On the contrary, the postcranial axial skeleton is represented exclusively from S.A., one individual, Sts 14, contributing no fewer than 28 of 34 S.A. bones from this part of the skeleton.

Table 3–5 shows the number and the percentage (in brackets) of items which have been recovered from each of the 14 S.A. and E.A. sites. The 14 sites have contributed very unequally to the stockpile of available items. Thus, Swartkrans and Sterkfontein, two of the three sites situated close together in the Krugersdorp District of the Transvaal, have between them contributed three-fifths of all the fossil items covered by this report. Their combined percentage of 59.7 percent of all fossil items is fairly closely paralleled by three of the anatomic categories (viz. Cranial Remains 57.7 percent, Deciduous Teeth 61.9 percent, Permanent Teeth 62.6 percent), while their share of the available sample in the fourth category—Postcranial Remains—lags somewhat with a combined percentage of 42.0 percent. Thus, a statistical analysis of almost any anatomical entity over the entire African sample is likely to be heavily biased by the characteristics and nature of the samples from Swartkrans and Sterkfontein.

Although Swartkrans and Sterkfontein have contributed respectively the biggest and second biggest shares to the overall total of fossil items (37.4 percent, 22.3 percent), and to the subtotals of cranial remains (34.2 percent, 23.5 percent) and of permanent teeth (41.5 percent, 21.1 percent), for two of the anatomical categories this numerical preponderance does not obtain. Of deciduous teeth, while Swartkrans has yielded nearly half (44.8 percent), Taung with its complete set of upper and lower deciduous teeth has given the second biggest share (19.0 percent), while the Sterkfontein contribution lies third with 17.1 percent. Of 138 postcranial items, Olduvai has yielded the most, with 52 fossils or 37.7 percent; the Sterkfontein share —44 items or 31.9 percent—is second biggest; and that from Swartkrans is third, with 14 bones or 10.1 percent.

At this stage, one well-preserved individual may make a major difference to the composition of any of the anatomic samples. Thus, the single Taung individual makes the second biggest site contribution to the 105 deciduous teeth available; similarly, two individuals from Olduvai (hominids 7 and 8) between them contribute over one-quarter of the total postcranial sample available (40 out of 138 bones), or 40 out of 89 limb-bones available from Africa. Similarly, one Sterkfontein individual (Sts 14) alone contributes 28 out of 138 postcranial items from Africa, or 28 out of 34 items from the postcranial portion of the axial skeleton. In other words, numerous as the available samples may seem to be on paper, there are anatomical parts which are poorly represented and where the remains of a small number of well-preserved individuals dominate the sample. We shall return to this point.

Table 3–5. Number of fossil items, and percentage of total, for each anatomical category from each of 14 sites

Site	Cranial Remains	Teeth			Postcranial Remains	Total Items
		deciduous	permanent	total		
Swartkrans, Tv1.	80 (34.2%)	47 (44.8%)	387 (41.5%)	434 (41.8%)	14 (10.1%)	528 (37.4%)
Sterkfontein, Tv1.	55 (23.5)	18 (17.1)	197 (21.1)	215 (20.7)	44 (31.9)	314 (22.3)
Olduvai, Tanzania	32 (13.7)	5 (4.8)	126 (13.5)	131 (12.6)	52 (37.7)	215 (15.2)
Omo, Ethiopia	6 (2.6)	5 (4.8)	99 (10.6)	104 (10.0)	0	110 (7.8)
Makapansgat, Tv1.	28 (12.0)	2 (1.9)	59 (6.3)	61 (5.9)	11 (8.0)	100 (7.1)
Kromdraai, Tv1.	7 (3.0)	8 (7.6)	27 (2.9)	35 (3.4)	9 (6.5)	51 (3.6)
Taung, Cape	5 (2.1)	20 (19.0)	4 (0.4)	24 (2.3)	0	29 (2.1)
E. Rudolf, Kenya	13 (5.6)	0	8 (0.9)	8 (0.8)	7 (5.1)	28 (2.0)
Peninj, Tanzania	1 (0.4)	0	16 (1.7)	16 (1.5)	0	17 (1.2)
Chesowanja, Kenya	3 (1.3)	0	6 (0.6)	6 (0.6)	0	9 (0.6)
Garusi, Tanzania	2 (0.9)	0	3 (0.3)	3 (0.3)	0	5 (0.4)
Lothagam, Kenya	1 (0.4)	0	1 (0.1)	1 (0.1)	0	2 (0.1)
Chemeron, Kenya	1 (0.4)	0	0	0	0	1 (0.1)
Kanapoi, Kenya	0	0	0	0	1 (0.7)	1 (0.1)
Grand Total	234 (100.1)	105 (100.0)	933 (99.9)	1038 (100.0)	138 (100.0)	1410 (100.0)

The spate of discoveries at Olduvai, especially since 1959, has brought the number of hominid individuals from that site to at least 36, represented by at least 215 fossil items, and has made Olduvai to date the third richest site in Africa. This position is true of the total stockpile of Olduvai fossil items (15.2 percent); of the cranial remains (13.7 percent) and of the permanent teeth (13.5 percent); on the other hand, Olduvai is richest in postcranial remains (37.7 percent) and fairly poor in deciduous teeth (4.8 percent). Omo lies fourth in total fossils (7.8 percent), although my data for the available teeth from here are somewhat incomplete. The 110 items are made up largely of permanent teeth (99). Kromdraai, however, exceeds Omo in cranial remains, deciduous teeth, and postcranial remains. Maka-pansgat lies fifth in total items (7.1 percent), but fourth in cranial remains (12.0 percent) and in postcranial remains (8.0 percent), but makes only a slight contribution (1.9 percent) to the sample of deciduous teeth. The special case of Taung and its big share of the deciduous teeth has already been mentioned; East Rudolf with 2.0 percent of all items is rapidly catching up after Richard Leakey's successful 1970 season.

Table 3–5 throws into relief the numerical smallness of the contributions from the remaining six sites—two in Tanzania and four in Kenya. Of course, the remains from these sites include some morphologically and geochrono-logically very important and interesting specimens, but in a statistical analysis, e.g. of the permanent teeth or the postcranial bones, their share would be negligible.

At least three general points emerge from this analysis: (a) the enormous total wealth of material; (b) the poor samples of certain parts of the skeleton; and (c) the need for studies by specialists as well as overall studies by "generalists." We shall deal with each point in turn.

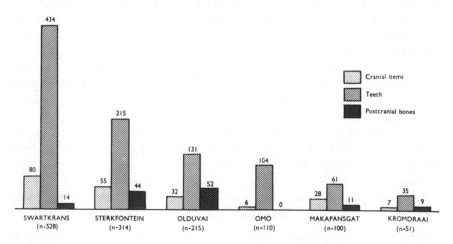

Figure 3–3. Hominid fossil items from the six richest sites in Sub-Saharan Africa; numbers of items in each anatomic category.

The Wealth of Remains

There is a tendency to underestimate the great wealth of materal available (see for example Coon 1963). The placing on record here of these lists of fossil items may assist workers from abroad to appreciate the immense quantity of material available and so to plan their study-visits to Africa more effectively, either by allowing enough time to enable a broad survey to be made of much of the material, or by concentrating on an in-depth study of one or other regional or functional complex.

The Poverty of Remains

Although overall sample size is impressive, a number of anatomical regions are still poorly represented. For example, despite Swartkrans having yielded the biggest site-sample of fossil items, only a single well-preserved *natural* endocast has been recovered from that site—probably though not certainly attributable to *A. robustus* (since *Homo* sp., too, has been identified from Swartkrans). Add to this that no other crania from Swartkrans are sufficiently well preserved to permit an artificial endocast to be made, and it is seen that we know pitifully little, if anything, of the cranial capacity and endocast morphology of *A. robustus* from S.A.

Similarly, the sample of deciduous dentition has important gaps. For example, only six upper incisors are known—four of them being the rather poorly preserved teeth in the Taung maxilla, the other two being upper central incisors of Swartkrans 839. Hence, only two individuals are represented in this sample, one, the type specimen of *A. africanus,* from Taung, the other from Swartkrans, commonly classified as *A. robustus crassidens.* No upper or lower deciduous incisors are known from E.A. From S.A., the sample of lower deciduous incisors numbers 13 teeth, belonging to one individual from Taung, one from Sterkfontein, two from Swartkrans, and one from Kromdraai.

Deciduous canines comprise three uppers and 12 lowers from S.A., representing one individual from Taung, two from Sterkfontein, three from Swartkrans, and one from Kromdraai; while there is one possible deciduous lower canine from Olduvai.

Deciduous upper molars are a little commoner, the total African sample numbering 24 and representing one individual from Taung, three from Sterkfontein, six from Swartkrans, two from Omo, and one or two from Olduvai.

No fewer than 22 individuals are represented among the 46 deciduous lower molars from Africa, these being drawn from site-samples as follows.

Taung	1 individual	Kromdraai	3 individuals
Sterkfontein	3 individuals	Olduvai	1 individual
Makapansgat	2 individuals	Omo	3 individuals
Swartkrans	9 individuals		

Thus, we require more representatives of all deciduous teeth with the possible exception of mandibular molars.

Of the permanent teeth, the number of individuals is much greater, except in respect of the following.

upper central incisors of *A. africanus* of S.A.
(number of teeth = 2, representing 1 individual from Sterkfontein).

upper lateral incisors of *A. africanus* of S.A.
(number of teeth = 5, representing 2 individuals from Sterkfontein and 2 from
    Makapansgat).

lower central incisors of *A. africanus*
(number of teeth = 7, representing 3 individuals from Sterkfontein and 1 from
    Makapansgat).

lower lateral incisiors of *A. africanus*
(number of teeth = 8, representing 3 or 4 individuals from Sterkfontein and 1
    from Makapansgat).

lower lateral incisors of *A. robustus* from S.A.
(number of teeth = 7, representing 6 individuals from Swartkrans).

upper incisors and canines from E.A.
(number of teeth: $I^1 = 6$, $I^2 = 5$, $\underline{C} = 7$, representing collectively 5 individuals
    from Olduvai, 2 from Omo and 1 from Chesowanja).

lower central incisors from E.A.
(number of teeth = 9, representing 4 individuals from Olduvai and 1 from
    Peninj).

Similarly, although 138 may seem a respectable total of postcranial remains, the student of any one structural or functional complex would experience the following limitations.

Scapula:	1 only from Sterkfontein (Sts 7)
Clavicle:	2 from Makapansgat (MLD 20 and MLD 36)
	1 from Olduvai (Old. hom. 8)
	1 from East Rudolf
Ulna:	1 from Kromdraai (TM 1517), and
	2 from Olduvai (Old. homs. 14 and 36)
Carpals:	4 altogether, made up as follows:
	Capitate 2 (TM 1526 from Sterkfontein and Old. hom. 7 from Olduvai).
	Scaphoid 1 (Old. hom. 7)
	Trapezium 1 (Old. hom. 7)
Hand: metacarpals:	9 (2 from Swartkrans, 1 from Kromdraai, *6 from Olduvai*)
Distal end of femur:	2 only, both from Sterkfontein (TM 1513 and Sts 34)
Tibia:	1 from Olduvai (Old. hom. 6), and
	1 from East Rudolf

Fibula:	1 only from Olduvai (Old. hom. 6)	
Talus:	1 from Kromdraai (TM 1517) and	
	1 from Olduvai (Old. hom. 8)	
Other 6 tarsals:	1 each, all from the same foot, namely of Old. hom. 8	
Metatarsals:	1 set of 5 from Old. hom. 8	
Toe phalanges:	2 from Sterkfontein, 2 from Kromdraai (TM 1517), and	
	1 from Olduvai (Old. hom. 10)	
Cervical vertebrae:	1 axis (C2) from Swartkrans (SK 854)	
Thoracic vertebrae:	9 from 1 individual (Sts 14) of Sterkfontein,	
	1 from Swartkrans (SK 3981 a)	
Lumbar vertebrae:	5 from 1 individual (Sts 14) of Sterkfontein, possibly 1	
	from Sts 65 also of Sterkfontein, 2 from Swartkrans	
	(SK 853 and SK 3981 b)	
Sacrum:	1 from Sterkfontein (Sts 14)	
Ribs:	4 complete ribs and 9 rib fragments, belonging to 1 in-	
	dividual (Sts 14) of Sterkfontein	

Thus, some bones which are most important for functional interpretations bearing on for example erectness, bipedalism, and manual grip, are in short supply.

On the other hand, humeri, radii, ossa coxae, and proximal femora are in better supply, *viz.*

	Humerus	*Radius*	*Os Coxae*	*Proximal Femur*
Sterkfontein	1	1	3	2
Makapansgat	2	3	2–3	1
Swartkrans	2	1	2	3
Kromdraai	1	—	1	—
Olduvai	1	1	1	1
Kanapoi	1	—	—	—
East Rudolf	2	—	—	1
Total	10	6	9–10	8

There has been a tendency to draw sweeping functional and even ecological interpretations from the girdle- and limb-bones. Yet the samples are small, the number of individuals represented is sometimes smaller, intrapopulation variability and the degree of sexual dimorphism are virtually unknown, and even the taxonomic assignment of individual postcranial bones is unsure, especially from sites such as Swartkrans and Olduvai, from each of which more than one taxon has been identified. All of these factors should be taken into account before the investigator makes claim for wide differences in function and ecology between various early hominid taxa, based on these fragmentary remains.

Enough has been said to justify the systematic search for more specimens being continued rigorously.

Specialized Studies

The wealth and diversity of material are now so great as to demand specialized studies of functional and structural complexes. It is no longer possible—and has not been for some time—for any one worker to be able to make definitive descriptions and analyses of all hominid fossils being considered. In 1959 and the early 1960's, when Dr. and Mrs. L.S.B. Leakey generously offered me all the hominid material they had exhumed from Olduvai, for study, reconstruction, description, and interpretation, I proposed that the postcranial remains be offered instead to a group of workers in London who had begun to specialize in limb-bones: in this way, the Olduvai postcranial fossils were entrusted to J. R. Napier, P. R. Davies, and M. H. Day.

More recently, we have planned a series of analytical studies upon anatomical regions, functional complexes, and other specialized aspects of the available australopithecine and other hominid material. These studies are being carried out by pre- and post-doctoral fellows from four or five countries, visiting scientists and other colleagues, in the Department of Anatomy, University of the Witwatersrand, in the Transvaal Museum, and elsewhere. Further studies are being carried out independently by J. T. Robinson and others. The following are some specialized studies at present underway:

MASTICATORY APPARATUS

A detailed study of the form and function of the masticatory apparatus by J. W. Wallace has been underway for over three years. This study includes a fine analysis of wear patterns, including scratch-marks, on the teeth; inferred vectors of force involved in masticatory thrust, bite pattern, tooth-tooth contact, crushing and grinding; reconstruction of muscle mechanisms operating on the jaws. The results are being considered in the light of the "dietary hypothesis," the "cultural hypothesis" on the use of the front teeth to account for differences between gracile and robust australopithecines, and other hypotheses.

DENTAL MORPHOLOGY

Many new teeth have been recovered since Robinson's dental monograph was published in 1956. In that study, according to von Koenigswald (1967), Robinson listed 78 deciduous and 448 permanent teeth from South Africa, or a total of 526 teeth. The list which has been compiled for this paper includes 95 deciduous and 674 permanent teeth from S.A., making a total of 769 teeth. This represents an increase of practically 50 percent. Furthermore, to this should be added 10 deciduous and 259 permanent teeth from E.A., giving a grand total of 1,038 teeth in place of the 526 upon which Robinson's monograph was based. This fact alone would justify a new odontographic study of the hominid teeth concerned. G. Sperber has already spent some two years on this task, devoting himself to the cheek teeth alone. At his disposal in S.A. are 61 deciduous molars, 113 maxillary and 85 mandibular permanent premolars, 204 maxillary and 160 mandibular permanent molars. It is hoped, too, that he will have the chance to study most of the available teeth from E.A. sites, comprising nine deciduous molars, 30 maxillary and 34 mandibular permanent premolars, 46 maxillary and 58 mandibular permanent molars, together with others whose existence is known of but whose identity has not yet been published. Furthermore, in his new statistical analysis, as in mine on the E.A. material, measurements on antimeres from the same individual will not be counted as separate items, but a left-right resultant will be used, thus preventing the teeth of any one individual from weighting the results unduly. Attention will be given to a search for evidences of sexual dimorphism of the teeth, especially in view of Campbell's suggestion that robust australopithecines are males and gracile ones females.

DETAILED DESCRIPTION OF NEW REMAINS

The new remains from Olduvai, Omo, and East Rudolf, as well as from Sterkfontein and Swartkrans, are being described and analyzed mainly by Clark Howell, Coppens, Day, Robinson, Mungai, Walker and Tobias. To do justice to the importance of these hominid fossils, their description and evaluation is of necessity arduous and protracted. Accordingly, often under pressure from colleagues, the discoverers and their co-workers have followed the usual practice of issuing preliminary reports in order to share some of the new data with confrères, even before the full monographic treatment of new discoveries has been completed. Such announcements are necessarily incomplete and sketchy and their conclusions provisional, pending the appearance of the larger works. Unhappily, harm has been done to paleoanthropology by more being read into such preliminary notices than was intended, and because their publication was considered by some as the occasion for indulging in lengthy and sometimes polemical rebuttals. Much premature and often needless controversy has thus been engendered, before the unavoidably time-consuming monographs have seen the light of day. The futility of such an exercise is obvious. A number of detailed monographic studies are at present in preparation. It is hoped they will provide paleoanthropologists everywhere with a corpus of factual data, which will enrich the sources of positive evidence and permit the more secure drawing of inferences.

CRANIAL CAPACITY AND ENDOCRANIAL CASTS

The evidence bearing on these problems was analyzed in detail by G.W.H. Schepers in the Transvaal Museum memoirs. Unfortunately, his functional interpretations went beyond what most paleoanthropologists and neurophysiologists were prepared to concede. A more objective, even conservative reanalysis of the material was clearly necessary. Accordingly, we invited G. von Bonin, a specialist on the brains of higher Primates, to visit our laboratories to restudy the casts, some 15 years ago. His cautious comments were summarized in his book, *"The Evolution of the Human Brain"* (1963).

Other studies have been made on the cranial capacities of early hominids more recently (Tobias 1963, 1964, 1968b). A detailed account of the published and some unpublished results was presented in the 38th James Arthur Lecture on the Evolution of the Human Brain, in April 1969 (Tobias 1971b). The results of these analyses indicated, inter alia, no appreciable difference in size or morphology between the endocasts of the robust and gracile members of the genus, *Australopithecus*.

Some new natural endocast material has come to hand, especially a particularly well-preserved and undistorted endocast from Swartkrans (Brain 1967a). Furthermore, several of the new, and largely unpublished calvariae from Olduvai have permitted artificial endocasts to be made. Accordingly, R. Holloway of Columbia University, who is a specialist in paleoneurology, was invited to our laboratories to make a detailed study of the African remains. He spent four months on this in 1969 and will return to Africa for a further period in 1972. His preliminary results suggest that the capacities measured or estimated for some of the australopithecine crania by previous workers are too great and that somewhat smaller values may in fact be correct. The outcome of his further analysis is awaited with interest, as well as his application to these data of paleoneurological principles of which he has for some time been engaged in study.

POSTCRANIAL BONES

Several workers are actively engaged on this subject. J. T. Robinson's monograph on the Sterkfontein skeleton and other postcranial remains is awaited following a lengthy gestation.

Meantime, A. Zihlman of Santa Cruz, California, and B. G. Campbell of Cambridge have both been re-examining the hindlimb and girdle remains; the former in regard to functional interpretations about bipedalism; the latter also on the anatomic correlations of bipedalism, as well as on the problem of body size of the gracile and robust australopithecines. Their studies are in addition to those of Napier, Davis, and Day, who have been describing and analyzing the Olduvai postcranial bones, and of Patterson and Howells (1967) on the humeral fragment from Kanapoi in northern Kenya.

DEMOGRAPHY OF AUSTRALOPITHECINE REMAINS

Following Dart's (1964) assessment of the minimum number of individuals represented among the S.A. australopithecines, Mann (1968) and Tobias (1968c) have both attempted analyses of the age of death among the australopithecines (Figure 3–4). Those S.A. australopithecine specimens

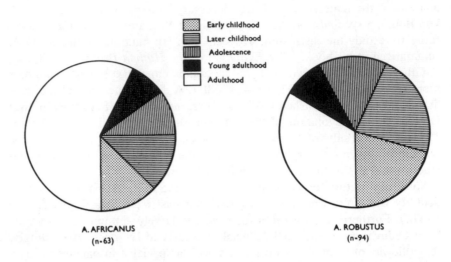

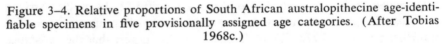

Figure 3–4. Relative proportions of South African australopithecine age-identifiable specimens in five provisionally assigned age categories. (After Tobias 1968c.)

to which it is possible to assign an age fall into two almost equal categories: 76 specimens (48 percent) are of immature individuals, while 81 (or 52 percent) are of mature individuals (Tables 3–6 and 3–7). An interesting contrast exists between the age-identifiable fossils assigned to *A. africanus* (from Taung, Sterkfontein, and Makapansgat) and those assigned to *A. robustus* (from Kromdraai and Swartkrans). Of 63 specimens of *A. africanus,* 35 percent are *immature* and 65 percent *mature;* whereas of 94 specimens assigned to *A. robustus,* 57 percent are *immature* and 43 percent are *mature* (Tobias 1968c). If we divide the data into two other categories, namely specimens of *pre-childbearing age* and of *childbearing age,* there is

Table 3–6. Provisional age assignment of australopithecine specimens from five South African sites

	Early Childhood	Later Childhood	Adoles-cence	Young Adulthood	Adulthood
A. africanus					
Taung	1	—	—	—	—
Sterkfontein	4	5	5	2	30
Makapansgat	3	3	1	3	6
Total	8	8	6	5	36
A. robustus					
Kromdraai	2	1	—	—	3
Swartkrans	17	20	14	8	29
Total	19	21	14	8	32

*Table 3–7. Provisional assignment of S.A. australopithecine specimens
to immature and mature categories**

	Immature Specimens	Mature Specimens
A. africanus	22 (35%)	41 (65%)
A. robustus	54 (57%)	40 (43%)
Total	76 (48%)	81 (52%)

*Immature category includes specimens classified as belonging to early childhood, later childhood, and adolescence.

Mature category includes specimens assigned to young adulthood and adulthood.

still a marked difference between the specimens of *A. africanus* and those of *A. robustus* (Table 3–8). If we regard the *specimens* in each category as representing *individuals,* it seems clear that, *as reflected in the cave deposits, A. africanus* individuals had a much better chance of surviving to childbearing age than did *A. robustus* individuals (Tobias 1968c). The *A. robustus* figures depend largely on the preponderance of juveniles at Swartkrans. In fact, the age composition of the Swartkrans sample differs markedly from that of any of the other S.A. sites. This casts into relief the question of how the bone accumulations developed, and whether the same mechanisms operated in different caves.

*Table 3–8. Provisional assignment of S.A. australopithecine specimens
to pre-childbearing and childbearing categories**

	Specimens of Pre-Childbearing Age	Specimens of Childbearing Age
A. africanus	16 (25.4%)	47 (74.6%)
A. robustus	40 (43.0%)	54 (57.0%)
Total	56 (35.7%)	101 (64.3%)

*Pre-childbearing category includes specimens classified as belonging to early and later childhood. Childbearing category includes specimens assigned to adolescence, young adulthood, and adulthood.

THE COMPOSITION AND NATURAL HISTORY OF THE CAVE DEPOSIT POPULATIONS

C. K. Brain (1968, 1969) has been making a detailed study of the habits of felines in relation to the Swartkrans deposit. He has suggested that leopards might have been responsible for much of the rich hominid accumulation at Swartkrans, an assemblage characterized by a high percentage of juveniles. In contrast, the accumulation at Makapansgat, including thousands of broken bones accompanying the hominid remains, accords more closely with the characteristics of a hominid habitation (Brain 1967b, 1967c, Tobias 1968d). Brain's major study of the characteristics and interpretation of various bone accumulations is now near completion.

Apart from different age distributions of hominid fossils, sites differ in the relative frequency with which various anatomical regions are represented. Since such differential composition may provide clues to the agency/ies responsible for the accumulation of hominid and other bones at each site, it is deemed worthwhile to place the data on record (Table 3–9). Here, the percentages cited are of the total fossil items recovered from each site. Only the six richest sites are included.

Most items from each site are teeth, which would be expected from the method of scoring adopted here: the frequencies range from 60.9 percent at Olduvai to 94.5 percent at Omo. Cranial parts, which average 16.6 percent of 1,410 items from 14 African sites, are relatively uncommon in the Omo sample (5.5 percent) and most frequent from Makapansgat (28.0

*Table 3–9. The percentage frequencies of items representing various anatomical regions of fossil hominids—from the six richest African sites**

	Cranial Items	Teeth	Postcranial Items	Total Items from Site
Swartkrans	15.2	82.2	2.7	528
Sterkfontein	17.5	68.5	14.0	314
Olduvai	14.9	60.9	24.2	215
Omo	5.5	94.5	0	110
Makapansgat	28.0	61.0	11.0	100
Kromdraai	13.7	68.6	17.6	51
14 African Sites	16.6	73.6	9.8	1410

*The percentages here refer to the proportion of the total number of items *from* each site.

percent). On the other hand, no limb-bone has yet come to light at Omo,* while no fewer than 24.2 percent of Olduvai items are of postcranial bones. Swartkrans, with its excellent sample of 528 items, has few postcranial bones (2.7 percent): any hypothesis to explain the accumulation of hominid remains must be able to suggest how 97.4 percent of items from that site are cranial and dental, and only 2.7 percent postcranial—in marked contrast to, say, Sterkfontein, Kromdraai, and Olduvai, where postcranial remains are as frequent as, or more frequent than, cranial remains.

Geochronology

The availability of volcanic materials such as anorthoclase and ignimbrite in E.A. has made possible a series of potassium-argon datings. In general, the dates from the southern part of the E.A. zone (eg. Olduvai and Peninj)

*Editor's note: The American team found a splendidly preserved ulna during the 1971 season.

are younger, those from the northern part (the Rudolf basin) older. The following is a summary of approximate dates from some E.A. sites.

Olduvai-Peninj:	0.7–1.8 m.y.
Baringo (Chemeron Beds):	4.0–4.4 m.y.
E. Rudolf (Koobi Fora):	2.6 m.y.
N. Rudolf (Omo):	1.8–3.1 m.y.
S.W. Rudolf (Kanapoi):	2.5 m.y.
S.W. Rudolf (Lothagam):	5.0 m.y.

An assiduous search is being made in the S.A. sites for datable material. The following avenues have been, are being or will be explored.

a. Paleomagnetism. To align the S.A. deposits with the sequence of reversals of the earth's magnetic field established in the United States of America and Australia. Prof. L. O. Nicolaysen, Director of the Bernard Price Institute for Geophysical Research, concluded that the Makapansgat bone-bearing breccias have been too altered to be dated by this procedure.

b. Fission track dating. It is believed that surface treating of the Transvaal samples in order to achieve the degree of planarity required for fission track dating will present extreme difficulty (Nicolaysen). A possible modification of this method to apply to our breccias and travertines has been suggested by Prof. E. Tongiorgi of Pisa.

c. Alpha-spectroscopic search for heavy radioactive-isotopes. Alpha-spectroscopic studies, using a large area ion-chamber, were made by Dr. R. Cherry and Mr. J. Hobbs in the Physics Department, University of Cape Town, on samples of breccia from Makapansgat. However, all that was seen was "a conventional alphaspectrum, showing the peaks as usually observed in an environmental sample containing the isotopes from the uranium and thorium series". No definite disequilibria were observed, "but fossil bone is such a 'natural ion-exchange medium' that it would be difficult or impossible to interpret disequilibria even if we did observe them."

d. X-ray diffraction studies. Studies were made on travertine and breccia from Sterkfontein, by Mr. I. H. Wright with the help of Dr. J. R. McIver and Dr. T. Partridge, in the Geology Department, University of the Witwatersrand. The treated samples contained no glauconite, but only quartz and, as an impurity, illite or hydroxymuscovite clay mineral. The latter substance may prove datable by the potassium-argon technique, if its presence is confirmed and if suitable concentrations of it can be obtained.

e. Microscopic examination. Dr. J. R. McIver of the Department of Geology made a thin section examination of travertine specimens from Makapansgat. The only recognizable authigenic materials he detected were calcium carbonate and iron oxide. He drew attention to the difficulty of differentiating between, and of separating, authigenic and allogenic materials.

f. Isotopic dating of carbonates. With the collaboration of Prof. Derek C. Ford of McMaster University, Ontario, Canada.

g. Thermoluminescence.

h. Analysis of coprolites.

i. O^{16}/O^{18}, Uranium/Thorium, and other isotopic tests on carbonates. Perhaps a series of carbonates obtained through the deposit could be tested in a number of ways (J. Vogel, E. Gill).

j. Relative dating through the deposit by a network of fluorine assays, nitrogen assays (collagen assessment), uranium (end-window beta counter), manganese. (In some Australian deposits E. Gill of the National Museum, Melbourne, Australia, has found that pyrolusite—manganese dioxide—accumulates as a function of time.)

In other words, what is needed is the application of the whole battery of modern tests to determine the palaeoecology and chronology of the S.A. australopithecine sites. We are indebted to many colleagues for their advice and assistance with the tests so far made.

Meantime, for relative dating of the five S.A. sites, we have been forced to rely upon faunal comparisons. This approach earlier suggested an age-sequence of the S.A. sites as follows (Cooke 1963, 1967).

Middle Pleistocene	Kromdraai
	Swartkrans
————	
	Makapansgat
Lower Pleistocene	Sterkfontein (Lower Breccia)
	Taung

However, a recent closer look by Wells (1969) at the faunal elements from the five S.A. sites has led him tentatively to suggest that Makapansgat could plausibly be regarded as earlier than Sterkfontein. Furthermore, of the age of Taung he states:

Those faunal elements most closely connected with the type specimen of *Australopithecus africanus* do not clearly favour equating this part of the deposit with Sterkfontein and Makapansgat Limeworks rather than with Swartkrans or even Kromdraai, and certainly do not warrant the view that the Taung child is the earliest South African Australopithecine. (Wells 1969, p. 94.)

On this reassessment, Makapansgat may be the oldest S.A. site and Sterkfontein (Lower Breccia) the second oldest. Clearly, we need many more faunal identifications from the S.A. sites; but regrettably we are no longer able further to clarify the faunistic position of Taung, as the site has been entirely destroyed by subsequent lime-working operations.

Space, Time, Species, and Sex

The newer evidence from Omo, added to that of Olduvai Beds I and II, shows that "gracile" and "robust" hominids coexisted in the same area for about 2 m.y. (from about 3.1 m.y. to about 1.00 m.y. B.P.). How much earlier this coexistence extends is uncertain, because of the paucity of

Pliocene remains in Africa. Only the Chemeron temporal bone and the Lothagam mandible date from the period immediately prior to the oldest Omo hominids (4.0 to 5.0 m.y.): both are compatible with gracile hominid structure. Thus, we have as yet no evidence that gracile and robust hominids were synchronic and sympatric earlier than about 3 m.y. B.P.

Paradoxically, the very duration of their synchronicity and sympatry has led some workers to infer a *closer* relationship, and others a *more remote* relationship, between the creatures represented by the two groups of fossils. For some, this long coexistence implies that the two groups must have been generically distinct. In fact, where J. T. Robinson previously regarded the gracile and robust fossils as representing the genera, *Australopithecus* and *Paranthropus,* he has lately seen fit—following the demonstration of the probable morphological and cultural status of *Homo habilis*—to lump the gracile australopithecines into *Homo,* while leaving the robust ones in *Paranthropus* (1965, 1966, 1967).

However, B. G. Campbell has suggested than an alternative explanation of the lengthy coexistence of the gracile and robust forms is that sexual dimorphism accounts for the differences between them. His revival of this old suggestion finds some support from the fact, pointed out in the latest *Yearbook of Physical Anthropology* (Tobias 1969), that "the two main streams within the early hominid group, robust and gracile, are represented in both the South and East of Africa—except that the divergent morphological streams seem to be further apart in East Africa. Thus, the E.A. robust forms are hyper-robust (*A. boisei*), while the E.A. gracile forms are ultra-gracile (*H. habilis*). The South African forms (*A. africanus* and *A. robustus*), on the other hand, seem to be closer together in morphology."

If, as now seems likely, the earliest E.A. deposits are older than the earliest S.A. ones, it would follow that the oldest synchronic E.A. hominids (hyper-robust and ultra-gracile) differ *more* from each other than do the later S.A. forms! I wondered then, "Whether we are presented here with a convergence in the later S.A. forms, or whether this apparent antinomy is the result of sampling error in highly variable populations, or whether some other explanation is to be sought " (Tobias 1969, p. 26).

Could "some other explanation" be that the robust forms were males and the gracile ones females, as Campbell has now tentatively suggested? There is, of course, precedent for this approach: as recently as 1962, Woo inferred of his *Gigantopithecus* teeth that his "large type" comprised male and his "small type" female members of the same species. If this were true of the robust and gracile australopithecines, it would follow that the degree of sexual dimorphism in the early E.A. populations was less than in the later S.A. forms (among which considerable overlap between robust and gracile forms has been shown, Tobias 1967). Since it seems that a reduction in sexual dimorphism has accompanied hominization (Le Gros Clark 1964, Campbell 1967), it would be entirely in keeping that, in such a lineage as that of *Australopithecus,* earlier populations would be characterized by greater sexual dimorphism and later ones by lesser dimorphism.

However, this is one of the few arguments which supports the sexual dimorphism hypothesis. There are powerful arguments against, such as the preponderance of "males" at Swartkrans and Kromdraai and of "females" at Sterkfontein! Another difficulty is that the most gracile of the E.A. hominids has a significantly larger cranial capacity than the robust, although males *exceed* females in mean cranial capacity in all well-studied higher Primates (Schultz 1965, Tobias 1971c).

Two compromise views, at present most widely held, are that the gracile and robust forms represent species of one genus, *Australopithecus* (Le Gros Clark 1964, Tobias 1968e), or at most, two subgenera (*Australopithecus*) and (*Paranthropus*), as proposed by Leakey, Tobias, and Napier (1964) and supported by Howell (1967).

Nevertheless, the very raising of the sexual dimorphism concept demands a reexamination of the australopithecine remains from each site for evidence of bimodal distribution curves. The possibility that the Swartkrans "Telanthropus" form (now widely regarded as *Homo* sp.) may represent simply the female of the other Swartkrans form had long ago been mooted, as mentioned by Robinson (1956, p. 152). Because of this, Robinson attempted to obtain evidence of sexual dimorphism among his dental samples. The only clear evidence seemed to be in the buccolingual diameters of the maxillary canines and of the mandibular first molars of the Swartkrans "Paranthropus" sample. However, his histograms include both left and right teeth from some individuals and hence are weighted. For example, the histogram for maxillary canine B-L diameters includes the left and right canines of SK55, and similarly those of SK65, while two isolated specimens, SK85 and SK93, are said to belong to opposite sides of the same individual (*op. cit.,* p. 41). Similarly, of the 18 M_1's represented, both left and right teeth of SK6, SK25, SK55b, and SK63 are plotted as separate items on the histogram. (Left and right teeth of the same individual have been treated as separate individuals in calculating means, throughout Robinson's dental monograph.) Further, the teeth plotted on the histograms are stated to be "Paranthropus" only: they specifically *exclude* teeth considered to be of "Telanthropus," eg. those in the "Telanthropus" jaws; on the other hand, they *include* great numbers of *isolated* teeth (eg. 8 out of the 15 maxillary canines plotted). *If* two taxa are represented by the jaws and calvariae of Swartkrans, they are almost certainly represented among the isolated teeth as well. Hence, the samples plotted in an attempt to find sexual dimorphism in one taxon may well include representatives of two taxa, while explicitly excluding some known representatives of the smaller taxon. Since a most important object was to determine whether or not sexual dimorphism could have been responsible for the two forms at Swartkrans, these studies are logically invalid and do not exclude the possibility. They should be repeated on unselected samples of Swartkrans teeth, such as G. Sperber is doing on cheek teeth.

For comparison, we need more information on the degree of sexual dimorphism in other hominids, including living man, and in other Primates. Such studies have recently been completed in the Anatomy Department on the crania (De Villiers 1968) and the teeth, jaws, and bony palate (Jacobson 1968) of the South African Bantu-speaking Negro, and on the cranial capacities of pongids and hylobatids (Tobias 1971c); while Professor J. F. van Reenen is engaged on a similar study of the teeth, jaws, and bony palate of Bushmen.

Comparative Primate Data

The need for more comparative data on living hominoids has just been mentioned. The inadequacy of some published lists of metrical data has been strikingly brought home by an experience with hominoid cranial capacities. The author was puzzled by the skewness of the distribution characteristics of a sample of cranial capacities ascribed by Vallois (1954) to 86 "gibbons proprement dit." Enquiries of M. Vallois and of Dr A. H. Schultz revealed that (a) the mean given in the paper by Vallois was wrong, as was the "corrected mean" he kindly communicated to me; (b) the sample range was given incorrectly, the lowest value being cited as 87 cc. instead of 70 cc.; (c) the sample of 86 gibbon capacities was based upon data for 77 gibbons published by Schultz (1933b) to which were added data for nine additional gibbons measured by Vallois himself; (d) male and female data had been lumped together; and (e) data for a number of species of *Hylobates*, recorded individually by Schultz, had been lumped by Vallois under the single rubric, "gibbons proprement dit." Thus, the 77 gibbon capacities which Vallois took over from Schultz comprise 28 of *H. lar,* 21 of *H. agilis,* 9 of *H. pileatus,* 6 of *H. concolor,* 10 of *H. cinereus,* and 3 of *H. hoolock* (but exclude 10 of *H. klossii,* which Schultz had shown to possess the smallest mean of the various species of *Hylobates*). The 9 gibbon capacities measured by Vallois comprised 3 of *H. concolor leucogenys,* 3 of *H. lar albimanus,* 1 of *H. agilis* and 2 of unknown species. The combined interspecific and bisexual sample of 86 capacities is clearly of no value for studies of sexual dimorphism, variability, skewness, or symmetry of distribution.

It has been necessary to start virtually from the beginning and to go back to raw data for animals of known sex and species. Data derived in this way with the kind help of Dr. Schultz, for samples of *H. lar* males, *H. lar* females, *H. agilis* (♂ + ♀), *Symphalangus syndactylus* males, and *S. syndactylus* females, have been presented in the 38th James Arthur Lecture (Tobias 1971b) and further elaborated at the Third International Primatological Congress (Tobias 1971c).

Similarly, recorded data for cranial capacities of all other hominoids are being reexamined and, where necessary, recomputed.

What has been found true of cranial capacity may well prove true, as well, of other metrical and indeed nonmetrical characters. In this respect, special tribute must be paid to Dr. Adolph Schultz, who has probably been responsible for putting more primate data on record than anyone else, and to Sir Solly Zuckerman and to Dr. E. H. Ashton, who, in their efforts to disprove the hominid affinities of the australopithecines, were jointly responsible for recording quantities of invaluable data on pongids in the Powell-Cotton collection. The only pity is that their dental measurements on pongids were made by techniques not the same as those usually employed by paleoanthropologists. Thus, many of their data are not directly comparable with published measurements on australopithecine teeth and other early hominid dentitions. Despite all that has been done toward standardization in anthropometry, there is still a great need for odontometric techniques to be standardized among primatologists and palaeoanthropologists.

Some More of the Many Unanswered Questions

Many problems in the analysis of the australopithecines remain unresolved. For sake of completeness, a brief list given in a recent review (Tobias 1969) is repeated here.

(1) What is the nature of the hominid in the "Middle Breccia" at Sterkfontein? Is it *A. africanus,* as Robinson (1962) has claimed; or a more advanced hominid (c.f. *H. habilis*), as I have been wondering?

(2) What is the nature of the Swartkrans hominid formerly known as "Telanthropus"? Is it really *H. erectus* as many assumed; or is it something closer to some of the ultra-gracile Olduvai hominids (c.f. *H. habilis*)?

(3) Do all of the scores of isolated teeth from Swartkrans which have been attributed to *A. robustus* really belong to this taxon; or do some belong to the other hominid/s at Swartkrans?

(4) Do the jaws dubbed originally "Telanthropus I" and "Telanthropus II" really belong to the same taxon; or are they members of different taxa, as R. Broom once thought?

(5) Is a second hominid represented among the existing sample of remains from Makapansgat, as I very tentatively suggested (Tobias 1968a)?

(6) The Taung skull is the type specimen of *A. africanus*. Are the Sterkfontein and Makapansgat fossils (as well as the fragment from Garusi in Tanzania) correctly assigned to the same species (the only permanent teeth shown by the Taung skull are the upper and lower first molars)?

These are only a few of many things we have tended to take for granted. Yet most of these are questions which can be answered by closer study and comparison of the remains themselves, as well as by the uncovering of more and better specimens.

1. Another important question is the degree to which mosaic evolution has been evident in hominid phylogeny during the last few million years. Evidence is accumulating from S.A. and E.A. that the evolution of the major structural and functional complexes of hominization did not proceed at the same pace. The dental and masticatory complex, for instance, may have outstripped the cerebral-behavioral complex at earlier stages in the process of hominization; subsequently, there is evidence suggesting a lengthy period of jaw and tooth stability, while cerebral expansion forged ahead.

2. If closer study validates the notion of mosaicism, a corollary would be that the characters of, for example, teeth and jaws would provide better evidence of taxonomic affinity and overall degree of hominization *at some periods than at others;* in fact, at certain stages, it would be rash to make a specific identification if only teeth and jaws were available. In other words, the taxonomic valency of teeth and jaws may not be the same at all periods in the last 5 million years of hominid evolution. Criteria which at one stage were useful might at other times cease to delineate effectively one taxon from another.

3. An interesting line of thought seeks to relate the variable features of the S.A. and E.A. hominids to their respective environmental niches. As mentioned before, the E.A. australopithecines are Equatorial; the S.A. ones are Subtropical. Even when allowance is made for shifts of the earth's axis over the last 5 million years, appreciable differences exist—and would have existed—between the two zones. What are the differences which normally obtain between Subtropical and Equatorial mammals? If such differences could be pinpointed, do they find a parallel in the differences between S.A. and E.A. mammalian forms? Are the differences between *A. robustus* of S.A. and the hyper-robust *A. boisei* of E.A., or those between *A. africanus* of S.A. and the ultra-gracile *H. habilis* of E.A., of the same kind and order as those obtaining between other E.A. and S.A. mammals? In other words, are the special characters—*if indeed such exist*—distinguishing the E.A. and S.A. australopithecines to be ascribed to ecotypic and geotypic variation, rather than to different stages in supposed phyletic lineages?

It is a comforting and stimulating thought that, as in other branches of scientific endeavor, each new discovery may solve a few problems, but poses great numbers of new questions. The windfall of new fossils has in a way brought new complexity into a simplifying picture: as our bubble of knowledge grows bigger, the interface with the surrounding sea of ignorance multiplies apace. Truly, it is no admission of defeat to claim that here, as elsewhere in growing branches of science, the more we know, the more we know we don't know!

Appendix

Latest lists of hominids from Omo and East Rudolf
(April 1971)

Since the above compilation was completed, more up-to-date information has become available on the hominid remains from East Africa. Mr. Richard Leakey has kindly allowed me to cite his unpublished list of hominids as at March 1971, from the two main East Rudolf localities, Koobi Fora and Ileret. Coppens (1971) has published a detailed list of those hominids from Omo recovered by the French Expedition (as at November 1970-January 1971).*

The following is a summary of the fossils now available from East Rudolf.

East Rudolf

	Koobi Fora	Ileret	Total
Calvaria	2	6	8
Face	0	3	3
Maxilla	1	3	4
Mandible	2	8	10
Cranial Parts Total	5	20	25 (13)
Deciduous teeth	0	0	0
Permanent teeth	4	?4	?8
Teeth Total	4	?4	?8 (8)
Upper limb bones	0	2	2
Lower limb bones	3	1	4
Postcranial Bones Total	3	3	6 (7)
Total	12	?27	?39 (28)

The figures in parentheses are the former totals as recorded in the chapter above. The total number of items for East Rudolf as a whole has risen by 11.

*Localisation dans le temps et dans l'espace des restes d'hominidés des formations Plio-Pléistocènes de l'Omo (Éthiopie). *C.r. Acad. Sci.* (1970) 271: 1968–1971.

In the following tabulation, Coppens' (1971) list of hominids recovered by the French Expedition is summarized. Alongside is cited a summary of Howell's list of specimens obtained by the American Expedition, and the total for Omo as a whole.

Omo

	French		American		Total	
Calvaria	0		1		1	
Face	0		0		0	
Maxilla	1		0		1	
Mandible	3		2		5	
Cranial Parts Total		4		3		7 (6)
Deciduous teeth	0		5		5	
Permanent teeth	44		60		104	
Teeth Total		44		65		109 (104)
Upper limb bones	0		0		0	
Lower limb bones	0		0		0	
Postcranial Bones Total		0		0		0
Total		48		68		116 (110)

The figures in parentheses are the totals as recorded in the chapter above. The number of items for Omo as a whole has risen by six.

Thus, the total of all specimens has risen from 1,410 to 1,427; to this total, East African sites have contributed 405 specimens or 28.4 percent, and South African sites 1,022 or 71.6 percent. Of 247 cranial parts, 175 come from South Africa and 72 from East Africa: with its 25 cranial parts, East Rudolf has the fifth largest site-sample after Swartkrans (80), Sterkfontein (55), Olduvai (32), and Makapansgat (28). The dental sample from Omo (109 teeth) is exceeded only by those of Swartkrans (434), Sterkfontein (215), and Olduvai (131).

II

Cranial Morphology

4

MATT CARTMILL
DUKE UNIVERSITY

Arboreal Adaptations and the Origin of the Order Primates

Segmentation of an evolving lineage into two or more temporally successive species is generally agreed to be an arbitrary procedure, depending in the last analysis on criteria of taste and taxonomic fashion. It follows that this must be true as well of temporally successive taxa at levels higher than the species. Certain artistic canons restrict the freedom of the evolutionary taxonomist. Taxonomic boundaries are expected to correspond as much as possible to major adaptive shifts, represented in the fossil record by phases of rapid morphological change that lead to the fixation of some new complex of traits underlying a subsequent evolutionary radiation. Such a complex of traits can be used as a diagnosis of the derived higher taxon. Several orders of mammals are defined in this way. The artiodactyls are defined by the acquisition of the double-pulley talus, a specialization for swift and efficient running; the rodents are defined by the dental specializations which have enabled them to become the dominant small herbivores in all recent mammal faunas; the bats are defined by the wings that give them access to flying insects; and so on. A few orders, such as the aardvarks and the South American edentates, are defined by the acquisition of traits for which no

I wish to thank Dr. John Buettner-Janusch for his advice and help throughout the preparation of this paper; Dr. R. H. Tuttle and Dr. C. E. Oxnard for directing the research on which it is based; and Vina Buettner-Janusch, Dr. L. B. Radinsky, and Dr. F. S. Szalay for their valuable comments and criticisms. I am also indebted to Dr. Szalay for providing me with the opportunity to photograph the femoral fragment attributed to *Hemiacodon*. The curators of the Field Museum of Natural History, the American Museum of Natural History, the Yale Peabody Museum, and the Museum of Comparative Zoology of Harvard University were generous and cooperative in permitting me to examine and photograph material in their collections. I am grateful to Dr. W. L. Hylander and Mr. Victor Lukas for taking the photographs reproduced in Plate 4–1. The study from which most of the conclusions presented here are derived (Cartmill 1970) was supported by a National Science Foundation graduate fellowship and by grants-in-aid from the Wenner-Gren Foundation for Anthropological Research and from the Society of the Sigma Xi.

adaptive significance is yet known; but there is no reason to think that these traits were not initially fixed by natural selection, or that their fixation was irrelevant to the subsequent evolutionary radiations of these orders.

It is not always possible to identify a major adaptive shift accompanying the appearance of a new grade of organization. The class Mammalia, for instance, seems to have been produced by the gradual accretion of small changes throughout the early Mesozoic, no one of which was sufficiently important to account for the mammalian radiations of the Cretaceous and Tertiary. In a situation like this—when, in Simpson's (1944) terms, there is no quantum evolution from one adaptive zone to another at the base of an adaptive radiation—placement of any taxonomic boundary between ancestral and descendant groups becomes largely a matter of caprice, subject only to the criterion of monophyly.

The insectivoran-primate transition is frequently described as a situation of this sort. Many students of the primates, from Huxley (1863) to Simpson (1955, 1961), Davis (1955), and Le Gros Clark (1959, p. 52), have asserted that the order Primates, both in the fossil record and in the Recent fauna, grades into the order Insectivora without any major shifts in adaptation. Simpson (1961) concludes that "no clear-cut diagnostic adaptation or 'heritage' distinguishes the order Primates as a whole, or specifically the primitive primates, from other primitive placental mammals."

If this is admitted, any proposed diagnosis of the order Primates must rest on traits which appeared as evolutionary novelties in the ancestral primate population and were fixed in that population, but had at that time no adaptive significance. The petrosal bulla may represent such a trait. In all living primates, the bulla surrounding the middle ear is formed by an extension of the petrosal bone, which also surrounds the inner ear. In most of the placentals, part of the bulla is formed by a separate entotympanic bone, not present in the primate skull. Accordingly, it is often suggested (Gregory 1915, Klaauw 1929, Van Valen 1965, McKenna 1966, Martin 1968) that the petrosal bulla can serve as a marker of the order Primates.

The significance of the petrosal bulla depends on its antecedents, which are still unknown. If the insectivorans ancestral to primates had an unossified bulla, the fixation of bullar ossification in the ancestral primate lineage may indeed have been involved with some adaptive shift, a possibility discussed by Frederick Szalay in Chapter 1. The absence of the entotympanic would then be a merely incidental feature reflecting the independent history of the primate bulla, and would require no further explanation. If primates evolved from insectivorans which already possessed an independent entotympanic ossification center, however, the loss of the entotympanic would be the evolutionary novelty marking the insectivoran-primate transition. This putative entotympanic loss cannot at present be explained in terms of natural selection. Until the antecedents of the primate pattern of bullar ossification are known, the adaptive significance of the petrosal bulla must remain unclear, and the use of this feature as a marker of the insectivoran-primate transition must render the boundaries of the order Primates essentially arbitrary.

Several students of the primates have attempted to *characterize* the order in terms of a distinctive complex of evolutionary trends, instead of trying to provide a list of traits which *define* the order. This complex includes trends toward enlargement of the brain, recession of the snout, convergence and approximation of the axes of vision, ossification of the walls of the orbit, atrophy of the olfactory apparatus, and specialization of the hands and feet for grasping. Most of these trends culminate in the catarrhine monkeys, apes, and men of the Old World. In these animals, the brain is relatively huge, the snout is generally rather short, the orbit is a complete bony cup, the eyes point directly forward, and the olfactory system is poorly developed. In all nonhuman catarrhines, hands and feet are effective grasping organs. These and related peculiarities of catarrhines determine our intuitive notions of what a primate is. The same suite of differences from other mammals is seen in the New World monkeys, and to a lesser extent in the living prosimians. None of the characteristic primate trends is manifested to any degree among the Paleogene forms grouped together by Van Valen (1969) as the superfamily Microsyopoidea of the suborder Prosimii. Known remains of *Plesiadapis, Phenacolemur, Palaechthon,* and *Microsyops* indicate that these mammals had large snouts, relatively small brains, unreduced olfactory apparatus, minimally convergent orbits, and no postorbital bar; in addition, *Plesiadapis* possessed clawed digits like those of a squirrel or tree shrew (Russell 1964, Simons 1967c, Szalay 1969a). The five microsyopoid families have been referred to the primates on the basis of dental resemblances to Eocene genera known to have been lemur-like in features of the skull and limbs; *Phenacolemur* and *Plesiadapis* are also thought to have lacked an entotympanic ossification center, unlike *Microsyops* (Szalay 1969a and Chapter 1).

The allocation to the order Primates of forms showing none of the adaptively significant features distinctive of the living primates is warranted if it is assumed that these features have been produced by selection pressures which were already affecting the microsyopoid populations. The primate evolutionary trends are generally held to represent the gradual accretion of adaptations to an arboral way of life. The inclusion of the microsyopoid families in the order Primates reflects a belief that no marked adaptive shift need be postulated to account for the evolution of a *Palaechthon*-like arboreal animal into the lemur- or tarsier-like primates of the Eocene, and that there is therefore no reason for posting an ordinal boundary between the two evolutionary grades. This paper will attempt to present some reasons for thinking otherwise.

The Arboreal Theory of Primate Evolution

Functional explanations of the primate evolutionary trends begin with the work of G. Elliot Smith and F. Wood Jones in the second decade of this century. In his Presidential Address to the British Association for the Advancement of Science in 1912, Smith (1924, pp. 17–46) attributed the origin and persistence of many of the primate evolutionary trends to the assumption of arboreal habits by Mesozoic primate ancestors. In the trees, Smith suggested, the sense of smell lost much of its usefulness, resulting in a gradual atrophy of the primitively elaborate olfactory apparatus. The dominant role that olfaction had played in guiding and directing the ancestral primates was taken over by vision and touch. The result was a great elaboration of visual and tactile receptors and associated cortical centers—and thus a general enlargement of the neopallium. The need for exceptional agility in running along unstable, discontinuous, and three-dimensional networks of branches favored a comparable elaboration of motor centers; and the demand for coordinated interaction of hand and eye in exploring the arboreal environment led to the development of extensive prefrontal association areas from secondary motor cortex. These trends, continuously maintained by the selective pressures of arboreal life, resulted in the appearance of primates of anthropoid grade by the end of the Eocene.

These ideas of Smith's were refined and expanded by his student Frederic Wood Jones. In a series of lectures before the Royal College of Surgeons in 1915 and 1916, Jones (1917) described most of the still unexplained primate trends as secondary effects of the functional differentiation of the fore and hind limbs. The ancestral primates, Jones conjectured, had become arboreal in the late Triassic, before terrestrial habits had subverted the primitive reptilian flexibility of their limbs to quadrupedal cursorial locomotion. Selection for efficient tree-climbing favored the specialization of the forelimb for reaching out and grasping new supports, and of the hind limb for supporting and propelling the body. Increasingly liberated from locomotor functions, the grasping primate forelimb gradually assumed the manipulatory functions primitively served by the snout. This, coupled with the olfactory atrophy induced by arboreal habits, led to the reduction and recession of the facial skeleton. As the snout dwindled between the orbits, the eye perforce were drawn together toward the midline. The liberation of the forelimb involved a trend toward upright posture, with corollary changes in the viscera and the skull base of the type earlier pointed out by Arthur Keith.

Smith and Jones saw orbital convergence as a passive by-product of the recession of the face. In 1921, Treacher Collins proposed that the need for accurate judgment of distance in leaping from one branch to another selected for overlap and stereoscopic integration of the two visual fields, and therefore favored convergence of the orbits. This is now generally accepted. The body of explanatory theory worked out by Smith, Jones, and Collins (and incorporating ideas from Klaatsch, Matthew, Bensley, and others) has been labeled the *arboreal theory* of primate evolution (Howells 1947).

Criticisms and Modifications of the Arboreal Theory

In its simplest form (sometimes encountered in introductory surveys of primate phylogeny), the arboreal theory is open to the objection that at least nine other orders of mammals include arboreal forms, none of which resemble anthropoids in relative brain size, olfactory regression, ossification of the eye socket, and so on. Realizing this, Jones took care to specify some hypothetical singularities of the primate phylum which accounted for the restriction of the primate trends to that order. Most arboreal mammals, Jones believed, had entered the trees secondarily after a prolonged period of adaptation to terrestrial life, which had inevitably brought about loss of digits, atrophy of the clavicle, and restriction of supination in the forelimb. These changes, in Jones' view, impaired the potential for that functional differentiation of the limbs which was necessary for any further evolutionary advancement.

Some fundamental criticisms of the arboreal theory were put forward by Hooton (1930). Parodying Jones' doctrines as the thesis that "the only prerequisites of intellect are prehensile limbs and a convenient tree," Hooton pointed out that the Malagasy lemurs had radiated into a variety of morphologically different types, none of which could be called markedly convergent with higher primates. "If tarsioid modifications are the result of arboreal life," Hooton asked, "why have not all arboreal lemurs undergone such changes?" Mistakenly believing that the extant Lemuriformes were all nocturnal and omnivorous, Hooton concluded that the primate radiation on Madagascar was not an adaptive radiation: that the morphological variety of the Malagasy lemurs, and the failure of any to attain an anthropoid grade of organization, proved that "the divergently adapted animals must have been . . . inherently different in their bodily organization and in their adaptive tendencies." The attack on Jones' explanations was thus followed by a retreat into teleology, and Hooton's objections were generally ignored.

Jones (1917) had tried to explain the persistent primitiveness of living strepsirhine prosimians by suggesting that their use of the hind limbs as suspensory organs had nullified the crucial trend toward postural uprightness and blocked the differentiation of the functions of the limbs. This same explanation was called on to account for the absence of upright, bipedal, culture-bearing ceboids. The difficulty here is apparent; if a habit of bipedal suspension is compatible with both prosimian and anthropoid grades of development, it cannot be invoked to explain the fact that some prosimians evolved into ceboids, while others did not.

A further contradiction exists between the original theory proposed by Smith and the elements added by Jones. If, as Smith had claimed (and Jones had agreed), arboreal habits lead to an atrophy of the useless olfactory apparatus, then why do a rhinarium, large olfactory lobes, and multiple ethmoturbinals persist in prosimian lineages which, according to Jones, entered the trees at the end of the Triassic? Further, why is it that mammals in general are characterized by the most elaborate olfactory apparatus of any vertebrate class if (as Jones asserted, following Matthew) the ancestral mammals were arboreal? Recent workers have attempted to resolve this inconsistency in different ways. Simons (1962b) asserts that the presence of well-developed organs of smell in prosimians implies that the primate ancestry was terrestrial. In reply to this, Napier and Napier (1967) point out that many arboreal nonprimates have retained elaborate olfactory apparatus in the trees, and infer that any trends toward olfactory regression in the primates must reflect not the uselessness of the sense of smell in an arboreal milieu, but rather the specialized nature of the arboreal adaptations of primates. They suggest that olfactory regression may be due to selection for orbital convergence, inverting the causal relationship proposed by Jones and Smith.

Evolutionary Explanations and Comparative Evidence

An explanatory theory of the type represented by the arboreal theory attempts to rule out teleological "explanations" like Hooton's by demonstrating that the evolutionary trends seen in a particular lineage result from selection pressures exerted on the evolving population by the sequence of environments through which it passes. The trends themselves are thereby eliminated as independent primitive terms. Such explanations can be refuted in two essentially different ways. The first is to demonstrate that one or more of the phenomena described is chimerical: that the specified lineage, for example, did not pass through the specified environmental sequence, had a different basal adaptation from that specified, was not characterized by the evolutionary trends described, or is not in fact a lineage. The second method is to demonstrate that the purported relationship between these elements does not obtain. This can be done by showing that there is some other lineage which resembles the first in all aspects specified by the explanation, but which does not exhibit the evolutionary trends in question. This indicates that the conjunction of basal adaptation and environmental sequence does not have the explanatory force claimed for it, and that the explanation offered is either wrong or insufficiently complicated. If the primate evolutionary trends have not been characteristic of other lineages of arboreal mammals, we may conclude that there is something wrong with the arboreal theory in its received form, and that any explanation of the primate trends must involve a more detailed description of the habitus of the ancestral primates; simply saying that they were arboreal (which is undoubtedly true) will not do the job.

The Varieties of Arboreal Adaptation

Most of the living orders of mammals have arboreal representatives. The adaptations of some—bats, dermopterans, gliding marsupials and rodents, arboreal anteaters, and sloths—are highly specialized and not comparable to those seen in primates. These animals will be largely ignored here. Tree-dwellers predominate in three families of marsupials: the didelphids (New World opossums), phalangerids (Australasian "possums") and phascolarctids (koalas). Although most Recent insectivorans are principally active on the ground, the tupaiids are usually described as arboreal. Of the several lineages of rodents that have developed arboreal adaptations, the sciurine squirrels are the most familiar and most widely distributed. The raccoons (Procyonidae) and civets (Viverridae) include the most specialized arboreal forms among living carnivorans. All of these taxa are represented in modern tropical forests which resemble the floral communities of the early Tertiary (Andrews 1961, Eyre 1963). The information presently available on the ecology, distribution, and behavior of these animals and of the nonhominoid primates (Cartmill 1970) suggests that most of them fall into eight broad adaptive categories:

1. *Forest-floor predators.* Most living insectivorans and carnivorans, many polyprotodont marsupials, and a few rodents feed on prey captured on the ground. Small terrestrial predators inhabiting modern tropical forests fall into two fairly distinct subcategories: those that feed on invertebrates concealed in the detritus of the forest floor, and those that rely on vision in detecting and capturing exposed invertebrates and small vertebrates. Animals in the first group, including *Monodelphis, Urogale, Tupaia tana, Rhinosciurus,* and most of the soricids, macroscelidids, and zalambdodont insectivorans, nose about through leaf litter, detecting prey by smell, hearing, and vibrissal contact, and seizing it with pointed anterior teeth (incisors or canines) borne at the end of an elongated snout. Animals in the second group, including *Felis, Cryptoprocta,* and *Genetta,* have transferred the function of prey detection from rostral organs to the visual system, and rely to a large extent on the retractile claws of the manus for seizing and immobilizing prey. These functional shifts are most pronounced in the felids, where the rostrum is virtually reduced to a visually-directed killing instrument.

2. *Shrub-layer insectivores.* The richest concentrations of insects in tropical rain forests (apart from the leaf-litter fauna) have been found in the dense shrub and undergrowth layers of clearings and forest margins

(Allee 1926, Paulian 1946, 1947). Mammals utilizing this resource, together with varying amounts of fruit, include *Microcebus, Phaner, Loris, Nycticebus,* most of the rain-forest populations of *Galago, Tarsius, Marmosa* and other small polyprotodonts, and the primitive diprotodonts (*Eudromicia, Cercartetus,* and possibly *Burramys*). Like the visually-directed predators of the forest floor, these shrub-layer insectivores rely on visual detection and ranging in the capture of exposed prey, which they stalk among the complex networks of terminal branches and seize with their grasping hands. This predatory technique has been described for *Microcebus, Loris, Nycticebus, Galago senegalensis, Galago crassicaudatus, Tarsius, Eudromicia,* and *Cercartetus* (Le Gros Clark 1924, Banks 1931, Phillips 1935, Hickman and Hickman 1960, Harrisson 1962, Bishop 1962, 1964).

3. *Canopy insectivores.* A few insect-eating arboreal mammals are active principally in the canopy layers of tropical forests, seldom descending into forest-margin undergrowth. This description seems to apply to *Caluromys, Ptilocercus, Galago elegantulus,* and *Perodicticus. Caluromys* and *Perodicticus,* at least, are known to track prey visually and seize it with their hands (Bishop 1964, Hall and Dahlquest 1963).

4. *Woodpecker avatars.* Woodpeckers (Picidae), which prey on woodboring insects in all the strata of the forests of the major continents, are armed with a stout bill for cutting wood and a slim protrusible tongue for extracting insects from their tunnels. In island or subcontinental faunas not colonized by woodpeckers, this niche is usually filled by other birds (e.g. the tool-using Galapagos finch or the huia of New Zealand). In Australasia and Madagascar, the woodpecker niche is filled by arboreal mammals which have developed enlarged wood-cutting incisors and have specialized one of the digits of the manus to serve the office of the woodpecker's tongue. This category comprises *Dactylopsila, Dactylonax,* and *Daubentonia.*

5. *Herbivores of the low canopy and ground.* This category includes those arboreal mammals which feed principally or exclusively on plant matter gathered in the lower canopy of tropical forests, but which also do a small amount of foraging on the ground: *Trichosurus, Lemur catta, Lemur fulvus,* at least some populations of *Propithecus verrauxi, Cebus, Callicebus,* and several other anthropoids.

6. *Canopy herbivores.* Mammals feeding on fruit and leaves in the higher strata of modern tropical forests include *Pseudocheirus, Phascolarctos, Lemur variegatus, Indri, Cercopithecus nictitans* (and many more specialized anthropoids), *Potos,* and *Ratufa.*

7. *Ground squirrels.* Terrestrial forest sciurids, including *Tamias, Eutamias,* some populations of *Xerus, Marmota,* and *Spermophilus,* and the long-nosed Asian ground squirrels (*Lariscus, Sundasciurus, Dremomys,* etc.), feed on fruits, seeds, and other plant matter gathered on the ground and in the shrub layer, together with varying amounts of leaf-litter insects.

8. *Vertically ranging herbivores.* This category comprises not only most of the tree squirrels (*Sciurus, Callosciurus, Heliosciurus,* etc.) but also the paradoxurine viverrids (*Paradoxurus, Arctogalidia, Arctictis,* etc.) and *Nandinia.* These mammals range through all the canopy layers of tropical forest and also forage extensively on the ground. Their diet consists principally of plant matter, together with some animal prey among the included carnivorans.

The adaptations of several species of arboreal mammals appear to be intermediate between these broad adaptive categories, suggesting possible evolutionary pathways from one kind of arboreal adaptation to another (Figure 4–1). The smaller tupaiids feed chiefly on insects encountered in nosing about through leaf litter, but they also enter the undergrowth and lower canopy layers to seek insects and fruit. These animals are intermediate between categories 1 and 2. This is also true of *Didelphis. Nasua,* which forages for prey in forest-floor detritus, also enters the canopy layers to gather fruit, and thus represents an approach to a paradoxure-like habitus. *Arctictis* is in some respects (dentition, prehensile tail) intermediate between *Paradoxurus* and the canopy herbivore *Potos. Cheirogaleus,* which is only slightly insectivorous and prefers larger branches as a locomotor substrate, is transitional in these respects between categories 2 and 5. This is also true of *Callithrix* and *Saguinus,* which forage diurnally for insects in marginal and riverine undergrowth but enter the canopy for nocturnal shelter and to some extent for fruit. The differences between categories 5 and 6 are merely quantitative.

Several lines of mammals which have adapted to life in the trees have not converged with primates to any significant degree. The giant squirrels of the genus *Ratufa,* common in many forest communities of tropical Asia, offer a particularly striking contrast with sympatric prosimians and monkeys. Although these squirrels are canopy occupants, seldom descending to the ground, they resemble other arboreal squirrels and differ from primates in many significant respects. Their eyes face almost directly laterally; the plane of the orbital margin falls less than 20° out of a parasagittal orientation (Cartmill 1970). The olfactory bulbs of *Ratufa* are large, filling the broad space between the medial orbital walls near their apices. The turbinal apparatus is complex and extensive. The thumb is greatly reduced, tipped with a flat nail, and apparently used only in holding food objects. The other digits of the hands and feet are armed with sharp claws. The first toe is not divergent. Despite these putative handicaps, these large squirrels compete successfully with many species of higher primates for the food resources of the rain-forest canopy. The same remarks could be made about any of the other tropical tree squirrels.

Two possible explanations for the absence of primate-like traits among tree squirrels are compatible with the general arboreal theory. The first is Jones' explanation: that the sciurid ancestry passed through a period of adaptation to quadrupedal terrestrial locomotion, which prevented later

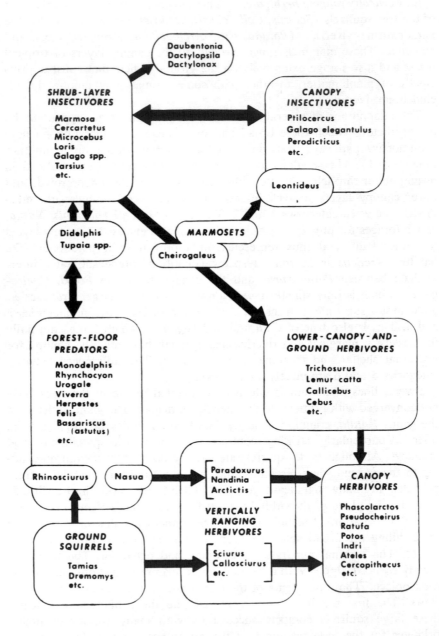

Figure 4–1. Diagram showing some of the major adaptive modes prevalent in the families of mammals discussed in the text. Arrows indicate probable evolutionary pathways from one kind of adaptation to another. It is not implied that any of these extant mammalian taxa is genetically or structurally ancestral to any other.

differentiation of the functions of fore and hind limbs. This explanation is unconvincing. Squirrels retain pentadactyl hands and feet, functional clavicles, and unfused long bones in the forearm and leg. They appear to have the morphological potential for developing grasping hands and feet like those of primates; they have simply not done so. If arboreal life favors these adaptations, there is no reason why they could not have appeared among the sciurids.

A second alternative might be urged by a partisan of the arboreal theory. It is possible that arboreal squirrels, viverrids, didelphids, etc., differ from close terrestrial relatives along vectors in morphological space parallel to the evolutionary vectors inferred for the primate lineages, although the magnitude of these vectors is less than in primates. Thus, although no squirrel has very convergent orbits, we might find that the orbits of tree squirrels are more convergent than those of ground squirrels. A relatively short period of arboreal adaptation might account for the imperfect expression of the primate evolutionary trends in these arboreal rodents—or similarly, in any other group of arboreal mammals which lack the characteristic primate specializations. Currently available behavioral and metrical data can afford us an adequate test of this explanation.

Grasping Extremities: Distribution and Adaptive Significance

As a substrate for locomotion, the branches of trees differ from the surface of the earth in at least four important respects: they are (1) discontinuous, (2) mobile, (3) variable (and always restricted) in width, and (4) oriented at all angles to the pull of gravity. Arboreal mammals must therefore be able to cross safely from one support to the next, hang underneath supports too narrow to balance on, and ascend and descend vertical supports—the last a competence required in no other environment.

In crossing from one support to the next, the safest procedure is to maintain a secure grip with the trailing extremity of the body while reaching and testing the next support with the forelimbs. Accordingly, the hind foot is more specialized for prehension than the hand in many arboreal mammals, including all primarily arboreal marsupials, *Daubentonia, Tarsius,* callithricids and other ceboids, and at least some tupaiids and arboreal mice. In forms lacking grasping specializations of the hind feet, a prehensile tail may serve the same function, as it does in *Potos, Arctictis,* and arboreal pangolins and South American anteaters. Although functional differentiation of the fore and hind limbs is common in arboreal mammals, it ordinarily works in the opposite direction from that specified by Jones.

Grasping hands and feet are advantageous to animals that habitually forage in terminal branches, since they permit these animals to suspend themselves by their hind limbs (and tail, if prehensile) while using the forelimbs to reach and manipulate food items. This can also be accomplished by squirrels and other forms which lack grasping cheiridia (squirrels do in fact sometimes forage in this way, hanging from narrow branches by their clawed hind feet), but their situation is more precarious. This is reflected in the sciurid habit of cutting fruits and seeds from terminal branches and then running down some trunk to retrieve them from the ground.

Per contra, locomotion on steeply sloping or vertical supports of relatively large diameter appears to be less hazardous for small mammals with clawed, squirrel-like hands and feet than it is for small mammals with grasping extremities and reduced claws. Whereas *Microcebus, Loris, Marmosa,* and *Didelphis* seek slender vines or terminal branches as locomotor substrates and are relatively incompetent on thick vertical trunks (Enders 1935, Subramoniam 1957, Petter 1962a, 1962b), arboreal sciurids and carnivorans move cursorially on large horizontal "highway branches" and run up and down vertical trunks with considerable agility, embracing the support with all four limbs terminally anchored by sharp claws. These clawed arboreal mammals descend trunks head first, which involves a retroversion or hyperinversion of the hind foot so that the plantar surface can be applied to the support with the digits pointing caudad.

Clawed, nonprehensile hands and feet are prevalent among relatively unspecialized terrestrial mammals, including the ground squirrels, forest-floor predators, and vertically-ranging herbivores listed above. Prehensile hands and feet like those of primates are universal among the shrub-layer insectivores and related herbivorous forms. The canopy occupants *Ratufa* and *Potos* lack grasping extremities, as do other sciurids and procyonids, but show certain analogical resemblances to primates; the pads of *Ratufa* are relatively enlarged (Pocock 1922),[1] and the tail of *Potos* is prehensile. The distribution of grasping extremities among arboreal mammals suggests that prehensile hands and feet represent an adaptation not to arboreal activity *per se*, but to activity on branches of relatively small diameter, whether in marginal undergrowth or in the forest canopy. For small mammals which habitually climb up and down large vertical trunks, claws apparently provide more effective traction than grasping digits tipped with enlarged pads. This inference is supported by the fact that the vertically-ranging woodpecker avatars have larger claws and less prehensile extremities than the related shrub-layer insectivores; *Daubentonia* differs from the cheirogaleines and other Malagasy lemurs in having true claws on all digits except the hallux, and *Dactylopsila* and *Dactylonax* lack the prehensile tail of the pigmy possums *Cercartetus* and *Eudromicia*.

Orbital Convergence: Definitions and Morphological Correlates

Visual convergence is not a simple univariate quantity. The orientation of the apparatus of vision can be estimated by several different procedures,

---

1. This is at least partly an allometric effect of the low surface-to-volume ratio in this giant squirrel.

no two of which yield uniformly comparable results. In higher primates, the optic axis (the axis of symmetry of the lens) and the visual axis (the line through the center of the cornea to the fovea or area centralis of the retina) roughly coincide; and primatologists therefore sometimes assume that visual and optic orientation are synonymous. However, in diurnal squirrels and tree shrews, the optic and visual axes are almost at right angles to each other (Polyak 1958, Rohen and Castenholz 1967). Nocturnal animals which lack cones in the retina have no visual axis; many birds have two foveas, and thus two visual axes, in each eye. Even in men and the great apes, the axis of the conical eye socket diverges from the mid-sagittal plane by about 23° (Broca 1873). Since the plane of the orbital margin is approximately perpendicular to the orbital axis in these giant primates, it follows that neither visual nor optic orientation can be directly inferred from the orientation of the orbits. This is true *a fortiori* for smaller mammals.

Most published estimates of the orientation of the eyes in primates are based on measurements of the orientation of the orbital margin, which is usually the only element of the optic adnexa that can be studied in fossil material. To describe orbital-margin orientation without ambiguity, three variables must be quantified: convergence (the dihedral angle between the mid-sagittal plane and the plane of the orbital margin), frontation (here defined as the plane angle between the nasion-inion line and the line along which orbital and mid-sagittal planes intersect), and relative interorbital breadth (expressed in orbital diameters). Convergence is a measure of the extent to which the two orbital planes face in the same direction. Frontation is a measure of the extent to which the two orbits face forward toward the end of the snout (as in an ape or man) rather than upward toward the skull roof (as in a crocodile). In higher primates, convergence and frontation are relatively greater than in more primitive mammals, and relative interorbital breadth is less. To assess the extent to which other arboreal mammals show functional convergences with higher primates in these respects, we must take into account some of the factors other than optic orientation which influence the relative position of the orbits.

In didelphids and other mammals with low frontal regions and wide, high zygomatic arches, the frontal margin of the orbit is depressed and the lateral margin of the orbit is elevated (Figure 4–2A). The orbits therefore face upward toward the roof of the skull; convergence is moderately pronounced, but frontation is slight. Here, orbital-margin convergence does not involve any reduction in relative interorbital breadth.

The frontal region of didelphids is low and flat because the braincase is small. The restriction of neurocranial surface area in these primitive marsupials demands a compensatory increase in the relative size of the zygomatic arches, to supplement the limited area available for the origins of the jaw-closing musculature (Hiiemae and Jenkins 1969). Dabelow (1929) plausibly suggests that the relatively small size of the braincase in

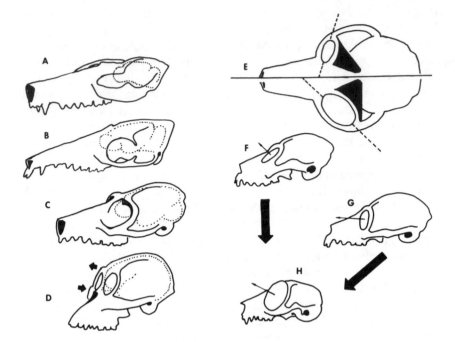

Figure 4–2. Diagrammatic representations of some of the factors influencing the orientation of the orbits.

A. Didelphid configuration, with low frontal region and high zygomatic arches. The orbits face upward toward the skull roof; convergence is moderate and frontation is minimal. B. Tupaiine (or sciurid) configuration. Increase in interorbital breadth associated with expansion of the brain (stippled outline), coupled with reduction in zygomatic height and bizygomatic breadth, results in laterally directed orbits; convergence is minimal. C. Prosimian primate with high convergence and moderately low frontation. D. Anthropoid primate; increased frontation reflects protrusion of the frontal region and recession of the maxilla (arrows). E. Mammals with moderate frontation and approximation of the periorbital cones (top) show an inverse relationship between convergence and relative eyeball size; in smaller species, the eye is relatively larger and the temporal musculature relatively smaller, and the posterior margin of the orbit is thus displaced backward along the zygomatic arch. F. Prosimian with high convergence and low frontation (e.g. *Nycticebus*). G. Prosimian with lower convergence and more pronounced frontation (e.g. *Lepilemur*). H. Smaller prosimian with increased vertical diameter of the orbit (e.g. *Galago senegalensis*); frontation is intermediate between (F) and (G).

adult marsupials is due to the precocious flattening-out of the chondrocranium in the pouch young; in a newborn marsupial, the pharynx must be open and the masticatory apparatus functioning at a stage of embryonic development equivalent to early fetal life in a placental. The low frontation and moderate convergence found in *Didelphis* can thus be seen as a consequence of the marsupial reproductive pattern. In placentals, the braincase is as a rule larger, the zygomatic arch more delicate, and the orbits correspondingly less convergent toward the skull roof (Figure 4–2B). Consequently, orbital-margin convergence in primates and other placentals involves orbital approximation (or else secondary increase in the height of the anterior root of the zygomatic arch, as in *Daubentonia*).

For a given degree of retinal summation and density of receptor cells, visual acuity depends on total retinal area in vertebrates with photopic retinas and on lens diameter in vertebrates with scotopic retinas (Walls 1942). Smaller animals must have relatively larger eyes to retain a degree of visual acuity comparable to that of larger relatives; therefore, relative eyeball diameter is inversely related to body size. This negative allometry influences convergence of the orbital margins. Other things being equal, orbital-margin convergence shows a direct correlation with body size; an increase in relative diameter of the eyeball must be accommodated by an increase in transverse orbital diameter, and this is most easily accomplished by caudad displacement of the posterior orbital margin along the zygomatic arch (Figure 4–2E). This allometric change in orbital-margin convergence does not necessarily reflect corresponding changes in the orientation of the optic axes. The notion that visual-field overlap is comparatively slight in prosimians (Zuckerman 1933, Hooton 1942, Brace and Montagu 1965) may be grounded in estimates of optic orientation made from skulls without correcting for allometry; the fallacy involved in doing this is demonstrated by any photograph of a living lorisid.

Allometric effects on orbital frontation are more complex. In small animals with relatively large eyes, the vertical diameter of the orbit must be relatively greater than in larger relatives. If frontation is pronounced in these larger relatives, it will be reduced in the smaller forms, since a relative increase in the vertical diameter of the orbits must involve elevation of the superior orbital margin (Figure 4–2G, 4–2H). If the larger relatives have minimal frontation, frontation will increase with depression of the orbital floor in smaller forms (Figure 4–2F and 4–2H). Increase in vertical orbital diameter therefore forces frontation to an intermediate value, whether it was large or small to begin with. Thus, orbital-margin orientation is nearly identical in *Tarsius, Galago senegalensis, Avahi,* and *Eudromicia,* although larger haplorhines, galagines, indriids, and phalangerids differ considerably from one another in this respect (Cartmill 1970).

Since the superior margin of the orbit lies on the frontal bone in most mammals, while the inferior orbital margin is formed by the anterior end of the zygomatic arch, orbital frontation depends on the position of the anterior zygomatic root relative to the frontal end of the neurocranium. A general reduction in the relative length of the face brings about a caudad displacement of the inferior orbital margin (or, more precisely, reduces rostrad displacement of the anterior root of the zygomatic arch during ontogeny), and therefore results in an increase in orbital frontation.

Anterior displacement of the superior orbital margin, through increase in the relative size of the frontal pole of the brain, also results in an increase in orbital frontation. Both these effects are seen in anthropoids (Figure 4–2C and 4–2D). Frontation can also be increased by pneumatization of the frontal bone, thrusting the superior orbital margin forward; this occurs to varying degrees in anthropoids, and is conspicuous in *Phalanger maculatus* among marsupials. None of these variables affect orbital frontation in sciurids or tupaiines; in these and other forms with minimal orbital-margin convergence, frontation is not a meaningful variable.

Orbital Convergence: Distribution and Adaptive Significance

If leaping about through the branches of trees selects for overlap of the two visual fields, we would expect to find more convergent orbits in arboreal mammals than in their closest terrestrial relatives. We would also expect to find that acrobatic arboreal mammals have more convergent orbits than their relatively slow-moving arboreal relatives. Quantitative measurements of orbital orientation in arboreal mammals (Cartmill 1970) suggest that neither of these expectations is justified. Orbital convergence is not significantly less in the terrestrial didelphid *Monodelphis* than in the more arboreal *Didelphis* and *Marmosa,* or than in the canopy dweller *Caluromys.* Orbital orientation in the slow-moving phalangerid *Phalanger maculatus* is more primate-like than in other phalangerids, including *Pseudocheirus lemuroides,* one of the most acrobatic of arboreal marsupials. When allometry is taken into account, the small insectivorous diprotodonts *Eudromicia, Dactylonax,* and *Dactylopsila* have more convergent orbits than other phalangerids or *Phascolarctos.* Among tupaiine tree shrews, convergence of the orbital margins is not significantly greater in *Tupaia minor* than in the less arboreal *Tupaia glis* or the largely terrestrial *Urogale.* The terrestrial sciurids *Cynomys, Tamias,* and *Xerus inauris* actually have more convergent orbits than the more arboreal forms *Sciurillus, Myosciurus, Sciurus carolinensis, Callosciurus prevosti,* and *Ratufa bicolor.* Corrected for allometry, orbital-margin convergence shows no correlation with arboreality among thirteen species of procyonids and viverrids, all of which have relatively less convergent orbits and greater interorbital breadth than the terrestrial *Felis bengalensis.* The slow-moving *Loris tardigradus* has more convergent orbits than any primate I have measured (including five species of New and Old World monkeys), when correction is made for allometric effects. Convergence and approximation of the orbital margins is only slightly less extreme among other lorisines. If the primate trends toward convergence and approximation of the orbits represent an adaptation for leaping from branch to branch, it is hard to see why these trends should culminate in *Loris* and *Nycticebus.*

The acrobatic locomotion of many sciurids demonstrates that orbital convergence is unnecessary for gauging distance in arboreal leaps. The visual axes are roughly parallel in diurnal sciurids and tupaiids, and the lamination of the lateral geniculate bodies of *Sciurus* and *Tupaia* (Diamond and Hall 1969) suggests that there may be a narrow anterior zone of overlap and neurological integration of the contralateral visual fields in these arboreal mammals. Further approximation of the orbits (with accompanying nasad shift of the retinal area centralis) would increase the amount of visual-field overlap, but decrease the amount of parallax, and thus reduce the distance over which stereoptic estimation of distance might be possible. For an acrobatic arboreal mammal, primate-like convergence and approximation of the orbital margins would appear to be as much a handicap as an advantage.

Outside of the primates, pronounced convergence of the optic axes is largely restricted to predators. Optic convergence is particularly marked in such animals as owls, hawks, and cats, which depend on vision for the detection of prey. Felids are the only nonprimate mammals whose optic axes diverge by less than five degrees (Johnson 1901). Cats also resemble primates in many neurological features of the visual system, including the lamination of the lateral geniculate, the high percentage of optic fibers which remain uncrossed at the chiasm, and the presence in the visual cortex of cells which respond differentially to stereoscopic differences between the two visual fields (Thuma 1928, Polyak 1958, Hubel and Wiesel 1970).

The relevance of these primate-like traits to the visually-directed predation carried on by most extant prosimians is obvious. Like cats, living cheirogaleines, lorises, galagos, and tarsiers feed on exposed and easily-alerted animal prey, which they carefully stalk and then strike suddenly with their prehensile hands. Stereoptic integration of the two visual fields improves the accuracy of the final strike; increase in visual-field overlap facilitates compensation for evasive movements of the prey.

Olfactory Regression

There is no evidence that the assumption of arboreal habits has been attended by loss of olfactory acuity or regression of the apparatus of smell in any nonprimate lineage. If the ancestral marsupials were arboreal, as the occurrence of hallucal nails and midline pedal contrahentes raphes among terrestrial marsupials suggests, then the didelphids and phalangerids have probably inhabited the trees since the Cretaceous without any atrophy of their large olfactory lobes and complex turbinal systems. Strepsirhine primates have presumably had as much time to adapt to life in the trees as haplorhine primates have; but they have much better-developed olfactory apparatus.

Jones (1917) asserted that scent glands and scent-marking behavior are reduced in arboreal mammals in general and disappear in higher primates. More recent studies show that olfactant-secreting glands and elaborate patterns of marking behavior are well-developed in many arboreal diprotodonts, tupaiids, procyonids, viverrids, lemurids, indriids, tarsiids, and ceboids (Pocock 1915, Bolliger and Hardy 1945, Story 1945, Andrew 1964, Schultze-Westrum 1964, Poglayen-Neuwall 1966, Epple and Lorenz 1967, Martin 1967, Moynihan 1967, Arao and Perkins 1969). Olfactory signals are fundamental to the maintenance of social organization and reproductive behavior in many arboreal mammals, including at least some Old World anthropoids (Michael and Keverne 1968). The thesis that the sense of smell is unimportant to arboreal mammals is no longer tenable.

The degree of approximation of the contralateral orbits necessarily affects the arrangement of the olfactory apparatus. The cone-shaped mammalian orbit has its apex at the optic foramen, on the ventral surface of the braincase. The orbit opens on the dorsal surface of the skull, bordering the interface between neurocranium and splanchnocranium. Approximation of the two orbital cones must therefore constrict the olfactory fossa and the neurological connections between the olfactory receptors and the brain. In lemurines, the large indriids, and *Daubentonia,* the approximation of the two orbits is not extreme (partly for allometric reasons). In these animals, ethmoturbinals may be more numerous than in smaller prosimians, the ethmoid is not exposed in the medial orbital wall, and the vertical height of the nasopharyngeal meatus is relatively great. In the cheirogaleines and most lorisiforms, greater orbital approximation results in increased constriction of the olfactory apparatus: turbinal counts are reduced, an "os planum" of the ethmoid appears in the medial orbital wall, and the vertically compressed nasopharyngeal meatus is progressively filled with projections of the displaced posterior ethmoturbinals (Kollman 1925, Kollman and Papin 1925). These trends culminate in *Loris,* where the orbital margins meet in the midline to form a marginal interorbital septum. This extreme approximation of the orbital margins accounts for the distinctive shape (Figure 4–3 and Plate 4–2) of the frontal cortex and olfactory lobes of *Loris* (Starck 1954, Radinsky 1968a).

A small interorbital septum often appears at the apex of the mammalian orbit if the basisphenoid is not pneumatized (Haines 1950). In *Tarsius,* extreme ocular hypertrophy coupled with marked optic convergence and orbital approximation results in a fusion of the medial walls of the two orbits to form an extensive apical interorbital septum (Spatz 1968). The olfactory fossa is accordingly displaced forward and reduced in size, and olfactory connections with the brain are limited to a slender tubus olfactorius which arches over the top of the interorbital septum (Figure 4–3A). A similar apical septum (and a corollary displacement and reduction of the olfactory fossa) characterizes small recent ceboids and was probably found in at least some Oligocene catarrhines as well

A. In *Tarsius,* the interorbital septum (stippled) forms near the apex of the orbital cone, and the olfactory process of the brain (outline) passes above the septum to reach the nasal fossa (black).

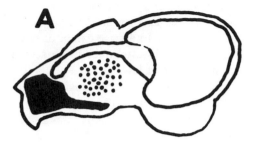

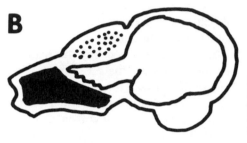

B. In *Loris,* the interorbital septum forms at the anterior margin of the orbital cone, and the olfactory bulbs lie beneath the septum in contact with a relatively large nasal fossa. Compare Plate 4–2.

Figure 4–3. Relations of interorbital septa in prosimians (diagrammatic).

(Simons 1959, Cave 1967). The regressive olfactory apparatus of· anthropoids evidently reflects not perfected adaptation to arboreal life, but derivation from some prosimian with a large apical interorbital septum like that of *Tarsius.*

The Onset of the Primate Evolutionary Trends

The thesis that the ancestral therian mammals were primitively arboreal is fundamental to the arboreal theory as formulated by Jones (1917). This thesis rests principally on the absence of "terrestrial" (i.e., cursorial) locomotor specializations in primitive and unspecialized mammals; on Matthew's (1904) inference from the shortness and divergence of the pollex in Paleocene condylarths that the therian pollex was primitively opposable; and on Morton's (1935) analysis of the tuber calcanei as having evolved to permit propulsive plantarflexion of the foot independent of flexion of the digits for grasping. It will suffice to say here that cursorial specializations are not evidently advantageous to small terrestrial mammals inhabiting floral communities providing a dense ground cover; that other explanations of the shortness and divergence of the mammalian pollex (Haines 1958) avoid the difficulty of accounting for the lack of an opposable thumb in didelphids; and that a tuber calcanei is found also in theriodont therapsids, thecodonts, and fresh-water crocodiles. Present evidence suggests that primitive therians were, like many extant insectivorans, forest-floor predators depending to a large extent on vibrissal contact and olfaction for locating leaf-litter invertebrates to be seized and killed by sharp anterior teeth in the elongated jaws.

The diversification of angiosperms in Cretaceous forest communities undoubtedly opened a variety of new ecological niches to insects and their predators. Several different lines of insect-eating vertebrates gradually developed adaptations for stalking and capturing exposed insects in undergrowth and lower canopy. Extant forms adapted to this way of life, including chameleons, small polyprotodonts, primitive diprotodonts, and most prosimians, possess grasping hands and feet facilitating locomotion on slender branches and extensive visual-field overlap (facultative in chameleons) facilitating prey detection and distance estimation.

Marsupials appear to have been the first mammals to occupy this habitus, and they still occupy it today in continental faunas lacking prosimians. During the Cretaceous, at least one and possibly two or more insectivoran lineages converged with the didelphids, acquiring clasping cheiridia (*sensu* Haines 1958) and pronounced orbital convergence. The low frontal region and high zygomatic arches of the primitive metatherian skull imply a considerable amount of orbital convergence, and this may have conferred an initial adaptive advantage on the didelphids.

The postorbital bar seen in primates is sometimes explained as a protective device or a strut for transmitting chewing forces, but it seems more likely that it originally served to prevent deformation of the orbital fossa by contractions of the masticatory muscles. In didelphids, the postorbital ligament lies in the plane of the superficial temporal fascia (of which it represents the free anterior edge), and tensile stresses in this fascia resulting from temporalis contraction or masseteric tension on the zygomatic arch do not tend to deform the lateral margin of the orbit. In small placentals with large eyes and relatively small temporal fossae, convergence of the orbital margins involves a deviation of the plane of the orbital margin from the "plane" of the temporal fossa. Contractions of the masticatory musculature are thus more likely to distort the orbital margin in prosimians than in marsupials. This potential interference with visual and oculomotor precision was countered in prosimians, as in small felids and herpestines, by the progressive ossification of the postorbital ligament.

The prehensile forelimbs necessary for stalking insects along thin branches serve also, among living insectivorous prosimians, as prey-seizing organs analogous to the tongue of a chameleon. The importance to primates of hand-eye coordination, which Smith was the first to stress, can be plausibly traced to an ancestral habitus in which the hand was used for striking prey. The transferral of prey-seizing functions from the anterior teeth to the manus may well explain the primate evolutionary trend toward recession of the snout. Corrected for allometry, the length of the dental arcade relative to prosthion-inion length is greater in *Tarsius* than in other extant prosimians (Cartmill 1970); it is somewhat less in lemuriforms, and still less in most lorisiforms. This pattern corresponds with the pattern of incisor use in these three groups; the enlarged incisors of tarsiers assist the imperfectly prehensile hands in capturing prey, whereas lemurines and indriids rarely, and lorisids seemingly never, employ their comparatively reduced incisors in retrieving food items (Le Gros Clark 1924, Bishop 1964, Jolly 1966). Jones' insight remains sound.

The shift of prey-seizing functions to the hand must also have made the vegetable component of the diet a more important factor in the selective pressures governing the morphology and occlusion of the cheek teeth. Like extant shrews (Hamilton 1930), living insectivorous prosimians eat a considerable amount of vegetable matter. However, among prosimians, but not among shrews, the grasping hands assist or replace the dentition in seizing, crushing, and dismembering insect prey, and efficiency in processing fruits and seeds assumes a correspondingly greater importance to the morphology of dentition. The fact that primitive primate molars display reduction of the stylar shelf, decrease in relative trigonid height, and other changes associated with herbivorous dietary specializations in some mammalian lineages (Szalay 1968a), does not contradict the hypothesis that the ancestral primates were primarily insectivorous. Comparable dental changes are seen in primitive diprotodonts like *Cercartetus,* whose quadritubercular, bunodont molars are well-suited to masticating the insects and fruits brought to the mouth by the prehensile hands. Among the diprotodonts, as among the prosimians, a shift to manual predation in the ancestral stock laid a pre-adaptive foundation for later radiations of more specialized herbivorous forms.

Were the Ancestral Primates Vertical Clingers and Leapers?

Napier and Walker (1967a) have proposed that the ancestors of all living primates can be placed, together with extant indriids, tarsiers, galagos, and *Lepilemur,* into a locomotor category of vertical clingers and leapers (VCLs), defined by a pattern of locomotion in which "the body is held vertically at rest and pressed to the trunk or main branch of a tree," and "movement from place to place is effected by a leap or jump from one vertical support to another" (Napier and Napier 1967). This thesis has been widely accepted. Napier and Walker advance two lines of reasoning in support of the claim that modern apes, monkeys, and quadrupedal prosimians are descended from vertically clinging and leaping (VCL) ancestors:

1. The VCL habit is an adaptation for escaping from predators. On Madagascar, there are no large predators, and vertical clinging and leaping

conveys no adaptive advantage; on the contrary, "the limitations imposed by the vertical clinging and leaping habit upon feeding behavior are . . . quite marked; the large forms, especially, are at a disadvantage when feeding in a small branch milieu" (Napier and Walker 1967a). The VCL habit therefore could not have evolved independently among Malagasy prosimians, and the ancestral lemuriform that colonized Madagascar must have been a VCL.

2. "Of all the fossil postcranial bones reputably assigned to Eocene prosimians, none show the morphological feature of quadrupeds: all the skeletal characters point to these animals being Vertical Clingers and Leapers." Nine skeletal characters are considered indicators of VCL adaptations in Eocene primates. Of these, the most important are the following: relatively elongated hind limbs; an elongated calcaneus and navicular; a tarsier- or galago-like femur with a cylindrical head and a high, narrow patellar groove; a "tarsier-like pelvis" with long ilium and short ischium; a centrally placed foramen magnum; and a reduced snout (reflected in the wide divergence of the mandibular rami).

The first line of reasoning is not sound. *Propithecus verrauxi* emit a variety of alarm calls on sighting raptorial birds and terrestrial predators (Jolly 1966); unless these calls are another Eocene survival, these Malagasy VCLs must be subject to predation. *Propithecus* feeding among small branches display more varied and flexible postures than *Lemur catta;* in moving through the canopy, they not only cling vertically and leap, but also walk quadrupedally or bipedally on branches, walk quadrupedally or bimanually under branches, hop bipedally, and hang by both hands or both feet (*ibid.,* pp. 34–38). If VCLs were at an adaptive disadvantage in the forests of Madagascar, they would surely have been replaced by competing quadrupeds during the course of the Tertiary and Quaternary. Quite the reverse: VCLs were the only prosimian locomotor type unaffected by the Pleistocene wave of extinctions in Madagascar (Walker 1967a).

Most of the supposedly diagnostic skeletal features cited by Napier and Walker do not separate extant VCLs from quadrupeds. On Madagascar, tarsal elongation is found among quadrupeds (*Microcebus, Phaner*) but not among VCLs; this is not wholly a matter of allometry, since *Galago crassicaudatus* has an elongated tarsus, while *Avahi,* an animal of the same general body size (Bauchot and Stephan 1966), does not. A cylindrical femoral head is not found in *Lepilemur* or the indriids; some cheirogaleines have an elongated femoral head (Plate 4–1). The ratio of ilium to ischium length does not separate *Nycticebus* from *Tarsius* and *Galago,* nor *Lemur* from *Propithecus* (Walker 1967b, Schultz 1969). The placement of the foramen magnum (Plate 4–2) shows no correlation with posture, as Biegert (1957) and others have repeatedly demonstrated. Divergence of the mandibular rami is a function not only of relative dental-arcade length (which itself is not correlated with posture), but also of relative dental-arcade breadth; thus, although *Tarsius* has an exceptionally long dental arcade for a prosimian, its mandibular rami diverge widely because the dental arcade is also extremely broad (Cartmill 1970), permitting the maxillae to give maximum support to the enormous eyeballs. The inferences about VCL habits in Eocene prosimians made by Napier and Walker, largely on the basis of published descriptions and pictures of fossils, are accordingly not persuasive.

The current evidence suggests that "vertical clinging and leaping" is an artificial category that includes two quite different kinds of adaptation. Saltatory modifications of the hind limb, including tarsal elongation and remodeling of the hip articulation (Grand and Lorenz 1968) are seen in many small insectivorous prosimians today, including quadrupedal forms (the cheirogaleines) as well as vertical clingers (*Tarsius* and the smaller galagos). This saltatory complex has been analyzed as an adaptation for leaping on animal prey (Walker 1969). Herbivorous prosimians on Madagascar have developed a leaping form of locomotion, possibly as an adaptation for avoiding predators; since canopy-feeding primates "must, in moving through the forest, either descend to lower levels where the canopy is *continuous* or pass from crown to crown in the *discontinuous* stratum by leaping or brachiating" (Napier 1963), the leaping adaptations of *Lepilemur* and the indriids may also be of advantage in feeding on leaves in the canopy. Herbivorous VCLs show no tarsal or coxal specializations, although they have long, powerful hind limbs.

During the Eocene, primates radiated into a variety of adaptive types. Several forms (*Tetonius,* the necrolemurines, and probably *Pronycticebus* and certain omomyids) are small animals with relatively large and convergent orbits; we are justified in inferring a persistence and further perfection of the ancestral *Microcebus*-like adaptation. The long hind limbs, small orbits, and seemingly herbivorous dentition of *Notharctus* (Gregory 1920) imply an adaptation like that of a diurnal indriid. Inconclusive evidence suggests that certain Eocene tarsiids had begun to develop saltatory specializations (Simons 1961a). Simpson's (1940) descriptions of postcranial remains assigned to omomyid genera show them to have resembled extant cheirogaleines as much as *Tarsius* or *Galago* (Plate 4–1).

Later Prosimian Evolution

At some point during the latter half of the Eocene, an as yet unidentified group of insectivorous prosimians with an apical interorbital septum and a corollary reduction and anterior displacement of the olfactory fossa (as in *Tarsius* or *Pseudoloris*) underwent an adaptive shift to a diurnal and largely herbivorous way of life. A slightly earlier shift of this sort had resulted in the production of forms like *Adapis* and *Notharctus* out of *Pronycticebus*-like ancestors. In animals with a tarsier-like orbit and rostrum, however, this shift had quite different consequences. Diurnal habits and increased body size implied a reduction in the relative size of the orbits. As the orbital roofs descended, the cranial space thus vacated was occupied by an expansion of the braincase over the persistent apical interorbital septum; the enlarged braincase thus regained a direct contact with the olfactory fossa, but the olfactory apparatus remained small and atrophic. Reduction in orbital diameter and expansion of the anterior part of the neurocranium narrowed the gap between the braincase and the lateral margin of the orbit. This permitted the spread of periorbital ossification (or the consolidation of a tarsier-like partial postorbital septum) to insulate the eyeball, with its increasingly photopic retina, from movements originating in the masticatory apparatus filling the temporal fossa. The resulting cranial configuration is seen in small ceboids today.

Although diurnal prosimian herbivores were supplanted in continental faunas by diurnal anthropoid herbivores, Old World prosimians continued to exploit the primitive primate adaptive zone. The ancestor of the lorisiform prosimians was probably a small insectivorous quadruped with slightly elongated tarsal bones, an internal carotid traversing the foramen lacerum medium, and a tympanic ring attached to the margin of the bulla. *Microcebus murinus* fits this description fairly well (Gregory 1915, Saban 1963, Charles-Dominique and Martin 1970). The early primate evolutionary trends have been carried to an extreme by the lorisines; *Loris tardigradus,* with its exaggerated orbital convergence and cheiridial prehensility, is more highly specialized than any other living strepsirhine for the mode of life whose adoption led to the differentiation of the order Primates from the other placentals. Under persistent competition from tarsiids and lorisids, the initially primate-convergent tupaiids (whose ancestral adaptation appears to be represented by the modern *Ptilocercus*) have tended increasingly to retreat to rostrally-guided predation on leaf-litter invertebrates. As a consequence, the tupaiine tree shrews display secondary increase in snout length and reduction of orbital convergence.

During the early Tertiary, the subcontinent of Madagascar was colonized by a small, nocturnal, insectivorous primate similar to the extant cheirogaleines; if this animal resembled *Microcebus,* it may have been derived from the African population that gave rise to the Lorisiformes. The adaptive radiation which ensued on Madagascar paralleled the continental Eocene prosimian radiations, the primitive visually-guided predatory lemuriforms giving rise to a variety of canopy herbivores. *Cheirogaleus* appears to represent a survival of an early stage in this adaptive shift.

Orbital convergence and displacement of the olfactory apparatus in *Microcebus* conforms to the pattern seen in lorisids, rather than that seen in tarsiers. The failure of an extensive apical interorbital septum to develop in the ancestral Malagasy lemurs meant that a wide separation between braincase and olfactory fossa was never effected; as a consequence, orbital reduction accompanying shifts to diurnal activity resulted in enlargement of the nasal fossa as much as of the braincase. Since there was little expansion of the anterior neurocranium, orbitotemporal confluence persisted, and masticatory movements continued to affect the eye; the retinas of diurnal lemuriformes accordingly remain photopic but afoveate, like those of diurnal sciurids or tupaiids.

The Basal Primate Adaptation and Ordinal Differentiation

The following points emerge from the preceding brief survey.

1. Primate-like morphology is not necessarily advantageous to arboreal mammals; the arboreal theory is untenable in its received form.

2. The distribution and functional significance of primate-like adaptations among extant mammals indicate that the ancestral primate adaptation involved nocturnal, visually-directed predation on insects in the terminal branches of the lower strata of tropical forests.

3. This presumed basal adaptation provides an adequate explanation of the origin, persistence, and variable expression of the primate trends toward cheiridial prehensility, periorbital ossification, orbital convergence and approximation, rostral recession, olfactory regression, and neurological integration of the contralateral visual fields.

4. Known cheiridial, cranial, and dental morphology of the Paleogene mammals grouped together by Van Valen (1969) as the prosimian super-family Microsyopoidea indicates that most of these forms could not have been visually-directed manual predators, and that none of their ancestors had undergone the adaptive shift proposed here to account for the primate evolutionary trends.

If these statements are correct, a monophyletic and adaptively meaningful order Primates may be delimited by taking the petrosal bulla, complete postorbital bar, and divergent hallux or pollex bearing a flattened nail as ordinally diagnostic. Such a diagnosis dictates the removal of the families Plesiadapidae and Microsyopidae to the Insectivora. The Paromomyidae must also be referred to the Insectivora if, as reported by Simons (1967c), the still-undescribed skull of *Palaechthon* lacks a postorbital bar. The dental specializations of the carpolestids and picrodontids suggest that these families are no more likely to meet these revised criteria of membership in the Primates. By this analysis, no Paleocene prosimians (except perhaps *Berruvius*) are known, and the primate fossil record begins in the Eocene. The five microsyopoid families appear to represent ultimately sterile side branches which diverged from the primate ancestry prior to the crucial shift to visually-directed predation.

Szalay (1968a), after considering a similarly restricted definition of the Primates, rejected it with these words:

Any attempt to impoverish an order such as the Primates from its early "side branches", to make it increasingly a vertical category, would be a taxonomic practice distinctly nonbiological. . . . I prefer to allocate doubtful groups to the orders to which they are strongly suspected to belong . . . rather than invariably burden the already much abused Insectivora.

Reluctance to burden the Insectivora by referring abortively specialized Paleocene placentals to this order is not warranted, because the order Insectivora is, and ought to be, a taxonomic catch-all. Within a higher taxon T of taxonomic rank j, it is customary to set up a "wastebasket taxon" of rank $(j-1)$ to receive those taxa of rank $(j-2)$ which have acquired the definitive characters of T but have undergone little further specialization. The order Insectivora provides such a taxonomic wastebasket in classifications of the placental mammals. In Ride's (1964) classification of the marsupials into four orders, the order Marsupicarnivora serves the same purpose. Forms lumped into these wastebasket taxa are sometimes more specialized than forms assigned to other taxa of the same rank: *Thylacosmilus* is a more specialized marsupial than *Caenolestes,* and *Talpa* is a more specialized placental than (for instance) most of the creodonts. The moles are not made a separate order, not because they are particularly primitive in their morphology, but because they have radiated into a rather narrow adaptive zone and have not acquired extraordinary specializations like those which compel us to draw an ordinal boundary around the pangolins. The order Insectivora is not restricted to ancestral placentals, or unspecialized placentals, or diminutive placentals, or insectivorous placentals; it comprises all placentals which need not be referred to any other order. Since a wastebasket taxon of this sort has no positive diagnosis, it cannot be overburdened. Any attempt to make the Insectivora a tight, satisfying taxon by referring aberrantly specialized Paleocene mammals to related extant orders must result in adaptively meaningless diagnoses for many orders, rather than only one. A shift in feeding habits, from nosing out forest-floor insects to spying and snatching insects among terminal branches, appears to account for the specializations distinctive of all Neogene primates. Given such an adaptive shift, we are entitled to mark it with a taxonomic boundary, relegating vaguely rat-like and multituberculate-like Paleocene primate relatives to the wastebasket category thoughtfully provided by earlier taxonomists.

5

W. W. HOWELLS
HARVARD UNIVERSITY

Analysis of Patterns of Variation in Crania of Recent Man

In interpreting morphology of man and primates, fossil and recent, one need has been for a body of comparative data, amenable to mathematical methods of analysis, to provide objective evaluations of a given problem. Here I wish to use measurements of recent human skulls, carefully assigned as to population, to show how two kinds of multivariate analysis (factor analysis, and multiple discriminant or canonical analysis) may be used to detect patterns of morphological variation.

Specifically I wish to enquire whether specific patterns of cranial variation found among individuals of a population—which we may assume to be variation primarily of genetic origin—are also the material on which population differences are based: whether the agents of microevolution or "subspeciation" are operating on coherent patterns of morphological variation and not merely having a taffy-pull with the skeleton of the head. Dobzhansky (1944) said races "are populations differing in the incidence of certain genes." Can we apply this to more complex entities in cranial variation?

The Role of Hypothesis and Methodology

Little is known about such patterns, and how they might be used in the interpretation of even the most recent human evolution. The problem just now is one of exploration; it differs from certain other sets of problems in which functional causes of difference can be postulated. This is much more the case for the postcranial skeleton, especially when different species can be compared by adequate methods.

Part of the research described was supported by National Science Foundation Grant No. GS-2465.

123

This may be epitomized by three stages in the study of the scapula:

1. classical dissection and comparison of muscle attachments. Oxnard, herein, refers to R. A. Miller's (1932) study, which supported only the hypothesis of increasing mobility in the shoulder joint from prosimians to man.

2. Direct experiment. Wolffson (1950) selectively severed or removed different muscles attached to the scapulae in newborn rats, a very good exploitation of Washburn's (1947) advocacy of such experiment, and done under his direction. This supported the hypothesis of shape dependent on muscle function, and negated older hypotheses of racial or genetic factors of shape variation in the case of man. This has not been carried to the point of trying to explain specific function-shape relations in primates.

3. Measurement of mechanical factors and multivariate analysis. Oxnard (Chapter 14) and associates devised measurements to reflect differences in essential mechanical function of the scapula and, using canonical analysis on data from many species of primates, demonstrated in simple form a pattern of these conforming closely to observed locomotor behavior of the species involved, also turning up a wealth of suggestive detail and hypotheses for further work.

In the skull of modern man, however, it is difficult to see how mechanical function might be related to variation except minimally. There are no muscles of any force other than those of the mandible, and these are certainly not adequate to explain much of the observed variation (although Hunt 1960 concluded that a hard diet increased the lateral growth of the palate and face of Australian aboriginal children as compared with Americans, and so dampened growth of the face vertically). Other special functional hypotheses have been addressed to climate and nose shape (see Weiner 1954, Wolpoff 1968), or intense cold adaptation and face shape (see Coon, Garn, and Birdsell 1950; Steegman 1970). These, of course, appeal to natural selection as the operative agent. Different, and more general, is the Weidenreich-Coon hypothesis (Coon 1962) that the populations of major geographic zones are to be ranked as subspecies; that is, their racial differences are to be traced, not to a recent adaptive differentiation within the species, but to different genetic heritages from different hominid ancestors, however the differences among the ancestors may have originally been established.

Some Craniometric History

At any rate, the beginning problem is analysis of "racial variation." Cranial configurations of some populations (Eskimos, Australians, Bushmen) are readily recognized by a student with moderate experience. They have been described in simple fashion: metrically, by such attributes as facial or nasal width; morphologically, by brow development or sagittal keeling. But these patterns have not been analyzed intensively for identification of their essential features, or for their interpopulation significance, which is really a first step in explaining them or using them in interpretations of fossil hominids.

The metrical study of crania has a history as old as anthropology itself, which I will outline here by using the cranial index as a whipping boy. The central position of this ratio in anthropology can hardly be overstated. Lapouge (1887, Revue d'Anthropologie, p. 151), a nineteenth century anthroposociologist, wrote that wars would be fought in the future over a few points in the cranial index—and he meant real wars, not scholarly scuffles. By the 1880s it was accepted as "an excellent means of distinguishing ethnoracial groups," and it is still being written about (for example, see Bunak 1969, from whom the latter quotation is taken). The trends, especially in Europe, toward brachycephaly, and some cases of debrachycephalization, have been considered and reconsidered. First a simple descriptive character, the index became so revered in its own right that the figure for the Trinil skull, in spite of its noncomparability to recent populations, was thought important; and in 1931 Keith wrote of the *Australopithecus* child that it was the first dolichocephalic ape. Brachycrany in the Krapina Neanderthals was a minor sensation. Eventually the index came to seem not a normally distributed variable differing on the average in human populations (its true character), but rather a trait with real categories (dolichocrany versus mesocrany versus brachycrany), thus allowing a population to be broken down into parts (and when combined with one or two other traits such as the nasal index, leading to "types") which could point to the different ancestral "races" merged in that population.

I recite this wayward history to emphasize how little real enquiry the index has received. It has mostly simply been accepted, almost as a God-given racial feature. There have also been absurd adaptive explanations related to riding yaks or climbing coconut trees. Some have viewed it as the interplay, or relative "dominance," of genes separately controlling cranial length and breadth (for a review, see Bunak, op. cit.). Perhaps the only student to inquire what the index really means was Weidenreich (1945), reviewing the whole history and evidence, and concluding that brachycephaly (that is, the emergence of "round" heads) was the concatenation and culmination of erect posture, expanding brain, reduction of the external architecture (supraorbital and occipital tori), and "rolling up" of the skull around the base into a more globular shape. But none of the long discussion has gone beyond the *a priori* assumption that this index is a special and significant trait in human variation. It may not be so; but that is not important here.

Multivariate Analysis

Oxnard (Chapter 14) covers the general nature and meaning of multivariate statistics so expressively that I shall restrict my own remarks to reinforcement of his on specific aspects. I refer particularly to what he says about the limitations of classical methods in attempting to synthesize information "by eye" and the general restraints of univariate analysis; about the way in which newer methods may suggest secondary insights and relationships not suspected from direct investigation or from classical anatomical deduction; and about forms of multivariate analysis generally.

One thing should be said. Multivariate analysis should not be looked on simply as a "technique," like gel electrophoresis, or split-line pattern analysis. It does call for the use of computer techniques, and covers a large number of specific mathematical procedures, but taken *in toto* it is an approach, not a technique.

To begin with, one aspect of multivariate studies is the use of *populations* to establish the nature and degree of variation. This is needed to supplement inferences, however sound, which can be drawn from nonquantitative, descriptive comparisons of a few specimens, whether the limit on numbers is imposed by availability or difficult methodology (for example, full-scale dissection as carried out with the precision insisted on by Tuttle, although this work is actually leading to valuable quantitative material) attaching to recent materials, or by rarity of fossils. However, even when specimens are few, students now universally apply the concept of population variation whether or not it can be sampled. (A well-known result is the decline of the practice of creating new genera or species on the basis of individual specimens—even among chimpanzees—or fossils, on the basis of minor differences from previously known material.)

Karl Pearson was much interested in exploring variation and correlation in human crania (Pearson and Davin 1924). Great though his contributions were, his statistics were addressed to the dispersion of single variates, or the correlation of two variables. The usual record of a set of P traits for a number (N) of individuals is as follows.

| Individuals | Traits | | | | |
	A	*B*	*C*	*D*	. . . . . . *P*
1	1a	1b	1c	1d	. . . . . . 1p
2	2a	2b	2c	2d	. . . . . . 2p
3	3a	3b	3c	3d	. . . . . . 3p
. .	. .	. .	. .	. .	. . . .
N	Na	Nb	Nc	Nd	. . . . . . Np

An individual (e.g. 1) becomes a vector of his *scores* on P traits. This is how he was treated in the earliest work; separated from his population, he was described integrally in this fashion, and "typed." Statisticians then turned to the vector of a *measurement* (e.g. A), consisting of its scores on N individuals, and concerned themselves with the parameters (e.g. mean, standard deviation) of that trait, in an attempt to describe the population as a whole. But the population is not contained in such single vectors, only in the matrix of numbers as a whole. This is not a new view. I mention it because anthropologists have long been reluctant to take account of it. Multivariate statistics were late on the scene, and later still in acceptance, being especially difficult to use without electronic computers. The above tabulation is simply meant to show that such statistics are nevertheless the natural and appropriate treatment of populations as such, or of several populations at once. Methodologically, they are the quantitative expression of "population thinking" (Mayr 1959), and the corrective, in structuring hypotheses, for "typological thinking."

Factor Analysis

The essence of this now well known group of methods is the attempt to explain—in fact to reproduce—the correlation or covariance among a set of measurements (or any variables) as simply as possible, by a transformation. A factor may be thought of as a "new" variable introduced among those actually used in measurement; it has loadings for, or correlations with, each of the original measurements. Figure 5–1 is an attempt to render the basic notion, minus various qualifications (the idea conveyed is actually that of a principal components analysis). Imagine plotting arm length and leg length on the usual bivariate scattergram: these two will obviously be strongly correlated, with the individual points forming a long ellipse. A new axis of measurement, laid along the long axis of the ellipse, would be a better way of registering what the correlation of leg and arm expresses: the axis constitutes a "factor" of generalized limb length. Both of the original two measurements will be even more strongly correlated with this factor than with one another (in fact they must be); and the individuals of the population can be measured, or scored, on this factor—that is, scored for "general limb length," instead of separately for arm and leg lengths.

There is, obviously, another axis of variation remaining, a much shorter axis of the ellipse at right angles to the first, which takes care of all the remaining variation of the individuals. In this instance it notes the lack of perfect correlation of arm and leg length, and this is a factor of differing proportion, or shape. Here is a choice. The individuals can be scored on it, which will retain all the original information. Or, as is more useful when things get complex, this smaller amount of information may be regarded as rubbish, and ignored.

When a larger number of measurements is used it becomes highly unlikely that all will co-vary harmoniously with arm and leg length, and that face breadths, for example, may form another cluster of correlation, a factor unrelated to the first suggested above. Without going into details, factors are extracted from a matrix of correlations by finding the latent roots (or eigenvalues) of the matrix—this is where computers make the

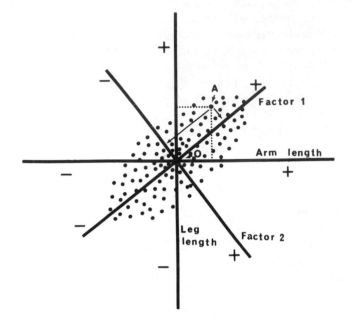

Figure 5–1. The relation of real measurements and factors. A hypothetical scattergram of individuals measured for arm length and leg length. The factors (or components) correspond, in two dimensions, to the major and minor axes of the ellipse of points and are uncorrelated, whereas the scatter of points shows that the measurements have a high correlation.

Factor 1 would represent "general limb length," in absolute size; + = long; — = short. Factor 2 represents *lack* of complete correlation, or "disproportion" in shape: + = arm relatively longer than leg; — = leg relatively longer. Individual A is well above average in both of the real measurements, but more so in leg length (the coordinates, or real measurement positions, are shown by dotted lines). This can be said directly in factor scores (coordinate positions shown by arrows): a high positive score, near the top end of the distribution, on Factor 1, for general limb length; and a negative score on Factor 2, within *absolutely* long limbs, *relatively* longer legs. An individual at 0, the mean point of measurements and factors, would be exactly average for the population in both absolute measurements. Factorially he would be exactly average in general limb length, and also in proportions.

difference. Each root has an eigenvector, here a "factor," associated with it; and the most important factor, in its association with the greatest proportion of the total correlation of the variables, is the first extracted. Meaning of factors is read from the loadings.

Succeeding factors, in this procedure, are uncorrelated with, or orthogonal to, the first and to each other; and it is a problem in factor analysis, to decide at what point dusk has come and further factors have dimmed into insignificance. In any case, it is the hope and purpose (generally) of the transformation that the number of meaningful factors will be small— much smaller than the number of measurements. Finally, factors (which are "new variables," or new reference axes) may be rotated to new positions relative to the original measurements, whose own positions remain fixed by their mutual correlation. In this rotation, the factors' mutual orthogonality, or zero correlation, may be retained or, to fit related groups of measurements, they may be placed obliquely, in which case the factors themselves become correlated positively or negatively.

PREVIOUS WORK ON CRANIA

This comprises studies by myself (1957, following a 1951 study on the living), Landauer (1962), Kanda and Kurisu (1967, 1968) and Brown (1967). All except the last were limited as to measurements used or completeness of coverage of the skull, due largely to the limitations of pre-electronic computation. Brown's study, much the most complete, used 77 measures of the dry skull and of X-rays of the same subjects. He went through five successive analyses (principal factor and maximum likelihood), first dropping many measurements because of overlap and "spurious" correlation, and dropping others, as well as factors, in an attempt to suit the surviving factors and measurements to one another in a form which could best be interpreted. The rotations were kept orthogonal (uncorrelated) instead of oblique. This has the advantages of reducing the subjective element in placing factors, and giving estimates of the relative contribution of each factor and measurement to the total communality.

Though comparisons are difficult to make, because of the limitations suggested, the following seem to be present as factors in more than one of the above studies:

Cranial-Skeletal Size. Found in my 1957 (and 1951) study, and probably also in that of Landauer, her V19 being saturated on length, but also such breadths as are present; and with the Japanese "length" factor, which has no associations with capacity or breadth. The skull base length, from basion to nasion, is associated with *none* of these factors. Brown, by contrast, found no specific cranial length or size factor. His "length" is confined to factor 4, "anterior cranial base length," as defined by the nasionsphenion length; and his measure of total length (endocranial) is distributed over four factors, with a sizable unique variance; thus the interpretation runs against a specific vault *length* factor.

Masticatory or Ruggedness Factor. Apparently common to Brown and Landauer, having loadings on bizygomatic breadth and infratemporal fossa depth, and on other indicators of strength of the muscles of mastication.

Brain Size. The Japanese study has cranial capacity loaded *only* on F1, together with loadings for height, breadth, and length in sharply decreasing order of magnitude. Present as described in my earlier study, and possibly in Landauer's.

Base breadth. Present as a minor factor, distinct by definition from cranial breadth, in my study; in the Japanese study, a breadth factor has its highest loadings in the basal part. No appropriate measurements in Brown or Landauer.

Lower Face. A not too specific agreement of measurements in all planes here, in the Japanese, Landauer, and Brown.

Face Height. Brown's no. 15, of general facial height (as interpreted), would correspond with the Japanese F5. The other studies lack the appropriate measurements.

Other possible correspondences exist. A common factor of breadth across the upper face may be present in the studies of Landauer, the Japanese, and myself, but here the comparisons become difficult to make. Different studies use different measurements, and results of a factor analysis are necessarily largely controlled by, and certainly limited to, the measurements used—no factors can be found where no measurements lie. Also, correlations between the same two measurements on different populations may differ considerably, and the reality and meaning of such differences badly needs investigating.

PRESENT ANALYSIS

This is a preliminary report of a new analysis, which should be more comparable to Brown's in method, though not in measurements used. The material is 834 skulls, known or judged to be male, belonging to 17 different populations, on which I took 70 measurements and angles. As in analysis of variance, the covariation has been decomposed; the correlations used are based not on the total group of skulls as such (which would increase the correlation by introducing the element of population differences), but upon the pooled within-group variances and covariances. This should furnish a good, generalized estimate of intrapopulation covariation for analysis. Unlike Brown, I did not discard measurements "spuriously" correlated (covering a similar region, or related like an angle and one of its sides), but decided to let the factors absorb and express all this. From an image-covariance common factor analysis[1] I took the first 18 factors

1. The previous explanation of factor analysis given, and portrayed in figure 1, actually was that of components analysis, which is not, properly speaking, factor analysis at all. In this, it will be recalled, all the information is retained: the matrix of correlations has 1.0's in the diagonals, which is just what you get in computing a table of correlations, so that any trait has a correlation of itself with 1.0, identical with its complete variance. Factor analysis proper is interested only in factors common to more than one trait; and it attempts to arrive at an estimate of the common variance for each trait, or communality, eliminating the unique variance which is the difference between the common variance and 1.0. In image covariance analysis, the off-diagonal correlations also undergo slight adjustment, for the special covariation of the two traits involved, so as more closely to reflect only common variance throughout the matrix. Needless to say, these refinements, and improvements in results, are a further compliment to computers.

for rotation and, like Brown, kept the rotations orthogonal. (The rotated factors are thus known to be independent for interpretation, but there is little point in saying, except in a general way, which are the most "important" since these are necessarily apt to be those with many measurements running in the same direction, the aspect Brown tried to correct for.) Unfortunately, there is not a broad correspondence between my list of measurements and Brown's. There are, however, some common results. For one, this time I, like Brown, did not find a factor of general size associated with cranial length; and, also like Brown, I found total cranial length to be of little or no importance in itself, and to be distributed over several different factors.

Of the original, unrotated factors, the first, associated with 23.3 percent of the common factor variance, did indeed carry the largest loading for glabello-occipital length (.70) of any factor; but the factor itself, though partly one of size, was primarily one of forward projection of the anterior part of the skull and face. In the rotations, the latter aspect became the essential nature of the factor, and skull length fell to minor importance. In general, rotation simplified and sharpened the factors very considerably; they make better "sense" by far in this form. The rotated factors appear to fall into two groups: the first generalized, the second anatomically local. The order of presentation below is more anatomical than expressive of their possible importance.

Forward Extension of the Facial Skeleton (especially the orbital margins, relative to the temporal region). Loadings are high from radii (originating at the meatus) to zygoorbitale, ectoconchion, frontomalare, and dacryon, that is, four points around the orbit; decreasing values for nasion and prosthion radii, and basion-nasion and basion-prosthion lengths. The correlations involved suggest the unity of the factor; the lower loading and correlations for glabello-occipital length—behind those of basion-

Measurements	Factor 1	Original Correlations						
		EKR†	FMR	DKR	NAR	PRR	BNL	GOL
Zygoorbitale radius	90*	84	76	84	80	76	65	48
Ectoconchion radius	90	..	89	77	73	65	60	45
Frontomalare radius	86		..	75	74	55	59	44
Dacryon radius	86			..	92	65	70	50
Nasion radius	82				..	64	75	53
Prosthion radius	76					..	58	45
Basion-nasion length	72						..	49
Glabello-occipital length	45							..

*Loadings and original coefficients of correlation are given without a preceding decimal point.

†Abbreviations used are defined in Howells (in press); here the measurements referred to are those listed at the left, in order, except the first shown.

nasion—imply that the association of cranial length is entirely secondary, and that there is an agreement between this factor and Brown's "anterior cranial base length." There is no agreement with "length" factors in previous analyses. Though criteria for judging are poor, this appears to be a prominent factor.

Vault Breadth. The loadings themselves suggest an uncomplicated factor of breadth centering on maximum cranial breadth and auricular breadth, diminishing in effect in all directions toward the frontal and occipital breadths; even palate breadth has a loading of .24.

Measurements	Factor 2	Original Correlations						
		AUB	XFB	STB	ZYB	ASB	WCB	JUB
Maximum breadth	70	61	62	54	44	46	35	32
Biauricular breadth	68	..	39	26	77	47	46	57
Maximum frontal breadth	60		..	81	37	27	38	37
Stephanic breadth	55			..	25	21	24	24
Bizygomatic breadth	54				..	35	48	77
Asterionic breadth	46					..	22	25
Minimum cranial breadth	36						..	46
Bijugal breadth	35							..
Palate breadth	24							

Such a breadth factor appears in Brown's study. The correlations, however, suggest that the question of a separate, if minor, factor of base breadth, indicated in my 1957 study and perhaps by the Japanese, is not settled. Although the loadings mask it, one suspects that if the small matrix above were factored in turn, it would give two factors, since there seems to be special correlation among biauricular, bizygomatic, and bijugal breadths. Now, as befits their mixed character as measurements, bizygomatic and bijugal also have good loadings on a facial factor (number 4, p. 133), while biauricular has a loading of consequence only here.

Face Height and Verticality. The prominence of the basion angle (of the facial triangle), and of facial height, suggests that it is a simple factor

Measurements	Factor 3	Original Correlations				
		NPH	NLH	NAA	OBH	WMH
Basion angle	93	78	56	−50	38	19
Nasion-prosthion height	89	..	72	−28	39	40
Nasal height	68		..	−36	42	33
Nasion angle	−46			..	−58	02
Orbit height	43				..	−03
Cheekbone height	27					..

of vertical maxillary growth; this is supported at a lower level by the height of the orbit and the narrow nasion angle. Cheekbone height, with a low loading and zero correlations with some of these, is obviously secondary, suggesting that malar development has nothing to do with the factor.

Brown, with quite different measurements, found an apparently similar factor, as did the Japanese study.

General Upper Facial Breadth. This is most highly saturated on biorbital and bifrontal breadths, diminishing through orbital, bijugal, and bizygomatic to palate breadth. Smaller loadings exist for breadths within the face (nasal, interorbital) and behind it (frontal, biauricular). As a factor it may be suspect, as being determined simply by the presence of a considerable number of transfacial measurements; at the same time, it has possible relationships to the "masticatory" factor seen in previous studies.

Measurements	Factor 4	Original Correlations				
		FMB	OBB	JUB	ZMB	ZYB
Biorbital breadth	84	91	68	77	51	59
Bifrontal breadth	81	..	62	73	46	47
Orbit breadth	74		..	52	23	41
Bijugal breadth	67			..	59	77
Zygomaxillary breadth	51				..	47
Bizygomatic breadth	48					..

General Midfacial Size. Loadings, which are not high, pick out bimaxillary, bijugal, and palate breadths, as well as cheek height and basion-nasion and basion-prosthion lengths. There is *no* loading for bizygomatic breadth. The factor does *not* reflect nasal or orbital breadth. If a single bone is pointed to, it seems like the maxilla. While not corresponding closely, factors of some kind of lower or midfacial expansion have appeared in the studies by Landauer (V9), the Japanese (F2), and possibly Brown's factor 2, "mandibular length." The particular rotations of different studies may have tended to put different emphasis on this nebulous factor, if indeed it is one common to the several studies.

Measurements	Factor 5	Original Correlations				
		BNL	WMH	MAB	BPL	JUB
Zygomaxillary breadth	62	29	41	47	34	59
Basion-nasion length	31	..	27	31	64	38
Cheekbone height	31		..	30	29	32
Palate breadth	27			..	35	47
Basion-prosthion length	26				..	34
Bijugal breadth	20					..

General Sagittal Lengthening.   Another low key factor suggesting total length (glabello-occipital has its highest loading here) which is contributed to by parietal length and flattening above and by length along the base (foramen and basion-nasion lengths) below. The original correlations show that it is not determined by broad correlation among the measurements involved, but emerges on rotation, which may be an argument for its validity.[2] It does not rely on occipital protrusion for lengthening (this is a different factor here), and so does not correspond with the Japanese "length" factor. It seems to represent a residual factor of cranial length after more specific contributions to lengthening (e.g. 1, 5, 14, 16, 17, 18) have been accounted for.

	Factor	Original Correlations				
Measurements	6	*PAA*	*FOL*	*PAC*	*PAS*	*BNL*
Glabello-occipital length	49	24	22	50	03	49
Parietal angle	49	..	10	−22	−89	11
Foramen magnum length	43		..	07	−06	14
Parietal chord	30			..	62	13
Parietal subtense	27				..	−04
Basion-nasion length	21					..

The remaining factors are fairly specific in nature and considerably more local in character.

Horizontal Profile of Orbits.   The high loadings express forward angulation of the medial margins of the orbits (dacryon), and of nasion, relative to the sides of the face. At a much lower level, the radii to nasion and dacryon reflect a degree of absolute prominence of this lower border of the frontal bone, while a negative loading for ectoconchion radius emphasizes the drawing back of the lateral orbital margin. This appears to be a

	Factor	Original Correlations				
Measurements	7	*DKA*	*NAS*	*NFA*	*NAR*	*BNL*	*DKR*
Dacryon subtense	89	−98	70	−66	36	27	40
Dacryon angle	−89	..	−68	28	−32	−23	−36
Nasio-frontal subtense	88		..	−96	47	36	37
Nasio-frontal angle	−88			..	−39	−27	−29
Nasion radius	34				..	75	92
Dasion-nasion length	31					..	70
Dacryon radius	31						..
Ectoconchion radius	−23						

2. This sounds illogical; the meaning is that, in spite of the mutual orthogonality or lack of correlation of, that is, PAA and FOL, the factor axis locates a region where these and the other measures all have positive projections on it (that is, correlations with it).

relatively important factor. (Certain other loadings, not shown, conform to the above interpretation.)

Horizontal Interorbital Profile. Quite distinct from the previous factor, this registers not relative flatness across the orbits, but across the nasal region between the orbits regardless of facial flatness on either side. There is, however, for obvious anatomical reasons, a low association of the prominence of nasion itself relative to the lateral angles of the frontal.

Measurements	Factor 8	Original Correlations			
		NAD	*NFA*	*SIS*	*NAS*
Naso-dacryal subtense	90	−77	−27	42	28
Naso-dacryal angle	−85	..	13	−30	−06
Nasio-frontal angle	−22		..	−22	−96
Simotic subtense	21			..	24
Nasio-frontal subtense	20				..

Prominence of the Nasalia. Again, a quite independent third factor of horizontal profile, restricted here to the nasal bones themselves. The essential feature is the absolute height of the nasal saddle (simotic subtense), and pitch of the nasal roof, not a pinching of the nasalia laterally, since the least nasal breadth has a *positive* loading, though a low one, and has a correlation of only .12 with the simotic angle. There is a low loading for the naso-dacryal angle, which measures the general prominence of the interorbital region as registered in the previous factor; but since nasal prominence is included in this, the loading is not unexpected, and does not negate the independence of nasal prominence suggested here.

Measurements	Factor 9	Original Correlations		
		SIA	*WNB*	*NDA*
Simotic subtense	83	−69	56	−30
Simotic angle	−76	..	12	42
Minimum nasal breadth	35		..	−30
Naso-dacryal angle	−24			..

This sorting out of different elements of midfacial flatness seems legitimate. Factor analysis may be expected to produce factors based on doubletons, or pairs of correlated variables having "spurious" correlation not strongly correlated with other variables. However, the angles involved in the last three factors (dacryal, nasio-dacryal, simotic) have relatively high communalities, especially naso-dacryal—that is, they are not strongly isolated—while their mutual original correlations are not high. They are, that is, indeed rather independent characters within the facial region.

Breadth of Interorbital Region. Still in the same region, this minor factor has its major loadings on interorbital and nasal bone breadth (with the secondary effect of a wider naso-dacryal angle). It is not related to facial breadth; orbit breadth is antagonistic, with a negative loading, and there is no loading for nasal breadth at the aperture. The secondary loadings are brought up in the rotation, evidently, judging by the original correlations.

Measurements	Factor 10	Original Correlations		
		WNB	OBB	NDA
Interorbital breadth	76	40	—06	54
Minimum nasal breadth	54	..	03	07
Orbit breadth	—38		..	—03
Naso-dacryal angle	34			..

Subnasal Flatness. Strictly a flatness (or the reverse) across the subnasal region, relative to the zygomaxillary border. There are low loadings for the radii from the auditory meatus to subspinale and prosthion, which are clearly a secondary effect and confirm the localization of the factor.

Measurements	Factor 11	Original Correlations		
		SSA	SSR	PRR
Zygomaxillary subtense	—89	—91	54	39
Zygomaxillary angle	89	..	—39	—24
Subspinale radius	—38		..	90
Prosthion radius	—22			..

Prognathism. A simple factor of classic prognathism, or relative projection of alveolar region and dentition as a whole (as registered by the radius to the first molar alveolus). The angles of the facial triangle (nasion and prosthion), rather than absolute measures, have the highest loadings. The factor is independent of the previous one of subnasal flat-

Measurements	Factor 12	Original Correlations				
		NAA	BPL	PRR	AVR	BBH
Prosthion angle	—88	—71	—46	—43	—33	25
Nasion angle	81	..	66	46	38	—27
Basion-prosthion length	58		..	84	74	12
Prosthion radius	43			..	88	26
Alveolar radius	33				..	26
Basion-bregma height	—29					..
Subspinale radius	24					
Nasion angle (ba-br)	—24					

ness; it is in a different plane, and the coefficient of correlation between prosthion angle, here, and xygomaxillary angle measuring subnasal flatness, is a negligible .13.

Malar Size. Isolated and simple, the factor is saturated only on measurements of the malar, and on its contribution to facial breadth (in the bijugal diameter) and to forward growth (in the zygomaxillare radius— both measurements with small loadings). Zygomaxillary breadth, the maximum breadth of the maxillae, is not involved; this factor is malar size independent of maxillary size.

Measurements	Factor 13	Original Correlations			
		XML	*MLS*	*ZMR*	*JUB*
Inferior malar length	73	70	40	50	39
Maximum malar length	69	..	52	40	46
Malar subtense	50		..	30	28
Zygomaxillare radius	27			..	37
Bijugal breadth	26				..

Frontal Bone Length and Size in the Midline. The highest loading is for the angle at basion subtended by the frontal chord, and for the frontal chord itself, so that the frontal length is both relative and absolute. This is not a matter of flatness or the opposite; the frontal angle has a zero loading. Secondary effects on the whole sagittal plane appear in the low loadings for total length and the radius to the vertex. Prominence of the glabella, however, is apparently related (allometrically?) to the size of the frontal itself. The matrix of correlations shows that this is a factor emerging as a result of rotation.

Measurements	Factor 14	Original Correlations					
		FRC	*NBA*	*FRS*	*GLS*	*GOL*	*VRR*
Basion angle (na-br)	84	64	−69	48	07	19	01
Frontal chord	75	..	−26	59	11	54	53
Nasion angle (ba-br)	−55		..	−19	−10	−24	30
Frontal subtense	32			..	−15	22	18
Glabella projection	32				..	22	03
Glabello-occipital length	28					..	46
Vertex radius	20						..

Frontal Bone Flatness in Profile. Distinguished from the previous factor, this is one of shape and of small size. The essential loadings are for flatness, as measured by both subtense and angle in the midline. The remaining smaller loadings shown, all of equivalent order of size, express shortness, narrowness, and glabellar projection. (In this analysis—and rotation—then, prominence of glabella is associated equally with general *large* size of frontal bone, and with flatness and small size. This is not easy to comprehend. Sex may be involved: although all these skulls are known or judged to be male, there is always an intrapopulation range of maleness; and small frontals may range between masculine flatness and supraorbital development and feminine angularity and suppressed glabella development; factor 14 is then a simple association of a tendency of glabella development to be associated with absolute size.)

Measurements	Factor 15	Original Correlations				
		FRS	FRC	GLS	STB	XFB
Frontal angle	94	−91	−22	24	−27	−26
Frontal subtense	−90	. .	59	−15	35	35
Frontal chord	−32		. .	11	29	33
Glabella projection	32			. .	−10	−08
Stephanic breadth	−31				. .	81
Maximum frontal breadth	−28					. .

Parietal Profile: Size and Curvature. This common factor seems to say that increased length of the parietal segment is related to increased curvature. In the height measures involved, the radius from the auditory meatus to vertex is, not surprisingly, a better measure of the development than is basion-bregma height. Also not surprisingly, this factor is one of the number contributing to glabello-occipital length.

Measurements	Factor 16	Original Correlations				
		PAC	PAA	VRR	BBH	NBA	GOL
Parietal subtense	91	62	−89	40	20	26	03
Parietal chord	82	. .	−22	50	40	22	50
Parietal angle	−68		. .	−22	−04	−21	24
Vertex radius	53			. .	66	30	46
Basion-bregma height	37				. .	02	34
Nasion angle (ba-br)	34					. .	−24
Glabello-occipital length	27						. .

Occipital Bone Profile: Curvature. A simple factor of occipital angulation. The subtense being almost a direct component of glabello-occipital length, it is not surprising that the latter has its second highest loading here, although of course it is a beneficiary, not a determiner, of the factor. Parietal curvature, with its low loading (as well as its negative original correlation with the occipital angle), reciprocates to a degree that of the occipital.

Measurements	Factor 17	Original Correlations		
		OCS	*GOL*	*PAA*
Occipital angle	—93	87	—35	—28
Occipital subtense	91	..	53	35
Glabello-occipital length	47		..	24
Parietal angle	23			..

Occipital Bone: Size and Height. This is distinct from the preceding factor of curvature—occipital angle has a near-zero loading. Rather, this factor points to the contribution of length of the occipital to general cranial height; basion-bregma (though not vertex radius) has much its highest loading on this factor. General cranial expansion associated with occipital generosity is even expressed by low loadings for total length, length of the frontal bone, and bistephanic breadth. Closer to home, however, this factor expresses growth in the midline only: the occipital, or biasterionic, breadth has a loading of only .16. Apparently this breadth is associated only with the factor (2) of lateral expansion.

An association of cranial height with sagittal occipital growth accords with previous suggestions or findings: Delattre's notion of cranial expansion in the phylogenetic sense by a sort of "rolling down" of the occipital (1958), or the finding of the lambdoid suture as the area of greatest frequency of wormian bone formation, suggesting special growth stress here (Bennett 1965).

Measurements	Factor 18	Original Correlations						
		BBH	*VRR*	*NBA*	*OCS*	*FRC*	*GOL*	*STB*
Occipital chord	82	41	48	24	49	23	46	15
Basion-bregma height	83	..	66	62	04	46	34	27
Vertex radius	53		..	30	17	53	46	37
Nasion angle (ba-br)	50			..	—08	—26	—24	11
Occipital subtense	33				..	16	53	02
Frontal chord	28					..	54	29
Glabello-occipital length	21						..	15
Stephanic breadth	20							..

SUMMARY

To summarize, we find some general factors: of forward facial extension, of lateral vault expansion at and above the ear, of upper facial broadening, of lower facial enlargement in no particular direction, and of vertical facial growth. In addition there are some 12 local factors, which in the vault are restricted to the midplane but in the face relate both to the vertical and the horizontal profile. Whatever their integration and harmonizing in the natural state, as computed here they are all independent: within this generalized population of skulls, individual emphasis on the pattern of facial extension (factor 1) has nothing to do with, for example, the degree of emphasis on lengthening of the frontal bone, the pattern of factor 14.

I have gone on about this at length because it is new information, though I have noted some encouraging correspondence with previous attempts. Let us now get to the main question: are these same patterns of within-group variation the material by which sharp differences between populations may arise?

Discriminant Analysis

In simplest form, a factor analysis will produce, for any one factor, an eigenvector, or set of factor loadings, which may be used as coefficients to multiply an individual's original measurements to give his factor score for that factor. Figure 5–2 shows the meaning of this graphically. Brown scored his Australian skulls on certain of his factors: the figure shows outlines of the skulls which scored highest and lowest on his "infra-temporal fossa depth" factor, and the differences reflect this particular scale of shape.

A discriminant analysis, by resolution of a different matrix, will produce an eigenvector—here a discriminant function—a set of coefficients which is used to multiply an individual's measurements to give his discriminant score for that function. By maximizing the multivariate differences between populations, it furnishes scores with the least among-group overlap for the populations. It thus also produces a discriminant axis (or

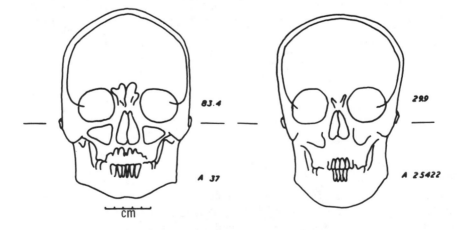

Figure 5–2. Australian aboriginal skulls scored high and low on Brown's "infra-temporal fossa depth." (Figure 19 from T. Brown, 1968; reproduced by permission of the author.)

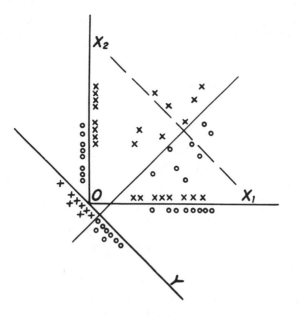

Figure 5–3. Representation of a discriminant function as a reference axis.

Two hypothetical populations, for which the individuals are symbolized by 0 and ×, are shown on a conventional bivariate scattergram for two measurements, X1 and X2. They are poorly discriminated by X1, the means of the groups in this case being obviously close. Segregation on X2 is obviously better. The best axis of discrimination (which is found as the discriminant function) is shown as the dashed line. The line Y is parallel to it and passes through the origin, 0, at an angle with the axes of the two measurements. The positions of the individuals on this axis are shown, giving perfect discrimination. The discriminant function acts as a set of direction cosines to transform the original measurements so as to project the individuals onto this new axis.

canonical variate) which, like a factor axis, passes through the space in which the information is distributed, the difference in the space being that the between-population element is present. Figure 5–3 is a diagrammatic rendering of a discriminant function as an axis of reference. Like factor axes, discriminant axes in the basic method are at right angles—uncorrelated—and serve as reference axes giving coordinates by which individuals, or populations, or measurements, may be placed relative to one another. (I hope, toward the beginning of this chapter, I made it clear that all this is present in the matrix of numbers—the total record of measurements—and simply requires appropriate multivariate treatment to bring out the aspects wanted. I shall not dwell on the methods of analysis, since these are being described in increasingly practical texts, e.g. Cooley and Lohnes 1962, or Jones, in press).

POPULATION COMPARISONS

Here the basic material is the same, but the 834 male skulls are now segregated into their 17 populations, and it is the difference among these we shall examine. (I shall not give results in detail; they are reported in full elsewhere—Howells in press). To resume, the matrix of covariation on which we operate here is specifically that *among* the groups, not that *within*. It is derived, in fact, by subtracting the matrix of within-group covariances, which was used in the factor analysis—call it W—from the matrix of covariances calculated on the whole body of material at once—call it T so that $T = A + W$. These components are separate and independent, and there is no logical or mathematical reason why the information in them should not be independent, as far as I can see, and according to advice from statistically expert friends.[3]

With 17 populations, 16 sources of difference are possible, that is, 16 simultaneous functions. (It is worth noting that the same analysis was carried out on the female skulls of the same populations. The very high degree of correspondence between all results of the two analyses is probably a better indicator of the stability of these results than are statistical tests of the significance of the functions themselves, even though these showed all functions to be "significant.") Since the functions are not rotated, and are less affected by the choice of measurements involved, it is safer to make statements as to their relative importance.

Figure 5–4 is an attempt to show the relative positions of the male populations when plotted simultaneously on the first three functions, which together account for about 54 percent of all the discrimination possible. (The balls are for recognition, and do not in any way suggest the actual variation of the individual scores around these population centroids.) There is seen to be a very satisfactory grouping of the populations by regions: four African groups are in the lower southwest corner, three European groups high in the middle-western region, three Southwest Pacific peoples in the east, and one Central Asiatic people (the Buriats

3. Discrimination among groups is maximized by establishing the ratio of the among to the within matrix, before extracting the functions, but it is not clear why this should have any effect on their basic independence. There is, of course, an analogy here with ordinary analysis of variance, in which the null hypothesis is based on the independence of the within and between components, the ratio between these having a distribution with the expected values for F given in all statistics books. As another possible disturbing factor, it is probable that the several within-group matrices of covariances for the different populations are not identical, and differential weighting might conceivably disturb the independence; however, the numbers in the populations are generally on a par, and I discount all these possibilities, especially in view of the strongly positive results which follow.

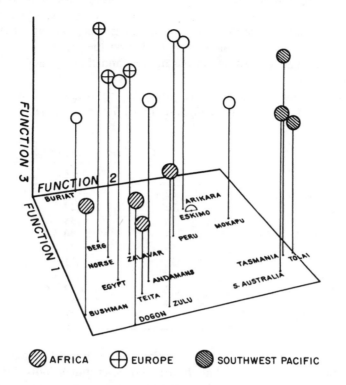

Figure 5–4. Model of distribution of 17 cranial populations on
3 discriminant functions.

of the Lake Baikal region) isolated at the extreme north "wall." It is only
further functions which (outside of this three-dimensional space) begin to
find important distinctions among other peoples.

COMPARISON OF FUNCTIONS AND FACTORS

Interesting though such information is for population relations, the important thing here is any correspondence of functions with factors. The nature
of the functions has been interpreted by correlating their scores with all
measurements, simply to find out which measurements most closely coincide
with the scores of a particular function, and thus point to morphological
distinctions which the function is registering. Now here is a choice: What
matrix of covariation is most appropriate in this case? In the tabulation
below, three sets of correlation coefficients are cited for the first function.
The "total" group correlations are just that: correlations from the total
group of 834 skulls; the "within" have the same basis as the correlations of
the factor analysis—that is the pooled within-group covariances, with the

element of group differences removed. The "among" correlations are computed from the among-group component of covariance, complementary to the within-group covariance. The "within" figures perhaps give the strictest basis for comparison with the factor results; on the other hand, *group* differences are what we look for in the figures, and the "among" may be the more indicative.

The correlations are ranked according to the "total" set, but all three columns include most of those highest for all 70 measurements.

	Correlations with function 1 scores		
Measurement	*total*	*within*	*among*
Biauricular breadth	.82	.53	.93
Cheek height	.72	.39	.93
Nasal height	.68	.38	.85
Maximum cranial breadth	.64	.40	.75
Bizygomatic breadth	.63	.37	.76
Nasion-prosthion height	. 59	.31	.78
Minimum cranial breadth	.58	.28	.83
Basion angle (na-pr)	.57	.29	.79
Nasion angle (ba-pr)	−.54	−.29	−.68
Maximum frontal breadth	.52	.28	.68

The among-group (and total) correlations are higher, which we might expect; but it is important that the function distributes individuals in the same fashion within groups. In any case, all figures say the same thing. This first function, carrying over a quarter of all the discrimination, passing through the multidimensional space set by the population differences, along its longest axis, separates the Buriats of Siberia, a Mongoloid people, at one end from both Africans and Southwest Pacific peoples at the other. Reading the meaning of the function, from the correlations of its scores with the original measurements, shows that this differentiation lies strongly in (a) breadth of the skull, above all at the base; and (b) height of the face. Now it is immediately apparent that these are the distinctions, among individual skulls, described by two important factors, in fact number 2, vault breadth, and number 3, face height. Though "genetically" independent by the factor hypothesis, they appear as important patterns in the greatest distinctions to be found among this set of modern populations. There are other differentiae to be found, of less importance, such as forward prominence of the lateral orbital borders (factor 1) and flatness of midface and subnasal regions (that is, factors 7 and 11), but these are less strongly accented.

The second function distinguishes between Africans and Southwest Pacific peoples—evidently the Australians and Oceanic "Negroids" have little to do with Africans in skull form. The function "pattern" is rather clear morphologically and does correspond to what one can see, in African and Australian skulls, as well as to what one can find in simple comparison of mean measurements of these same series. Note, however, that these earlier methods—use of eyeball and of univariate means—do not pick out the essential pattern as efficiently as the function. In terms of factors, the pattern is prognathism (factor 12) and subnasal horizontal flatness (factor 11)—here *non*flatness in the Australians. Again, although definitely independent in the factor analysis, these are pattern elements which combine to distinguish the Australian-Melanesian profile from that of Africans (perhaps surprisingly but truly). One highly important difference on the function is glabella protrusion, which is of low importance factorially, making itself felt only in the frontal bone factors of length (factor 14) and flatness (factor 15), which appear as secondary components of the function. The among-group correlations in particular point to another difference, reflected by the function: Australian nasal roots are *not* flat, but seem recessed because of the overhanging supraorbital region. This fact has, I am sure, escaped general notice by the eyeball system of analysis.

A final component is factor 4, upper face breadth. Now the function combines this with factor 15 (frontal flatness and narrowness) so that Australians are *narrow* in the frontal and *wide* in the zygomatic arch, and so have a high infratemporal fossa depth (which, compared to Africans, they have indeed). Thus the *function* is doing the work of Brown's "infratemporal fossa depth" factor, which therefore may here be distributed on two different factors. We should remember that factor analysis still has ground to cover.

Function 3 serves to pick out Greenland Eskimos, distinguishing them principally from Europeans and American Indians (while leaving middle ground to Africans and Pacific people). Here the factorial picture is almost a cookbook recipe. Take the general factor 1, forward extension of the facial, especially the orbital, skeleton; the other ingredients are: factor 9, flat nasalia; factor 8, a nonprominent nasal region relative to the orbits; factor 13, large malars[4]; and factor 11, a flat subnasal region. (A minds-eye view of a European skull, or an American one, will confirm this from the opposite side). Factor 7, another horizontal profile character, relating to a sweeping back of the sides of the frontal and the orbits relative to the midface (nasion and dacryon) is *not* in the recipe; such distinctions among the populations involved does not seem of consequence, and instead it is the absolute swinging forward of the eyes (and of the malars) which is crucial; and, relative to the eyes, the retrusion of the nasal space and of the subnasal skeleton along with it.

4. A description of American Indian skulls which implies small malar size may be surprising, and of course only two Indian populations are represented. But in fact, malars of the Peruvian skulls are definitely small, and those of Arikara of only moderate size.

Thus it seems clear that the patterns of difference between populations are compounds of certain specific patterns of variation among individuals—and of a limited number of such patterns—rather than an independent set of differences having nothing to do with such circumscribed entities of covariation. The correspondence might be shown in other ways as well, which I have not had the opportunity to carry out, such as scoring individual skulls, or whole populations, on the factors (see Figure 5–2). Matrix transformations of combinations of factors into function scores might be attempted. It must, of course, be remembered that the factor patterns discerned here are limited by the measurements taken, that is, by the description of shape achieved, and also, as to their true biological or genetic independence, by the mathematical imposition of orthogonality on the factors. It is quite possible that further work would modify the present factors or their interpretation somewhat.[5]

Nevertheless, the specificity of function-factor correspondences bespeaks a reality for the nature of the factors. We have seen this set of relations (based on my reading of figures not reproduced here—see Howells 1970c).

Function 1: Mongoloid Buriats vs.   Factor 2 Vault and base breadth
    Africans and Australo-Melanesians   3 Face height, verticality
                                               1 Facial (orbital) prominence
                                               7 Midface flatness
                                              11 Subnasal flatness

Function 2: Africans vs. Australo-   Factor 12 Prognathism
    Melanesians   11 Subnasal prominence
                                              15 Frontal flatness
                                               4 Upper face breadth

Function 3: Eskimos vs. Europeans,   Factor 1 Facial (orbital) prominence
    American Indians   9 Nasal bone flatness
                                               8 Interorbital flatness
                                              11 Subnasal flatness
                                              13 Malar size

Consideration of these most important functions should suffice to make the point. Subsequent functions, associated with progressively smaller proportions of total possible discrimination, also become less clear as to patterns of differentiation, and thus as to associations with factors. Such associations nevertheless continue; function 4, for example, primarily dis-

5. In another analysis not completed, I have combined an equal number of populations which are all East Asiatic, Oceanic, and American, except for the Norwegians. The pooled within-group correlations, which I have not factor-analyzed, are close throughout to those forming the basis of this analysis, so that resulting factors should be much the same as those herein.

tinguishes the Mokapu Hawaiians, through a high skull and a flattish occiput; the correlations clearly show a correspondence with factors 17, for a flat, open-angled occiput, and 18, for a long occiput and high skull, irrespective of occipital angulation.

There are other possibilities in interpretation. I have scored three Neanderthal skulls (or rather casts and restorations thereof) on the various discriminant functions. On four out of ten functions tried, they give values at the very limit or outside the limits of modern crania. One is function 3, on which excessively low Neanderthal scores imply prominence of the facial midline and a lack of prominence of the lateral orbital margins, as well as prominence of the subnasal region relative to the malars. On function 10, which corresponds well to factor 7, the extremely low Neanderthal scores are highly appropriate to the swept-back character of the sides of the face. Function 5, however, seems equivocal. An important element is factor 8, for a high naso-dacryal angle, that is, a nasal saddle not prominent relative to dacryon, which hardly seems characteristic of Neanderthals, although they seem better described by factor 18, for occipital angling and vault lowness, a less important association of the same function.

The functions, although confirming some of the already recognized distinctions of "classic" Neanderthals, may serve mainly to place them taxonomically outside the range of recent man. The lack of fuller concordance with what the functions express, in spite of extreme discriminant scores, suggests that actually individual factors might be a more informative way of analyzing Neanderthal cranial features than the functions which distinguish among the populations of modern man.

General Considerations

LIMITATIONS OF METHOD

Multivariate transformations have their hazards; we want measurements that will "stay put when our backs are turned," and both mathematically and intuitively it is correct to feel that direct measurements are the solid citizens in this respect. Cohen covers this well in Chapter 19 of this text (which contains the quote). In 1939 Wilson and Worcester wrote about factor analysis: "There is perhaps nothing more likely to convince one that he has something of value than the ability to execute a mechanical arithmetical procedure," and Ehrenberg (1962) said: "Its practitioners seem to be largely unaware of the technical and methodological problems they have let themselves in for," pointing out that the computation only shows that a factor *can* exist in such and such a situation, and that mathematically even the same loadings allow an infinity of such factors.

These strictures are important, even though the climate that gave rise to them was the application of the methods to psychological variables, themselves less determinate than cranial measurements. The transformed variates, as reference axes, may indeed be placed in or rotated to any desired position, and so are essentially indeterminate—this has been pointed out above. Since factors and functions are transformations, there is nothing absolute about them; they depend on the choice of measurements and subjects, and no factor, we noted, will place itself in a region where no measurement vectors lie. In the case of discriminant functions the populations at least set the pattern of axes, so one knows where one stands at first. But the pattern will change whenever the populations change, and the same is true for other kinds of multivariate statistics: generalized distances, and clusterings based on these. This is the reason for Oxnard's advocacy of neighborhood limited classification.

Another liability of discriminants, in their primary form, as generally applied heretofore, is this: their application to the cases on which they were based, and their application to further test cases, are not strictly comparable, since they overdiscriminate in the case of the former. That is to say, a function will discriminate what it is asked to; given groups of Hawaiian and Eskimo skulls, for example, it will discriminate according to the essential morphological differences of these (which is what you want), but it will further discriminate according to the specific and unique aspects of these particular population samples—a component which you certainly do *not* want (Howells 1970c). But this can be taken account of and corrected for in newer discriminant procedures (Dempster 1969).

To take the meanest view, these statistics are most successful when they tell us what we already know. But this is too mean: they tell us much more, and arrange material as we cannot do by intuition. They give convincingly real results at many points, and by successive approximation may be expected to go on giving information, as I have tried to show by comparing the results of factors and functions, and as may be done better in the further comparison of different factor analyses. Also, it is now apparent that arrangements of populations, by discriminants or generalized distances, will give the same arrangement when the material submitted is measurements on the one hand and blood gene frequencies on the other (Friedlander et al. 1971), or, if this is not the case, pose a problem for solution by the absence of concordance (Hiernaux 1956).

ADVANTAGES

Multivariate methods, then, are not automatic friends of the user. A simplistic application of them can be misleading, and the cookbook approach may be dangerous. But then, in univariate statistics, has a worker considered the job done when some hireling has computed means and standard deviations? Or has something important been found out when it is shown, by the tables in the back of the book, that two population samples differ "significantly"?—a popular booby trap in many anthropological situations. The comparable case in multivariate statistics would be something like testing a Zulu skull on a function to discriminate Eskimos and Hawaiians, a situation which might have relevance for such treatment of Neanderthal skulls. (Either Neanderthals *are* no more than a racial variant of modern man, in which case they can legitimately be tested by multiple discriminants developed on modern populations, or else they are not, in which case there are limits on such application.)

Thus, limitations of methods may have a reverse face of advantage, if they are known and understood. Because of their very flexibility (the Good Fairy side of their nature) they afford a richness of possibility by way of re-forming and re-testing material which is infinitely greater than standard anatomical or univariate methods will allow. A suitable case is given by Oxnard: testing the scapula of *Daubentonia* as an "unknown" gave cryptic results, the suggestiveness of which remained fugitive until data on *Daubentonia* itself were introduced into the canonical analysis, as a partaking population. In a more primitive study I (1966) arrived at an equivocal result in assessing late Jomon period skulls of Japan: certainly not ancestral Japanese, they were more like Ainu, by a function to discriminate Japanese and Ainu, but were not acceptable outright as such, and have had to be left in the suspense account, for want of fuller data. (Here again is a comment on Neanderthals.)

In addition, I suggest that typical methods now in use still rest on the relatively simple algebra of their original formulations—certainly this is true of discriminants. As biologists use them more intensively for real problems, and find theoretical and actual hitches in such application, as well as specific needs for new and refined forms, feedback to the mathematical statisticians will result in such refinements, a highly important matter. A reading of Cohen (Chapter 19) should put this in perspective. I have a feeling that the rather squishy nature of the variables and subject matter (psychological data) on which factor analysis and discriminant analysis have been developed in the last generation has had something to do with slowing progress, although certainly the recency of really capacious and fast computer hardware also is responsible.

FUTURE APPLICATIONS

It is evident that the derivation of suggestions, for hypotheses and work, from multivariate analyses, and the further use of transformed data, have only begun to be exploited. Future possibilities include the following.

1. Comparison of factors of the skull in other studies, including some specifically designed for testing (for example, specific sites of growth) should lead to more stability of results. This study has shown some correspondence of factors and discriminant functions. This needs checking, including fuller development of among-group correlations than I have been able heretofore to get through. It could include direct testing of

factors and functions, by arithmetically attempting to transform one into the other for goodness of fit, etc., a package I leave to mathematicians.

2. Retesting of population differences in many ways, such as regrouping, and closer study of anatomical aspects suggested by preliminary work such as that herein. (An interflow between this and anatomical and experimental work seems to be called for.) Even in the present material, the discriminant analysis gave immediate and useful guidelines for reinspection and interpretation of the simple univariate mean differences among the populations involved.

3. The use of discriminants which are not, as in the standard form, generalized and orthogonal, but specific for pairs of populations within a space of more than two populations, and thus not uncorrelated, might be useful in many studies. These would be analagous to rotated, especially oblique, factors in a factor analysis.

4. More use of different methods of grouping and clustering populations beyond the simple stages of most current usage. Some algorithms may be mathematically appealing, but fail to give biologically helpful results. Here is where experience with a variety of techniques would be useful.

The above suggestions relate strictly to expansion of what I have covered in this paper. They disregard the kind of mathematical refinement I mentioned, and do not touch on the kind of model-testing and model-building described by Oxnard and Cohen. Another entire field is interpretation of isolated fossil parts by any combination of weapons. Examples are Day's work, especially the spectacular example of a distal phalanx of the hallux from Olduvai (1966, 1967), Knussmann on the ulnae of *Oreopithecus* (1967), and Patterson and Howells on a humeral fragment from Kanapoi (1967). This brings us, among other questions, to that of more light on hominid fossils. How, for example, may we better specify and quantify the differences among Neanderthal specimens, Broken Hill, and the Ngandong series?

Such questions will doubtless be better attacked after further work on control data (that is, on populations large enough for proper analysis) in order to define the significance of measurable anatomical differences as well as possible. The work would converge on that described by Oxnard. I mentioned the usefulness of building up a body of comparative data, which would allow attempts to assess fragmentary remains, even from cremations. I have done some highly preliminary work on the probably Pleistocene frontal bone from the Tabon Cave, Palawan, Philippines, through the kindness of Dr. Robert Fox. Actual assignment is not possible just now, but the individual appears to be excluded from affiliation with any Far Eastern "Mongoloid" population, as represented by Japanese, Chinese, Filipinos, or Atayals of Formosa. This is largely on the basis of nonflatness of the interorbital region, the sort of feature specified by factor 8 above.

In this paper I have reported specific results from applications of specific techniques. The real message, like that from many aspects of the symposium, is the use of broadened analytical approaches, as ways of looking and problems and materials. As a non-mathematician, it seems to me that anthropology and primatology stand to gain much from appropriate mathematical procedures. We need only to remember how their introduction by Fisher and Wright into agricultural genetics and general evolutionary theory actually revolutionized those fields.

Summary

1. Application of two forms of multivariate statistics (factor analysis and discriminant analysis) to the same body of cranial measurements indicates that interpopulation differences in recent man involve the same morphological patterns as does individual variation within populations.

2. These patterns do not follow such classic distinctions as cranial or nasal indices. They consist rather of general differences in: facial height, vault and base breadth, forward prominence of lateral borders of the face, upper facial broadening, etc.; as well as more anatomically local differences in such features as: horizontal profile of the orbital margin of the frontal bone, midfacial flatness, occipital sagittal contour, etc.; to a total of not less than 18 factors from the measurements used. Variations in total cranial length and breadth of nasal aperture are notable for their lack of importance in these cranial patterns.

3. Limitations in this type of method lie in the essential indeterminacy of the transformed variates, which must be allowed for in interpretation.

4. Taking account of these limitations, the great flexibility and breadth of possible uses of transformed variates should make them an increasingly useful set of tools in examining data and testing hypotheses of variation, form, and function.

5. The methods have already been used in judging the affiliation of primate and hominid fossils, and analyzing the relation of form and function of the scapula in different primate species. Further accumulation of data and refinements of method should widen understanding of cranial and other features of shape in modern and fossil man.

III

Comparative Neurobiology and Endocasts

6

HEINZ STEPHAN
MAX-PLANCK-INSTITUT FÜR HIRNFORSCHUNG
NEUROBIOLOGISCHE ABTEILUNG, NEUROPRIMATOLOGIE

Evolution of Primate Brains: A Comparative Anatomical Investigation

Paleontological studies have shown that the Primates evolved from Insectivores. Thenius (1969) pointed out that the initial attribute of the Primates may have been the development of prehensile hands and feet in relation to arboreality. Further, the beginnings of arboreality were accompanied by enlargement of the eyes, a rostrally directed shift of the optical axis, shortening of the facial skull, and the development of a more voluminous brain case. The enlargement of the brain is thought to have been an important and, to a certain degree, specific characteristic of primate evolution.

The present paper will discuss the extent to which this enlargement is more pronounced in Primates than in other mammalian orders and how far various parts of the brain behave differently during this increase.

Direct information concerning the size and certain characteristics of primate brains from the past may be obtained from natural or artificial endocranial casts of fossil skulls. The papers in this volume by Radinsky and Holloway present informative results obtained from such "fossil brains." It is impossible, however, to acquire any knowledge about the size of brain structures that are not represented near the surface and about neuronal differentiation without additionally employing the indirect method of comparing the brains of recent species. How far this latter scientific approach, known as *"Comparative Neuroanatomy,"* can contribute to the knowledge of phylogenetic processes shall be discussed. Comparisons are

I would like to thank the following staff and colleagues: Misses C. Roberg, H. Keiner, A. Rehbein, and M. Stichling, who skillfully prepared long series of microscopic sections; Miss H. Grobecker, who took care of the hundreds of photographs necessary for the volumetric estimations and for publication; Miss H. Lehmann and Mrs. I. Stephan, who patiently typed the text; Mr. J. Kampe, who executed the figures with great care; and finally to Mrs. B. Harrisson, who checked the English text.

thought to be especially effective in this respect when based on the brains of primitive Insectivores.

The recent representatives of the Insectivores are not uniform with regard to brain development. With the aid of quantitative methods, we have tried (since 1956 together with Andy, Bauchot, and Spatz) to identify the species of Insectivora with the most primitive cerebral pattern.

We have grouped together these primitive forms as *"Basal Insectivores."*[1] To this group belong representatives of the tenrecs (Tenrecidae), hedgehogs (Erinaceidae), and shrews (Soricidae). The primitiveness of the brains in the basal Insectivores is reflected in the fact that not only the total brain but also all progressive structures,[2] and complexes of cerebral structures respectively, are quantitatively the least developed. With regard to the quantitative composition of the brain, the basal Insectivores represent a fairly uniform type. In contrast, the remainder of the Insectivores reveal distinct marks of higher development, for which reason we call them *"Progressive Insectivores."* This group contains predominantly specialized types (e.g. those adapted to burrowing or digging, or to the search for food in water). In comparison, the basal Insectivores must be designated as being unspecialized or less specialized (Stephan 1967).

Of all placental mammals, the basal Insectivores may soonest be expected to show the least degree of change since their first appearance (permanent types). Hence, they can be expected to be still comparatively similar to the early forerunners of the placental mammals, and therefore to represent a good base of reference for evaluating evolutionary progress.

Materials and Methods

BRAIN WEIGHT

Comparisons of brain weights considered in conjunction with the differences in body weight by aid of the allometry method give information about "encephalization." Investigations of encephalization in Insectivores and Primates based on extensive material from our own collection and from literature were published mostly in collaboration with Bauchot (Stephan 1959, Bauchot and Stephan 1964, 1966, 1968, 1969, Stephan and Bauchot 1965). Data from Chiroptera also were collected by Pirlot and Stephan (1970).

The allometry formula[3] $h = b \cdot k^\alpha$ was found empirically by Snell

---

1. Basal, basic form, and basic group connote taxonomical units "occupying the lowest place among the *recent* placental mammals according to characteristics of the brain." In the course of phylogenetic development the fossil relatives of the recent basal Insectivores, which may have played a role in the evolution of the placental mammals, represented transitional stages. For these fossils, the terms *layer* or *level* would be more appropriate.

2. Progressive structures are those which become enlarged during a progressive brain development (e.g. in an ascending primate scale, see further).

3. $h =$ brain weight. $k =$ body weight.

(1892) and Dubois (1897). The relation of brain weight to body weight is found to be linear on a double logarithmic scale (Figure 6–1), if closely related species are compared (log h = log b + α · log k). The slope of the straight line (α = angle with the abscissa) indicates the rate at which a given structure enlarges with increasing body weight. The relationship between the size of the body and the size of the brain is remarkably uniform in interspecific comparisons of related groups, even if these groups belong to very different orders of mammals. It is found to come close to 0.63. This value is obtained by comparing as many related groups as possible in all available species of Insectivora, Primates, and Chiroptera. It is the expression of the most typical size relationship between brain weight and body weight.

A line with the gradient thus derived is drawn through the point of the mean values of the basal Insectivores (Figure 6–1). If the typical relationship between brain and body weight is known, it is also possible to detect those differences in brain size which depend on factors other than

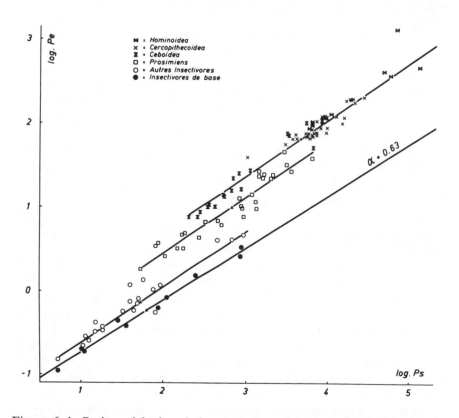

Figure 6–1. Brain weight in relation to body weight in a double logarithmic scale. (From Bauchot and Stephan 1969.) Reference base is the "Basal Insectivore Line." (See text.) α = slope of the line, Pe = brain weight, Ps = body weight.

body size. Such differences in size are expressed on the double logarithmic scale by the different levels at which the straight lines are situated or by the distances of different points from a base line. These differences in position are usually thought to be due to differences in evolutionary level. Using the basal Insectivore line (log $h = 1.632 + 0.63 \cdot \log k$), one can calculate the probable brain weight in basal Insectivores with different body weights. We designate these values as "basal sizes" (BG). The actual brain weights of the different species are designated as "progression sizes" (PrG) because they are in general higher than those of the basal forms, that is, progressive. Comparisons of the progression sizes with the basal sizes gives a value of the respective degree of progression. The "Index of Progression" ($IP = PrG/BG$) is a numerical value expressing the degree of enlargement of the brain in one species by comparison with that of a typical basal Insectivore of equal body weight. Thus this index is an expression of the encephalization as understood by us. The progression indices are indicative values which, although not precise, are certainly adequate for our purposes.

VOLUMES OF VARIOUS BRAIN COMPONENTS

Investigations on the volumes of various brain parts are very time consuming, since they are performed on serial sections of whole brains. Therefore comparisons cannot be based on extensive materials as like the studies on brain weight have. We have so far measured a total of 84 brains from 63 species in our collection, including 22 Insectivores, 20 Prosimians, and 21 Simians. The volumes were estimated from enlarged photographs of serial sections at equal intervals (60–80 per brain). The volume of the sum of all measured structures (= total brain) obtained in this manner is markedly smaller than the volume of the fresh brain. This is due to the shrinkage resulting from the fixation and embedding procedures. The degree of shrinkage is different for each brain, even if the technique of preparation is the same. In order to obtain comparable values, we have corrected all figures to the volume of the fresh brain. The method is described in more detail in Stephan, Bauchot, and Andy (1970).

Subsequent comparisons are based on the same principle described above for total brain weight. The volume of each component is related to body size and compared to the size of the corresponding structures in the basal Insectivores. The volume comparisons have been performed for a variety of structures, quite different in size and complexity: medulla oblongata, mesencephalon, cerebellum, diencephalon, corpus geniculatum laterale, various nuclei of the diencephalon (by Bauchot), subcommissural organ and subfornical body, epiphysis and hypophysis (the latter by Bauchot), olfactory bulb, accessory olfactory bulb, paleocortex + amygdaloid complex, Ncl. tractus olfactorii lateralis, schizocortex (regio entorhinalis + praesubicularis), hippocampus, septum, various nuclei of the septum, striatum, neocortex and neocortical area striata. I will discuss here only some of these structures.

A further subdivision of the more complex cerebral regions and the measurement of smaller and more circumscribed structural units are now in progress.

Results

TOTAL BRAIN WEIGHT (= ENCEPHALIZATION)

In Figure 6–1, data on all available Insectivores and Primates are presented. It is evident that (1) typical relationships between brain weight and body weight exist, which are similar in the various groups (parallel lines in Figure 6–1); and (2) a progression in brain size is present ascending from basal Insectivores to progressive Insectivores, Prosimians, and Simians. The distances of the various species from the basal Insectivore line, expressed by progression indices, are represented in figures 6–2 and 6–3. The scales provide information about the degree of encephalization of each species. Only some general results will be discussed here.

The total range of encephalization seems to be larger in Primates than in any other order (Figure 6–4). The bottom of the scale is occupied by *Lepilemur* which has an encephalization index of 2.4. The top of the scale is occupied by man with an index of 28.8. Thus encephalization is 12 times larger in *Homo* than in *Lepilemur*. However, the distance between man and

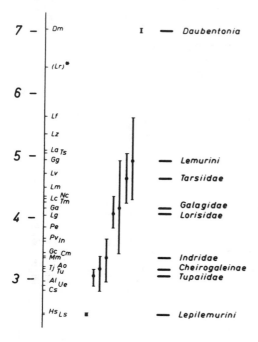

Figure 6–2. Encephalization indices of the Prosimians. Reference base (=1; not indicated) is the average encephalization in basal Insectivores. Vertical bars represent the range of variation within each systematic group with the mean values of the indices shown by a dot. The mean points form an ascending pattern from left to right. The horizontal bars to the right correspond with these means and the name of the individual systematic group is given.

*Value Lr (*Lemur rufiventer*), taken from literature, deviates so much from all other Lemur-species, that it is given only with reservation.

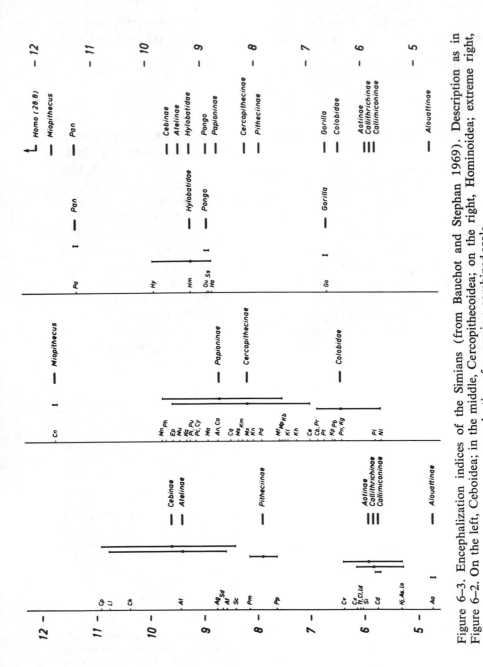

Figure 6–3. Encephalization indices of the Simians (from Bauchot and Stephan 1969). Description as in Figure 6–2. On the left, Ceboidea; in the middle, Cercopithecoidea; on the right, Hominoidea; extreme right, reproduction of means in a combined scale.

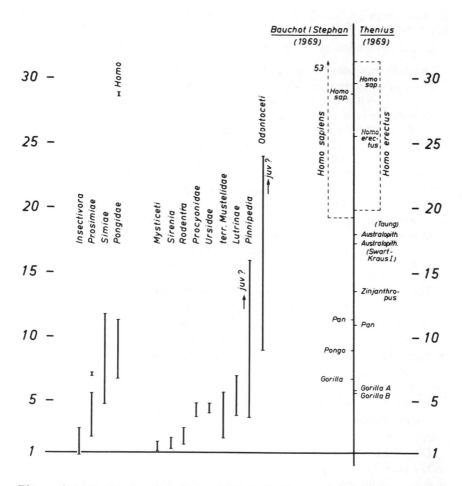

Figure 6–4. Encephalization indices. On the left, values of various systematic groups taken from Bauchot and Stephan (1966, 1968, 1969). Possibly in Pinnipedia and Odontoceti (data taken from the literature), part of the upper range represents indices from juveniles. On the right, brain weights of some apes and fossil forerunners of man (Thenius 1969) compared with data given by Bauchot and Stephan (1968). The range of variation of the indices in *Homo sapiens* is calculated from data given by Gjukic (1955). Broken lines: range of variation in *Homo sapiens* and *H. erectus*. Reference base as in Figure 6–2.

the highest encephalized nonhuman Primate is larger than the distance between the latter and the basal Insectivores. Thus the encephalization of Primates gets its wide range mainly from the very high position of man.

Other surprising results that are evident in figures 6–2 and 6–3 include (1) an overlap between Prosimians and Simians; and (2) a closely similar degree and range of encephalization in all of the three superfamilies of

Simians (man excluded). In particular, no clear difference exists between the Ceboidea and the Cercopithecoidea and the nonhuman Hominoidea. The encephalization of the gorilla is even lower than that of *Daubentonia*.

The encephalization of *Daubentonia* with an index of 7.0 exceeds by far that of the other Prosimians. The next highest index is 5.6 in *Lemur fulvus*. The very high position of *Daubentonia* within the Prosimians is thought to be the consequence of a secondary reduction in body size. Such *dwarfism* sometimes occurs and the diminution in brain size in relation to the reduction in body size appears to follow an *intra*specific exponent (generally close to 0.25) which is clearly lower than the *inter*specific one (see Bauchot and Stephan 1969).

The possibility of a secondary body size reduction in *Daubentonia madagascariensis* is supported by the existence of a large form (*Daubentonia robusta*) in the Quaternary of Madagascar (Thenius 1969, p. 176).

Another sample of dwarfism seems to be the talapoin monkey, *Cercopithecus* (*Miopithecus*) *talapoin*. This species again shows an unexpectedly high encephalization, higher than that of the chimpanzee.

The *encephalization in Primates compared with that in other orders* is shown in Figure 6–4. A Prosimian stage is reached by many nonprimate mammals. A Simian stage is evidenced by the seals, whereas the Odontoceti exceeds by far the Simians(inclusive the Pongidae) but without reaching human values. All data for Odontoceti are taken from literature. The upper part of the scale for Odontoceti may be occupied by juvenile specimens, the encephalization of which is generally higher than that of adults.

It must be concluded from these results that a high encephalization is *not* a specific characteristic of Primates and that even the highest nonhuman Primates are not characterized by unique outstanding brain size.

Only the encephalization of man is higher than that of all other mammals so far investigated. The large gap between recent nonhuman Primates and man can be filled by *fossil forerunners of man*. Data on brain weights and body weights given by Thenius (1969, taken from Jerison 1963, and Tobias 1967) for *Zinjanthropus, Australopithecus,* and *Homo erectus* suggest that these forms clearly comprise intermediate stages between *Homo sapiens* and other Simians in their encephalization (Figure 6–4). We may infer that the high encephalization of recent man has developed gradually during human phylogeny.

The encephalization of *Homo erectus* varies very widely and in the upper part of its range it exceeds the mean of *Homo sapiens*. We infer that this wide range is only partly due to the effect of progressive evolution in successive phases, but that a high variability already existed in *H. erectus* populations. It cannot be expected, however, that this variation was as extreme as that in recent man, in whom we can observe encephalization indices from 19.2 up to 53.0. The high variability in modern man is probably in part due to the suppression of natural selection.

The scales given in figures 6–2 and 6–3 may be considered to be *"Ascend-*

ing Primate Scales" and may be employed to evaluate the evolutionary level reached by the various species. The theoretical background to this consideration is the following: A fairly reliable and characteristic feature of directed progressive evolution in the whole animal kingdom is the concentration, enlargement, and differentiation of the nervous system. Originating from diffuse neural nets, the nerve cells become more and more concentrated in ganglia and finally in central nervous systems characterized by increasingly complex brains. This development culminated in mammals. The brain is generally believed to represent the most important indicator of evolutionary progress in mammals.

But I believe that the brain as a whole does not represent the best possible criterion for evolutionary progress, since its enlargement is expected to be nonproportional, that is, not uniform with respect to its individual parts. The evolutionary level is expected to be best indicated by those structures, which are most notably enlarged by comparison with similar parts in basal forms.

Therefore we will postpone the determination of evolutionary level and the discussion of its phylogenetic interpretation, until the progression scale of the most progressive structure has been represented.

VOLUMES OF VARIOUS BRAIN COMPONENTS[4]

In order to obtain a general indication of the trends (whether progressive or regressive) and intensities of the size modifications from basal Insectivores to Primates, we averaged the progression indices of all available Primates (Figure 6–5). The neocortex, which contains the most highly developed structures in the central nervous system, shows the strongest progression. Its extraordinary supremacy in comparison to all other structures is remarkable. In Prosimians the neocortex is on the average 14.5 times and in Simians it is 45.5 times larger than those in basal Insectivores. Therefore its size represents the best cerebral criterion (at present available) for the classification of a given species in a scale of increasing evolutionary level (= "Ascending Primate Scale").

The Ascending Primate Scale. The graph based on the indices of neocortical progression (Figure 6–6) shows that within the Primates many different evolutionary stages are preserved in extant species. The lowest position is occupied by the tree-shrews (Tupaiidae, the systematic classification of which is still debated[5]) and the Lepilemurini (including the

4. Parts of the given results have been published previously in the Annals of the New York Academy of Sciences (Stephan and Andy 1969).

5. According to some investigators, they are classified as Insectivores, whereas others consider them to be Primates. It is, however, commonly admitted that they have to be placed in a transitional position between Insectivores and Primates. Regarding the indices of neocorticalization, their position is closer to the Prosimians. This fact supports the opinion, which the majority of the investigators are inclined to accept, that they should be classified as Subprimates together with the Prosimians. Recent suggestions to separate the tree-shrews entirely from both Insectivores and Primates are premature.

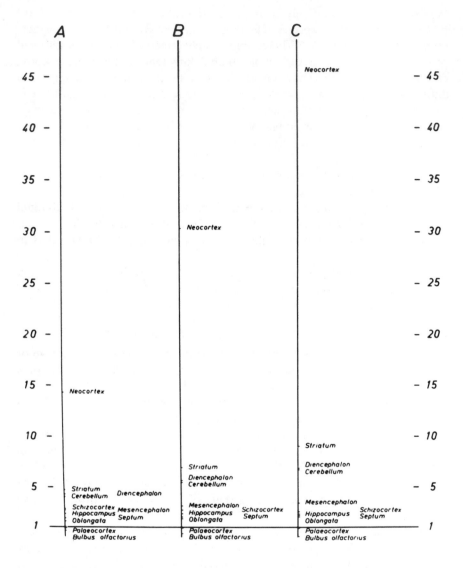

Figure 6–5. Average progression indices of measured brain structures in 20 Prosimians (A), 21 Simians (C), and the total of these 41 Primates (B) so far investigated. Reference base (=1) is the average progression index of these structures in basal Insectivores.

genera *Lepilemur* and *Hapalemur* from Madagascar). The latter stand only slightly higher than the Tupaiidae (Index of the Neocortex = IPN = 8.3 and 7.7 respectively). The mouse and dwarf lemurs (Cheirogaleinae; IPN = 9.0 − 12.0) are also comparatively low in neocorticalization. The highest indices of neocorticalization within the Prosimians are found in the true lemurs (Lemurini; IPN = 17.5 − 23.3), tarsier (*Tarsius;* IPN = 21.5), and the aye-aye (*Daubentonia;* IPN = 26.5). The position of *Daubentonia* is especially uncommonly high. It exceeds considerably that of all other Prosimians and advances into the province of the Simians. Within the Simians, the howler (*Alouatta;* IPN = 20.8) and the gorilla (*Gorilla;* IPN = 32.1) occupy uncommonly low positions. Both genera are obviously more primitive in their brain development than hitherto assumed. The very high position of the talapoin (*Miopithecus;* IPN = 60.1) is probably related to secondary "dwarfing," as already mentioned in connection with the encephalization. The remarkable distance between man and all other Primates, mentioned in connection with encephalization, is even greater in neocorticalization. The neocortex of *Homo* (index = 156) is approximately 19 times larger than that of *Lepilemur* (index = 8.3), whereas in encephalization *Homo* is only 12 times greater than *Lepilemur*. As in encephalization also man's neocorticalization is farther from the highest nonhuman Primate than the latter is from the basal Insectivores (Figure 6–6). Man's exceptionally large neocortex undoubtedly represents the morphological substrate for the very high and complex functional capacities of his central nervous system.

Phylogenetic Interpretations of Neocorticalization. It must be stressed that the degree of neocorticalization is not a measure of *systematic relationship* between species. Related forms may be at different levels, whereas very distinct species may be at similar levels of neocorticalization. This distribution is not unique to Primates since it occurs also in other highly diverse groups of mammals, such as bats. The same is true of encephalization, as illustrated in figures 6–2 and 6–4.

Furthermore, it must be emphasized that the scale of increasing neocorticalization, which we have interpreted as an Ascending Primate Scale, represents *no direct sequence of evolutionary stages* in the sense that the more progressive forms have passed through all the individual lower stages. How, then, can our ascending scale contribute to the problem of phylogenesis? We think it can, under the following assumptions: (1) It is likely that the lower extant forms under consideration correspond well in their brain patterns with phylogenetic stages, through which higher forms have passed during phylogenesis. Therefore, we may expect to get some information about the brains of the extinct forerunners of living species. (2) It is improbable that the development of the highest centers of integration in the neocortex revert to lower levels under normal wildlife conditions (in contrast to domestication and capitivity). The size of the neocortex is therefore considered to be an effective guide in the discussion of phylogenetic sequences.

According to these assumptions, we do not believe that forms with lower levels of neocorticalization were derived from forms with higher levels. For example, the very low position of the gorilla may be interpreted as

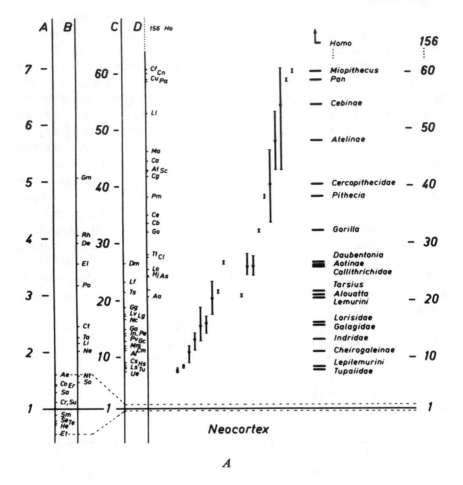

Figure 6–6. Progression indices of the neocortex.

Indices express how many times larger the neocortex is than that of a typical basal Insectivore of equal body weight.

A. On the left (vertical columns) the different species: (A) Basal Insectivores, (B) Progressive Insectivores, (C) Prosimians, (D) Simians; in the center (vertical bars) systematic groups of Primates according to their respective variabilities; on the right (horizontal bars) averages of the indices within these systematic groups. Order of vertical bars (middle) from left to right, according to increasing neocortical progression (Ascending Primate Scale). Broken line = range of the Basal Insectivores.

B. A reduced scale to demonstrate the actual position of man. P = Prosimians, Dm = Daubentonia.

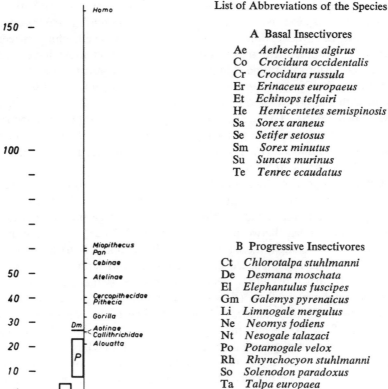

List of Abbreviations of the Species

A Basal Insectivores

Ae   *Aethechinus algirus*
Co   *Crocidura occidentalis*
Cr   *Crocidura russula*
Er   *Erinaceus europaeus*
Et   *Echinops telfairi*
He   *Hemicentetes semispinosis*
Sa   *Sorex araneus*
Se   *Setifer setosus*
Sm   *Sorex minutus*
Su   *Suncus murinus*
Te   *Tenrec ecaudatus*

B Progressive Insectivores

Ct   *Chlorotalpa stuhlmanni*
De   *Desmana moschata*
El   *Elephantulus fuscipes*
Gm   *Galemys pyrenaicus*
Li   *Limnogale mergulus*
Ne   *Neomys fodiens*
Nt   *Nesogale talazaci*
Po   *Potamogale velox*
Rh   *Rhynchocyon stuhlmanni*
So   *Solenodon paradoxus*
Ta   *Talpa europaea*

B

C Prosimians

Al   *Avahi laniger*
Cm   *Cheirogaleus major*
Cs   *Cheirogaleus medius*
Dm   *Daubentonia madagascariensis*
Ga   *Galago senegalensis*
Gc   *Galago crassicaudatus*
Gg   *Galago demidovii*
Hs   *Hapalemur simus*
In   *Indri indri*
Lf   *Lemur fulvus*
Lg   *Loris gracilis*
Ls   *Lepilemur ruficaudatus*
Lv   *Lemur variegatus*
Mm   *Microcebus murinus*
Nc   *Nycticebus cougang*
Pe   *Perodicticus potto*
Pv   *Propithecus verreauxi*
Ts   *Tarsius syrichta*
Tu   *Tupaia glis*
Ue   *Urogale everetti*

D Simians

Aa   *Alouatta seniculus*
As   *Aotus trivirgatus*
At   *Ateles paniscus*
Ca   *Cercopithecus ascanius*
Cb   *Colobus badius*
Ce   *Cercopithecus mitis*
Cf   *Cebus albifrons*
Cg   *Cercocebus albigena*
Cl   *Callicebus moloch*
Cn   *Ceropithecus (Miopithecus) talapoin*
Cu   *Cebus sp.*
Go   *Gorilla gorilla*
Hj   *Callithrix jacchus*
Ho   *Homo sapiens*
Ll   *Lagothrix lagotricha*
Lo   *Leontocebus oedipus*
Ma   *Macaca mulatta*
Pa   *Pan troglodytes*
Pm   *Pithecia monacha*
Sc   *Saimiri sciureus*
Tt   *Saguinus tamarin*

follows: The Hominoidea probably radiated from forerunners with a level of neocorticalization that was less than that of the gorilla. After separation of the two lineages leading to recent gorillas and recent chimpanzees, the further development of higher centers of integration must have occurred at a much faster rate in the chimpanzee than in the gorilla.

Volumetric Comparisons of Non-neocortical Brain Structures. Compared with the progression of the neocortex, that of all other structures is low (Figure 6–5). The second highest progression after the neocortex is found in the striatum with an average of approximately 5 in the Prosimians and 9 in the Simians. This progression is somewhat larger than that of the diencephalon and cerebellum.

Lesser progressions are found in a group of structures and structural complexes, respectively, consisting of mesencephalon, schizocortex, hippocampus, septum, and medulla oblongata. The average progression index of these structures lies between 2 and 3. The mesencephalon and medulla oblongata show a slight increase from the Prosimians to the Simians, which is not noticeable in schizocortex, hippocampus, and septum.

The only structure with a clear-cut regression is the olfactory bulb, the average indices being 0.66 (thus corresponding to an average reduction by ⅓) in the Prosimians and 0.09 (corresponding to an average reduction of more than $\frac{9}{10}$) in the Simians. The paleocortex, including the amygdaloid complex, shows no quantitative deviation from the basal Insectivores to the Simians. The reduction of the olfactory structures of the paleocortical complex seems to be balanced by an enlargement of the higher centers of the amygdaloid complex. A further subdivision of this complex is in preparation (Stephan and Andy).

The progression indices of the non-neocortical structures have been arranged in scales (see for example Figure 6–7). Such scales provide information about the comparative size of these structures in individual species and in other systematic groups. In order to determine whether there are definite tendencies in size alterations of various structures in the Ascending Primate Scale and in order to discover which species show an especially strong deviation from a given trend, we have compared the progression indices of the individual structures directly with those of the neocortex (Figure 6–8).

Striatum (figures 6–7 and 6–8): The scale of the progression indices of the striatum (Figure 6–7) shows strong similarities to that of the neocortex (Figure 6–6). However, the degree of progression is much less and the position of man comes closer to the other Primates. The progression in the striatum is almost constantly in an equal proportion to that of the neocortex. The enlargement of the neocortex is generally five times that of the striatum. These close parallels in size seem to indicate a functional relationship between the two structures.

Diencephalon and Cerebellum: Trends in the diencephalon and cerebel-

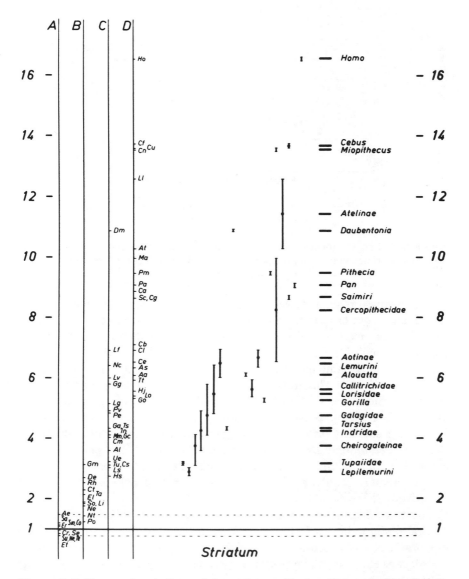

Figure 6–7. Progression indices of the striatum. Explanation and abbreviations as in Figure 6–6.

lum in the Ascending Primate Scale are similar to that of the striatum with regard to evidence for steady progressions. However, in contrast to the rather uniform and linear ascent in the striatum, the increase in the size of the cerebellum and still more so that of the diencephalon are most accentuated in the earlier phases (from the basal Insectivores up to the Prosimians) and then become increasingly less pronounced. During the later stages, the

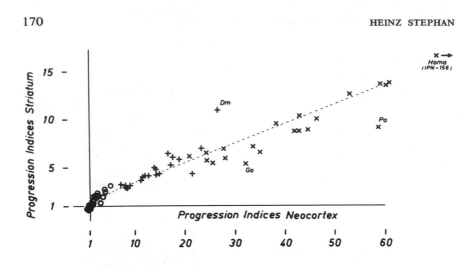

Figure 6–8. Progression indices of striatum in relation to those of neocortex. Broken line represents general course of size development of striatum in the Ascending Primate Scale.

general increase remains larger in the cerebellum than in the diencephalon. In the late stages leading from higher Primates to man, the cerebellum still shows a large increase, whereas the enlargement of diencephalon is comparatively small.

Hippocampus: Among the Prosimians a clear trend in hippocampal enlargement is found. This trend does not continue in the Simians in which group the mean values remain unchanged with an average progression index of 2.3. Comparatively low values are found in *Loris* among the Prosimians and in *Gorilla* among the Simians. Comparatively high values occur in *Lemur* and *Daubentonia* among the Prosimians and in *Homo* among the Simians. Man has a progression index of 4.2. Thus man's hippocampus is more than four times larger than that of a hypothetical basal Insectivore of the same body-weight.

The wide distribution of the hippocampal indices at similar levels of corticalization indicates further that there is no close size relation between hippocampus and neocortex. Very high indices, up to 3.8, are reached by the Macroscelididae, which are progressive Insectivores with poorly developed neocortices.

Size alterations very similar to those of the hippocampus are found in the septum and schizocortex. These three structures do not obviously vary independently of one another. The interdependence of these structures becomes very clear, when their volumes are correlated directly with one another. The exceptionally close correlations among them are entirely independent of the very distinct evolutionary levels of the various species from the primitive Insectivores up to man.

The well-known and very close functional relationships between these three structures are thus clearly reflected in their size interdependence. One would expect also to find similar size correlations with the olfactory centers, if the limbic and olfactory systems are in close functional interdependence.

Bulbus olfactorius: In most *Prosimians* the olfactory bulb is more or less clearly reduced, although there exist some forms in which the size of the olfactory bulb is still insectivore-like. This is the case in the tree-shrews, in the dwarf galago, and in the aye-aye. All other Prosimians have smaller olfactory bulbs than the average basal Insectivore. The strongest reduction is found in the *Indri*. In this genus the size of the olfactory bulb is only one-eighth that in a basal Insectivore of equal body weight.

In the *Simians* the olfactory bulb is always markedly reduced. It is largest in the night monkey (*Aotus*), yet only one-fifth the size of that in an average basal Insectivore. Man shows the most marked reduction of the olfactory bulb among available Primates (index = 0.023). The size of the human olfactory bulb is less than one-fortieth of that of basal Insectivores.

The trend for olfactory bulb diminution in Primates is most notable during the Prosimian phase. In the beginning of the Simian phase (average index of neocorticalization = 25) the olfactory bulb is in general reduced by four-fifths. In the higher Simians the values approach the zero-line but without reaching it. No primate species lacks olfactory bulbs. There is no obvious dependence between the development of the neocortex and simultaneous reduction of centers in the olfactory system. The reduction of the olfactory centers appears to be a functional adaptation, which is prominently paralleled in the Primates by a progressive neocorticalization. From surface measurements it was established that the secondary olfactory centers are reduced in exactly the same proportion as the primary olfactory bulb (Stephan 1961). Here again, very clear size relationships can be observed between structures which are functionally interdependent.

If the trend of the olfactory bulb in the Ascending Primate Scale is compared with that of the hippocampus, very distinct tendencies emerge. The reduction of the olfactory structures stands in clear contrast to the progression of the limbic centers. There exist highly diversified developmental trends, and we believe that convincing reasons for a clear separation of the two systems (olfactory and limbic) are indicated by our comparative anatomical results.

Bulbus olfactorius accessorius: The relatively small accessory bulb shows no clear trend in the Ascending Primate Scale. In most Prosimians it is better developed than in basal Insectivores, but less developed in New World monkeys and absent in Old World monkeys. These results suggest that a progressive phase was followed by a regressive one, which ultimately lead to its total loss in major groups of Primates.

Medulla oblongata and *mesencephalon:* Neither of these structures show a clear trend towards gradual enlargement in the Ascending Primate Scale.

Nevertheless, these structures show a distinct progressive development, because (apart from a few exceptions) in the Primates and also in the progressive Insectivores they are larger than in basal Insectivores. Obviously, the weak tendencies of an increasing enlargement are obscured by stronger, presumably functional influences.

Area striata and corpus geniculatum laterale: The delineation of visual structures in Insectivores is very complex and investigations are still incomplete. Therefore, comparisons between visual structures of Insectivores and those of Primates are not yet possible. Comparisons within Primates show a clear progression from Prosimians to Simians. In relation to the neocortical progression, the increase of the area striata is most accentuated in the earlier phases (up to the lower Simians) and less pronounced in the higher ones. Especially well developed optic centers are found in species of *Cebus,* whereas in *Homo* they do not exceed the average for Simians. In *Alouatta* and *Gorilla* the progression indices are especially low and do not exceed a level characteristic for Prosimians (Stephan 1969a).

Periventricular organs (Stephan 1969b): (1) *Subfornical body:* A clear trend cannot be observed in the development of the subfornical body. Generally the trend is progressive in the Prosimian phase, but regressive in the Simian phase, especially in the Hominoidea. In man the index is less than 0.5.

(2) *Subcommissural body:* This structure is clearly progressive in the Prosimians compared with basal Insectivores, and especially large in *Daubentonia* and *Tarsius.* Simians show no further progression. In the New World monkeys it is generally larger than in the Old World monkeys. In the Hominoidea, the progression indices decrease. In man, the subcommissural body is generally regressive by comparison with that of basal Insectivores (index < 1).

(3) *Epiphysis:* This structure shows neither progression nor regression in the ascending scale. It seems to have a strong interdependency with body weight.

Phylogenetic development of the non-neocortical structures. The preceding size comparisons indicate prevailing trends in the phylogenetic development of different parts of the brain, that is, whether progressive or regressive developments have occurred. Even subtle changes can be reliably determined in structures that evidence no or opposite evolutionary trends without quantitative analysis. One such structure is the septum, which on the basis of simple macromorphological studies is generally believed to undergo strong reduction in the Ascending Primate Scale, especially in man. Allometric investigations demonstrate just the opposite, namely a clearly progressive development (Andy and Stephan 1968). The apparent reduction is suggested by the very strong enlargement of other parts of the brain, especially that of the neocortex.

We also believe that it is possible to give information concerning different intensities of an increase or a decrease in successive phases of evolution. For example, our results suggest that the progression of the optic centers and the regression of the olfactory centers have been more accentuated in the Prosimian phase than in the Simian one.

More rarely the original progression seems to revert into a regressive phase in the course of evolution. This is indicated for the accessory olfactory bulb and the subfornical body. Both structures are clearly progressive in most Prosimians and small or absent in many Simians. It also seems likely that a regressive phase may be followed by re-enlargement. Size modifications of the olfactory bulb in Chiroptera suggest this possibility (Stephan and Pirlot 1970).

Thus it is possible that forms similar to recent Prosimians may have played a role during the phylogenetic development of the Simians, even though certain characteristics now set them apart, outside the general trend in the ascending scale. For example, the strong development of the olfactory bulb in the tree-shrews does not argue against the possible importance of tree-shrew-like forms in primate phylogeny. During the initial evolutionary phase (from insectivores to tree-shrew-like forms), the olfactory bulb may have become enlarged but subsequently, in a second phase, markedly reduced.

Comparative neuroanatomical results must be interpreted with considerable caution, especially when employed to elucidate questions on the way the sizes of the structures have been achieved in recent species (which represent the end points of individual evolutionary lines). It can be assumed that structures, which show a distinct and steady increase in the Ascending Primate Scale, also have undergone a more or less continuous enlargement during their phylogenetic development (for example, the striatum). By contrast, this type of continuity seems unlikely in structures such as the hippocampus, in which the size varies very strongly in the Ascending Primate Scale. I suggest that during the development of such a structure, strongly progressive phases have alternated with less marked or even regressive ones. It can be assumed further that after divergence into various distinct phylogenetic lines, the continued evolution of structures became independent within those lines, not only concerning the intensity of size alterations but also the lapse of time. This is suggested by the marked scattering of the indices of progression, even within closely related groups.

Summary and Conclusions

All comparisons are referred to the relationships in primitive (="Basal") Insectivores.

The size of the brain, considered in conjunction with the differences in body weight by aid of the allometry formula (=encephalization), is not more pronounced in Primates than in other orders. A Prosimian stage is reached by many mammals and a Simian or even Pongid stage is matched by several semi-aquatic and/or aquatic groups. Only man has encephalization, which exceeds that of all animals. He is the only Primate with an outstanding brain size.

Brains of fossil forerunners of man represent intermediate stages and fill the gap between man and recent nonhuman Primates. The extraordinary brain size of *Homo sapiens* seems to have developed gradually.

The enlargement of the brain is not proportional; that is, all parts do not develop at the same rate. The neocortex is by far the most progressive structure and therefore used to evaluate evolutionary progress (=Ascending Primate Scale). Under the assumption that neocortical development has never been retrograde during phylogeny, the degree of neocorticalization is considered to be an effective guide in the discussion of phylogenetic sequences.

Directed size alterations of the non-neocortical structures in the Ascending Primate Scale allow conclusions on prevailing trends in their phylogenetic development. Most of the structures or structural complexes so far investigated show progressive trends. The olfactory structures, by contrast, evidence regressive ones. In several small structures the trend can revert, so that a progressive phase in Prosimians may be followed by a regressive one in Simians.

On the basis of such prevailing trends, it seems possible to obtain information about the composition of a fossil brain and the progression of its various parts, when brain and body sizes are known.

Up to that point, the given phylogenetic interpretation of the results of our comparative neuroanatomical investigations based on recent material seems to be well established. It must be stressed, however, that because our scientific approach is indirect, it can provide only inferences, not proofs.

7

LEONARD RADINSKY
UNIVERSITY OF CHICAGO

Endocasts and Studies of Primate Brain Evolution

Two sources of information are available for studies of primate brain evolution. One source, comparative studies of brains of living forms, provides access to a wealth of anatomical detail, the functional significance of which may be determined by experimental work and by correlation with observed behavioral patterns. For example, the greater degree of differentiation of the lateral geniculate body and greater area of striate (=visual) cortex in the brains of monkeys compared to lemurs (LeGros Clark 1962) suggests a more elaborate processing and use of visual information in the former group. That hypothesis is open to elaboration and verification by experimental ablation studies, visual discrimination testing, and observation of the use of vision in natural populations of monkeys and lemurs. The disadvantage of this approach for evolutionary studies is that no series of living species constitutes an evolutionary sequence, and therefore comparisons between living forms may be misleading sources of information for reconstruction of evolutionary histories. Thus, while the series *Tupaia-Lemur-Macaca-Pan* provides an example of increasing complexity in many brain structures, it does not necessarily reveal stages in the evolution of the chimpanzee brain, since in the absence of control from the fossil record or from very extensive comparative studies, it is not safe to assume that any one of those forms preserves a representation of an ancestral stage of any other form. All are the successful products of natural selection over an equal amount of time for their respective niches.

The second source of information for brain evolution studies is the actual fossil record of brains of extinct species. In birds and mammals the inside of the braincase is molded by the brain so that an internal cast of the braincase (=an endocast) duplicates, with more or less clarity, the external anatomy of the brain. Thus an endocast from a fossil skull can reveal the external morphology of the brain of an extinct species. The main limitation

175

of this approach is that it supplies at best a restricted amount of information (external brain morphology), the functional significance of which is not susceptible to direct testing. The great importance of this approach is that it provides the only source of direct evidence of brain evolution. Three reservations commonly expressed about the use of endocasts concern the feasibility of making them, the degree to which they reproduce external brain morphology, and the functional significance of external brain morphology.

Endocast Preparation

Endocasts can be made from fossil skulls either by cleaning out the braincase and making an artificial cast, or if the braincase is filled with a hard matrix, by removing part of the braincase to expose the natural fossil endocast. The first method has the advantage of leaving the braincase intact, although it is usually necessary to open the braincase to allow complete cleaning of the inside. A good casting material is liquid latex, which is strong and elastic when cured, and when pulled through a small opening (such as the foramen magnum) will return to its original shape. The latex casting technique is ideal for recent skulls, since it does not require sectioning or otherwise damaging the skulls (for details see Tugby 1953, or Radinsky 1968a).

With the second technique, exposure of the natural endocast, a cast should first be made of the outside of the braincase, to minimize possible information loss. It is often possible to remove the braincase wall in pieces large enough to allow their reconstruction apart from the endocast. Even when this is not possible, it should be obvious that the advantages in terms of information to be gained from study of the endocast far outweigh the possibility of loss of a small amount of information about the external surface of the braincase.

Endocasts and External Brain Morphology

Examination of endocasts of about 500 different species of mammals revealed that in the great majority of mammals, most details of external brain morphology, including the pattern of cerebral convolutions, are reproduced on the endocast. In addition, endocasts include impressions of intracranial blood vessels, the stumps of cranial nerves, and often the pattern of braincase sutures. Impressions of sulci become less distinct with increasing brain size, and in mammals such as cetaceans, proboscideans, large ungulates, and the largest primates, few sulci are visible on the endocast. This size-related phenomenon appears to hold within but not between major groups of mammals. Sulcal details are blurred on endocasts of the largest-brained rodents and monkeys, but are sharp on equally large endocasts of ungulates and carnivores.

Among primates, endocasts of all genera of living prosimians reproduce all of the cerebral sulci (Bauchot and Stephan 1967, Radinsky 1968a). In *Megaladapis edwardsi,* an extinct giant lemuroid with a cranial capacity of about 140 cc., sulcal details are indistinct; but in another giant lemuroid, *Archaeolemur majori,* (cranial capacity about 95 cc.), the endocast reproduces every sulcus (Radinsky 1970). In the New World monkeys, endocasts preserve fine cerebral details in most of the genera, but sulci are blurred in the largest-brained forms, such as *Cebus* and the atelines. However, carefully selected skulls (for maximum ossification of the cerebral juga) of even the largest ceboids will provide endocasts that reproduce all of the sulci. Only the smallest of the living Old World monkey endocasts clearly preserve the sulcal pattern, but exceptional skulls of even the largest (baboons) show all sulcal details. Natural fossil endocasts of the Pliocene cercopithecids *Mesopithecus* and *Libypithecus* appear to reproduce all of the cerebral sulci (Piveteau 1957, Edinger 1938). All cerebral convolutions are reproduced on some gibbon endocasts, but most details are lost on all great ape and human endocasts (Connolly 1950, and personal observations). The widely publicized lack of sulcal detail on human and great ape endocasts (Symington 1916, Le Gros Clark et al. 1936) may have been responsible for the general neglect of this area of study by primatologists.

Significance of External Brain Morphology

RELATIVE SIZE

Since endocasts closely reproduce external brain morphology, they can be used to measure brain volume and thus provide data for studies of relative brain size. For extinct species, body weights can be estimated from associated postcranial elements or represented by foramen magnum area (Radinsky 1967a, and in preparation). It is well known that primates differ in relative brain size (Stephan and Bauchot 1965, Bauchot and Stephan 1969), but it is difficult to make precise statements about the significance of such differences. Rensch (1956) summarized some of the data showing correlation between brain size and intelligence. Jerison (1963) estimated number of neurons from brain and endocast volumes to derive a coefficient of "brain efficiency," but Holloway (1966b) pointed out serious problems with that approach. Sacher (1959) demonstrated a high positive correlation between brain size and longevity in mammals, and the implications of that relationship in terms of evolutionary strategy should be investigated for the primates.

GROSS PROPORTIONS

Even where details of cerebral morphology are indistinct, endocasts reveal the relative size of some major parts of the brain, such as the olfactory bulbs, cerebellum, frontal lobes, and occipital lobes, and allow general functional interpretation. For example, relatively reduced olfactory bulbs and expanded occipital lobes in the endocast of *Tetonius homunculus,* a 55-million-year-old prosimian, indicate reduction of olfaction and increased importance of vision at least that far back in primate evolutionary history (Radinsky 1967b). Similarly, comparison of endocasts of modern gibbons and monkeys reveals a relatively larger cerebellum in the former, which suggests more highly developed muscular control in gibbons compared to monkeys.

TAXONOMIC CHARACTERS

Anatomical features revealed by endocasts may be indicative of phylogenetic relationships between taxa and therefore of taxonomic significance. For example, the complex sulcal pattern displayed by the endocast of *Archaeolemur* was considered by Le Gros Clark (1945) as anthropoid in appearance, and by Piveteau (1950) to indicate an evolutionary level comparable to that of advanced monkeys. However, comparison of the *Archaeolemur* endocast with those of living monkeys and lemuroids reveals the sulcal pattern to be more like that of indriid lemuroids than any anthropoids: the coronolateral sulcal complex (=rectus and intraparietal sulci) is more prominent than the central sulcus, and there is no inferior precentral (=arcuate) sulcus (Figure 7–1). The different appearance of the lower part of the frontal lobe from that of other lemuroids results from the larger size and smaller orbital impressions in *Archaeolemur.*

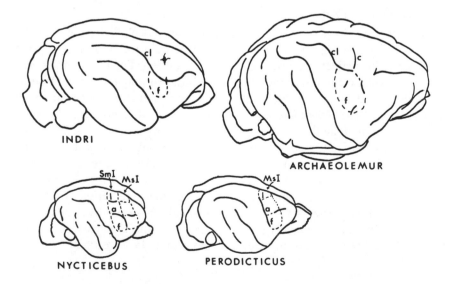

Figure 7–1. Prosimian endocasts. Cortical maps from references cited in text. *Abbreviations:* c, central sulcus; cl, coronolateral sulcus; l, a, f, cortical representations of hind limb, fore limb and head respectively; MsI, primary motor cortex; SmI, primary somatosensory cortex.

Anthony (1946) divided cebids into two groups based on differences in sulcal pattern (which can be seen on endocasts). In *Aotus, Saimiri, Alouatta, Lagothrix, Brachyteles,* and *Ateles* the intraparietal sulcus is continuous with the sylvian sulcus, while in *Callicebus, Pithecia, Cacajao,* and *Cebus* it is not (for examples, see Figure 7–2). In separating *Aotus* from *Callicebus,* and *Saimiri* from *Cebus,* Anthony's division differs from other proposed subfamilial classifications of the Cebidae, and before undue reliance is placed on that single character, its functional significance should be investigated. That such a character might not have profound phylogenetic significance is suggested by the occurrence of the same feature in the sulcal pattern of *Nycticebus* (intraparietal sulcus continuous with sylvian) but not in *Loris, Perodicticus,* or *Arctocebus.* Hirsch and Coxe (1958) have suggested that the confluence of intraparietal and sylvian sulci in *Ateles* may have resulted from a lateral shift of the parietal (postcentral) gyrus

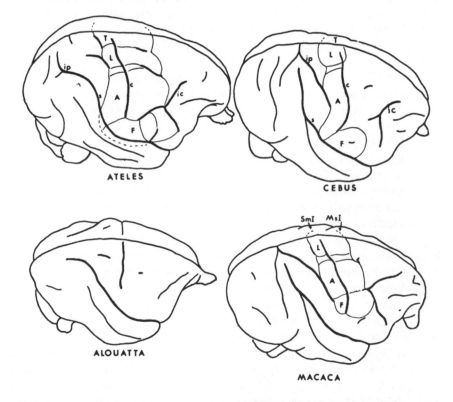

Figure 7–2. Monkey endocasts. Cortical maps from references cited in text. *Abbreviations:* c, central sulcus; ic, inferior precentral or arcuate sulcus; ip, intraparietal sulcus; s, sylvian sulcus; T, L, A, F, cortical representations of tail, hind limb, fore limb, and head, respectively; MsI, primary motor cortex; SmI, primary somatosensory cortex.

correlated with the expansion of the somatic sensory tail projection area
(pp. 181–182).

SULCI AND CORTICAL LOCALIZATION

Electrophysiological mapping experiments on a variety of mammals, including the dog (Pinto-Hamuy et al. 1956), cat, and macaque (Woolsey 1958),
procyonids (Welker and Campos 1963), sloth (Meulders et al. 1966),
and loris (Krishnamurti and Welker 1965, and Sanides and Krishnamurti
1967) indicate that the majority of neocortical sulci delimit functional or
somatotopic areas. For example, Krishnamurti and Welker's work on
Nycticebus (*op. cit.*) showed that the short sulci of the frontal lobe form a
partial boundary between primary somatic sensory and motor cortices, and
also separate face from arm from leg projection areas within the somatic
sensory cortex (Figure 7–1). More dramatically, Welker and Seidenstein
(1959) demonstrated that the secondary sulcal complex within the somatic
sensory hand projection area of the raccoon separates representations
from individual digits and palmar pads. LeGros Clark (1962, p. 245)
called attention to the presence of longitudinally oriented sulci that appear
to cut across histologically defined cortical areas in lemur brains, in contrast
to the predominantly transversely arranged sulci that delimit cortical areas in
anthropoid brains. However, electrophysiological mapping studies (Krishnamurti and Welker 1965, Zuckerman and Fulton 1941, Vogt and Vogt
1907) indicate that the longitudinal sulci separate somatotopic areas (face
from arm representations) within primary motor and somatic sensory cortex; and detailed cytoarchitectonic studies (Sanides and Krishnamurti 1967)
reveal that those sulci separate histologically different areas in the frontal
and parietal association fields. Thus in prosimians as well as in anthropoids
and other mammals, most sulci delimit cortical areas as defined by histologi-
cal or electrophysiological methods.

Because cortical areas are delimited by sulci, and because sulci are reproduced on endocasts, it is possible to interpret functional areas on endocasts,
providing that sulci can be confidently homologized from a mapped species
to an unmapped one. Fortunately this is possible for most of the sulci in
most of the primates; only *Daubentonia* presents major problems for interpretation of some of its sulci (on the frontal lobe). The ability to delimit
functional areas on endocasts is important because behavioral or sensory
specializations may be reflected in differential enlargement of particular
cortical areas. For example, the great sensitivity of hands in raccoons, which
is correlated with a behavioral pattern involving extensive manipulation
and palpation of objects, is reflected in the neocortex in a great enlargement
of the hand representation within the primary somatic sensory area (Welker
and Campos 1963). Similarly, the tactile pad and prehensile ability of the
tail of the spider monkey *Ateles* is indicated on the cortex by an expansion
of the tail representation within the primary somatic sensory and motor

areas (Fulton and Dusser de Baranne 1933, Hirsch and Coxe 1958). In both the raccoon and spider monkey, the differential enlargement of particular somatotopic areas could be interpreted from the external morphology of the brain or endocast.

Recognition of differential enlargement of face and hand somatic sensory projection areas on otter endocasts led to hypotheses on sensory specialization that appear to correlate with observed anatomical and behavioral characters, but remain to be confirmed by neurophysiological experiments (Radinsky 1968b). In the case of the otters, fossil endocasts allowed tracing back of the various sensory specializations 12 and 25 million years.

Few primate brains show sensory specializations comparable to those of otters or raccoons. A more expanded visual cortex in monkeys compared to prosimians is indicated by the more rostral position of the lunate sulcus, and endocasts of the ancient anthropoid *Aegyptopithecus* indicate that the visual cortex had expanded to the degree seen in higher primates at least 25 million years ago (Radinsky, in preparation).

The presumed face area of the primary somatic sensory cortex in *Archaeolemur* is expanded and elaborated by one or two secondary sulci (Figure 7–1), suggesting specialization for increased tactile sensitivity in the facial region of that extinct giant lemuroid. However, the infraorbital foramen in *Archaeolemur* is not enlarged, which one would expect to find with such a specialization. Perhaps one or both of those secondary sulci mark part of the boundary between primary and secondary somatic sensory cortices, as in some *Nycticebus* brains (intersomatic sensory sulcus of Krishnamurti 1966, Sanides and Krishnamurti 1967). Cortical mapping studies of *Lemur* and *Indri* (or *Propithecus*), specifically to delimit the boundaries of the somatic sensory face representation, might help resolve this problem of interpretation.

The expansion of the tail area within somatic sensory and motor cortex in *Ateles* is reflected in several features of the sulcal pattern (Figure 7–2). The somatic sensory tail representation, contained in most monkeys within the medial wall of the hemisphere, has expanded onto the lateral face of the hemisphere, and its caudal boundary is often marked by a short sulcus. The remaining somatic sensory body representation has been displaced laterally, which may account for the confluence of intraparietal and sylvian sulci. The face area has been displaced laterally into the sylvian sulcus, and also rostrally to a precentral position. This has displaced the motor face area rostrally, although there appears to be some overlap between the two face areas. This displacement of the motor face area is indicated by the broad separation of the inferior precentral (=arcuate) sulcus (which marks the rostral boundary of the motor face area) from the central sulcus. An unusually long caudal spur, extending back from the inferior precentral sulcus, separates the face representation from the more medially located motor arm area. The motor tail representation is also expanded onto the lateral face of the hemisphere, and its boundary with the leg area is marked by a short sulcus, incorrectly identified by Anthony (1946) and Connolly (1950) as the superior precentral sulcus. The true superior precentral sulcus, which marks the level of a thin band of trunk representation separating motor leg and arm areas, is located further laterally on the brain of *Ateles*. The relatively great anteroposterior width of the motor cortex results in part from the rostral displacement of the face representation and in part, according to Hirsch and Coxe (1958), from increased cortical control of proximal limb musculature, presumably related to brachiating.

Most of the sulcal features described for *Ateles* are present also in brains of *Lagothrix* and *Brachyteles* (see Anthony 1946), both of which have comparable tail specializations. The overlap of intraparietal and sylvian sulci in *Saimiri* is also apparently correlated with a lateral displacement of the somatotopic representation (Benjamin and Welker 1957, Hirsch and Coxe 1958), although in that genus the tail is less specialized than in the atelines. In *Cebus,* which like *Saimiri* has a slight amount of tail prehensility and tactility, the intraparietal sulcus is not contiguous with the sylvian sulcus, although there has been enough lateral displacement (from a slight expansion of the tail representation and enlargement of the hand and face representation relative to the condition in *Macaca*) of the somatic sensory face area to move some of the face area rostrally past the lateral end of the central sulcus, with a consequent rostral displacement of the inferior precentral sulcus (Figure 7–2). The difference in position of the intraparietal sulcus in *Cebus* compared to *Saimiri* may result from the larger size of species of the former genus, since, as an apparent general allometric phenomenon, primary sensory and motor areas occupy relatively less of the total cortical area in larger brains. Total amounts of cortex in square millimeters to which the tail projects are: 8 for the macaque, 15 for *Saimiri,* 30 for *Cebus,* and 110 for *Ateles* (Hirsch and Coxe 1958, Welt 1962).

Of the other two monkeys with confluent intraparietal and sylvian sulci, *Alouatta* has a prehensile tail with a bare tactile area, but *Aotus* does not. Except for a few motor cortex experiments reported by Vogt and Vogt (1907), the brains of neither of these genera has been mapped, either electrophysiologically or histologically. Expansion of the tail representation with lateral displacement of the remaining somatotopic representation may account for the confluent intraparietal and sylvian sulci in *Alouatta,* but another explanation must be sought for *Aotus. Alouatta* lacks the short sulcus above the superior precentral sulcus that in *Ateles* separates motor leg and tail areas and indicates that the tail area has expanded onto the lateral face of the hemisphere. Its absence in *Alouatta* may be due to the smaller size of its brain compared to *Ateles,* since as another general allometric phenomenon, smaller-brained mammals have fewer cerebral sulci than their larger-brained close relatives.

The above discussion indicates some of the potential for interpreting specializations from fossil monkey endocasts. A prehensile or highly sensitive tail would be indicated by the presence of sulci delimiting the tail area of the primary motor or somatic sensory cortex, providing that the brain was large enough (that is, larger than that of *Alouatta*). For smaller brains, the confluence of intraparietal and sylvian sulci may prove to be an indicator of tail mobility or sensitivity, at least to the degree seen in *Saimiri*. However, further neurophysiological experiments will be necessary before that character can be used with confidence.

EXTRINSIC FACTORS

While the development of various cortical areas is a major determinant of external brain morphology, the relative size and shape of other parts of the head may also effect sulcal pattern and brain shape. For example, the lack of orbital impressions in *Daubentonia* and the long-skulled giant lemuroids (*Megaladapis, Palaeopropithecus*) suggests a lack of rostrocaudal compression which may be correlated with the less opercularized sylvian sulcus in

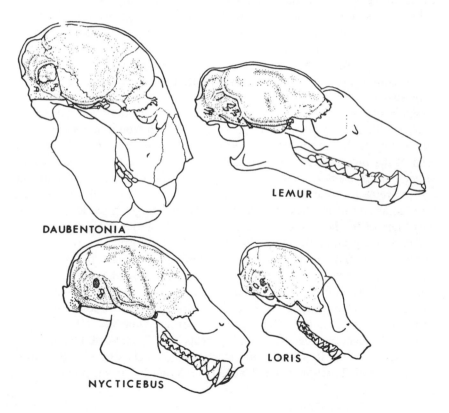

Figure 7–3. Prosimian endocasts superimposed on skull outlines. From Radinsky 1968a.

those genera (Radinsky 1970). Another example is the high, rounded frontal lobe profile of *Daubentonia,* with its peculiar sulcal pattern, which apparently results from the extreme flexion of the facial part of the skull relative to the braincase (compare *Daubentonia* with *Lemur* in Figure 7–3). A third example is the more convex dorsal profile of the frontal lobe of *Loris* compared to the condition in *Nycticebus* (and other lorisines), which appears to have resulted from the small size and relatively large orbits of *Loris* (see Figure 7–3); similar differences in brain shape are seen in comparisons of brains of small and large breeds of domestic dogs.

Future Endocast Research

Detailed comparative studies of external brain morphology are needed for many groups of primates. There are still no studies of prosimian or cercopithecoid brain anatomy comparable in depth and extent of coverage to Anthony's (1946) and Hershkovitz's (1970) works on ceboid brains. In addition, all of the older studies could be reinterpreted profitably in light of what is now known about the functional significance of external brain morphology. Endocasts could be helpful in such studies since for many genera they are much easier to obtain than are real brains. Such studies are important not only for what they may reveal about the biology of the animals, or about phylogenetic relationships, but also because they may suggest promising directions for future research. For example, the confluence of intraparietal and sylvian sulci in *Aotus* and *Alouatta* suggests differential expansion of some cortical areas, which would make them interesting subjects for experimental neurophysiological work. Or, the forelimb area of the motor cortex in *Indri* and *Propithecus* appears to be relatively larger than in *Lemur,* which suggests greater cortical control of hand and arm movement in the indriids (Radinsky 1968a). This could be tested by behavioral and neurophysiological methods, and might provide new insight into some of the factors involved in lemuroid evolutionary radiations.

Endocasts remain the only direct source of information on brains of extinct forms and the only primary evidence for primate brain evolution studies. There are fossil primate skulls now in museum collections which could be prepared to yield endocasts. In addition, some of the fossil primate endocasts that have already been described should be restudied and interpreted in light of modern knowledge of cortical function. Ultimately the major significance of endocast studies rests on the state of knowledge of brain function, and it is hoped that neurophysiological studies will continue to elucidate the functional significance of external brain morphology.

8

RALPH L. HOLLOWAY
COLUMBIA UNIVERSITY

Australopithecine Endocasts, Brain Evolution in the Hominoidea, and a Model of Hominid Evolution

Each living species of primate has a species-specific interdependent organization of anatomical structures and behavioral repertoire, modifiable within unknown limits of environmental influence. The same statement must apply to fossil primates. Obversely, but in addition, there is still an underlying core of adaptive commonality involving locomotion, prehensility, social behavior, and intelligence, that has been derived through millions of years of evolution within a basically arboreal environment. This core is probably not species-specific; rather, it is shared across several taxa within the Primate order. In general, our knowledge about the relationships of neuroanatomical patterns and processes to this latter set of commonalities is grossly, but fairly clearly, understood.

However, were we to measure our knowledge of the relationship between neuroanatomical parameters and species-specific behavioral attributes, we would have to admit our almost total ignorance. Recently, (Holloway 1969a), I raised two questions which were meant to serve as a rough gauge of our knowledge of primate behavior and neurological parameters: (1) can we predict behavioral qualities (social and sensorimotor) from a description of brain morphology?; (2) can we predict brain morphology or other neural parameters from a description of primate behavior? This is not a problem unique to primatology. The structure-behavioral differences be-

I am greatly indebted to Professor Phillip V. Tobias and to Dr. C. K. Brain for allowing me to undertake the study of the australopithecine endocasts, and for their hospitality and encouragement. I am also grateful to the National Science Foundation, whose support through Grant GS-2300 enabled me to study the fossils. Finally, I am very grateful to the organizer of this effort, Dr. R. Tuttle, for inviting me to participate, and to the Wenner-Gren Foundation, particularly Mrs. Lita Osmundsen and her staff, for their marvelous hospitality, and to Mrs. Sue Gould, Columbia University, for her help with the editing and typing of this paper.

tween basenji hounds and cocker spaniels are not known any more than are the differences between chimpanzee and gorilla, or between black and Norway rats. Some areas of both behavior and neural structure are easier to deal with in terms relevant to species-specific behavior, as for example in sensorimotor representation and cortical maps (e.g. Welt 1962a) as contrasted with social behavior. Part of our problem, but a major part, is that we have not yet successfully defined the appropriate units of behavior or neural structures which will permit a causal understanding between behavior and structure at the neural level.

Elsewhere (Holloway 1964, 1966a and 1966b, 1967, 1968, 1969a), I reviewed the problems of parameters (and their quantification) used for structural-behavioral and evolutionary analysis. Here I will concentrate more on some concrete evidence for brain reorganization resulting from my studies on the early hominid endocasts from South Africa. Then I will construct an admittedly speculative model of early hominid evolution, which I hope might have some feedback effect in structuring our perceptions on current research in primatology. First, however, it is appropriate to make some general comments on hominoid brain evolution.

Problems in the Study of Brain Evolution

There are three ways by which one may approach the evolution of the human brain: (1) the study of the only empirical evidence of brain evolution, paleoneurology, or the study of endocasts; (2) neurological study of an extant comparative series; (3) study of neurological parameters and behavior, which can include neurophysiology, neurochemistry, inbred strains, and neurological genetics. Few if any of us study brain evolution by any one method independently, since each method has severe limitations, and our goal of finding syntheses between structure, behavior, and evolutionary adaptations requires us to cast our nets as widely as possible.

Few anthropologists fully understand the enormity of the problems of studying brain evolution. The brain is the most complex organ (or interacting sets of organs) in the primate body. It has billions of parts, (if we count cells) and a stunning number of interconnected nuclei and fiber systems. A thorough study of *one* brain can be a matter of lifetime effort. As any other organ in a species, the brain is variable. Thus, one little appreciated fact (except by neuroanatomists) is that our masses of data seldom contain information about neural variability in a species, a rather basic datum for most other morphological complexes, particularly bones and teeth. Tilney's (1928) interesting cross-sectional measurements for a comparative primate series modeling phylogenetic trends are based on single specimens. Blinkov and Glezer's (1968) compilation of quantitative studies on human and other primate brains provides data of a similar nature. Aside from brain weights and volumes, the variability of almost any neural subsystem one wishes to choose is not known. There is no knowledge about sexual dimorphism in the brain (not even in the hypothalamus); and even with rats and mice, where highly inbred genetically-selected strains have been investigated for years, there is very little to link behavior with neuroanatomical variables. A few exceptions exist: work relating to specific neurological disorders such as waltzing, wheeling, spastic, choreoid mice, and a recent investigation of inbred mice strains and heritability of relative and absolute amounts of cortex and hippocampus (Wimer et al., 1969). Behavioral analysis or correlation with neural parameters have not yet been published in the above study.

The usual parameters utilized, such as neuron size and density, neural/ glial ratios, even nuclear volumes, are not available for more than a few specimens and few species, a fact particularly true for the pongids. Dendritic branching parameters are unknown for any primate other than man (Conel 1939–1963, Schade et al., 1962), yet synaptic connectivity is surely an important parameter to any understanding of intelligent problem-solving behavior. There are almost no studies on subcortical fiber systems, except for Daitz (1953) and Powell et al., (1957) on the fornix and mammillo-thalamic systems, and these compare monkey and man. Not even a be-ginning comparative picture exists for synaptic junctions or vesicles based on electron microscopy.

Indeed, even for that most crude of parameters, cranial capacity, we are not in the very best shape. Tobias' (1970) recent paper has clearly shown the serious lack of dependable data on this measurement in modern man. In fact, it is not possible even today to extract a sample of cranial capacities for pongids to compare statistically with our present australopithecine sample. We have averages, ranges, s.d.'s, c.v.'s, but no basic enumerative data, as Tobias pointed out in his discussion of skewed samples (1968b). Even our current empirical claims regarding the cranial capacities of fossil hominids are, as far as I am concerned, totally suspect and in need of serious restudy. Throughout the United States and across the world, there are numerous laboratories and Primate Centers where hundreds of primates are studied each year, but there is no coordinated plan for saving tissues, skeletons, or collecting volumes which might be profitably studied in the future.

Neurochemical differences between hominoid brains are just beginning to be investigated, but I have never found any serious attempts to think about the relationships between neurochemical differences, neuroanatomical variables, and behavior. Lists of transmitter substances or even their quanti-fication is interesting and potentially useful information when coupled with neuroanatomical data, but I have found few examples of any concern with hominoid materials. Goodman et al. (1969) have provided figures on heart vs. muscle-type lactic de-hydrogenase in primates, which shows an interest-ing trend within the Primate order, but has yet to be related specifically with either neuroanatomical or behavior variables. For example, this study shows that there is a shift from M-LDH to H-LDH in primate phylogeny. H-LDH is more suited to aerobic metabolism and sustained production of energy, increasing as we approach man, suggesting a possible tie-in with the decreasing neural density in the cerebral cortex and increase in the glial-supporting cells. Perhaps the H-LDH increase is associated with this latter parameter.

The point is that unless some effort is made to relate these biochemical variables to real anatomical structure, we cannot hope to find meaningful taxonomic or functional relationships between neurochemistry and behavioral adaptations. Similarly, isolated published figures for different neurotransmitters are often meaningless without knowledge of the neuro-anatomical subsystems these work in, their roles in terms of inhibition and facilitation, and the quantitative knowledge of the fiber systems mediating impulses from one nucleus to another. Thanks to the work of Stephan and Andy, we are beginning to get good data on nuclear volumes of different neural systems.

By way of review, the following general statements sum up most of the problems facing us regarding hominoid brains and their evolution.

1. Each primate brain bears a specific stamp of qualitative and quantitative features gained through natural selection, reflected in quantitative parameters such as brain size (both absolute and relative), cortical differentiation, neural densities, glial/neural ratios, nuclear volumes, fiber cross-sectional areas, neurophysiological (functional) homologous mappings of sensorimotor capacities, and various amounts of different neurotransmitters, etc.

2. Each primate brain is therefore *reorganized* according to the evolutionarily oriented adaptations of social, manipulative, and locomotor behaviors. Different primate brains are never simply smaller or larger versions of each other. Fossil endocasts reflect only poorly, if at all, the different internal organizations of neural structures, and an extant comparative series is not an evolutionary sequence.

3. Mass parameters, such as brain weights, volumes, brain/body or encephalization coefficients, *mask* species-specific adaptations and do not bear on strictly causal analyses between form, function, and behavior except at exceedingly gross or molar levels.

4. Any item or unit of observable behavior is a product of the interaction of multiple systems of neural fibers originating from neurons in different nuclei of many neural systems, and the programming of these interactions depends on the inherent genetic information about their formation and operation in conjunction with all past interactions of these subsystems in the individual's lifetime and environments.

5. Quantitative parameters of neural subsystems indicate each species has a particular specificity of organization; and that in general, what is

largest reflects high levels of intensity of past forces of natural selection. These in turn reflect different rates of mitotic divisions and precise instructions about their durations. This means that in addition to changes of genetic instinctions for the brain, there were also changes involving wider physiological systems, (e.g. the endocrine system and growth).

6. Since behavior is the result of complex interactions of many neural subsystems through time, we cannot expect to make simple relationships between complex behavior and single neural parameters. Similarly, we must avoid reductionistic explanations that relate ablation or stimulation studies of one neural area to an item of behavior. Localization of function to one particular area of the brain is not possible, since functions are "localized" in the interactions of many components in a multitude of subsystems within systems within the large system of brains, environments, and past ontogenetic histories.[1] We can only say that an area *x* or *y,* or *x + y,* is *involved* in a function, i.e., rage, language (!), etc.

7. Brains, and their neural systems, are variable just as are other morphological systems, and data for one brain is not sufficient to make relationships between neural parameters, behavior, and species differences in both.

8. Different patterns of morphological form in the musculoskeletal systems between different primates, whether fossil or extant, requires reorganization of nervous elements involved in the expression of motor behavior, and the latter's integration within the whole social, perceptual, and material environment of the particular primate species. This means that regardless of the cranial capacity or external form of a fossil species, the brain of that species had undergone evolutionary changes to match and service the anatomical patterns selected. Statements regarding the evolution of brains first or last are oversimplistic and mask reality, leaving us with unnecessarily crude perspectives of our own evolution.

The Usefulness of Cranial Capacity

Elsewhere (Holloway 1966a, 1968), I have criticised cathecting on mass variables, e.g. cranial capacity, as an explanatory variable in behavioral and brain evolution. I have emphasized that in hominids, at least, cranial capacity is but an outward manifestation of multiple internal organizational changes that are better suited to correlation with behavioral variables. Most simply, comparisons based on cranial capacities alone are not comparisons of equal units, and indeed, the extreme variability of this parameter without demonstrated correlation or causal relationship to behavioral parameters undermines its effectiveness in discussing human evolution.

This position does not at all mean, however, that this parameter is useless,

---

1. This sentence is purposely complex, almost unreadable, and is given without apology.

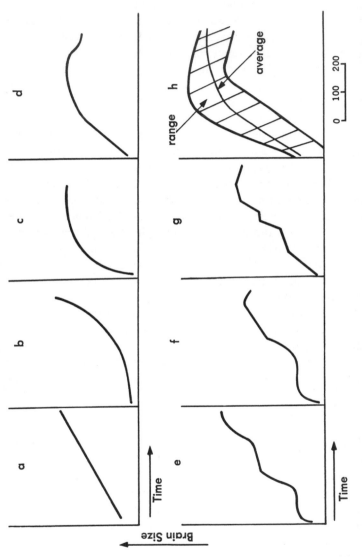

Figure 8–1. Hypothetical curves showing possible brain growth rates with time, during hominid evolution. *a* shows constant increase with time, suggesting a constancy of selective forces; *b* is increasing acceleration with slow initial phase; *c* is high initial phase with decreasing acceleration; *d* through *g* indicate different combinations suggesting a variety of selective forces through time; *h* shows both average and range, with skewed distributions at different times, suggesting different adaptive dynamics. It is doubtful that our data will ever allow such relationships to be drawn.

or should be ignored or treated lightly. We need much more reliable data on cranial capacities for a variety of reasons. It is a good parameter for characterizing morphological variability, and for providing a statistical basis for taxon comparison and elucidation. Were enough data available, it might be sensitive enough to distinguish between variability attributable to sexual dimorphism, subspecific, or specific variation. It could tell us something about skewed samples, and provide a basis for reasonable speculation about selection dynamics. It could provide us with extremely interesting data about differing rates of human evolution, given that our techniques of absolute dating, such as K/A, improve and are extended. These in turn could enlighten us about more molecular dynamics of hominid evolution and natural selection. For example, consider the rich speculative possibilities *if* we had the data to draw the graphs shown in Figure 8–1.

Another basic reason for wanting precise data on cranial capacities is the feedback effect between these data and finding new methods of reliably reconstructing fragmented fossil volumes. It will become clearer later in this chapter that one of the problems I faced in reinvestigating the australopithecine endocasts was the utilization of different methods for different casts to get precise volumes. More precise data on *Homo erectus* and Neanderthal endocasts are needed to make accurate estimates on recent finds from Olduvai Bed II cranial remains, and the new finds by Clark Howell and Richard Leakey at Omo and Lake Rudolf.

Data on the non-neural contributions to cranial capacity (that is, blood vessels, tentoria, meninges, fluid, etc.) are needed for all hominoids to provide better estimates of neural contents.

Cranial capacities, particularly in the fossil record, might even be used to estimate body weights and heights, parameters useful for analyzing past ecological relationships such as biomass, carrying capacity, hunting efficiency, and other demographic variables.

While there are many uses for this parameter, regardless of its crudity, there are also abuses of which we must be constantly aware. The most obvious examples are facile attempts to correlate it with variables such as intelligence, race, and culture. "Rubicon" models of language are another abuse, but are seldom seriously considered these days. Another abuse involves oversimplified cathection on the concept of mosaic evolution, whereby one builds up a whole concatenated theory of evolutionary dynamics to explain hominid adaptations based on cranial capacities alone, overlooking evidence for reorganization, and ending up with statements that relegate the brain as the final target for evolutionary change.

Early Hominid Cranial Capacities

On the basis of my studies thus far on the australopithecines, I believe that the endocranial evidence offers support for the views that two forms,

Table 8–1. Old and new volume determinations of the australopithcine endocasts

Specimen	Volume, C.C. (This Study)	Volume, C.C. (Previous)
Taung child	404 (440 adult)	525–600
STS 60 (Ples. 1)	428	436
STS 5 (Ples. 5)	485	480
STS 19 (Ples. 8)	436	550–570
STS 71 (Ples. 7)	428	480–520
MLD 37/38	435	480
Old. Hom. 5 ("Zinj")	530	530
SK 1585	530	—

a robust and gracile one, existed. The precise taxonomic position of these forms relative to each other and other hominids is surely an area for debate, and the small sample of endocranial volumes available will not provide closure on these problems.

Before going further, it is necessary to briefly describe some of the methods used to arrive at the final total volumes given in Table 8–1. Fuller descriptions, and analysis of morphological form and measurements, are still under study.

The simplest method of reconstructing a precise volume is to add the missing parts to an already reasonably complete endocast (Plates 8–1 to 8–4). The Taung child, STS 60, and SK 1585 endocasts allowed such simple reconstruction, for the added parts made up only a small portion of the whole. Thus small errors of interpretation, such as the degree of angling of the frontal lobe, whether the poles are rounded or more pointed, the size of its rostral portion, the cerebellar-temporal cleft formed by the petrous portion, or the brain stem to the level of the foramen magnum, probably contain small errors in interpretation which could cancel out. If they are not included in the reconstructions, an additive error of under-estimation is very likely. The rather large difference between the previous volume for the Taung child and my own reconstruction is a matter of correctly finding the most reasonable midsagittal plane, and adding a frontal portion which matches the matrix still imbedded in the facial fragment. Of course, one must assume perfect symmetry between hemispheres when arriving at the total volume. (For details on this reconstruction, the reader is referred to Holloway 1970.)

STS 5 and the Olduvai Hominid #5 ("Zinj") were already complete enough and free of distortion to be measured without reconstruction.

STS 19, however, required the utilization of a partial endocast method, based on that used by Tobias (1967, and in press) in his analyses of the Bed I and II hominids.

This involved the making of a hemiendocast of the base of the brain, including the cerebellum, the sigmoid vein, or sinus, and as much of the temporal lobe as was available. The same constructions were made from the Taung, STS 5, SK 1585, and Old. Hom. #5. Obviously a certain amount of "eyeballing" the upper border of the temporal portion was necessary. It was gratifying to discover that in all of these pieces, the partial or percent volumes to their known totals varied by less than two percent, a result which Tobias can corroborate from his own earlier utilization of this method. Extrapolation on the basis of a mean percentage provided the total volume given for the STS 19 specimen.

The STS 71 cranium is highly distorted, and the endocast portions taken from it were likewise distorted. The reconstruction attempted to rid the endocast of that distortion, and so provide a final form that bore closer resemblance to an australopithecine brain. Obviously, this was a matter of some "feel," and I hope that my familiarity with these specimens has led me to make the proper decisions.

MLD 37/38 required an entirely different method, utilizing the MLD 1 occipital fragment to ascertain probable bone thickness in the regions of the occipital pole, lower cerebellum, and temporal regions; estimates of bone thickness on the MLD 37/38 solid piece, and a reconstruction of probable total distance between frontal and occipital poles. Once these were obtained, it was possible to obtain four measurements and apply a formula to them. The four diameters were: (1) frontal pole to occipital pole; (2) maximum biparietal breadth; (3) bregma to lowest cerebellar region; (4) vertex to deepest projection of the temporal lobes. The formula used was a modified version of that given by MacKinnon et al (1956): $V = \frac{1}{2} (L \times W \times B) + (L \times W \times H)$, which was used to calculate volumes based on radiographs of human skulls whose volumes had been previously determined by the seed method. The formula used was:

$$V = c_1(L \times W \times B) + c_2(L \times W \times H),$$

where B = bregma-lowest cerebellar diameter, and H = vertex to lowest temporal lobe projection. The constants c_1 and c_2 were obtained by comparing the actual volumes for the other endocasts with the diameters and obtaining a best fit. A pilot study first made on a series of ten chimpanzees and nine gorillas indicated that the formula could predict brain capacity within 2 percent error of the actual water-displacement determinations. Since the bregma region is intact on the MLD 37/38 specimen, the anterior projection from a plane through bregma and perpendicular to a frontal to occipital pole axis was taken on the basis of the mean distance of such measurements from the other intact endocasts.

It should be immediately apparent that at least some bias is introduced in any analysis of the statistical properties of this sample of endocasts by the very methods to arrive at the total volumes, since some of the methods had to rely on reconstructions based on such endocasts. This probably helps to partly explain the very low degree of standard deviation given in the table for the gracile volumes.

Naturally, the search for total volumes is only the first step in reappraisal

Table 8–2. Some statistical results based on data in Table 8–1

	Gracile Alone	Combined
Average	442	464
Standard deviation	21.6	44.71
Coefficient variation	4.88	9.63

Student-t test

Test	t	p
(1) gracile vs. robust, d.f. = 6	6.69	.001
(2) combined vs. Hominid 7 (657 c.c.), d.f. = 7	4.07	.01
(3) gracile vs. Hominid 7 (657 c.c.), d.f. = 5	9	.0005
(4) *robust vs. Hominid 7 (675 c.c.), d.f. = 10†	3	.01
(5) 752 c.c. gorilla vs. average 550 c.c. male sample (Ashton and Spence), d.f. = 62	3.26	.01
(6) Sexual dimorphism between gracile and robust = 16.6% Sexual dimorphism between male and female gorilla = 16.3% Sexual dimorphism between male and female chimps = 7–11%		

*Assumes robust average = 530, N = 6, S.D. = 53 (c.v. = 10%).
†Assumes Hominid 7 average = 657, N = 6, S.D. = 65 (c.v. = 10%).

of the australopithecine endocasts, and further study on the dioptographic tracings, tape and caliper measurements on the actual casts, and analysis of morphological features, including definition of gyri and sulci, are still in progress.

Based on my reconstructions and estimates (see Table 8–1), the gracile forms have an average volume of 442 c.c., and the robust specimens (that is, Olduvai Hominid 5 and SK 1585) both are estimated at 530 c.c. Table 8–2 provides some basic statistics, which are only suggestive, keeping the small sample sizes in mind. The differences can be interpreted at the sexual, subspecific, or species level. Brace (1969) believes the evidence of dental and skeletal remains of robust and gracile forms can be explained by assuming sexual dimorphism. I would think that the dental evidence of pattern differences between anterior and posterior teeth, the dimensions, and the different localities would argue against such a position. The difference between robust and gracile australopithecine volumes, based on tables 8–1 and 8–2 is 16.6 percent, a figure somewhat higher than either the Ashton and Spence (1958) or Randall's (1943) gorilla samples, both providing a sexual dimorphism of 16.3 percent. Surely the canine length reduction existing in both forms of australopithecine indicates a general reduction of sexual dimorphism relating to skeletal and dental components, although there may well have been increases in the epigamic features such as breasts, fat distribution, etc. With respect to the first set of features, the gorilla is by far the most dimorphic hominoid, and I find it very difficult to imagine that early hominids had more dimorphism in these features than the gorilla.

In addition to this large difference of 16.6 percent between gracile and robust averages, the form of the endocasts shows other differences that are not concordant with differences at the level of sex. The expansion of the lateral part of the cerebellum, the shape and smallness of the occipital poles, the increase in height in the parietal region, all suggest differences of a greater magnitude than sexual dimorphism, i.e., either subspecific or specific. Examination of shape in male and female endocasts of both chimpanzee and gorilla does not show differences of this order in the australopithecines.

I believe that the differences are more readily explained at the species than subspecific level. This opinion is made on the basis of the endocasts, the dental and skeletal morphology, and the site distributions.

Considerable controversy has surrounded the designation of certain of the East African discoveries as a new species, *Homo habilis,* particularly the Olduvai Hominid #7 (Wolpoff 1969, Pilbeam 1969c). The statistics provided in Table 8–2 suggest that the volume of 657 c.c. for this hominid (this is not an adult value; Tobias 1971b) is significantly different from either gracile or robust form of australopithecine, and is certainly well under the published values for the Javanese *Homo erectus* volumes or that of Olduvai Hominid #9, from the LLK site (Tobias 1967). Certainly the teeth of several of these Bed I and Bed II finds, as well as the parietal fragments of Hom. #7, and those of Hom. #13, suggest a form distinct from either gracile or robust form of australopithecine. There is really not enough morphological completeness in these fragments to be definitive, but the shape of the parietals in #7 and the occipital portion of #13 suggests a lateral and height expansion of cortex above the level of the australopithecines.

While it would be premature to speculate very far on the new finds discovered by Clark Howell at Omo, or those of Richard Leakey at East Rudolf, my brief examination of the original of Omo and the casts of the East Rudolf finds leads me to expect that the brain sizes were within the ranges of the australopithecines described in Table 8–1. The Omo specimen would best be considered a gracile australopithecine, and the East Rudolf finds would be examples of the robust species. (These speculations are based on measurements taken through the courtesy and invitation of Clark Howell, and are unpublished pending his and Leakey's descriptions.)

The South African forms are without chronological dates, and the East African forms have dates extending back to a possible 2.4 to 2.6 million years B.P. The *Homo habilis* materials are dated at somewhat greater than 1.75 million years, whereas the LLK Hominid #9, with a volume of ca. 1000 c.c. is around 900,000 years B.P. This suggests a period of perhaps one million years during which brain size was fairly constant, but which was rapidly becoming larger near the end of this period (c.a. 1.5 million years?). Unfortunately, we do not know very much about the body sizes of these hominids, although a recent estimate by Lovejoy and Heiple (1970) suggests a stature of 42 to 43 inches and a weight of between 40 and 50 pounds for the gracile form, based on extrapolations from femoral fragments. This would give a brain:body ratio of 1:46 for the gracile form, a figure very similar to the average for modern man.

Using a ratio of 1:46, and an average brain weight of 530 grams for the robust forms, the expected body weight would be approximately 54 pounds, which seems considerably lower than expected. Assuming a body weight of between 90 to 100 pounds, (i.e. 95 pounds), the brain weight to body weight ratio for these robust forms is 1:77, which is quite low for hominids; indeed, for primates. This small exercise should indicate how necessary it is to have much firmer estimates of body size to arrive at the true ratios. One might ask at this point, what is the value of such ratios? As I see them, their main use is in giving us estimates of changes with time, which might be fitted into a framework which attempts to synthesize growth rates and durations with ecological, dietary, and behavioral efficiency, and thereby give us added insights to possible dynamics involved in hominid evolution and the adaptive relationship between different contemporaneous taxa within the hominid family.

It seems likely, considering the large cranium of LLK, and those of the Javanese specimens, that body sizes greatly increased during this period, and that part of the increase in cranial capacity must be associated with this increase. There is no way at present to know, in fact, whether there was any significant increase in cranial capacity not associated with increase in body size from *Australopithecus* to *Homo erectus*. We are in great need of surer absolute dates and better indications of body size than we have at the present time. This suggests that the brain had undergone reorganization prior to and during the early australopithecines which did not involve mass as much as it did the relationships between neural units, and that subsequent brain enlargement might be related to increasing body size up to the *Homo erectus* period.

Evolution of Hominid Behavior Based on Brain Sizes

My own view, based on these considerations and those of the increasing complexity of stone tools and greater dependence on hunting, is that natural selection was favoring extended dependency times, i.e., prolonged growth, leading to both hypertrophy and hyperplasia of neural elements, resulting in stronger dependency and thereby cooperative ties, with concomitant increase in cultural learning and adaptability. The features I discussed in my 1967 paper on positive feedback and "initial kick" were surely underway, resulting in increased complexity-management. In other words, the significant variable that ties together expansion of cranial capacity and cultural complexity is a change in social behavior, the main effect of which is prolonged dependency and growth. I fully believe that the most significant neural changes occurring during this period are those leading to increased memory functioning and changes in the nature of set and attention variables. Krantz's (1968) suggestions regarding memory and persistence hunting are easily accommodated in this framework, which I will detail in a succeeding section.

This leaves a long anterior time period, perhaps going back to *Ramapithecus,* during which the "initial kick" is getting started. In other words, I believe that whatever we define as specifically human (e.g. symbolizing behavior, decreased dental and skeletal dimorphism, increased epigamic dimorphism, close familial ties with a cooperative economic division of labor and sharing of foods, incipient and developing bipedalism, possibly associated with ventral-ventral copulation and signifying a social behavioral change to reduced aggressiveness and more permanent male-female associations and permanent sexual receptivity of the female, decreased intragroup aggression, and beginning stone tool cultures based on standardized rules)—these are developing during the pre- and actual australopithecine phase, extending into *Homo habilis.* The effect of these changes in species-specific behavioral patterns is to set the adaptive stage for the effective cultural programming of offspring to utilize a cultural adaptation, and to set in motion a feedback relationship between increased cultural efficiency, prolonged dependency and postnatal learning and growth, and increase in cortical reorganization manifested outwardly as a great increase in cranial capacity. Increasing body size attends these changes, reflecting new growth rates and durations, and an increase in utilization efficiency of ecological niches.

Other Evidence from the Endocasts

Another reason for the above view regarding dynamics of hominid evolution is based on my assessment of that infamous sulcal landmark, the lunate sulcus. Schepers' (1946, 1950) publications dealing with sulcal and gyral markings have not received an enthusiastic welcome from most paleoanthropologists. I think that this is unfortunate, for however enthusiastic his work, his assessment of those casts as basically human may well be correct. I must admit my own inability to feel secure about the identification of all the sulci and gyri which Schepers was able to identify; and even if correct, I do not believe it is possible to extrapolate to behavioral qualities to the extent that he did. I am sure that Le Gros Clark's (1947) careful assessment of these patterns cannot be reproached, and that our knowledge of the actual brain morphology will remain forever in doubt. Nevertheless, the lunate sulcus, at least on the Taung specimen, can safely be attributed to its posterior position as originally claimed by Dart (1926) and corroborated by Schepers (*op cit.*). As Le Gros Clark noted, no other interpretation makes any sense, since a more anterior placement must necessarily carry the lunate to a position forward to that known for extant pongids.

The minimal interpretation of this fact is that by the time of the Taung child, the hominid brain was already reorganized in a human direction, regardless of the chimpanzee-like size, and that the expansion of superior and inferior parietal cerebral cortex, with attending thalamic (pulvinar) increase, had already taken place. The STS 60 and the type 2 (crushed) endocasts showing fracture depressions strongly suggest a greater degree of cortical fissuration than found on chimpanzee and gorilla (although this an admittedly subjective appraisal which I hope to quantify at a later date). The Hom. #5 and SK 1585 endocasts both indicate an expanded and elevated parietal breadth and height, showing a distinctly different pattern than one finds on pongid endocasts. Similarly, the cerebelli show morphological advances in shape and relationship to the overlying cortex, although not necessarily in absolute or relative size. We already know from the remaining skeletal evidence of the dentition, skull, pelves, and limb bone fragments that these hominids are not pongids. Thus the human neuroanatomical form had already been established (albeit not completed) regardless of cranial capacity by australopithecine times. This evidence is clearly concordant with the views on dynamics suggested above, although it does not yet offer any elaboration on specifics.

The question has been raised several times by colleagues as to whether the endocranial casts of the fossils give any indication of differential enlargement of particular lobes (frontal, parietal, temporal) or regions of the cerebral cortex, or the cerebellum. As I have reviewed this question elsewhere (Holloway 1964, 1968), a full study of this question is not likely to be profitable. First, aside from the well-known diminution of occipital primary visual cortex, as anteriorally bounded by the lunate sulcus, and the concomitant expansion of posterior parietal and temporal cortex, the literature suggests that the frontal lobe shows no proportional increase from between apes, microcephalic humans, fossil hominids, and modern man. Second, there are serious methodological problems associated with delineating the appropriate boundaries of such arbitrary divisions on the endocasts. Associated with this problem is an additional one of variability. That is, even assuming an accurate delineation of boundaries, there are so few specimens that the information obtained would be almost meaningless. This is particularly true of the cerebellum, which is very similar within living species of the Hominoidea, in which one can easily separate the cerebellar hemispheres from the overlying cortex. The dorsal surface of the cerebellum can only be guessed on the endocasts. Finally, would such an effort really provide any useful additional information beyond what one can see through visual inspection? The important point is that the cortex of the australopithecine brain was apparently organized toward the modern human condition; knowing the precise amount of increase in relative size of temporal or parietal cortex would not add anything essential to our understanding of this fact. (This author does plan to make such attempts in the near future, but is dubious about the value of such studies.)

The Expansion of Parietal Cortex and Language

Both Geschwind (1965, 1964) and Lancaster (1968) have offered some very simplistic accounts of the relationship of language behavior to cortical areas. Language behavior, as Lenneberg (1967) notes, involves a complex cognitive and neural reorganization that cannot be explained away either by "object-naming" or an additional increment of inferior parietal cortex. Lancaster's suggestion that language ("object-naming") would be an advantageous adaptation is simply banal, without relating the cognitive properties of such behavior within a matrix of social behaviorial evolution. Similarly, Geschwind's theories of non-limbic to non-limbic associations does not appear to be wholly supported by either clinical or neuroanatomical evidence. This does not detract from his position that infraparietal cortex has an important role in language behavior, but it is likely to be important to more general cognitive behavioral abilities that underly language and involve subcortical nuclei and fiber systems as well, particularly the pulvinar.

To return to the hominid endocasts, I have shown elsewhere (Holloway 1964, 1968) that "rubicon" models will not suffice to explain the emergence of language, and Lenneberg's (1964, 1967) data supports such a position. It is not possible to look at the endocasts and say, "aha, here is the evidence that these hominids possessed language because the parietal area is expanded." Even if the parietal area is expanded, as well as the inferior frontal convolution making up part of Broca's area, and the posterior temporal region, a part of Wernicke's area, this cannot prove the presence of a behavior as complex as language. These patterns can only show that there is nothing in the brain morphology of the australopithecines incompatible with such a behavioral ability. This is obviously quite a different kind of statement, and places the burden of proof on a future understanding of brain anatomy and language, and an elucidation of the behavioral significance of other kinds of non-neural analyses, such as the appearance of stone tools made to a standard pattern. My own position on this latter matter is clear (Holloway 1969b). I believe any hominid capable of making stone tools to such a pattern was capable of imposing arbitrary form upon the environment, and thus was capable of producing language by arbitrary symbols, regardless how primitive such behavior might have been, linguistically.

Recently, Geschwind et al (1968) have found neuroanatomical evidence for a larger development of cortex on the left side of the brain, a fact which associates well with current notions about handedness and laterality of language capabilities, although the matter is far more complex than simply attributing speech to one side. Alas, the endocasts are not of much help in this respect, unless the interesting fact that most of the endocasts are of the right side of the brain indicates that our australopithecine ancestors preferred to lay down and die on that side! Nor have I seen any statistical analyses of early hominid stone tools which can be taken as an indication of laterality in use.

In sum, there is nothing in the brain morphology of the endocasts which denies these abilities, or which militates against the kind of hand-eye co-ordination which could produce stone tools. This all means, as I have stressed elsewhere, that the brain was hardly the last organ to evolve, and was clearly a focus for important and stage-setting selective pressures early in hominid evolution, associated with changes in locomotion, manipulative abilities, food-getting activities, and most importantly, social behavior.

A Speculative Model of Early Hominid Evolution

I would like to present a three-stage model, continuous in nature but also a progressively differentiating one, based on the endocasts, skeletal and dental evidence, associated faunal remains and cultural artifacts, and logical relationships expected from comparative physiological and psychological processes. The model is built up from many sources too numerous to list here, such as Hockett and Ascher (1964), Hallowell (1961), Etkin (1954, 1963), Holloway (1967), Rensch (1959), Caspari (1961), and Maruyama (1963).

If the *Ramapithecus* fragments are indeed hominid, one can only suggest, as have Pilbeam and Simons (1965) and particularly Jolly (1970), that a different adaptation based on a savannah environment, utilizing seeds, grass, and other vegetable foods, was leading to a new evolutionary adaptation with bipedalism under strong positive selection. I do not think we can go beyond this without additional materials. Consequently, my proposed model starts *after* this *Ramapithecus* level of adaptation.

Stage 1. End of *Ramapithecus* and Early Australopithecine Phase, with major emphasis on social behavioral adaptation, involving bipedalism and endocrine organization.

Stage 2. Late Australopithecine–Habiline phase, with major emphasis on consolidation and refinements of Stage 1.

Stage 3. Late Habiline–early *Homo erectus* to Neandertal-Sapiens phase, with emphasis of elaboration of cultural skills in a positive-feedback relationship with brain enlargement.

The essential features of each stage follow in abstracted form.

STAGE 1

Incipient development of a more cooperative, sex-role-separated social group, based on endocrine changes involving hormones and target-tissues, with selection for a reduction of sexual dimorphism in dental and skeletal size, and an increase in epigamic features of such secondary sexual characteristics as permanent breasts, fat distribution, and other possible changes toward facilitating more permanent sexual receptivity of the female. These developments are seen as a parcel of correlated anatomical, physiological, and behavioral changes leading to more prolonged sexual and social control between the sexes, in turn associated with increasing postnatal dependence of the offspring on the mother, and increased growth and learning time. In mind are changes in the organization of interactions between hormones and target-tissues, such that reduction of aggressive components of behavior and sexual dimorphism are linked with increased growth and delay of maturation of skeletal development. Associated with this complex is the incipient development of symbolic language abilities, and incipient hunting and/or scavenging combined with bipedal mobility to secure protein sources, plant foods, and water. The same endocrine changes leading to the dimorphic features cited above may have played an important role in decreasing intragroup aggression, thereby permitting groups to live more densely, where cooperative behavior among males and between males and females would be a strong protection against predators, and other hominid groups of the same or other species. The stress on development of secondary sexual characteristics such as permanently enlarged breasts is for two reasons. First, they had an important attraction function in intensifying social and sexual behavior involving foreplay and the development of ventral-ventral copulation as normative. Second, the female permanently-enlarged breast is both an erogenous zone and one that provides warmth and security to offspring and satisfaction to a mother.

The enlargement of the permanent human female breast might be seen in the following light: that is, as a method of increasing the surface area per body size of a ventral zone important to the warmth and security of the infant, an increase in the surface area for erogenous foreplay between female and male, an enhancement of satisfaction to a nursing mother, and an increase in an epigamic feature of sexual dimorphism with a signaling function associated with permanent sexual receptivity of the human female.

I regard the development of language abilities more bound up with social affect and control than with hunting behavior involving signaling, although this does not mean that such a development would not have had strong positive selection favoring it. In addition to these reorganizations of the social-behavioral and bipedal adaptations, the brain became reorganized, minimally involving a decrease in primary visual cortex on the convex cerebral surface, and an increase in parietal and temporal association cortex allowing for greater discrimination among complex cues of the environment, and the extension of foresight and memory to more effectively cope with savannah-type environment. Associated is the beginning manufacture of stone tools to extend an efficient economic base. This may have involved the breaking of bones to secure marrow, detach pieces of flesh or skin, and to drive off carnivores from their kills by using the stones as missiles.

STAGE 2

Refinement and elaboration of social behavioral changes discussed under Stage 1, with increasing dependence on social cohesion, language, and stone tools for adaptation. Bipedal locomotion is now essentially of a fully human type. Expansion of the brain both relatively and absolutely, the major increment of which is associated with increased body size. Fuller efficiency of economic sharing and complementation between the sexes, providing the necessary basis for increased postnatal dependency, learning, and the beginning of a feedback system between better nurtured brains and cultural behavior, mediated through increased time of postnatal dependency and stimulation. Language behavior is more strongly developed, and there exists cognitive behavior of a more fully human type, where language and tool-making arise from the same psychological structuring. There are true stone tool cultures at this stage, and language has prime importance both in maintaining social cohesion, control, and the "programming" of offspring. Dependence on hunting increases, and there is more success with stalking and hunting of larger game. Increased body size, bipedal agility, predictive abilities, and associative skills are selected for as essential to more successful hunting. The social behavioral changes outlined in Stage 1 permit longer male-male association for persistent hunting, and the protection of secure home-base for females and young, who are also providing small game and vegetable foods. The "initial kick" or "human revolution" is fully set, and brings in Stage 3.

STAGE 3

Positive feedback between development of the brain and cultural complexity mediated through a continuing trend to increase dependency time and postnatal learning of cultural behavior in a more complex and stimulating material and social environment. The major neural changes are those of size and refinement of the basic human reorganization developed in stages 1 and 2, (that is, sensorimotor, associative, extrapyramidal modulation, and cerebellar involvement in manual skills. Not a stage of behavioral innovation, but elaboration of "complexity-management," that is, fineness of sensory discrimination and associations between larger sets of past memories and skills (Holloway 1967).

Comments

It must be emphasized that these stages are seen as gradual and continuous, but certain developments are stressed more strongly in one stage than another. The major point of this framework is to insist that social behavioral mechanisms were really the initial kick, and have had a long evolutionary development in early hominids. In a sense, the brain expansion changes are minor compared to the underlying social matrix evolving to the point where the brain expansion rests on a solid behavioral foundation. This framework also provides a base for synthesis of anatomical, behavioral (social and individual), physiological, adaptational, and ecological variables, and fits both the skeletal and cultural remains.

It is possible, I believe, to tie in more molecular analyses within this model. At the level of neuroanatomy, one can offer many brain regions which can be synthesized with behavioral attributes such as set and attention, concentration, "memory" (permanence, quantity, facility and strategy of recall), hand-eye-and-running coordination, mother-infant affect, babbling and reticular core reorganization, play, curiosity, prolongation of prepubertal vividness of experience and memory. To do so, however, is far beyond the limits of this paper.

Naturally, it must be understood that the analysis of endocranial casts alone cannot play more than a limited role in elaborating these hypotheses, or in supporting some of these speculations. The external morphology of endocasts only provides clues about past selection pressures, not proof, and those clues are fairly gross or molar. The judicious use of endocasts, both as clues to neural reorganization and to changes of growth variables which suggest more molecular views of the interplay of selection forces, physiological process, and adaptations, must await further discoveries with firm chronological dates. I hope that this contribution has suggested a number of ways in which the present data might be utilized to focus our attention on interconnected sets of processes that led to our present evolutionary position and condition. While further studies of the australopithecine endocasts are in progress, it should be apparent that these specimens have potential use, both as clues to general evolutionary events in hominid evolution, and as morphological patterns for taxonomic purposes. The analysis given thus far of the australopithecine endocasts shows, I believe, that the evolution of the brain has always been an important part of hominid evolution, and not something that took place following other changes in different morphological sectors of the hominids. The concept of mosaic evolution is surely a useful one in understanding hominid evolution, but it should be used with some fuller understanding, so that gross application of the concept does not mask important interacting dynamics of various selection pressures during our evolution.

IV

Postcranial Morphology

9

OWEN J. LEWIS
ST. BARTHOLOMEW'S HOSPITAL MEDICAL COLLEGE

Evolution of the Hominoid Wrist

It is now widely accepted that one of the most fundamental characteristics of the primitive arboreal mammal—prosimian or marsupial—is a prehensile lower limb (Lewis 1964a). Characteristic features of this dominant grasping and propulsive member are a divergent hallux and some capacity to rotate the distal segment of the limb so that the modified plantar surface may grasp in varying attitudes. This latter function is usually achieved by inversion and eversion—properties of the subtalar joint complex. In *Galago* and *Tarsius,* however, elongation of the navicular and calcaneus has produced a functional complex analogous to the pronating-supinating forearm bones (Hall-Craggs 1966); and in marsupials a meniscus at the lateral side of the ankle joint facilitates considerable rotatory movement.

Apparently independently in a number of primate lines the forelimb has taken on a more important role and a divergent, grasping pollex has been elaborated. In arboreal quadrupeds little more is required: the wrist joint is well adapted for weight bearing with the ulna transmitting weight directly to triquetral and heel-like pisiform (Figure 9–1A) but nevertheless accommodating pronation-supination of about 90°. However, when the forelimb assumes the dominant role in suspensory locomotion, wrist joint structure is apparently subjected to new evolutionary pressures.

The Wrist Joint in Extant Apes

All living Pongidae possess fully elaborated synovial inferior radio-ulnar joints incorporating a neomorphic ulnar head and with the inferior capsule modified to form the descriptive triangular articular disk. This diarthrosis arises at the site primitively occupied by a syndesmosis; other Primates may show a varying and imperfect synovial modification in the interior of this ligamentous union.

It has also been recognized for the past century that the Hominoidea differ

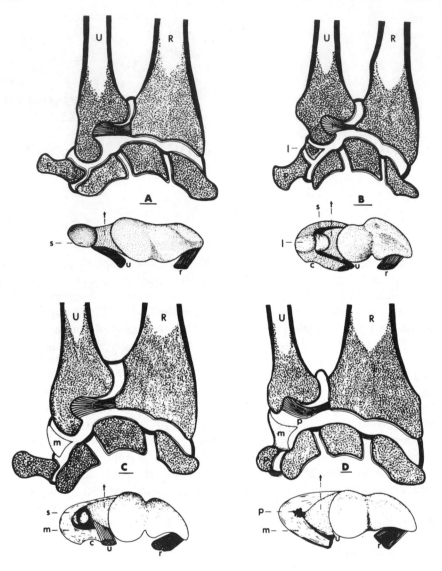

Figure 9–1. The right wrist of: A, *Cercopithecus nictitans;* B, *Hylobates lar leuciscus;* C, *Pan troglodytes;* D, *Homo sapiens.* In each case the upper diagram represents an idealized frontal section of the articulation. Features that, due to the concavity of the carpus, do not occupy one plane are nevertheless included here; thus the pisiform is represented together with the three bones of the proximal carpal row—triquetral, lunate, and scaphoid, from left to right in each case. The lower diagram is a view of the proximal articular surface of the same articulation, with its varying form and components.

The following features are shown: radius (R); ulna (U); pisiform (P); triangular articular disk (t); meniscus (m); intrameniscal lunula (l); palmar from other Primates in withdrawal of the ulna from its primitive articulation with the triquetral and pisiform bones, the original carpal extremity becoming the so-called styloid process. Only recently (Lewis 1965, 1969) has it become apparent that striking and far-reaching anatomical modifications are correlated with progressive exclusion of the ulna from the wrist joint. Certain comparative observations furnished the clue: it is well known that in gibbons a new carpal element—the bony os Daubentonii—occupies the interval created by retreat of the ulna and that at a comparable site in the marsupial ankle lies the so-called "os intermedium tarsi." The realization that the latter bone is no more than a lunula (a sesamoid-like ossification within the ankle meniscus noted above) has demolished a popular scheme of tarsal homologies (Lewis 1964b) but has also prompted the notion that the os Daubentonii might likewise be a lunula. This presupposes the presence of a meniscus within the gibbon wrist, and this has now been confirmed. The elaboration of this evolutionary novelty can be shown to have played a key role in the evolution of the hominoid wrist.

In *Hylobates lar* (Figure 9–1B) the meniscus (with a radially directed, concave, free margin) partially intervenes between the cartilage-clothed ulnar styloid process and the triquetral adjacent to its pisiform articulation. At the anterior margin of the meniscus, the pisotriquetral cavity freely communicates with the radiocarpal cavity, and also extends proximally as a short blind diverticulum related to the periphery of the meniscus; in effect, therefore, the pisiform articulates largely with the meniscus. The meniscus, with horns attaching posteriorly to the radius and anteriorly to the lunate, effectively excludes the ulnar styloid process from contact with the pisiform but not entirely from the triquetral. An intrameniscal bony lunula—the os Daubentonii—is apparently constantly present in adult specimens.

Pan troglodytes (Figure 9–1C) presents a similar meniscus which, however, lacks a lunula. Again the pisotriquetral cavity communicates with the main radiocarpal cavity at the indented anterior border of the meniscus. In some specimens, the cartilage-clothed ulnar styloid process may still contact the triquetral through the aperture bounded by the meniscus. In others, this aperture may be constricted, thus more completely excluding the ulnar styloid process from the wrist joint cavity within its own proximal synovial compartment. In such cases the meniscus effectively becomes the deeply concave ulnar portion of the proximal articular surface by merging with the triangular articular disk. A small aperture located at the apex of the triangular disk then leads into a proximal diverticulum of the wrist joint lodging the still articular ulnar styloid process.

---

ulnocarpal ligament (u); palmar radiocarpal ligament (r); opening into prestyloid recess (p); site of communication between pisotriquetral and wrist joint cavities (c); ulnar styloid process (s). (From Lewis, Hamshere and Bucknill, 1970.)

In *Gorilla,* isolation of the ulnar styloid process from the wrist joint is even more complete. Incorporation of the meniscus with the triangular articular disk into a smoothly concave articular surface is well established and the cartilage-clad ulnar styloid process is totally excluded from direct participation in the radiocarpal joint, being lodged in its own capacious synovial cavity communicating with the wrist joint by a small irregular opening. This arrangement has been observed in *Gorilla gorilla gorilla* and in *Gorilla gorilla beringei.* A modified pattern has, however, been observed in an example of the latter subspecies where the proximal synovial cavity was found to have become completely sealed off from the wrist joint proper.

Quite comparable arrangements have been noted in a specimen of *Pongo pygmaeus* (Lewis 1971), which possessed a similar closed synovial cavity investing a cartilage-covered ulnar styloid process, and demarcating the upper surface of what was clearly the homologue of the typical hominoid meniscus. In other specimens, the proximal synovial cavity may be obliterated leaving a nonarticular ulnar styloid process embedded in a mass of fibrous tissue representing the meniscus homologue; fusion of this tissue to the triquetral, sealing off the pisotriquetral articulation, may further obscure its true identity as a derivative of a primitively free intraarticular meniscus.

In all these hominoid genera, the essential mammalian ligamentous apparatus (consisting of thick palmar intracapsular ligaments converging from radius and ulna to the lunate) is conserved, but with certain modifications: the palmar ulnocarpal ligament is merged with the triangular articular disk and the massive palmar radiocarpal ligament is bifascicular, the divergent bands attaching distally to the lunate and capitate. In all genera the radial and ulnar collateral ligaments are mere indefinite capsular thickenings, scarcely worthy of special note, but a rather more substantial dorsal radiotriquetral ligament is invariably present.

Function in the Hominoid Wrist Joint

Such a major reconstruction of the ape wrist must be a response to changed functional requirements. There can be no doubt that retreat of the ulna from the carpus is an essential prerequisite for an increased range of supination, for if the radius, carrying with it the whole carpus, is to rotate about the ulnar head, the eccentrically situated styloid process must be freed from its restricting articulation with the triquetral and pisiform. Interposition of a meniscus effects this. Accurate biomechanical data are lacking, but it appears that monkeys are capable of only about 90° of pronation-supination, compared with about 180° in pongids.

Avis (1962), in a most important study, has critically analyzed the essential features of brachiation. She has shown that the arboreal activities of all pongids include certain fundamental features. All the Pongidae can indulge in arm-swinging in which the body swings forward, suspended from one hand, rotating as it goes, towards a grip with the other hand; this rotation is achieved by supination of the grasping limb. Avis has suggested that the term "brachiation" should be restricted to such activity. A strictly comparable locomotor pattern is never seen among monkeys, even among those most favoring suspensory locomotion. Of course, apes demonstrate a wide range of other arboreal locomotor activities, and brachiation as above defined might best be thought of as an indicator of versatility in suspensory locomotion which also finds its expression in these other activities. This newly acquired capacity apparently permitted the invasion of an ecological niche not previously available to Primates—perhaps the thin flexible branches—and was presumably a key factor in the divergence of the hominoid line. There is a case then for believing that improvement of a key morphological component, the wrist joint, opened up a whole new phase of primate evolution.

Paradoxically, Avis demonstrated that the gibbon, supposedly the most accomplished brachiator, habitually uses incomplete bodily rotation of about 90°, but this is coupled with abrupt elbow flexion propelling the animal through the air to a grip with the other hand. Gibbons can, however, use a full 180° turn in slower swinging. Their frequent use of limited rotation may have its structural basis in the less complete exclusion of the ulna from the carpus in this species; although data are lacking, there may possibly be some restriction to free supination in this species.

Apart from radical reconstruction involving the ulnar side of the wrist, certain other changes are characteristic of the so-called brachiators. Some limitation of wrist extension would be a logical consequence of aligning hand and forearm, and this is most manifest apparently in *Pan* and *Gorilla* (Tuttle 1969a). In response to this, all show some degree of volar inclination of the distal radial articular surface. The very thick palmar radiocarpal ligament, part of the primitive mammalian palmigrade heritage, has become bifascicular in all the Pongidae, extending its distal attachment from the lunate to embrace the capitate as well; this subdivision is least apparent in *Hylobates*. This extension spanning the mid-carpal joint may be associated with diminished mobility at that joint, which is perhaps also reflected in the early or late fusion of the pongid centrale to the scaphoid (Jouffroy and Lessertisseur 1960). A potent factor in limitation of wrist extension, however, is the relative shortness of the long flexors. These various modifications coupled with a new orientation of the pisiform (see below, "Osteological Features at the Primate Wrist") must play their part in determining the posture of the hand during terrestrial activities.

Wrist extension is apparently little restricted in at least some gibbons, which may then adopt a palmigrade posture. The wrist joint in *Hylobates* is, of course, still well adapted for weight bearing with a still long ulnar styloid process capable of transmitting weight through the often large os Daubentonii to a pisiform which retains something of a heel-like character.

The fist-walking of orangutans is doubtless consequent upon the shortness of their long flexors, and their occasional adducted fist-walking posture (Tuttle 1967)—with even 90° of wrist adduction—is rendered possible by the extreme withdrawal of the ulnar styloid process from the wrist joint. The knuckle-walking habitus of the African apes seems likewise to result from limitation of wrist joint extension which precludes palmigrade weight bearing. The attitude is, in effect, an expression of the position of rest of the hand; even man, on occasion, uses the hand in this way as a steadying strut. The author has been unable to confirm the suggestions made by Tuttle (1967, 1969b) that there are other special modifications of the wrist associated with knuckle-walking. Thick palmar radiocarpal ligaments are not the exclusive property of knuckle-walkers, nor do these apes have especially strong radial collateral ligaments attaching to a scaphoid tubercle whose length might then be correlated with this activity; this latter notion results from wrong identification of the palmar radiocarpal ligament and confusion of its bony attachments. It thus seems that knuckle-walking requires no especially striking modifications of the wrist joint.

The Human Wrist Joint

It is of obvious phylogenetic significance to consider whether *Homo* has shared similar wrist joint modifications with the Pongidae. Consultation of traditional textbook accounts would tend to give the impression that the construction of the human joint has little in common with those of the Pongidae (except perhaps *Pongo pygmaeus*). But these texts seldom include reference to two significant features, well documented in the literature but apparently usually dismissed as irrelevant minutiae: a radiographic opacity is sometimes (½ to 1 percent) demonstrable in the human wrist at the site occupied by the gibbon os Daubentonii; a synovial cul-de-sac (the prestyloid recess) extends from the wrist joint cavity into relationship with the ulnar styloid process.

Arthrograms show a striking similarity between this prestyloid recess and that compartment of the chimpanzee wrist joint proximal to the meniscus. These two diverticula are similar in size, position, and in the location of their communication with the cavity of the wrist joint proper. Moreover, the arthrograms clearly delineate the meniscus in *Pan* and its homologue in *Homo*—the most ulnar part of the proximal articular surface (Figure 9–1D). This most medial component of the proximal articular surface, contacting the triquetral, has not hitherto been satisfactorily accounted for. The human meniscus homologue is thus well integrated into the proximal articular surface, just as it often is in *Pan* and normally is in *Gorilla*. The joint of *Homo* is distinguished mainly by the commonly more extreme withdrawal of the ulnar styloid process, beyond the confines of the synovial prestyloid recess with consequent loss of articular character. The human joint shows a considerable range of variations (Lewis, Hamshere, and Bucknill 1970), all of which are clearly derived from a basic hominoid type of structure, and which include varieties comparable to those normally found in chimpanzees and gorillas.

The prestyloid recess (and so a meniscus homologue) is a constant feature of the human wrist joint. Rarely, (2 percent) the meniscus may be quite free (Plate 9–1) as it may be in *Pan*. More commonly the meniscus is well merged with the triangular articular disk (Plate 9–2) thus constricting the opening into the recess which may, however, be easily located adjacent to the apex of the triangular disk. The size of the diverticulum varies considerably but always approaches the anterior aspect of the ulnar styloid process, which varies greatly in length. When long, the process protrudes into the recess and is then clothed by articular cartilage (30 percent); in about 8 percent, it is even sufficiently long to contact the triquetral through the recess opening, at least in ulnar deviation. Short styloid processes lie beyond the confines of the synovial recess and are nonarticular. One example of a separate ossicle has been observed in a dissection adjacent to the ulnar styloid process (at the site occupied by the opacities known to radiologists); it was embedded in the tissue of the meniscus homologue and was clearly comparable to the gibbon lunula, the os Daubentonii.

In 34 percent of human wrists, the radiocarpal cavity communicates with the pisotriquetral joint just as in *Hylobates, Pan,* and *Gorilla.* In the remainder, the communication is obliterated by attachment of the meniscus homologue to the triquetral (Figure 9–1D), subdividing the articular surface of that bone.

The only ligaments worthy of mention are quite comparable to those of the Pongidae. Again, the palmar ulnocarpal ligament has lost its separate identity by incorporation with the triangular disk, giving the impression that the anterolateral angle of this structure is attached to the lunate. The palmar radiocarpal ligament is massive, bifascicular, and with the usual hominoid attachments. The dorsal capsule is relatively thin but presents a thickened band—the dorsal radiocarpal ligament joining radius to triquetral.

Development of the Human Wrist Joint

Developmental changes in the human wrist (Lewis 1970) further emphasize affinities with the Pongidae. For a time during embryonic life, the ulnar styloid process contacts both triquetral and pisiform; but by the end of the embryonic period (60 days; 30 mm C.R. length), a considerable mesenchyme filled interval appears between it and the pisiform. During the early part (30–60 mm C.R.) of the fetal period, joint cavities are formed. Three separate cavities soon develop: radioscaphoid and radiolunate cavities separated by a thin mesenchymal septum, and a more widely

separated ulnotriquetral cavity where the cartilaginous anlagen of the
ulnar styloid process and the triquetral articulate. At this stage the radiolunate and ulnotriquetral cavities are separated by a considerable extent of
loose mesenchyme which is situated distal to the condensed tissue of the
developing triangular disk and commonly contains a cartilaginous nodule,
the so-called "intermedium antebrachii." After 60 mm C.R., the ulnar
styloid process retreats from the triquetral and a mass of condensed
mesenchyme, triangular in section, comes to intervene. Subsequent development leaves little doubt that this mass is the human homologue of the
pongid meniscus; when the cartilaginous "intermedium antebrachii" persists
into this stage it lies in the interior of the mass. The ulnar styloid process is
initially embedded in the upper surface of the meniscus homologue whose
lower surface is separated from the triquetral by the persistent ulnocarpal
cavity. As development proceeds, cavitation extends from here into relationship with the styloid process, thus forming the prestyloid recess and
delaminating the meniscus as a free entity, triangular in section, projecting
into the joint. The initially separate radioscaphoid and radiolunate cavities
soon communicate and persistent vestiges of the intervening septum give rise
to the synovial folds (large anterior and small posterior) found here in all
Primates. The ulnotriquetral cavity soon joins this common radiocarpal
cavity by dissolution of the intervening mesenchyme. Persistent partitions
here (complete or incomplete) doubtless represent the primitive condition
and have been observed in marsupials, carnivores, lemurs, platyrrhine
monkeys, and even in a single gibbon specimen.

The appearance of a pisotriquetral cavity is delayed until about 60 mm
C.R. It is situated between the fetal pisiform and the triquetral and meniscus
homologue. It commonly communicates with the wrist joint cavity proper.
The fetal pisiform is located more proximally (overriding the periphery of
the meniscus homologue) than is the case in the adult condition. In later
fetal life, the pisiform is distally displaced and the communication between
radiocarpal and pisotriquetral cavities may be obliterated. The derivation of
the varying adult arrangements is here quite apparent.

In some fetuses, the retreat of the ulna is considerably retarded; variations
in the rate of this process adequately explain the frequent adult variations.

Persistence and ossification of the cartilaginous nodule ("intermedium
antebrachii") almost certainly account for the occasional adult radiographic
opacities, and there can be little doubt that these structures are homologous
with the gibbon lunula.

Thus, in a quite striking way, human development recapitulates the postulated phylogenetic stages, obscured somewhat by altered time relationships
(heterochrony), e.g. the precocious development of the cartilaginous lunula
before clear delineation of a meniscus.

Embryology also provides convincing proof that the only true accessory
ligaments of the wrist joint are the anterior radiocarpal (radius to lunate

and capitate), posterior radiocarpal (radius to triquetral), and the variable palmar ulnocarpal (ulna to lunate) ligament. In even the earliest fetal stages, these ligaments are readily identifiable and distinguishable from the fibrous capsule by their characteristic structure.

Evolution of the Hominoid Wrist Joint

Thus, the human wrist, both in adult structure and in developmental history, bears the unmistakable imprint of a pongid heritage. But the hominoid modification of the wrist, imperfect in *Hylobates* but more advanced in *Pan* and *Gorilla,* may reasonably be supposed to be correlated with the emergence of a new locomotor pattern (see above, "Function in the Hominoid Wrist Joint"). This implies that hominid evolution has included a phase of adept suspensory locomotion characterized by an accomplished ability for rotation of the body from the grasping hand. Brachiation is one facet of such an arboreal repertoire. It is not suggested that early hominoids indulged in a stylized form of gibbon-like arm-swinging as their exclusive, or even customary, mode of locomotion. In fact, the gibbon propensity for completing its bodily rotation by propelling itself through the air perhaps reflects the biomechanical imperfections of an incompletely remodeled wrist joint. Freedom of movement would certainly be restricted in those gibbon specimens having a persistent septum between the primitive ulnar and radial compartments of the joint. The suggestion by Avis (1962) that this new method of locomotion (now seen to be dependent on a major anatomical change) unlocked a new ecological niche, the small flexible branches, is worth serious consideration. The evolutionary novelty of a remodeled wrist joint might then be seen as a key modification leading to the emergence of the Hominoidea and eventually to the Hominidae. There are other well documented examples where the rapid evolution of a new character complex has led to invasion of a fresh adaptive zone (with subsequent emergence of other modifications); this is apparently the common course by which new higher taxonomic categories emerge. One of the best known examples concerns the rapidly evolving changes in the tarsal joints which led to the probable invasion of a plains environment and to the emergence of the order Artiodactyla (Schaeffer 1948). The hominoid wrist joint changes seem to present an evolutionary breakthrough of comparable magnitude.

The living pongids exhibit a progressive sequence of wrist joint specialization, and the probability is that the surviving stages of a true phylogenetic sequence are here represented, rather than varying grades of parallel evolution. Simons (1962b) has maintained that brachiation has evolved in parallel from three to six times, and Napier (1963) and Straus (1964) have speculated that brachiation has evolved independently in gibbons and African anthropoid apes (it must be remembered, however, that these authors were using forelimb elongation as their criterion for brachiation). While experimentation with suspensory locomotion has undoubtedly occurred several times during primate evolution, there is no sound evidence that profound wrist joint modifications have been realized repeatedly also. For instance, this has not occurred in semibrachiating monkeys (Lewis 1971). The appearance of a transient cartilaginous nodule in the human fetal wrist (and its occasional persistence and ossification) at a site comparable to the gibbon os Daubentonii suggests that a gibbon-like stage of organization of this functional complex has been represented in human phylogeny. Parallelism might have been a likelier proposition if embryology had indicated that the essential morphological changes were mere developmental consequences triggered off by retreat of the ulna from the carpus. Whillis' (1940) work on joint development has, after all, indicated that intraarticular disks (and menisci) result from persistence of the embryonic mesenchymal joint disk, when the growth of two bones together is retarded. The facts reported above ("Development of the Human Wrist Joint"), however, show that for a time in development the ulnar styloid process articulates through a joint cavity with the triquetral, and the meniscus is later fashioned anew. The genetic control must here be more complex than in the cases cited by Whillis and, presumably, therefore less likely to be duplicated in evolution.

It has frequently been pointed out that brachiation could have provided an excellent apprenticeship for erect bipedal locomotion. It could now be further suggested that wrist joint modifications, consequent upon brachiation, strikingly preadapt the hands for manipulative, feeding, carrying, and tool-using functions, with all their profound implications for the progressive evolution of the hominid phase of primate evolution.

Osteological Features at the Primate Wrist

For the reasons given above, the anatomical arrangements at the wrist should carry considerable weight in assessing the locomotor habits and taxonomic status of fossil primates. The application of these findings to fossil material necessitates a clear awareness of the osteological features which mirror the joint changes. These bony features are diagrammatically shown in figures 9–2 and 9–3B.

In the Cercopithecoidea and Ceboidea (Figure 9–3B), the triquetral is a large bone in accord with its weight-bearing role. It is rather cuboidal in shape, bears a tuberosity projecting forwards and distally from its medial surface, and is set with moderate obliquity at the medial aspect of the carpal condyle. It bears a large, gently concave facet on its proximal surface; the pisiform articulates at the anterior margin of this facet, forming with it an articular cup for the ulnar styloid process. In the habitually pronated quadrupedal posture, the pisiform projects backwards into the heel of the hand. The ulnar styloid process presents a facet on its carpal aspect (facing the interior of the joint); the opposite peripheral aspect of the process is non-

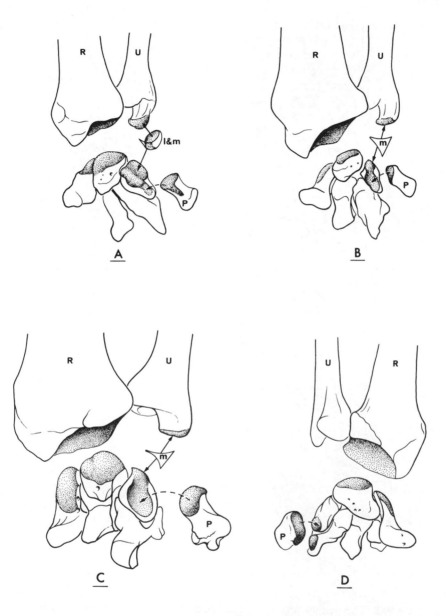

Figure 9–2. The palmar aspects of the bones entering into the wrist joint (with the articular surfaces participating in this joint stippled in each case), together with the capitate and hamate, in: A, *Hylobates lar leuciscus;* B, *Pan troglodytes;* C, *Gorilla gorilla gorilla;* D, *Pongo pygmaeus.* In each case the pisiform (P) has been rolled away from its articulation with the remainder of the carpus. Other lettering is as in Figure 9–1.

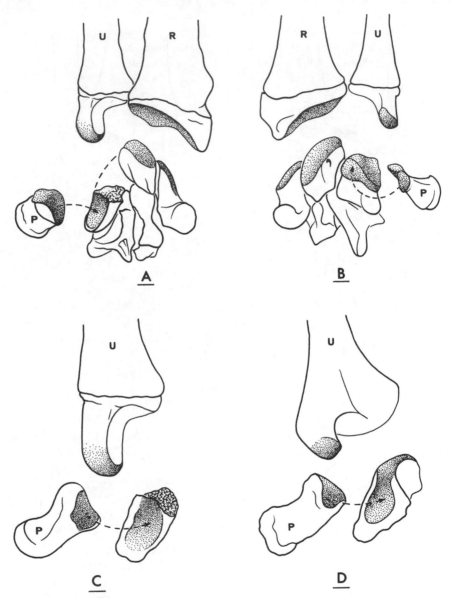

Figure 9–3. The palmar aspects of the bones entering into the wrist joint (with the articular surfaces participating in this joint stippled) together with the capitate and hamate in: A, *Dryopithecus (Proconsul) africanus;* and B, *Cebus nigrivittatus.* The ulna, triquetral fragment and pisiform of *Dryopithecus (Proconsul) africanus* are shown in C; and the corresponding bones of *Pan troglodytes* in D. In each case the pisiform (P) is shown rolled away from its articulation with the triquetral. Other lettering is as in Figure 9–1. In A, the pisiform appears rather foreshortened due to the effect of perspective.

articular. In the weight-bearing position, the styloid process is comparable to the lateral malleolus of the hind limb; this arrangement has been well illustrated for the howler monkey by Grand (1968b).

In *Hylobates* (Figure 9–2A), the triquetral closely resembles that of monkeys and has a large slightly concave proximal articular surface contacting the lunula-containing meniscus. The upper aspect of the latter plays upon the distal and peripheral aspects of the somewhat hook-shaped ulnar styloid process; compared to monkeys, the articular surface here has shown a shift from the surface of the process directed towards the interior of the joint, where primitively it contacted the pisiform and triquetral. The pisiform is more distally directed than that of monkeys and bears an inconspicuous concave facet on its dorsal surface for the triquetral; the flattened proximal extremity of the pisiform abuts against the periphery of the meniscus with a synovial pocket of the pisotriquetral joint intervening. The two articular surfaces on the triquetral are, of course, in continuity.

In *Pan* (Figure 9–2B), the triquetral is more distally directed, lying along the ulnar margin of the carpal condyle. The pisiform is strikingly reorientated into the palm and its distal extremity is firmly tethered to the hook of the hamate by the pisohamate ligament. It bears a greatly enlarged convex facet found on its dorsal surface for the triquetral; in contrast the facet found at its proximal extremity is much reduced and articulates with the periphery of the meniscus at the site where the radiocarpal and pisotriquetral joints communicate. The shallow concave facet for the pisiform on the anterior aspect of the triquetral is confluent with a convex facet on the summit of the bone for the meniscus, reflecting this communication. The ulnar styloid process is usually quite large, often rather flattened at its extremity, and may show features somewhat reminiscent of the hook-like character noted in *Hylobates*. The process bears an articular surface on its distal and peripheral aspects where it contacts the meniscus.

Essentially similar bony arrangements exist in *Gorilla* (Figure 9–2C), although the blunt truncated ulnar styloid process articulating with the meniscus homologue is usually quite short.

In principle, the arrangements in *Homo* differ little from those in *Pan* and *Gorilla*. The pisiform is, of course, greatly foreshortened and bears a concave facet for the triquetral. This articulation has usually undergone a greater distal migration than in *Pan* or *Gorilla*, and a nonarticular area commonly separates the pisiform and meniscus facets on the triquetral. As noted above ("The Human Wrist Joint"), the pisotriquetral joint is usually sealed off in man by attachment of the periphery of the meniscus homologue to the triquetral.

The most aberrant derivative of the basic hominoid pattern is seen in *Pongo* (Figure 9–2D). Distal migration of the reduced pisiform has here progressed to the point where it may articulate, not only with the very apex of the triquetral, but also with the hook of the hamate. The very short, conical, ulnar styloid process is also far withdrawn from the joint. The triquetral bears no meniscus facet.

In essence, the following osteological trends are shown by the living Hominoidea. First, where the ulnar styloid process bears an articular facet it has migrated from the surface directed towards the interior of the joint (where in monkeys it articulates with pisiform and triquetral) onto the external surface of the process (where the meniscus is interposed into the joint); this leaves a large nonarticular area on the process for attachment of the triangular articular disk. These changes, in the early stages, give the ulnar styloid process a rather hook-like form, as seen in *Hylobates* and some specimens of *Pan*. A similar appearance is also strikingly apparent during fetal life in *Homo sapiens*. Second, the triquetral becomes reduced in size and obliquely orientated. Apparently associated with this is a distal migration of the pisiform which is also directed markedly distally into the palm. In *Hylobates* a large facet is present on the proximal extremity of the bone for articulation with the periphery of the meniscus; this facet is much reduced in *Pan* and *Gorilla,* while the triquetral facet on its dorsal surface is greatly enlarged. No facet for the meniscus is found in *Pongo* and *Homo,* but during human fetal development the pisiform is more proximally located and largely articulates with the periphery of the meniscus.

The characteristic hominoid angulation and migration of the pisiform distally into the palm (only incipient in *Hylobates*) is in strong contrast to its disposition in monkeys. By virtue of its strong ligamentous attachment to the hook of the hamate, it thus comes to form a splint spanning the volar aspect of the midcarpal joint and presumably contributing to the limitation of extension here.

The Wrist of Dryopithecus (Proconsul) africanus

Currently there is a widely held view that the origin of the hominid line is to be found among the Dryopithecinae of the early or middle Miocene. The presumptive ancestral form has been dubbed "a dental ape"—a primate combining a hominoid dentition with limbs of quadrupedal monkey-like morphology perhaps not unlike those of living semibrachiating forms. The discovery of the almost intact forelimb bones of *Dryopithecus* (*Proconsul*) *africanus* and their description by Napier and Davis (1959) has formed the keystone of this hypothesis. A forelimb such as that of *Dryopithecus* (*Proconsul*) *africanus,* said to preserve a quadrupedal type of wrist articulation (the ulna articulating with pisiform and triquetral), but showing incipient brachiating features in other parts, is suggested as a reasonable prototype for that of terrestrial hominids and of arboreal pongids. It must be remembered that the description of these fossil remains, now apparently universally accepted, was carried out without appreciation of the uniquely hominoid changes in the wrist articulation.

There seem to be considerable difficulties implicit in this view of human evolution. The studies reviewed here demonstrate that a uniquely modified type of wrist joint is shared by the Pongidae and Hominidae but is quite unlike the monkey type of morphology suggested for *Dryopithecus* (*Proconsul*) *africanus*. Parallel and similar modifications must then be postulated in the two families, one arboreal (at least primitively) and the other terrestrial; but if suspensory locomotion provides the effective selective pressure, this seems unlikely. If the notion of parallel evolution is rejected, then the origin of bipedal man must be sought at a later stage from a primate more advanced in the evolution of the brachiating habit. Alternatively, the fundamental proposition that the architecture of the wrist joint region in *Dryopithecus* (*Proconsul*) *africanus* is essentially that of an arboreal quadruped could be in error.

This later possibility has been tested by a reexamination of the casts of the relevant forelimb bones of *Proconsul africanus* held in the British Museum (Natural History). A brief summary of the conclusions will be presented here; a full report with supporting evidence will be published elsewhere.

The casts of the individual bones are shown articulated in Figure 9–3A. In the light of the osteological data described above ("Osteological Features at the Primate Wrist"), certain features appear to be noteworthy. The ulnar styloid process has a somewhat hook-like form and is clearly articular on its distal and peripheral aspects, not on the carpal aspect, facing the interior of the joint as in monkeys (Figure 9–3B). This is at least suggestive that the joint possessed a meniscus; it is not conclusive, however, for this appearance can occasionally be mimicked in the macerated limbs of monkeys which were known to have facets in the usual interior situation in the fresh state.

The disposition of the pisiform and triquetral is much less equivocal and clearly seems to indicate a quite advanced hominoid type of organization. The reduced obliquely-lying triquetral with its large concave anterior facet for the pisiform is quite unlike that of monkeys (Figure 9–3B). Similarly, the pisiform has all the hallmarks of a quite advanced type of hominoid organization (figures 9–3C and 9–3D). The small proximal facet on the pisiform (interpreted by Napier and Davis 1959 as being for the ulnar styloid) is quite comparable to that seen in *Pan* and *Gorilla* for articulation with the periphery of the meniscus. The proximal fragment of the triquetral is lacking but it is suggested that it would probably have possessed a smooth facet for the meniscus, as in the African anthropoid apes. All this suggests a wrist joint organization more advanced along hominoid lines than that of *Hylobates* and approaching in some respects the structure seen in *Pan*. These suggestions go far towards resolving the paradoxical idea (proposed by Napier and Davis 1959) of a primate possessing a monkey-like wrist joint interposed between elbow and hand regions not unlike those of *Pan*.

Thus it seems that by the middle Miocene at the latest, there were already present Primates possessing a key locomotor specialization unique to the Hominoidea and which had opened up to them a new arboreal habitat. The possibility that locomotor changes, accompanying the invasion of a new habitat, preceded dental modifications must be considered. Further, by the middle Miocene, apes had evolved to a stage where they were preadapted by brachiation for a terrestrial life and for efficient use of the forelimbs in manipulation; such apes could have been ancestral to *Ramapithecus punjabicus* of the late Miocene or perhaps early Pliocene, which has been considered to be an early hominid (Simons and Pilbeam 1965).

10

FRIDERUN ANKEL
YALE UNIVERSITY

Vertebral Morphology of Fossil and Extant Primates

How closely related are form and function in animals? The answer to this question is critical for understanding the biology of contemporary animals and for the interpretation of fossil remains. A considerable literature is available on teeth, skulls, and limb bones. The results of dental studies—far-ranging over many years—support numerous taxonomic conclusions. Functional studies on the anatomy of limbs can rely to some degree upon descriptions of free-ranging animals. In many cases, locomotor categories can be applied easily to limb anatomy. Limb bone fragments often are preserved both quantitatively and qualitatively. All this information on correspondences represents a tremendous base for the interpretation of fossils.

In addition to limb anatomy, one must also know what can reasonably be deduced from single vertebrae or from fragments of vertebrae. Fossil vertebrae are rare. They often are small, easily lost, and casually disregarded. Thus descriptions of them are even more rare than the specimens. It is not known how many specimens of the primate vertebral column are hidden in boxes and cabinets of museum collections in Africa, Europe, and America. What functional and phylogenetic interpretations do they contain? What interpretations can be based on them? In the evaluation of fossil remains, what can be deduced from the entire column and from individual segments? By attempting to answer these questions I regard my research to be a contribution to evolutionary primatology. This paper is also an essay on the functional anatomy of an interesting and complicated region of the animal body.

General Morphology

The vertebral column—the structural core of the vertebrate body—is composed of numerous linked segments that share several common features:

Dr. Th. Grand of the Oregon Regional Primate Research Center corrected the manuscript and made helpful suggestions. I thank him for his assistance.

1. The body or centrum, which has weight-bearing, force-transmitting functions.
2. The neural arch, which protects the spinal cord dorsally.
3. Various spines and projections.
4. Several joint surfaces by which each vertebra articulates anteriorly and posteriorly.

The vertebral body and neural arches exhibit processes or apophyses, some of which are paired, and to which muscles, tendons, and ligaments are attached. Articular facets on the neural arch, situated cranially (anteriorly) and caudally (posteriorly) on paired apophyses, are called zygapophyses. Prezygapophyses are directly cranially; postzygapophyes are directed caudally (Figure 10–1).

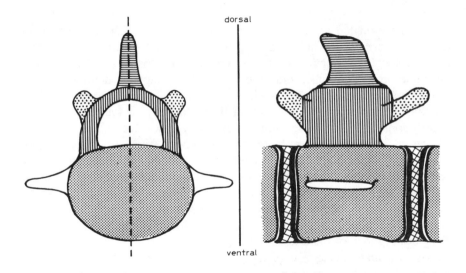

Figure 10–1. Vertebra, schematic, left seen from cranial, right from lateral. (From Ankel 1967.)

Because of their different orientations in each region, the paired pre- and postzygapophyses direct and limit movement. Between adjacent bodies, the intervertebral discs permit limited turning and bending in all directions. Thus, vertebral segments commonly articulate at three points in front and three points behind. The number of articulations is reduced at particular segments in the column: between the first (atlas) and second (axis) cervical vertebrae, within the sacrum, and in the distal tail segments.

From segment to segment and region to region, general morphological characteristics change slowly. Thus, vertebral unit characters and character-istics contribute collectively to distinctive features of a region. The func-

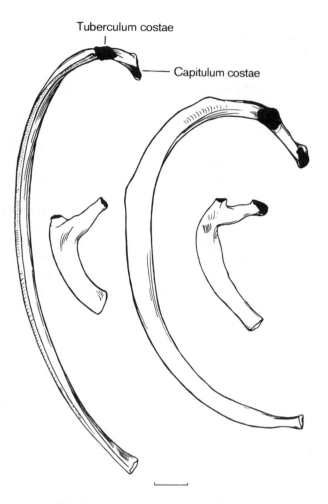

Figure 10–2. Sixth and first rib from *Cercopithecus* and *Hylobates* (from left to right). (From Ankel 1967.)

tional morphology of vertebral pairs summates in action so that one observes movement potential or leverage functions for the whole region. Since the paired articulations along the neural arches do not change at all from one segment to the next, functional movements within each region remain distinct.

These articulations have different positions within the different regions of the vertebral column (Figure 10–3). The paired articular facets, however, are an exception to the slowly changing characteristics; the joints change abruptly between the cervical and thoracic regions on the first thoracic vertebra and on the "anticlinal" vertebra between the thoracic and lumbar regions. At these points, the proximal articulations (prezygapophyses) show the position of the anterior (cervical; thoracic) region; the posterior articulations (postzygapophyses) display the position of the distal adjoining (thoracic; lumbar) region.

Morphology of Different Regions

In addition to the general features mentioned above, each region exhibits certain structural distinctions. Cervical vertebrae are marked by foramina passing craniad to caudad through each transverse process. The thoracic vertebrae bear ribs and display one or two articular facets for each rib. Lumbar vertebrae have prominent transverse processes. Sacral vertebrae fuse and form foramina intersacralia, dorsally and ventrally, within each broadened pair of segments. The proximal tail vertebrae have small neural arches and ventral arches or haemapophyses; distally, the caudal vertebrae consist of vertebral bodies only (Remane 1936, Schultz 1961, Ankel 1967).

Since specializations in primate vertebrae are not marked, it is usually difficult to determine from which genus single vertebrae are derived. Of the seven living prosimian families, only two (Lorisidae and Tarsiidae) are particularly specialized in the thorax and vertebral column. There are minor differences (e.g. in positions of the transverse processes and the craniocaudal length of the lumbar spines) in the lumbar region of Cercopithecoidea and some Ceboidea. Otherwise, the species of Old and New World monkeys are relatively unspecialized. The Hominoidea manifest certain trends which will be discussed later.

In an "unspecialized primate," the average number of segments in each region is 7 cervical, 12 to 13 thoracic, 6 lumbar, 3 sacral, and 25 caudal. In man, we find considerable reduction but reductions are even more evident in *Pan* and the other Pongidae (Table 10–1). By comparison with other primates, in the Hominoidea reduction of the number of vertebrae and the shortening of each segment in the presacral region reduce potential movement and provide a more stable trunk. These reductions in length and number of presacral vertebrae approximate the shoulder and hip girdle more closely in the Hominoidea. But the approximation of the shoulder girdle and hip articulations is partly equalized by the length of the ilia in the Pongidae.

In prosimians and monkeys, vertebral bodies increase in length caudally. In Hominoidea and especially in *Homo,* the diameter of the vertebral bodies likewise increases caudally. In the human sacrum, the lateral articulations with the ilia are also relatively enlarged. This increase in diameter and the expansion of the sacroiliac articulation in man correspond with the increase of weight that accompanies truncal erectness (Schultz 1961). However, presence of the same trend to increase the diameter of vertebral bodies in the Hylobatidae and Pongidae somewhat invalidates this interpretation.

In prosimians and monkeys, the vertebral column functions as a tie-beam

Table 10–1. Numbers of vertebrae*

Genus	Cervical	Thoracic	Lumbar	Sacral	Caudal
Tupaia	7	12–13	6	3	25
Perodicticus	7	15–18	6–7	6–9	7–12
Callithrix	7	13	6–7	3	27
Ateles	7	14	4–5	3	31
Macaca	7	12–13	7	3	17
Pan	7	13	3–4	5–6	3
Homo	7	12	5	5	4

*Averages, from Schultz 1961.

and pressbeam, distributing and absorbing tensile, compressive and shearing forces (Kummer 1959), as manifested in neck lordosis (dorsally concave) and rump kyphosis (dorsally convex). In man, however, the vertebrae, supported from below, function as a true column. This elastic column has a threefold S-shaped curve (neck lordosis, thoracic kyphosis, lumbar lordosis). Because of the mechanical curves, it can adjust to sudden shock, as when running and jumping. Were the column straight and unjointed, damage would surely result.

I will compare now the different regions of the vertebral column and define their specializations.

Morphological Specializations

The first and second cervical vertebrae are unlike any other vertebrae. The first vertebra, the atlas, articulates with the skull. The occipital condyles of the skull are bound to the atlas, and nodding is accomplished at this joint pair. It is more or less ring-shaped and lacks a vertebral body.

Phylogenetically, the first cervical vertebra lost its centrum. The centrum of the atlas fused with the body of the second vertebra, the axis, to form the cone-shaped odontoid process (dens). The odontoid process is elongated cranially into the atlas. Fossil series demonstrate that the odontoid process is homologous with the vertebral body of the atlas. But some authors believe that the odontoid is a phylogenetically new structure in mammals (Meienberg 1962, Jenkins 1969).

The ventral portion of the atlas, a thin bony band, articulates with the odontoid process. On the ventral aspect of the odontoid process lies an articular facet, forming a swivel joint for the head upon the neck. The odontoid process is locked into place by a strong dorsal ligament (ligamentum transversum), which prevents injurious contact between it and the spinal cord. Caudalaterally on each side of the tooth, paired articular facets join the axis to the atlas. Taxonomic differences appear when the orientation and inclination of the odontoid process and its basal articular surface are considered.

In prosimians and monkeys, the odontoid process and its facet are sharply angled. The ventral surface is inclined about 30° dorsally, and the paired lateral articulations angle about 45° caudalaterally (to the sides of the vertebral column). In the Hylobatidae and the Pongidae, both surfaces are less sharply inclined and their angles are reduced. In the Hylobatidae and in *Pan,* the odontoid process tends to be straight; in *Pongo* and *Gorilla* it is more inclined. In man, the odontoid process is straight following the longitudinal axis of the body of the vertebra; the paired articular facets incline 10° or less. Some variability exists in prosimians, monkeys, and apes, but has not yet been detected in man. The inclination of the paired facets varies less, but both characters should be evaluated together.

The paired facets are almost flat in man; in monkeys they are concave/convex. In man the atlas and axis are simply superimposed; in prosimians and monkeys, the atlas encloses more of the cranial portion of the axis. It is also less enclosed in apes than in prosimians and monkeys. Comparisons between man and other primates on the anatomical basis of man's upright posture and the steepness of the cervical region relative to head postures in nonhuman primates should be made with caution.

The articulations between the axis and the third cervical vertebra are typically like those along the rest of the column: intervertebral disc and paired joints along the neural arches. Paired cranial facets face dorsally, paired caudal facets face ventrally (Remane 1936).

These articulations in the cervical region are situated lateral to the vertebral body (Figure 10–3). The centra are saddle-shaped and interlock, thus preventing turning. The paired neural arch articulations permit bending ventrally and laterally and less movement dorsally. Therefore, little motion is possible between the second and third and posterior cervical vertebrae. In *Tarsius,* the paired articulations of the cervical vertebrae do not lie flat against the vertebral bodies as in other primates. They are oriented steeply, in almost the same planes as the neural arches and the vertebral bodies are not as saddle-shaped in *Tarsius* as in other primates; this permits greater

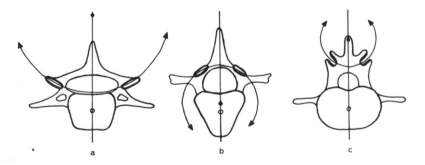

Figure 10–3. Position of zygapophyses within: (A) cervical; (B) thoracic; and (C) lumbar region. (From Ankel 1967.)

turning movements. In man, the spinal processes project caudally and usually the tips are double.

In the lorisoid species *P. potto,* the lower cervical and upper thoracic vertebrae ($C_{2..7}$, $T_{1..3}$) exhibit elongated dorsal spines, protruding beyond the body contour and into the fur between the last two cervical vertebrae and the first two or three thoracic vertebrae. Surrounded by the long hairs of the back, the spines are covered with a thick cornified epithelium. Walker (1970) has clarified the nature and function of these structures. The hairless tubercles of skin covering the elongated spines are sensitive. Pottos in captivity are recorded to present this region to other pottos and humans for tactile stimulation. Although Hofer (1957), Schultz (1961), and Ankel (1967) interpreted these structures to be defensive organs, Hofer (1957) expressed some doubts about this hypothesis. In the Pongidae, the cervical and the first thoracic dorsal spines are elongated, especially in adult male gorillas and orangutans. These spines end in knobs and are embedded in the enlarged dorsal musculature of the neck. The muscles might be correlated with the heavy heads of these pongids (Schultz 1961). Specializations like those in *Perodiciticus* and the Pongidae can reduce or even prevent certain movements.

The thoracic vertebrae in all four genera of the Lorisidae are specialized. Foramina open ventrally into the base of each transverse process. In *Arctocebus* the ribs are broadened posteriorly (craniad to caudad) and cover succeeding ribs in a shingle-like arrangement (Schultz 1961, Jenkins 1970). No adequate functional interpretation has been given for this structural complex.

The barrel-shaped hominoid thorax is broader than deep. Its configuration depends on the shape of the ribs and the breadth of the sternum. The ribs are deeply flexed near their vertebral articulations (Figure 10–2). The position of the vertebral column is altered, so that it lies more centrally within the trunk. In smaller animals, the ribs hang from the vertebrae and the column is positioned dorsally and more or less peripheral to the body outline (Schultz 1961). However, a relatively broad thorax can occur also because of craniocaudal shortness (compression) of the thoracic vertebrae, short distances between the capitulum and tuberculum costae (rib articulations with vertebrae), or particular directions of the abutments of the ribs.

The vertebral bodies of Hominoidea are broader than long, and this is most pronounced in human thoracic and lumbar vertebrae (Schultz 1961, compare p. 226).

In all other primates, the thorax is deeper than broad. However, the genera *Perodicticus, Tarsius, Ateles* (*Brachyteles*), and *Colobus* show some broadness in the thorax, but they do not have broad sterna.

In the thoracic vertebrae, the neural arch articulations rise dorsally above the vertebral bodies. They are oriented from dorsomedially to laterally and ventrally. These facets can be visualized as part of a circle, the center of which is in the middle of the cranial or caudal surface of the centrum and intervertebral disc (Figure 10–3B; Rockwell, Evans, and Pheasant 1938). Thoracic turning motions would be considerable were it not for the ribs, the ligaments, and the muscular insertions.

The "anticlinal" vertebrae in various species, defined by directional change of its pre- and postzygapophyses, have different positions within the rib-bearing region (Gottlieb 1914). In monkeys, the spine of the "anticlinal" vertebra is usually straight. The cervical and upper thoracic spines project caudally; in the lumbar region they project cranially.

The paired articular facets in the lumbar region lie as if on a small circle whose center point is the dorsal spine (Figure 10–3C). Such an arrangement obstructs rotary movements, but lateral bending and flexion in a ventral direction are not influenced. The lumbar region is longer in prosimians and monkeys than in the Hominoidea, especially the Pongidae. The lumbar vertebrae of the prosimians and monkeys are also longer and more numerous than those in hominoids; they are usually keeled ventrally in prosimians and monkeys because of the ligamentum longitudinale ventrale. Additional spines (processes accessorii) are situated ventral to the postzygapophyses in nonhominoid primates. In some joints, these spines lock the articulations of succeeding vertebrae (Figure 10–4). In some of the Ceboidae, the dorsal lumbar spines are long cranio-caudally and interlock; negligible dorsal movement is possible between them (Erikson 1963). Some platyrrhine genera (Alouattinae and Atelinae) differ from the other platyrrhines and from the Cercopithecoidea in the position of the lumbar transverse processes. In these advanced ceboids, the lumbar transverse processes arise from the roots of the neural arch, while in the cercopithecoid monkeys they arise from the broadest part of the vertebral body (Figure 10–5).

In adult man, the last two lumbar vertebrae are positioned so that their ventral borders are considerably higher than their dorsal borders. The cranial

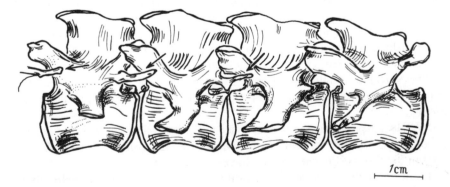

1cm

Figure 10–4. First to fourth lumbar vertebrae of *Lagothrix* seen from left side. Arrows = Processus accessorii. (From Ankel 1967.)

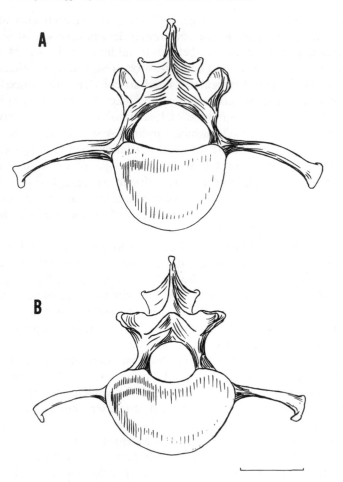

Figure 10–5. Different position of lumbar transverse processes.
A. *Ateles*, B. *Macaca*   (From Ankel 1967.)

aspect of the sacrum shows the same tendency. The lower lumbar and first sacral vertebrae together form the promontory, a point of major stress during human truncal erectness. In some other primates, one can detect promontories, but they are not as distinctive as that in adult man (e.g. *Papio*, Schultz 1961).

Since the sacrum lies at the structural intersections of the anterior vertebral column with the caudal segments, and of the pelvis with the lower extremities, it is called upon to perform a variety of functions. As a "keystone" in the pelvis, it transfers body weight and propulsive force from the lower legs to the vertebral column. The pelvic inlet, including the sacrum, must be studied in the female in relation to the head and body size of the neonate (Schultz 1949, Leutenegger 1970a). The sacrum also reflects its primary association with the tail, the morphological diversity of which provides significant taxonomic characters. In addition, it is one of the most likely elements of the vertebral column to be preserved by fossilization.

In all sacra, the intervertebral discs and articulations are reduced, and the lateral processes fuse and connect with the iliac blades. The sacroiliac joint varies with the relative magnitude of locomotor stresses. It is expanded in the prehensile-tailed monkeys, which can hang from their tails with no other support (Leutenegger 1970b). In man, it is larger than in Pongidae of comparable weight since it must transmit the weight of the trunk, arms and head to the lower extremities. In East African women, accessory sacroiliac joints are correlated with carrying burdens and children on the lower back (Trotter 1964). Thus the sacral articular surface directly reflects posture and locomotion (Schultz 1961).

Sacral vertebrae number varies from two to nine (Schultz 1961). More numerous sacral segments are usually correlated with caudal reduction. Increase in number is associated with specialization; unspecialized sacra have three segments. Tail reduction is possible without concomitant increase in sacral number (*Indri, Papio sphinx, Macaca mulatto*), but sacra with more than four segments are always followed by less than the unspecialized number (25) of caudal segments.

Additional specializations appear by comparing the sacral and caudal regions of (1) hominoids; (2) long-tailed monkeys; (3) *Cebus,* having a long tail with prehensile function; and (4) the prehensile-tailed Alouattinae and Atelinae, which exhibit a tactile pad of friction skin on the ventrodistal one-third of the tail.

In hominoids, the few small caudal vertebrae show no traces of neural arches or apophyses. Haemapophyses can be present, but this is rare.

In monkeys, if there is an outer tail, two regions of variable length are distinguished: the proximal caudal and distal caudal regions (Schmidt 1886, Ankel 1962). The proximal or first caudal region has neural and ventral arches. The neural arches articulate with zygapophyses as in the lumbar region, but bony prominences are reduced from segment to segment. Succeeding vertebrae of the second region have neither arches nor processes. Here the vertebral bodies are joined by intervertebral discs.

The vertebrae in the proximal region are short, but lengthen successively to the longest caudal vertebra. Then the length of each segment decreases caudad to the tip of the tail. The distal vertebrae are long and rather rounded in cross-section. The ventral arches, the haemapophyses, are short craniocaudally and arch- or V-shaped, forming chevron bones. They articulate anteriorly on each vertebra in the proximal caudal region, and in succeeding distal segments lie farther beneath the intervertebral discs. Chevron bones are reduced to small paired bony nuclei, which are clearly revealed on roentgenograms.

Table 10–2. *Average numbers of sacral and caudal*
segments and sacral indices (Ankel 1965)

Species	Sacrals	Caudals	Sacral index	Specimens
Lemur	3	26	70.2	1
Perodicticus	6	11	21.0	1
Galago	3	25	51.0	1
Tarsius	3	27	77.0	1
Saimiri	3	28	77.1	3
Cebus	3	25	80.7	7
Alouatta	3	27	94.4	2
Ateles	3	31	121.0	7
Lagothrix	3	26	112.0	1
Macaca mulatta.	3	15	46.3	20
M. fascicularis ..	3	25	68.2	10
Papio hamadryas ⎱ P. papio .... ⎰	3	20	65.3	6
P. sphinx	3	8	33.0	1
Cercocebus	3	25	69.2	6
Cercopithecus ..	3	26	61.0	6
Pliopithecus† ...	3 ?	?	63.2	1
Hylobates	5	3	11.5	20
Symphalangus ..	5	2	4.3	8
Pongo	5	3	15.3	8
Pan	6	3	4.3	14
Gorilla	6	3	11.4	20
Homo	5	4	15.1	10

In long-tailed monkeys that lack caudal prehension, the first caudal segments are comparatively long craniocaudally. Only four of them bear low
neural arches (Figure 10–6C). The longest segment is situated near the
sacrum. Thus, mobility of the proximal region is reduced, compared with
that of prehensile-tailed monkeys.

In *Cebus,* a prehensile-tailed monkey lacking a specialized caudal tactile
pad, the proximal caudal region contains six segments. The longest caudal
segment is situated away from the sacrum so the mobility of the tail-root in
Cebus is greater than that in nonprehensile-tailed forms. The neural arches
in *Cebus* are more elevated than in monkeys with nonprehensile tails. The

A
B
C

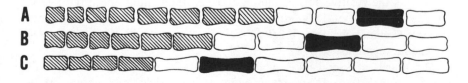

Figure 10–6. (A) *Lagothrix;* (B) *Cebus;* and (C) *Cercopithecus.* Gray = first
caudal region; black = longest tail vertebra. (From Ankel 1962.)

segments of the tail-tip in *Cebus* are shorter than in nonprehensile forms. In all these characteristics *Cebus* is intermediate between the nonprehensile- and the prehensile-tailed monkeys (Figure 10–6B).

In the Alouattinae and Atelinae, the proximal caudal region has high neural arches and is comparatively long. It consists of eight segments. Since the single segments are short, there are more articulations within the same tail length of these subfamilies than in those of other long-tailed primates. The neural arches are positioned high above the vertebral bodies and the ventral arches are depressed below them. The longest caudal segment lies far from the sacrum (Figure 10–6A). Dorsal movements (elevating, dorsi-flexing) are more prominent than ventral movements (depressing, flexing) in the proximal caudal region. Movements of the proximal tail region are facilitated by lumbar-like facets on the vertebrae.

While dorsal movement is primary at the base of the tail, ventral movement (flexing) is relatively more important at the tip of the tail, especially for picking up objects and hanging from branches.

The distal tail segments are very short and dorsoventrally flat in ateline and alouattine monkeys. The tail musculature reflects these differences. Proximally, the dorsal muscle bundles are thicker than the ventral ones; distally, the reverse is true. In long-tailed monkeys that lack caudal pre-hensile functions, the diameter of all muscle bundles (the two dorsal and two ventral) are similar at any position along the tail.

Specialization occurs in the caudal vertebrae of *Saimiri*. Lateral longi-tudinal ridges extend along the entire length of each vertebra in *Saimiri*. I cannot explain this structure functionally.

Comparisons of sacra and the proximal caudal region in different forms show a significant correlation between the length and height of the neural arch and the neural canal. In tailless forms, the canal ends within the sacrum; in long-tailed primates, the plane of the distal (caudal) sacral opening is smaller than the plane of the proximal (cranial) opening. In prehensile-tailed forms, particularly in the Alouattinae, the distal opening of the sacral canal is as broad and as high as the proximal opening. In the Atelinae, the distal opening is higher than the proximal opening. Direct correlation exists between the plane and other dimensions of the proximal sacral opening, those of the distal sacral openings, and the size of the caudal region. This correlation between dimensions of the neural canal is ex-pressed as the relative size of the cranial and the caudal apertures (Ankel 1965). Samplings of 152 individuals from 19 genera and 21 species show strong correlation between this "sacral index" and the number of succeed-ing caudal segments.

The ranges of sacral index values are 0 to 20 in primates with totally reduced tails, 20 to 50 in forms with reduced tails, and above 80 for prehensile-tailed monkeys. *Cebus,* with a sacral index of 80.7, is intermedi-ate between nonprehensile- and prehensile-tailed species. The Alouattinae

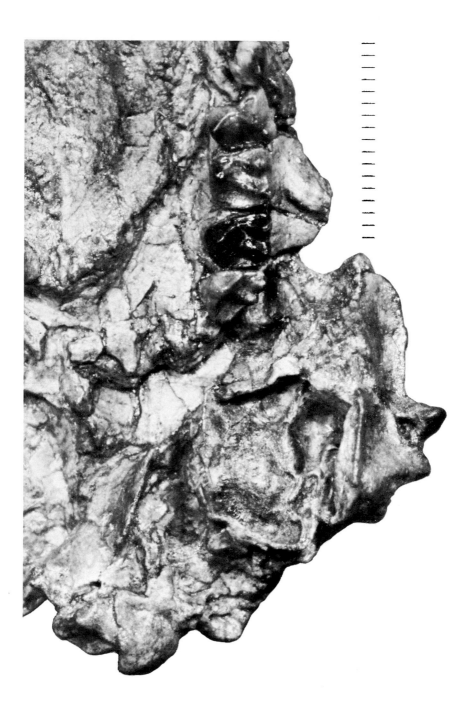

Plate 1–1. Left ear region of crushed skull of *Phenacolemur jepseni* (AMNH No. 48005), early Eocene. Subdivisions on the scale are 0.5 mm.

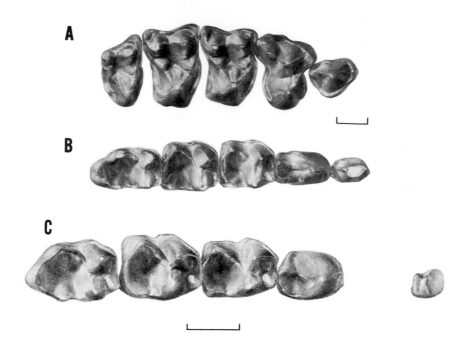

Plate 1–2. *Palaechthon alticuspis* (A and B), P^3–M^3 and P_3–M_3, medial Paleocene; *Palenochtha minor* (C), canine and P_4–M_3, medial Paleocene. Occlusal views; scales represent 1 millimeter. (After Simpson, 1955.)

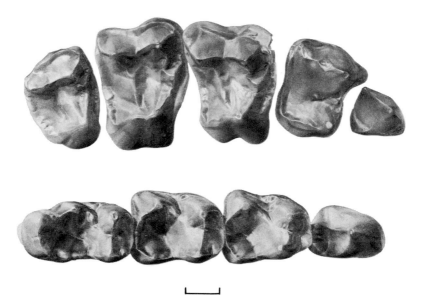

Plate 1–3. *Paromomys maturus,* medial Paleocene, P^3–M^3 and P_4–M_3. Occlusal views; scale represents 1 millimeter. (After Simpson, 1955.)

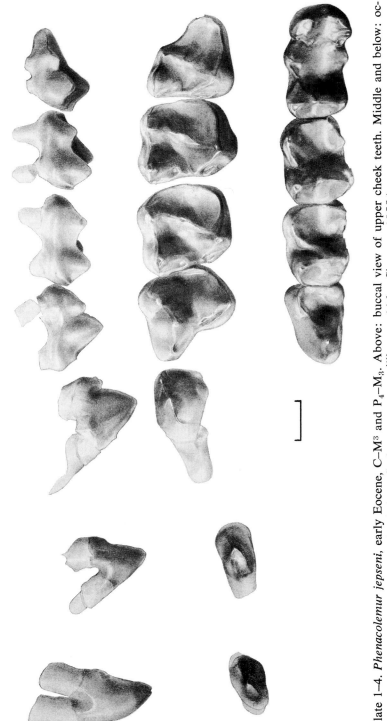

Plate 1–4. *Phenacolemur jepseni*, early Eocene, C–M³ and P₄–M₃. Above: buccal view of upper cheek teeth. Middle and below: occlusal views. Scale represents 1 millimeter. (After Simpson, 1955.)

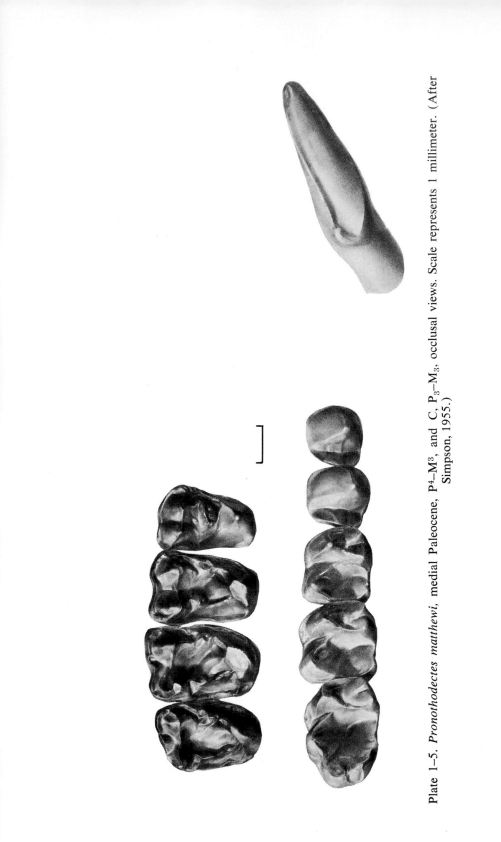

Plate 1–5. *Pronothodectes matthewi,* medial Paleocene, P⁴–M³, and C, P₃–M₃, occlusal views. Scale represents 1 millimeter. (After Simpson, 1955.)

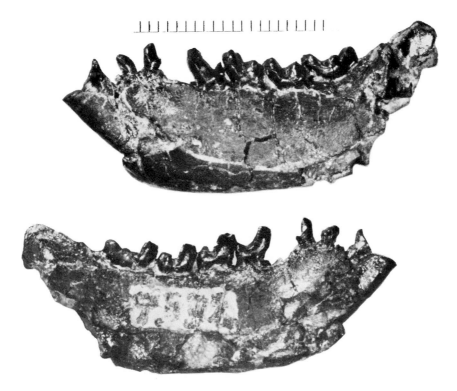

Plate 1–6. *Pronothodectes matthewi*, USNM No. 9332, lateral (above) and medial (below) views of mandible fragment to show the robust enlarged incisor and the deep horizontal ramus. Subdivisions on the scale are 0.5 mm.

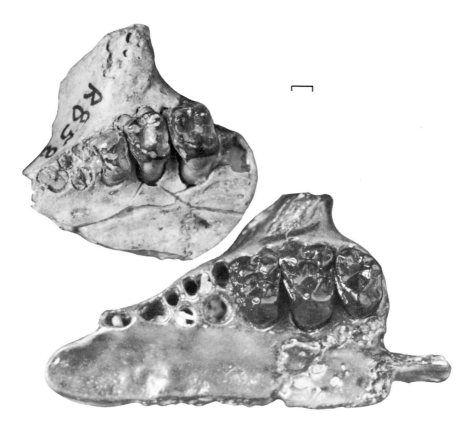

Plate 1–7. *Chiromyoides campanicus* and *Plesiadapis tricuspidens*, comparison of maxilla fragments (CR 858 and WA/288) to show relative position of the base of the zygoma. (Courtesy of Donald E. Russell.)

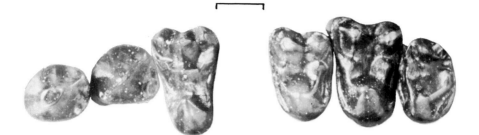

Plate 1–8. *Saxonella crepaturae,* composite, partial, upper dentition, [C–P^2 (WA/395), P^3 (WA/396), and M$^{1-3}$ (WA/394)], late medial Paleocene. Scale represents 1 millimeter. (Courtesy of Donald E. Russell.)

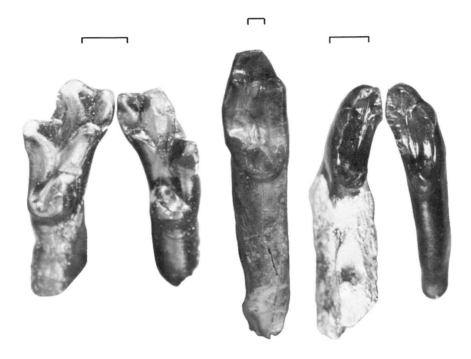

Plate 1–9. Enlarged upper incisors (from left to right) of *Saxonella crepaturae* (WA/393), *Chiromyoides campanicus* (CR 357), and *Plesiadapis walbeckensis* (WA/297 and WA/391), occlusal views. Scale represents 1 millimeter. (Courtesy of Donald E. Russell.)

Plate 2–1. New specimen of *Aegyptopithecus* from Quarry I, Oligocene of the Fayum, Egypt. YPM 23944, left ramus, P_3–M_3.

Plate 2–2. Internal view of the type mandible of *Oligopithecus savagei* from the Fayum Oligocene of Egypt. Note the sectorial P_3.

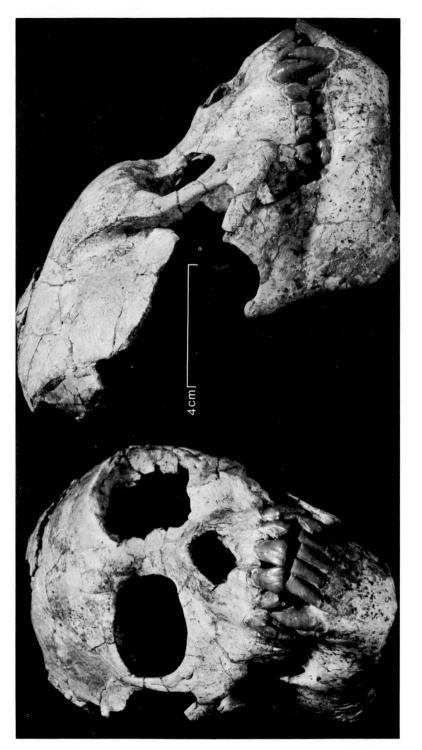

Plate 2–3. Facial and lateral view of the 1948 skull of *Dryopithecus africanus* from Rusinga Island, Kenya, East Africa (of early Middle Miocene age) found by Mary Leakey. This is the most complete fossil ape skull known. (KNM—RU 1829.)

4cm

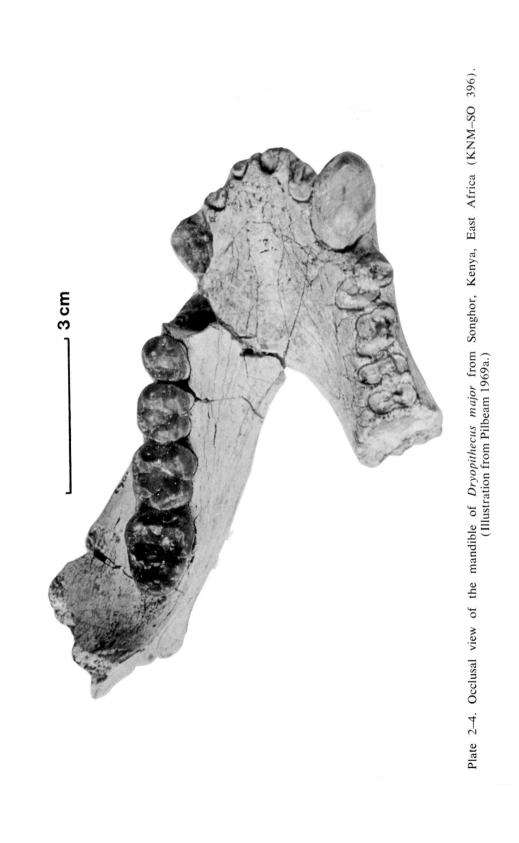

3 cm

Plate 2–4. Occlusal view of the mandible of *Dryopithecus major* from Songhor, Kenya, East Africa (KNM–SO 396). (Illustration from Pilbeam 1969a.)

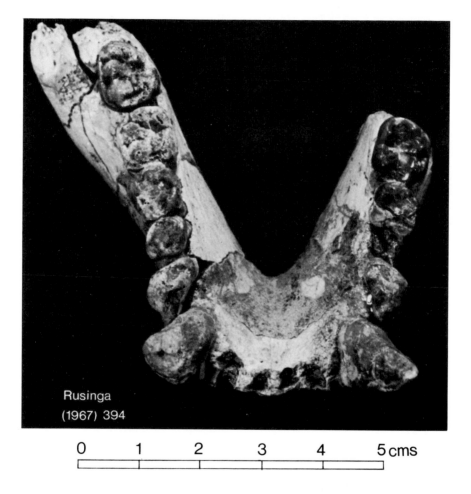

Plate 2–5. Crown view of the dentition and mandible of *"Kenyapithecus afri-canus"* from Rusinga Island, Kenya. (Photo courtesy of L. S. B. Leakey.)

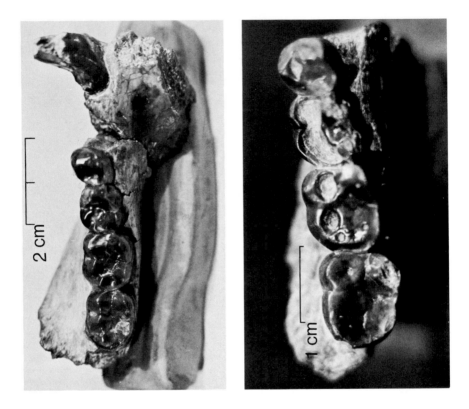

Plate 2–6 (left). Crown view of the left mandibular ramus of *Dryopithecus indicus* from latest Miocene deposits at Haritalyangar, India. This mandible shows the symphyseal fragment found by Rao in 1915 and the portion of horizontal ramus recovered in 1935.

Plate 2–7 (right). Occlusal view of the Calcutta mandible of *Ramapithecus* found near Haritalyangar, India. Notice the broad, flattened tooth crowns and the sharply decreasing wear gradient posteriorly.

Plate 4–1. Left proximal femora of prosimian primates, posterior aspect (not to same scale). Left to right: *Galago crassicaudatus* (FMNH No. 53077), *Hemiacodon gracilis* (AMNH No. 12613), *Cheirogaleus major* (FMNH No. 5656), *Microcebus murinus* (AMNH No. 185629), *Lemur fulvus* (FMNH No. 85136), *Propithecus verrauxi* (FMNH No. 89204), and *Avahi laniger* (FMNH No. 5654). The fragment attributed to *Hemiacodon* bears a subcylindrical articular surface somewhat like that seen in *Galago;* but it is not greatly different from *Microcebu*s in this respect, and it resembles *Cheirogaleus* more closely in the general shape and proportions of the three trochanters and the trochanteric fossa.

(*over*)

Plate 4–2. Radiographs of prosimian skulls, from vertical (left column) and lateral (right column) aspects. Top to bottom: *Avahi laniger* (YPM No. 317), *Hapalemur griseus* (MCZ No. 44917), *Nycticebus coucang* (FMNH No. 89467), *Loris tardigradus* (FMNH No. 95027), and *Galago senegalensis* (FMNH No. 81756). The lateral radiographs are here oriented on a horizontal passing through anterior and posterior nasal spines. Orientation of the foramen magnum with respect to this horizontal axis of the splanchnocranium is shown by a black line passing through basion and opisthion. The quadrupedal *Hapalemur* has a significantly less vertical foramen magnum than the vertically clinging *Avahi; Loris* and *G. senegalensis* do not differ significantly.

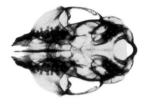

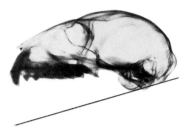

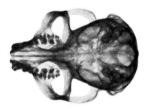
1 cm

Plate 8–1. Lateral view of a plaster replica of the Taung endocast, toned with charcoal.

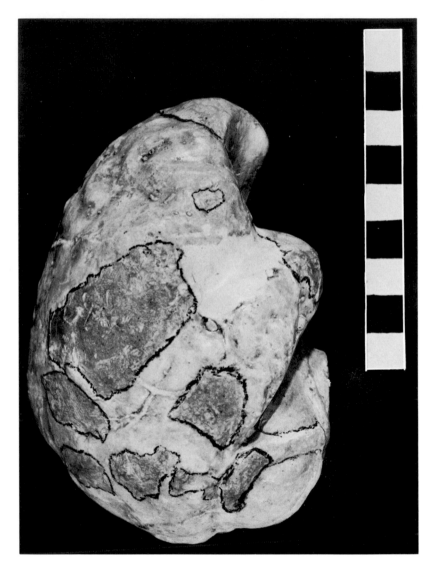

Plate 8–2. Reconstruction of Taung hemiendocast.

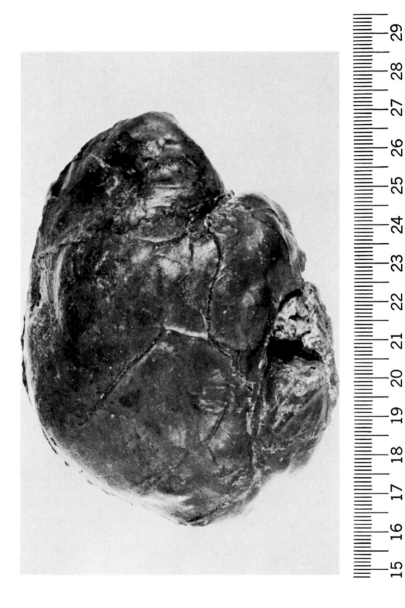

Plate 8–3. Lateral view of actual endocast from Swartkrans, SK 1585.

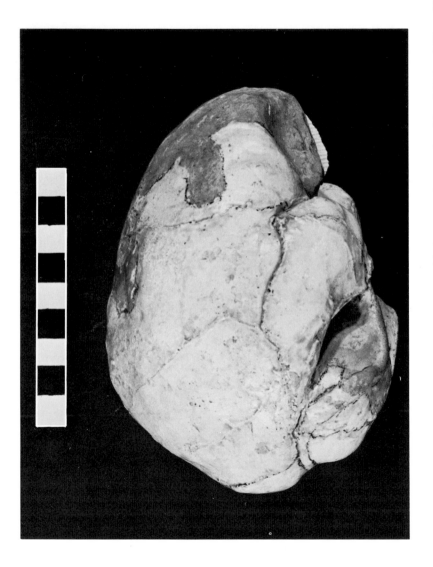

Plate 8–4. Reconstruction of SK 1585 (small white segment on orbital rostrum is opposite side of Olduvai Hom. No. 5 cast).

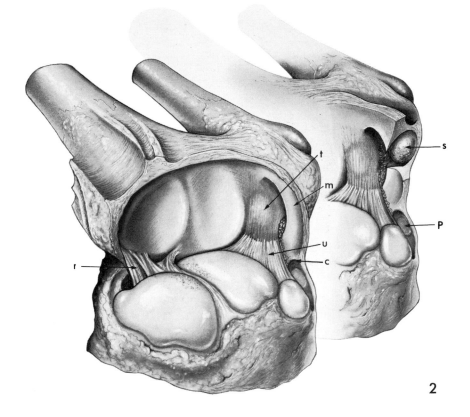

2

Plate 9–1. The left wrist joint of a human cadaver opened dorsally and flexed. This specimen is unusual in retaining a virtually free semilunar meniscus (m) independent of the triangular articular disk (t); an arrangement very comparable to that shown in the chimpanzee illustrated in Figure 9–1C has here been retained. The wrist joint cavity communicates above the free radially directed margin of the meniscus with another synovial compartment—the prestyloid recess. In the detail shown to the right, this recess has been opened by incising the meniscus to reveal a cartilage-clothed ulnar styloid process (s) protruding into it. The wrist joint cavity is also in communication (c) with the pisotrique-tral cavity; the articular surface of the pisiform is shown (P). The palmar ulnocarpal ligament (u) is shown, as is the palmar radiocarpal ligament (r).
(From Lewis, Hamshere, and Bucknill, 1970.)

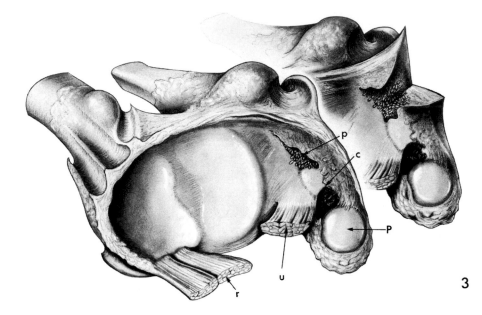

3

Plate 9–2. The upper articular surface of the left wrist joint (together with the pisiform (P) of a human cadaver, with the two bands of the palmar radiocarpal ligament (r) cut from their attachment to lunate and capitate, and the palmar ulnocarpal ligament (u) cut from the lunate. This specimen illustrates the arrangements when the prestyloid recess fails to extend up to the ulnar styloid process (which is not, therefore, clothed with articular cartilage). The opening (p) into the prestyloid recess is masked by a protruding mass of synovial villi, and the meniscus, as is usually the case, has lost its separate identity, becoming the most ulnar part of the proximal articular surface. In the detail shown above, the prestyloid recess has been opened by an incision revealing a luxuriant growth of synovial villi arising from the lining membrane. In this specimen there was continuity (c) between the radiocarpal and pisotriquetral joint cavities. The pisotriquetral joint cavity, as is often the case, is ballooned proximally to the pisiform bone (P) itself. (From Lewis, Hamshere, and Bucknill, 1970.)

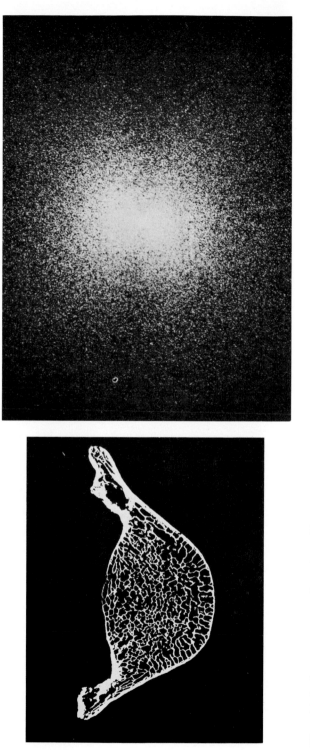

Plate 14–1. A transverse section of the fifth lumbar vertebra of man and (right) its power spectrum as derived by optical data analysis.

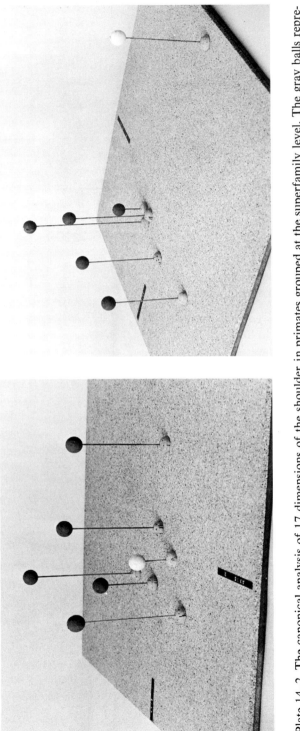

Plate 14–2. The canonical analysis of 17 dimensions of the shoulder in primates grouped at the superfamily level. The gray balls represent the positions of the major superfamilies; the white ball represents the position of *Daubentonia*. The three dimensions of the models are the first, second, and third canonical axes. The general scale is the same for both models and such that the major dimension of the right model is about 15 standard deviation units. The left model represents, then, the analysis where *Daubentonia* is interpolated indirectly (as though it were a fossil). It suggests that *Daubentonia* "belongs" with the other forms. The right model represents the analysis where *Daubentonia* is added directly (as though it were an extant form). It shows the true picture: *Daubentonia* is uniquely different (in its shoulder) from the other primates.

Plate 15–1. A thin section of a rock.

Plate 15–2. A power spectrum of the rock section.

Plate 15–3. The contoured power spectrum of the rock. The arrow demonstrates an obvious departure from circularity.

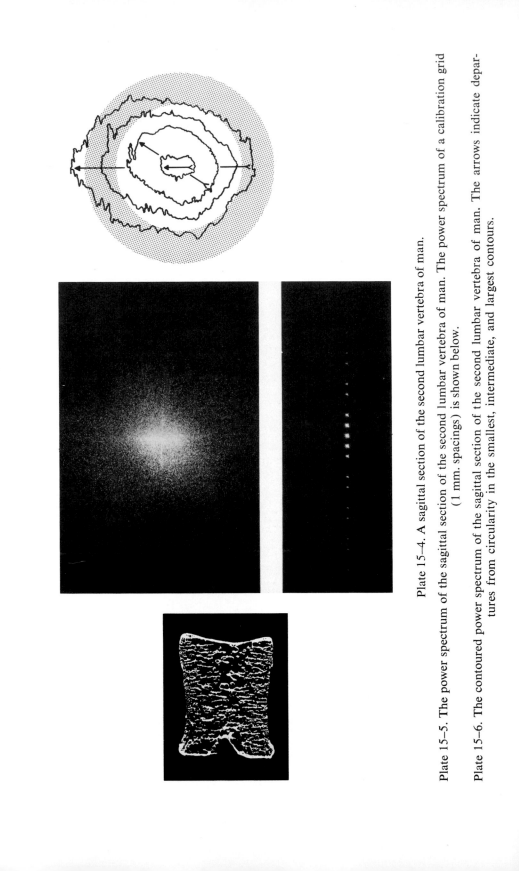

Plate 15–4. A sagittal section of the second lumbar vertebra of man.

Plate 15–5. The power spectrum of the sagittal section of the second lumbar vertebra of man. The power spectrum of a calibration grid (1 mm. spacings) is shown below.

Plate 15–6. The contoured power spectrum of the sagittal section of the second lumbar vertebra of man. The arrows indicate departures from circularity in the smallest, intermediate, and largest contours.

have a sacral index of 94.4; the Atelinae exhibit values greater than 100.0 (*Lagothrix* 112.0; *Ateles* 121.0). Thus, it is shown that one morphological complex of the sacrum is directly correlated with tail length and function.

An additional explanatory and logical base underlies this observation. The prehensile tails of the Alouattinae and Atelinae have the functional capacity of a fifth extremity (Ankel 1962) and must be supplied appropriately by neurovascular connections. In man, the filum terminale of the spinal cord lies in the first lumbar segment. In *Ateles,* the filum is situated in the eighth caudal segment. The nerves in the prehensile tail are also comparatively thick (Hofer and Tigges 1964). The lumen of the neural canal in the fourth lumbar vertebra is much larger in *Ateles* than in *Macaca mulatta* (Figure 10–5). The vascular system in *Lagothrix,* a representative prehensile-tailed form, is more highly developed than in nonprehensile long-tailed forms (Wrobel 1966). Two systems of arteries serve the proximal tail root and the prehensile tip. In the skeleton, these facts are reflected in the high neural arches and deep haemapophyses.

Discussion

The phrase "vertebral specialization" sounds innocent enough, but it presents two inherent difficulties. First, the vertebral column in different regions mirrors different adaptations of the organism, and therefore has not evolved as one unit: the cervical vertebrae evolve in relation to the mobility of the neck; the thoracic segments are associated with the shape of the thorax and anchoring of the ribs; the flexibility of the lumbar region is associated with climbing, jumping or walking; the sacrum and intervertebral discs are related to truncal orientation; the sacrum and pelvic canal must permit childbirth in the female; the tail is integrally involved in many locomotor and behavioral patterns; the vertebrae reflect nervous and vascular distributions. This enumeration of features is more arbitrary than real. (Such division in the marsupial pelvis is considered by Elftman 1929, and in several areas of the human body by Washburn 1951b).

Second, we might ask, what is "specialization"? With regard to locomotor phenomena, it may mean restriction of potential motion in the same region in *X*'s ancestor or in animal *Y;* or it may refer to the fact that *X* shows some increased range of motion in one region, that X shows increased leverage in one segment, or that *X* resists increasing weight bearing in some position. Yet, this specialization "of motion" is not clear-cut. *Tarsius, Callithrix, Cebuella,* and *Propithecus* are able to turn the head directly backwards while clinging to vertical supports. Some structural features to facilitate this increased motion may be seen in *Tarsius,* but none have been detected in the other forms. Thus, locomotor specialization is not always clearly reflected in structures. As described throughout this review, some structural features certainly possess taxonomic or functional significance;

for example, the cervicals in *Tarsius* long cervical and upper thoracic spines in *Perodicticus* and the Pongidae, the axis vertebra of *Homo,* the broadening of the hominoid thorax, the foramina in the thoracic transverse processes of the lorises, the relative broadening of the last lumbar vertebrae and expansion of the sacroiliac joints in the hominids, the marked reduction of the caudal region of man and the apes, the caudal vertebrae of *Saimiri,* and the complex changes in the sacrocaudal area of the prehensile-tailed South American monkeys.

But for the most part, vertebral morphology of the Old and New World monkeys remains uniform. The number of presacral elements varies between groups and between thoracic and lumbar regions and thus indicates certain trends. But these minor differences in form and size are not taxonomically significant. Identification of single vertebral elements at the family level is sometimes difficult; taxonomic identification at the generic level is nearly impossible. With these perspectives, the primate vertebral column must be judged to be "relatively unspecialized." From many attempts to correlate function and morphology in primates, we learn that it is not possible to deduce function from morphology alone. We must know as much as possible of the locomotor-behavior of our subjects in the natural habitat. This is often totally ignored.

The question, "How much is the vertebral column of primates involved in locomotion and all kinds of movements?" can partly be solved by studies on film sequences of these actions. Observations of locomotion in the natural habitat of baboons (*Papio anubis,* Kenya) and evaluation of a film on gray langurs (*Presbytis entellus*) demonstrate that the vertebral column is not involved in much "action," even during high speed running and climbing.[1] Thus, it is not difficult to understand why the vertebral column is relatively uniform in morphology in these species.

What Can Be Said about Evolutionary History?

Stewart (1962) showed that the cervical region in Neanderthal men does not differ fundamentally from that in modern man.

The "sacral index" suggests that *Pliopithecus vindobonensis,* a middle Miocene hominoid from Europe, had a pendant tail. *Pliopithecus* is regarded commonly to be an ancestor of the gibbons. The sacral index of *Pliopithecus* (63.3) falls within the range of long-tailed monkeys. If *Pliopithecus* had four sacral segments instead of the three found, the sacral index value of 50.0 would still indicate (Figure 10–7) a tail of 10 to 15 caudal vertebrae (Ankel 1965). The sacrum of other fossils should also be

1. My gratitude is extended to the Wenner-Gren Foundation for Anthropological Research, New York for a grant (No. 2605), making it possible for me to visit South Africa and Kenya. I thank Drs. B. B. Beck and R. Tuttle for lending me their film on Ceylonese gray langurs.

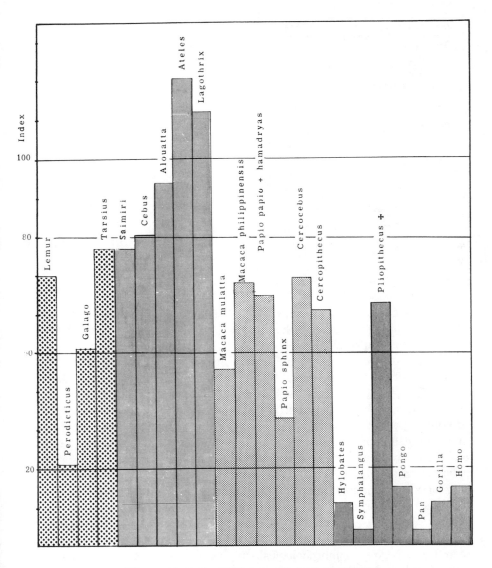

Figure 10–7. Sacral index. (From Ankel 1965.)

Table 10–3. Morphology of vertebral column and
thorax in extant primates*

Generalized	Specialized	Specialization of
Tupaiidae		
Lemuridae		
Indriidae		
Daubentoniidae		
	Lorisidae	Entire vertebral column
Galagidae		
	Tarsiidae	Presacral column
Cebidae except	Alouattinae ⎫	Sacral and caudal
	Atelinae ⎬	regions
	Saimiri ⎭	Caudal region
Cercopithecoidea		
	Hylobatidae ⎫	
	Pongidae ⎬	Entire vertebral column
	Hominidae ⎭	

*From Ankel 1967.

considered. *Oreopithecus* (Straus 1963, Schultz 1960), a hominoid from the lower Pliocene (Pontian), has a sacral canal with tapering configuration, an indication of outer tail reduction. The lumbar vertebrae are broad and short, but with longitudinal ventral keeling as in monkeys. The *Australopithecus africanus* sacrum (Sterkfontein 14) exhibits similarities in shape and tapering sacral canal like that of modern man. Robinson (1970) described two vertebrae from Swartkrans, attributed by him to *Paranthropus*.

With respect to fossils then, the axiom of few deductions from few fragments must be applied even more strictly. If such deductions about vertebral parts are combined with whatever else remains of the foot, leg, or pelvis of an animal, or of an extinct type, far more can be deduced about form and function; and the evolutionary emergence of the vertebral column in the array of primates will be illuminated. The analysis of sacrocaudal relations provides an example of function deduced from morphology. Sometimes, however, primates are divided into locomotor categories. The search then begins for morphological support for these categories. Caution toward such comparisons must be exercised.

A barrel-like trunk characterizes the Hominoidea. But the hominoids represent a spectrum of locomotor "types": the Hylobatidae and *Pongo* are "true brachiators": *Pan* and *Gorilla* are knuckle-walkers (Tuttle 1970); *Homo* is erect and bipedal. Since the tendency to truncal broadening occurs in other primates, one cannot say how closely correlated this is with truncal erectness. The barrel-shaped thorax may favor erect posture by positioning the center of gravity more within the thorax than is found in most monkeys.

In springing and brachiating primates, thoracolumbar movement is fundamentally different. Ribs and thorax are rather rigid, and the lumbar region does allow dorso-ventral springing movements. The "length of the lumbar region" may therefore be functionally important. The "anticlinal" vertebra, where the thoracic and lumbar articulations change orientation, lies within the rib-bearing thorax (Gottlieb 1914, Erikson 1963). But this position cannot be the sole functional adaptation. The adaptive complex must be more diffused, as Erikson (1963) notes, to include regional lengthening, increase in segment number, and shape of vertebral bodies as well as craniad extent of the lumbar joint pattern. But neither thoracic nor lumbar joints influence dorsoventral "spring movements."

These seem like clear adaptations for springing movements, yet the "anticlinical" vertebra is found in the rib region of mammals incapable of springing movements (as in *Erinaceus*). Therefore, animals can do more functionally than may be indicated by structure; and not all morphological features have a functional explanation.

Anthropocentric conclusions must be avoided (Edinger 1961). Even man is not so specialized in the vertebral column that major deductions about locomotion can be made. Man is specially adapted neither for spectacular high-wire performance in the circus nor for swimming through the water with ease or grace. Yet he exhibits a broad spectrum of activities in which the vertebral column is employed (Ankel 1966, 1967).

Locomotion is more than a particular set of morphological features such as relative tail length or scapular shape. It is also related to habitat, flight response, availability of food, body size, age, and temperament. On specific occasions animals can function more efficiently than their daily life usually demands. Long-term observations of wild primates in their natural habitat reveal locomotor variations far beyond the traditional categories (Ripley 1967b). Whether one considers juvenile flexibility, functional potentiality or adaptability, or Prost's (1965) "toti potentiality," the vertebral column answers a broad set of requirements.

If the "functional potency" of a morphological feature is high, this structure is not likely to undergo morphological specialization. This remains true for the vertebral column (Ankel 1967, p. 117). Leutenegger (1970b, p. 96) states the same for the primate pelvis.

Future Approach to the Problems

Can future research probe more deeply into the relationships between the form and the function of vertebrae? In addition to the morphological approach (vertebral size and number, size and orientation of the joint surfaces, relative region length comparisons) used by Schultz and Ankel, attempts have been made to find other significant measures. Masali (1968) has constructed "rachigrams" by superimposing sagittal sections of vertebrae, but these do not reveal function or work well for single vertebrae. Demonstating caudal increase in weight for man and other mammals, Delmas *et al.* (1958) determined the relative weight of single vertebrae within columns. Delmas and Pineau (1959) calculated the coefficients of all vertebral weights within the entire human column. But this does not say anything to illuminate the interpretation of fossils. It cannot reveal functional potentials of the column in living forms.

It would be helpful to study the internal architecture of vertebrae by split-line techniques and to section the centra to examine the internal bony trajectories (as in M. C. Hall 1965), or to study stress patterns with plexiglas models, as Pauwels (1965) and Kummer (1965) have done. This would clarify the stresses on individual segments. Studies of the back musculature should be correlated with studies of the flexibility and potential movements of the segments in anesthetized or freshly dead animals, as Slijper (1946) has done (Ankel 1967). Observations through x-ray cinematography have begun (Jouffroy, Jenkins, Grand, personal communication). Selected muscles and nerves should be sectioned to determine how the coordination and position sense of the trunk and tail are affected within the motor pattern of the animal. We need more observations and films of locomotor behavior in natural and artificial habitats to elucidate functions of the vertebral column. Further research concerning the diameter of neural canal in primates, perhaps comparing other mammals, should be done.

11

DONALD R. WILSON
CALVIN COLLEGE

Tail Reduction in *Macaca*

Tail reduction has occurred independently in various primate taxa. This reduction is accompanied by a number of corresponding changes in other features of pelvic anatomy. In this investigation I use the genus *Macaca* as a natural model of tail reduction in order to elucidate this process and to examine its relationship to behavioral and ecological differences between the species.

Methodological Perspective

This paper is designed to be not only a contribution of substantive nature, but also a test, or possibly an illustration, of a methodology. This methodology is based on the perspective that morphological parameters are, to a greater or lesser degree, functionally explainable in terms of behavioral and/or ecological characteristics. A logical extension of this perspective is that the morphological differences between various taxa are consequently explainable in terms of behavioral and/or ecological differences. It is this latter procedure (that is, the morphological comparison) that I will illustrate here.

I thank the other participants of this symposium for the discussions, criticisms, suggestions, and general stimulation given during the conference. The contributions of Russell Tuttle merit special recognition for the instruction given me in a research perspective, and for the valuable insights and helpful criticism, which he generously offered during the entire course of this research.

I acknowledge the support of the National Science Foundation with a Science Faculty Fellowship for the year 1965–1966, the Danforth Foundation with Danforth Teacher Fellowships for the years 1965–1966 and 1966–1967, and the Wenner-Gren Foundation with a research grant for the summer of 1968. I thank Dr. Louis S. B. Leakey, Curator of the National Museum in Nairobi, Kenya for inviting me to be a guest investigator at the Tigoni Primate Research Center in Limuru, Kenya, summer of 1966; Dr. Leon Schmidt, former Director of the National Center of Primate Biology, Davis, California, for allowing me to study there in August, 1968; and Mr. L. Pulchritudoff, Curator of Osteology at the National Center of Primate Biology, for his generous assistance.

My methodological procedure involves the following series of steps: first, the quantification of morphological parameters; second, the determination of internal consistencies within and between these parameters; and third, the construction of possible functional explanations for these morphological differences in terms of behavioral and ecological differences between the taxa.

Admittedly each of these steps has certain inherent difficulties. For example, in step one, any morphological characteristic has numerous quantifiable parameters, and any single parameter will give only a very incomplete picture of that characteristic. Consequently one must determine which of the possible parameters measures significantly the most meaningful dimension of that characteristic. Because similar difficulties are inherent to each of the steps, the tentativeness and debatability of the resulting explanatory models must clearly be recognized.

Specific Objectives

In view of the methodological perspective of this investigation, the specific objectives are threefold: first, to examine and define as completely as possible the parameters of (a) tail size and structure, (b) tail mobility, (c) pelvo-caudal musculature, and (d) ischial callosity and tuberosity size and shape; second, to determine correlations and interrelationships within and between these parameters; and third, to examine possible functional explanations for these morphological differences in terms of behavioral parameters such as tail use, sitting and sleeping postures, and locomotion, and in terms of differences in ecology.

In order to control closely parameters which differ for reasons unrelated to these functional factors, it is necessary to compare species related as closely as possible, but which show contrasts with respect to the specific variables. Therefore, the species *M. fascicularis, M. mulatta, M. nemestrina,* and *M. arctoides* were selected because of their close taxonomic relationships and wide variations in tail and ischial callosity morphology.

Materials

The species of *Macaca* may be divided into three groups based on length of tail. My study includes representatives of each of these three groups: *M. fascicularis* from the long-tailed forms, *M. mulatta* and *M. nemestrina* from the medium-tailed forms, and *M. arctoides* from the short-tailed forms. Two forms with medium length tails were selected because their tails differ significantly in structure and mobility even though they are of similar length.

In general the long-tailed macaques are arboreal and inhabit lower latitude areas under 7,000 feet altitude. In contrast, the medium-tailed

macaques, which tend to inhabit more northerly, rocky and montane regions, are more terrestrial and stockily-built monkeys. Two of the short-
tailed species live in the coldest areas inhabited by macaques.

The subjects used in this research were of three types: (1) cadavers for
dissection, (2) live animals for physical examinations and radiographs,
and (3) skeletal material. The approximate number of specimens used
were five males and five females of each of the four species for each of the
three types of subjects. The exact distribution of research specimens is
given in Table 11–1. The total number of specimens used was 136.

Table 11–1. Number of research specimens

	Cadaveric			Live			Skeletal			Total for Species
	M	F	Total	M	F	Total	M	F	Total	
M. fas .	5	5	10	5	5	10	2	10	12	32
M. mul.	6	5	11	5	5	10	6	5	11	32
M. nem.	7	4	11	5	5	10	14	7	21	42
M. arc.	3	6	9	5	5	10	5	6	11	30
Total for Type of Subject			41			40			55	136

From the cadaveric material the following data were recorded: tail length
and volume (by water displacement), crown–rump length (C–R ht.), and
diagrams of the callosities from which surface areas were calculated.

With these measurements various ratios were constructed which would
make comparisons between the species more accurate by eliminating differences in size or weight.

Then, each of the muscles of the pelvocaudal muscle set were removed
individually by dissection, cleaned of external fascia and tendons, labeled,
and immersed in a 10 percent solution of formalin for at least 48 hours
prior to weighing. As each muscle was removed, observations on the nature
and extent of its origin and insertion were recorded. The muscles were
subsequently dried at 40°C. for 24 hours. After removal from the oven,
they were allowed to cool for 10 minutes and were then weighed on a
Mettler Type H analytical balance. Weights were recorded for each of the
three muscles to the nearest mg. (0.001 grams). Reweights were taken on
selected muscles to insure the consistency and the accuracy of the initial
weights. Ratios were calculated expressing the relative mass of each muscle
to the pelvocaudal muscle set. The methods of muscle weight analysis are
similar to those used by Tuttle (1967, 1969a). In addition, caudal vertebrae counts were taken.

From the live subjects, data were recorded on tail dimensions, C–R
height, callosity shape and area, and tail mobility. Caudal vertebrae counts

were made from both the live subjects (by radiographs) and the osteological subjects.

Limitation of Research Data

Admittedly the data recorded on each of the variables under investigation supply us with a rather incomplete picture of that parameter. For example, muscle weights do not give us information about the functional efficiency of the muscle, nor do callosity areas instruct us concerning the internal features of that structure. No single selected quantitative measure is capable of presenting an adequate description. These measures are, however, desirable inasmuch as they avoid some of the limitations of general verbal descriptions, are more useful in comparative studies, and are more subject to statistical evaluation. The data recorded is, I hope, a measure of some meaningful dimension of that characteristic which can in turn be related to behavioral and ecological differences between the species.

One of the limitations encountered in the use of the research material was the differences in age and size between the cadaveric and live specimens, which placed limitations on the types and reliability of comparisons between species.

Step 1: The Quantification of Morphological Parameters

TAIL SIZE

The three parameters of tail size examined are: (1) number of caudal vertebrae; (2) tail length; and (3) tail mass. A summary of the data on each of these parameters is given in Table 11–2.

The mean number of vertebrae for *M. fascicularis* is 26.33 based on a sample of 12 animals (5 males, 7 females) with complete tails. This mean agrees very well with the mean of 26.1 given by Schultz (1938, p. 6). One unusual characteristic of *M. fascicularis* is the very high frequency of incomplete tails. Of 31 crab-eating monkeys which were examined for vertebrae number, only 12, or 40 percent, had complete tails. Of the 19 with incomplete tails, all had at least 18 vertebrae, so the missing section was comparatively short, generally less than 5 cm. The variability of the vertebrae number is expressed for comparative purposes by the coefficient of variation (that is, the standard deviation divided by the mean). In *M. fascicularis* the variability is .024. This indicates relatively small variation. In fact, 11 of 12 specimens had either 26 or 27 caudal vertebrae. The one exceptional specimen possessed 25 caudal vertebrae.

For *M. mulatta* the number of caudal vertebrae is 18.33 based on a sample of 31 animals (16 males and 15 females), which value is higher than Schultz's figure of 16.6 based on a sample of eight (1936, p. 6). The

Table 11–2. Summary table—raw data

Line No.		M. fas.	M. mul.	M. nem.	M. arc.
1	Caudal vert. number	26.33	18.32	17.21	7.72
2	Tail length	50.12	22.76	16.64	3.81
3	Tail volume (cm.³)	77.0	33.1	18.3	2.6
4	Tail length ratio	1.252	.482	.354	.079
5	Tail volume ratio	39.1	7.1	4.1(est.)	.7
6	V on vert. no.	.024	.056	.109	.155
7	V on tail lgth. ratio	.039	.118	.130	.177
8	Extension at base	110°	100°	155°	—
9	Abduction at base	60°	100°	90°	—
10	Extension distally	15°–25°	30°–45°	0–10°	—
11	Abduction distally	15°–25°	30°–45°	15°–30°	—
12	Flexion distally	15°–25°	30°–50°	30°–50°	—
	Muscle prop.—male				
13	Pubocaudalis	.178	.333	} .583	.483
14	Iliocaudalis	.537	.365		.148
15	Ischiocaudalis	.284	.302	.417	.369
	Muscle prop.—female				
16	Pubocaudalis	.198	.315	} .663	.548
17	Iliocaudalis	.514	.400		.101
18	Ischiocaudalis	.288	.285	.337	.351
	Muscle wt. ratios				
19	Male—Totals	70.2	28.1	29.4	15.9
20	Pubocaudalis	12.4	9.4	} 18.4	7.7
21	Iliocaudalis	37.8	9.3		2.3
22	Ischiocaudalis	20.0	9.4	11.0	5.9
23	Female—Totals	82.1	40.7	—	26.3
24	Pubocaudalis	16.3	12.6	—	14.5
25	Iliocaudalis	42.2	16.6	—	2.7
26	Ischiocaudalis	23.6	11.5	—	9.1
27	Callosity area	11.7	14.1	16.1	24.6
28	Tuberosity area (skel.)	4.5	11.3	10.5	13.0
29	Callosity ratio	2.39	2.14	2.20	2.71
30	Tuber. ratio (skel.)	1.61	1.63	1.79	1.98
31	Call.–tuber. ratio	1.44	1.29	1.15	1.41

range in my sample is from 17 to 20 and the coefficient of variation is .056. Thus, whereas the mean is approximately eight less, the measure of variation is greater than that in *M. fascicularis*.

M. nemestrina has approximately the same number of vertebrae as *M. mulatta,* but a considerably larger range, 14 to 20. In *M. nemestrina* the mean number of caudal vertebrae is 17.21 based on a sample of 29 (19 males and 10 females). This mean is about 1.5 less than that recorded by Schultz (1938) whose figure was based on a sample of eight. Thus *M. nemestrina* contrasts principally with *M. mulatta,* not in mean values, but rather in variability. Its range is clearly greater than *M. mulatta,* and its coefficient of variation is approximately twice as large, (.109) as that of *M. mulatta* (.056).

M. arctoides has a mean of 7.72 with a range of 6 to 10 caudal vertebrae based on a sample of 29 animals (12 males and 17 females). The coefficient of variation has increased to .155.

In summary, there is a very substantial and significant difference in caudal vertebrae number between these species, and this reduction is accompanied by an increase in the variability of vertebrae number as measured by the coefficient of variation.

Between each pair of taxa, the significance of the difference of the group means is <.001 except in the case of *M. mulatta:M. nemestrina,* where it is .01–.001.

Previous vertebrae counts are inadequate because they lumped together different species of macaques into a single sample. Keith (1902, p. 30) merely gave a range from 4 to 21. Schultz (1961, pp. 5–19) gave a mean of 17.0 based on a sample of 162 in which the range was from 5 to 28. Only one study (Schultz 1938) gives the mean and sample size for various species.

The contrast between the tail lengths is greater than between the number of caudal vertebrae because the vertebrae become not only fewer but shorter as well. The maximum tail length of *M. fascicularis* was close to 60 cm. with a mean of 50.12 cm. The mean tail length ratio (that is, tail length divide by crown-ischial callosity height) is 1.252. Sitting height and tail lengths are recorded by Davis (1962), Kellogg (1944), Lyon (1908), Miller (1907a, 1907b), and Sody (1949). The ratios calculated for their data range from 1.10 (Miller, 1907b) to 1.40 (Kellogg 1944), with the majority falling between 1.20 and 1.30. The smaller ratios possibly reflect incomplete tails which are very frequent among *M. fascicularis.* Furthermore, Sody (1949) notes that there are small regional differences in tail lengths. Also, inasmuch as the tail length ratio is partially a function of age (Schultz 1938), some of the differences may merely reflect differences in the ages of the specimens.

The mean tail length for *M. mulatta* is 22.76 with a ratio of .482. These figures are very similar to those of other researchers (Burton 1962, Allen 1938, Schultz 1933a). The length of the tail in *M. mulatta* is only 40 percent the length of the tail in *M. fascicularis* when measured relative to body length.

Although the tail of *M. nemestrina* has only one less vertebra in it than the tail of *M. mulatta,* its relative length is only 75 percent that of *M. mulatta.* This is due to the shorter lengths of its caudal vertebrae. The tail length ratio of *M. nemestrina* is .354. The tail of *M. arctoides* has a mean length of 3.81 and a ratio of .079.

As with caudal vertebrae number, the coefficient of variation increases as tail length is reduced, increasing from .039 in *M. fascicularis,* to .177 in *M. arctoides.*

In a comparison of the pairs of taxa, the significance is <0.001 in all cases.

Although caudal vertebrae number and the tail length ratio show a significant difference between the species, the tremendous difference that does exist in tail size has not yet been seen because these measures have not taken into consideration the differences that exist in the mass of the tails of the different species. In order to assess differences in tail mass, the volume of the tails on the dissection specimens was measured by water displacement. The mean tail volumes for each of the four species range from 77 cm.^{3} for *M. fascicularis* to 2.6 cm.^{3} for *M. arctoides* (Table 11–2, line 3). In order to compensate for the differences in the gross size of the different species, a ratio for tail mass was calculated by dividing tail volume by the weight of the specimen in kilograms. Because weights of the dissection specimen were not available in *M. nemestrina* the ratio is calculated for estimated weights. In this comparison (Table 11–2, line 5) the great differences in tail size become more evident. Whereas the tail of *M. fascicularis* has only 1½ times more vertebrae than the tail of *M. mulatta* and is 2½ times as long relative to C–R height, it is five times as large in volume in relation to body weight. The contrast between the more muscular tail of *M. mulatta* and the more spindly tail of *M. nemestrina* is also shown. Whereas they differ only by one vertebra and in tail length by about 30 percent, they differ in relative tail mass by a factor of 1.75.

In summary, the contrast in tail size becomes more distinct as one moves from caudal vertebrae number to tail length ratio, and then to tail volume ratio. The coefficient of variation is inversely related to tail size.

TAIL MOBILITY

In considering the range of mobility in the tail, it is necessary to divide the organ into two parts. First, mobility at the base of the tail involves flexion, extension, and abduction by a unit of the first four or five caudal vertebrae. Second, there is mobility between adjacent vertebrae distal to the first four or five caudal segments.

Lines 8 and 9 of Table 11–2 give a summary of mean mobility at the tail base for each of the species except *M. arctoides.*

With respect to extension, two ranges are evident: first, the range of 100° to 110° characteristic of *M. fascicularis* and *M. mulatta;* and second, the range of 150° characteristic of *M. nemestrina.* The only comparable reference in available literature gives a figure of 85° for maximum extension of the tail of *M. mulatta* which may occur either in standing posture with extended limbs or in presenting posture (Deutsch 1966, p. 14). With respect to abduction, again two ranges are evident, but with a different distribution between the species: one range, that of 60° characteristic of *M. fascicularis;* and the second range, that of 90° to 100° characteristic of *M. mulatta* and *M. nemestrina.* Thus, these three species of *Macaca* present three different patterns of potential tail mobility at the base:

1. *M. fascicularis:* restricted extension—restricted abduction.
2. *M. mulatta:* restricted extension—extensive abduction.
3. *M. nemestrina:* extensive extension—extensive abduction.

The tail of *M. arctoides* presents a unique situation, so this species is not included in this section of Table 11–2. In the young stump-tailed macaque, the tail has considerable mobility, both in extension and abduction. However, in mature forms there is very little movement because of the monkey's habit of sitting directly on the tail. Consequently it tends to become permanently fixed in one position. Sometimes this is an extended position, but more commonly it is in an abducted position. Thus, in adult stump-tailed macaques, the tail generally evidences the capacity to abduct only to one side.

Each available species of *Macaca* displays its own unique pattern of movement between vertebrae distal to the base of the tail. Further the degree of movement varies from one section of the tail to another. Lines 10 to 12 of Table 11–2 give the number of degrees of movement possible between adjacent vertebrae for various sections of the tail and for three types of movement: extension, abduction, and flexion.

M. fascicularis has fairly limited movement in all distal sections of the tail. The average amount of movement between adjacent vertebrae is approximately 20°.

By contrast, *M. mulatta* has an extremely flexible tail. Beyond the eighth vertebra, mobility between adjacent vertebrae averages 40°. This, coupled with the much shorter vertebrae in *M. mulatta,* enable its tail to circumscribe areas of very small diameter.

M. nemestrina contrasts sharply with the rigid tail of *M. fascicularis* and the highly flexible tail of *M. mulatta.* It has an ability to flex which is comparable to *M. mulatta,* an ability to abduct comparable to *M. fascicularis,* and its ability to extend is much more restricted than either of them. So whereas the tail has great ability to extend at its base, it has extremely limited ability to extend at any point distal to the base.

There appears to be no significant sexual difference in tail mobility in any of these species, either at the base or within the distal extremity of the tail.

PELVOCAUDAL MUSCULATURE

The basic investigations of the pelvocaudal musculature of Cercopithecoidea are those of Eggeling (1896), Elftman (1932), and Hill (1966). Eggeling (1896) surveyed the entire range of pelvic and perineal musculature for the Primates. His observations on the genus *Macaca* were based on only two dissections, one male and one female, from different species. Elftman's investigation (1932) of Old World monkeys was based on two specimens, neither of which was a macaque. Elftman's study was limited further by including only males. Hill (1966) reports on the musculature of *Cercopithecus,* but states that it is little different from that of other genera of the Cercopithecoidea. A basic description of the musculature of *Macaca* is provided by Howell and Straus (1933) on *M. mulatta.* No previous study has compared the musculature of various species of *Macaca.* The impression given by previous studies is that variations between various genera of the Cercopithecoidea, or between species within these genera, in the pelvocaudal muscle set are not extensive. My investigation does not support this generalization.

Lines 13 through 18 of Table 11–2 contain a summary of the proportion that each muscle contributes to the set of three pelvocaudal muscles in each species and sex. The table shows that in male *M. fascicularis,* the iliocaudalis is by far the largest (.538) and the pubocaudalis is the smallest (.178), with ischiocaudalis (.284) intermediate between them. In fact, the iliocaudalis muscle is three times as large as the pubocaudalis muscle and almost twice as large as the ischiocaudalis muscle. In *M. mulatta* the three muscles are fairly similar in size, but in *M. arctoides,* pubocaudalis and iliocaudalis have reversed their positions in relative size, with pubocaudalis now being the largest (.483) and iliocaudalis the smallest (.148) and with ischiocaudalis (.369) again intermediate between them. In fact, the reversal is so complete that pubocaudalis is now three times the magnitude of the iliocaudalis muscle. Separate figures for pubocaudalis and iliocaudalis are not given for *M. nemestrina* because they are fused in this species. In females, iliocaudalis is two and one-half times the mass of pubocaudalis in *M. fascicularis.* By contrast, in *M. arctoides,* the pubocaudalis muscle is five times the mass of the iliocaudalis muscle. In short, the shift in the proportional size of pubocaudalis and iliocaudalis is very distinct and radical in both males and females.

The differences between any pair of the three species is significant at the .001 level for pubocaudalis and iliocaudalis.

Lines 19 through 26 of Table 11–2 show the sizes of each muscle and of the muscle set relative to the weight of the specimen. The index is computed by dividing the muscle weight in grams by the specimen weight in kilograms and multiplying the result by 100. The larger the index, the larger is the size of the muscle or the muscle set in relation to the size of the animal.

This section of the table indicates that the pelvocaudal muscle complex is far larger relative to body size in *M. fascicularis* than in either *M. mulatta* or *M. arctoides*. Table 11–2 also demonstrates that in each species the muscle complex is relatively larger in females than in males. In each species there is a difference of 10 to 12 points on the index. This larger size in the female is possibly explained by their special functional importance as a support for pelvic viscera in the female.

In addition, Table 11–2 shows the relative change by sex for each of the three muscles. From male *M. fascularis* to male *M. arctoides,* pubocaudalis decreases 38 percent, ischiocaudalis decreases 70 percent, and iliocaudalis decreases 94 percent. In females, decreases have not been as large (10 percent for pubocaudalis, 62 percent for ischiocaudalis, and 93 percent iliocaudalis). Thus, whereas all three muscles evidence significant decrease in the male, only iliocaudalis and ischiocaudalis exhibit decrease in the female. This suggests a special functional role for the pubocaudalis in the female.

ISCHIAL CALLOSITIES AND TUBEROSITIES

Ischial callosities function as sitting pads. They are present in all Old World monkeys, gibbons, simians, and in some great apes (R. A. Miller 1945, Washburn 1957, Clark 1959). They are highly variable in size and shape, not only between genera, but also within certain genera (Hill 1967, Crook 1966, Wickler 1967). This is particularly true in *Macaca* where they vary from close conformance in size and shape to the underlying ischial tuberosities to being considerably larger than the tuberosities, especially in males (Pocock 1925). It is necessary first to note various differences as determined from anesthetized and dead specimens.

Comparison of Callosities.   Three basic shapes are expressed in the callosities of the four available species of *Macaca*. First, there is an oval shape which is characteristic of *M. fascicularis* and *M. mulatta*. It is greatly modified in the male of M. *fascicularis* where the two callosities are fused into a single continuous bar. Second, is the triangular shape of *M. nemestrina.* Here the medial edges are roughly parallel and the dorsal edges form a straight line. In comparison with *M. mulatta,* which has callosities similar in area, the callosities of *M. nemestrina* are more projecting medio-dorsally and reduced in projection latero-ventrally. Thus each callosity forms a triangular area, and the two together also form a single triangular area. Third, is the more circular shape of *M. arctoides,* both in the male where they are fused and in the female where they are separated.

Another contrasting feature between species is the angle that is formed by the extension of the longitudinal axes of the two callosities. This angle is greater in *M. fascicularis* (120–130°) than it is in *M. mulatta* and *M. nemestrina* (95–105°). In *M. fascicularis* this greater angle permits a greater distance to remain between the dorsal and lateral extremities of the callosity and the base of the tail. The base of the tail in *M. fascicularis* is distant from the callosities, whereas in *M. nemestrina* and *M. arctoides* they are closely related. *M. mulatta* is intermediate between *M. fascicularis* and *M. nemestrina* in this feature.

Lines 27 and 29 in Table 11–2 give the mean callosity area and ratio for each of the four species. With the callosity ratio, I attempt to indicate callosity size in relation to weight of the specimen. In order to compensate for two measures of different dimensional natures, the ratio is defined as the square root of the area of the callosity divided by the cube root of body weight. The larger the numerical expression of the ratio, the larger is the area of the callosity relative to specimen weight.

From this section of the table, a difference in mean callosity area is evident, ranging from 11.7 cm.^{2} for *M. fascicularis* to 24.6 cm.^{2} for *M. arctoides*. However, a different picture emerges when the ratios are examined. Although *M. fascicularis* has absolutely the smallest callosities, it evidences larger ones in relation to body weight than either *M. mulatta* or *M. nemestrina*. Ratios of the latter two species are fairly similar. *M. arctoides* exhibits absolutely the largest callosities and the highest ratio.

Comparison of Tuberosities. There are three basic shapes of the tuberosities in the four species corresponding to the three shapes of the callosities. First is the oval shape characteristic of *M. fascicularis* and *M. mulatta*. Second is the triangular shape corresponding to the callosity shape in *M. nemestrina*. Third is a more circular shape characteristic of *M. arctoides*. There is, however, much less distinction between *M. nemestrina* and *M. arctoides* in tuberosity shapes than in callosity shapes. In fact, the overlap that exists between the tuberosity shapes is sufficiently large to make these three shapes merely general tendencies rather than discrete and stable types. Although this overlap exists, they are still significantly dissimilar that a prediction of the species based on shape of tuberosity alone would probably be accurate most of the time.

Lines 28 and 30 in Table 11–2 constitute a summary of the tuberosity areas and ratios for the four species. The columns on tuberosity area indicate that there is an increase in tuberosity size through the series of four species. This is partially an artifact of specimen size. The tuberosity ratios indicate similarities between *M. fascicularis* and *M. mulatta,* relative increases in *M. nemestrina,* and still greater increases in *M. arctoides*.

There are slight differences between the species in the size of the callosities relative to the size of the tuberosities. Because the sample sizes are small and the variation due to age and size is great, the differences are not statistically significant. The results of a comparison in relative size for the four species is given in line 31 of Table 11–2. These results, which must be understood as being very tentative, suggest that there may be more correspondence between tuberosity and callosity sizes in *M. mulatta* and *M. nemestrina,* and a greater differential in this feature in *M. fascicularis* and *M. arctoides*.

Step 2: The Determination of Correlations

The relative extent and direction of change in the different parameters in Table 11–2 are difficult to envision because different bases and ranges are used to present the data. Consequently, in Table 11–3 the value of 1.00 for each parameter in M. fascicularis is given, in order to make the extent and direction of changes through the series of four species more apparent.

CORRELATIONS WITHIN THE
FOUR MAJOR PARAMETERS

Each of three measures (caudal vertebrae number, tail length, and tail volume) indicate an aspect of tail reduction. Caudal vertebrae number shows the least contrast (1.00, .70, .65, .29) and tail volume the most (1.00, .43, .24, .03). The contrasts in terms of the ratios on tail length and tail volume are even more distinct. As the tail reduces, its variability increases, both in terms of the vertebrae number and tail length ratios.

The caudal vertebrae numbers by themselves do not clearly indicate the radical reduction in tail size that has occurred in Macaca. The important point to note here is that these measures are not only different ways of expressing tail reduction, but also that they form an internally consistent cluster.

Regarding tail mobility, it is necessary to distinguish between mobility at the base of the tail and mobility more distally, and to discern between potential mobility (that is, that which is structurally feasible) and actual mobility. Potential mobility at the tail base may be characterized generally to be inversely proportional to tail length, except in M. arctoides. In other words, the long-tailed forms have less potential mobility of the tail at the base than do shorter-tailed forms. Mobility at the tail base may be viewed in terms of three types of movement: extension, flexion, and abduction. There is a differential degree to which each of these movements is possible. Since the degree of flexion is limited by the perineal area, all three forms appear to have the same potential for flexion at the base. However, with respect to extension and abduction they differ significantly. The comparative degree of abduction is small in M. fascicularis (1.00) and large in M. mulatta (1.66) and M. nemestrina (1.50), whereas the comparative degree of extension is small in M. fascicularis (1.00) and M. mulatta (.90) and large in M. nemestrina (1.40). Thus, during tail reduction the degree of potential abduction increases sooner than the degree of potential extension. In summary, potential extension and abduction at the tail base is comparatively small in M. fascicularis; abduction is large and extension is small in M. mulatta; and both movements are extensive for M. nemestrina.

Probably a more fruitful way of viewing this mobility is by comparing the degree of stability with the degree of flexibility. In M. fascicularis the tail base exhibits high stability with relatively low flexibility; in M. mulatta,

Table 11–3. Summary table—relative data

Line No.		M. fas.	M. mul.	M. nem.	M. arc.
1	Caudal vert. no.	1.00	.70	.65	.29
2	Tail length	1.00	.45	.33	.08
3	Tail volume	1.00	.43	.24	.03
4	Tail length ratio	1.00	.38	.28	.06
5	Tail volume ratio	1.00	.18	.10	.02
6	V on vert. no.	1.00	2.33	4.54	6.46
7	V on tail length ratio	1.00	3.06	3.33	4.54
8	Extension at base	1.00	.90	1.40	—
9	Abduction at base	1.00	1.66	1.50	—
10	Extension distally	1.00	2.00	.50	—
11	Abduction distally	1.00	2.00	1.00	—
12	Flexion distally	1.00	2.00	2.00	—
	Muscle prop.—male				
13	Pubocaudalis	1.00	1.85	} .81	2.71
14	Iliocaudalis	1.00	.68		.27
15	Ischiocaudalis	1.00	1.06	1.46	1.29
	Muscle prop.—female				
16	Pubocaudalis	1.00	2.59	} .93	2.76
17	Iliocaudalis	1.00	.78		.20
18	Ischiocaudalis	1.00	.99	1.17	1.22
	Muscle wt. ratios				
19	Male—Totals	1.00	.40	.42	.22
20	Pubocaudalis	1.00	.76	} .37	.62
21	Iliocaudalis	1.00	.25		.06
22	Ischiocaudalis	1.00	.47	.55	.30
23	Female—Totals	1.00	.50	—	.33
24	Pubocaudalis	1.00	.77	—	.91
25	Iliocaudalis	1.00	.40	—	.06
26	Ischiocaudalis	1.00	.49	—	.39
27	Callosity area	1.00	1.21	1.37	2.10
28	Tuberosity area	1.00	2.51	2.33	2.89
29	Callosity ratio	1.00	.89	.92	1.13
30	Tuberosity ratio	1.00	1.01	1.11	1.23
31	Call.–Tuber. ratio	1.00	.90	.80	.98

stability and flexibility are more balanced; and in *M. nemestrina,* flexibility has been greatly increased at the expense of stability. Mobility may be osteologically possible although it might not be myologically possible. The reverse situation is not to be anticipated. In other words, the presence of certain types of articular facets and processes on the caudal vertebrae may impose limits on mobility and the absence of them may make greater mobility possible. But if the presence or absence of these osteological characteristics is not accompanied by corresponding myological components, active movements are limited. For example, in *M. nemestrina* the tail is, to a large degree, permanently hyperextended and it has become semi-vestigial. In contrast with *M. nemestrina,* the range of actual mobility corresponds well with the range of potential mobility in *M. fascicularis* and *M. mulatta.* In these two forms the full range of potential mobility is facili-tated myologically.

A third major variable between species of *Macaca* is the development of the pelvocaudal musculature. The features examined here are: (1) the relative proportions of each of the three muscles; and (2) the relative development of the muscle set to body weight. The comparative weight ratio of the muscle sets are: 1.00 in *M. fascicularis,* 0.40 in *M. mulatta,* 0.42 in *M. nemestrina,* and 0.22 in *M. arctoides.* Thus the pelvocaudal muscle sets in male *M. mulatta* and *M. nemestrina* are equivalent in size and they are 40 percent of the size in *M. fascicularis.* The muscle set of *M. arctoides* is approximately one-half the size of that in *M. mulatta* and *M. nemestrina.* The figures for the females indicate a reduction that is not quite as marked as that in males. In *M. arctoides,* reduction of the ischiocaudalis muscle is 0.30, reduction of the pubocaudalis muscle is 0.62, but reduction of the iliocaudalis muscle is 0.06. In other words, in *M. arctoides* pubocaudalis is two-thirds the relative mass of that in *M. fascicularis,* ischiocaudalis is one-third as large, and iliocaudalis is only one-sixteenth as large.

Another way of approaching such comparative changes is to look at the proportion that each muscle contributes to the pubocaudalis set. In Table 11–3 we see that during tail reduction, the relative proportion of pubocau-dalis increases 171 percent in the males (to ratio 2.71) and 176 percent in the females (to ratio 2.76), that the relative proportion of ischiocaudalis remains essentially the same, whereas the relative proportion of iliocaudalis decreases 73 percent in the male (to ratio .27) and 80 percent in the female (to ratio .20).

The major changes in the pelvocaudal musculature are summarized as follows: (1) a consistent but radical decrease in the relative size of the total muscle set; and (2) a reversal in the relative sizes of pubocaudalis and iliocaudalis.

A fourth major variable to be considered is callosity and tuberosity size. There is a general increase in callosity size from *M. fascicularis* (ratio = 1.00) to *M. mulatta* (1.21), *M. nemestrina* (1.37), and *M. arctoides* (2.10). The same situation exists with respect to tuberosity size, being 1.00, 2.51, 2.33, and 2.89 respectively. However, some of the variation in gross tuberosity and callosity size is accountable in terms of specimen size. When size is considered, as it is in the callosity and tuberosity ratios, the evidence pointing toward larger gross size does not correspond with larger relative size as one might anticipate. A close examination shows that in fact *M. fascicu-*

laris does not fit into the expected sequence. The tuberosities and callosities of *M. mulatta* and *M. nemestrina* are similar and are significantly smaller than those of *M. arctoides*. Although *M. fascicularis* has absolutely the smallest callosities and tuberosities, they are not as small as one might expect in relation to gross body size.

In summary, there is a general increase in the sizes of the callosities and the tuberosities. There is also an increase in their relative sizes, with the exception of the callosities of *M. fascicularis*.

CORRELATIONS BETWEEN THE FOUR MAJOR PARAMETERS

Two basic trends evidenced through a sequence consisting of *M. fascicularis, M. mulatta, M. nemestrina,* and *M. arctoides* are a decrease in tail size and an increase in potential tail mobility. The reason for this correlation is that the larger a tail is, the greater must be the strength of its internal structure. This internal strength is created by a highly developed myological system and an osteological system characterized by stability. Stability is achieved osteologically by the development of large articular surfaces and highly developed vertebral processes and articular facets, which limit mobility.

Conversely, the shorter the tail, the greater the mobility. But this axiom remains true only as long as the tail remains a fully functioning tail. When the tail atrophies, as it has in *M. nemestrina* and *M. arctoides,* its potential mobility is not fully employed. The atrophy of the tail of these two species is reflected by the relatively greater variability in vertebrae number and in the tail length ratio.

The relative size of the tail and the relative size of the pelvocaudal muscle set are clearly and positively correlated. For example, the tail length ratios in the four species are 1.00, .45, .33, and .08, while muscle weight ratios are 1.00, .40, .42, and .22. As tail size decreases, the relative importance of the iliocaudalis muscle decreases, and the relative importance of the pubocaudalis muscle increases.

As tail size decreases, the area of the callosities (1.00, 1.21, 1.37, 2.10) and of the tuberosities (1.00, 2.51, 2.33, 2.89) increases.

One correlation that appears strange initially is that as potential tail mobility increases, pelvocaudal musculature decreases. This potential mobility is realized only in *M. fascicularis* and *M. mulatta,* and in fact between those two the relative tail length ratios (1.00 and .43) are almost identical to the relative muscle weight ratios of the male (1.00 and .40). So, because the potential mobility is not realized in *M. nemestrina* and *M. arctoides,* the musculature development is small.

As potential tail mobility increases, there is a corresponding increase in the sizes of the callosities and tuberosities.

The relative development of the pelvocaudal muscle set is inversely correlated with relative areas of the callosities and the tuberosities.

Step 3: The Construction of Functional Explanations

RELATION OF THE TAIL
TO BEHAVIOR AND ECOLOGY

Relation of the Tail to Sitting Posture.   Various classes of sitting postures
have been described for *M. mulatta* by Hinde and Rowell (1962) and by
Deutsch (1966). These classes include upright sitting, forward sitting,
relaxed sitting, cat-like sitting, and hunched sitting (Deutsch 1966, pp.
7–10). The primary difference between these postures is the degree to
which the animal sits on the ventral or the dorsal parts of the callosities.
In positions called upright sitting and forward sitting, the animal places its
weight primarily on the ventral edge or portion of the callosity. In relaxed
sitting, the weight is more on the entire surface, and in cat-like sitting and
hunched sitting, the weight is more on the dorsal section of the callosity.
It should be noted that not all these types of sitting are equally practical
or possible to the various species of macaques because of interference by the
tail. The tail of *M. fascicularis* is the largest and least flexible of the four
available species. It is not possible for a *M. fascicularis* to extend or to
abduct its tail to accommodate cat-like sitting or hunched sitting except
when the tail is hanging over the edge of the branch or other support. The
tail of *M. mulatta,* having significantly greater extension at the base, permits the animal a greater range of sitting postures. Consequently *M. mulatta*
tends to sit farther dorsally on its callosities. The tail of *M. nemestrina* is
so permanently hyperextended that it does not interfere at all with sitting
postures. Consequently, *M. nemestrina* is capable of sitting even farther
dorsally on its callosities without interference by the tail. The tail of
M. arctoides is so small, and so tucked between its callosities, that it has no
influence on sitting postures. Thus, cat-like and hunched sitting postures
are much more frequent in *M. nemestrina* and *M. arctoides* than in
M. mulatta and *M. fascicularis.*

In summary, the size of the tail and its mobility at the base affects sitting
postures in various macaques, restricting *M. fascicularis* the most (that is,
to forward sitting), restricting *M. mulatta* somewhat less, and not restricting
M. nemestrina and *M. arctoides.* The effects that these differences have on
the size and shape of the callosities are considered below.

Relation of Tail Function to Ecology.   By far the most important
functions of the tail in the long-tailed macaques relate to locomotor activities,
especially to aspects of balance. Although the tail often serves communica-
tive functions it never exists as the only possible means by which communication can take place. Tail gestures are always accompanied by other postural and facial gestures. Visceral support can be fulfilled by the pelvocaudal
musculature in the absence of a sizable external tail.

There are two factors that make arboreal quadrupedal locomotion significantly different from terrestrial quadrupedal locomotion: (1) the smaller

surface area of the substrate; and (2) the more discontinuous nature of the substrate. Factor (2) means that locomotion often occurs over open spaces (that is, from limb to limb). Consequently, leaping becomes a more important part of the locomotive repertoire in arboreal macaques. Both of these factors require acute balance, and this is the major function of the tail in the long-tailed forms (Campbell 1966). As macaques penetrated more terrestrial niches, the tail became less important, and consequently other selective pressures on tail form could become more effective to produce changes in it.

Climatic Adaptation. Allen's rule states that "protruding body parts, such as tails, ears, bills, extremities, and so forth, are relatively shorter in the cooler parts of the range of the species than in the warmer parts." This postulation generally agrees with the geographic distribution of the species of *Macaca,* with certain exceptions. Although all species of *Macaca* do not fit exactly the distribution anticipated by application of Allen's rule, correspondence is sufficiently close to make the rule useful as a description of the general distribution of the macaques.

Allen's rule is based on the principle that shape of the protruding parts of the body facilitates or inhibits heat loss and heat retention. A long tail, with its large surface area, is ineffective in preventing heat loss; a short tail, with its correspondingly smaller surface area, suffers less heat loss. A long tail is dysfunctional in a cold climate because its large surface area exposes it to frostbite. Thus long-tailed forms would be selected against in colder regions.

Another ecological rule which is applicable to the macaques is Bergmann's rule which states that "the smaller-sized geographic races of a species are found in the warmer parts of the range, the larger-sized races in the cooler districts." So if tail loss is an adaptation to the cooler habits, it should be accompanied by a corresponding increase in the size of the species. All four of the long-tailed species of *Macaca* have body weights ranging from approximately 4 to 8 kilograms in the males and from 3 to 6 kilograms in the females. The weight for the medium-tailed macaques ranges up to 14 kilograms, and for the short-tailed forms the weight ranges up to 18 kilograms. The geographic distribution of the macaques closely follows the expected pattern as predicted from Bergmann's rule. It should be noted that the increase in the weight of the short-tailed species is not accompanied by a corresponding increase in the linear dimensions of the species, but rather is reflected in a stockier and more robust physique. So body physique also appears to be a converging argument that the short-tailed forms are adapted to cooler climates.

One other indication of adaptation to the cold in short-tailed forms is the length and thickness of hair. Although data are not available for each species, some observations indicate that the short-tailed forms are also the heaviest haired (Sanderson 1957). Such evidence is of little value alone, but it may be corroborative in association with evidence of heavier bodies and shorter tails.

Visceral Support. The pelvocaudal muscle set performs visceral sup-portive functions in the macaques. The amount of visceral support required is directly proportional to the amount of intra-abdominal pressure. This pressure may be of two types which differ in the way they originate. The first type, static pressure, is the result of a decrease in the volume of the abdomi-nal cavity which occurs during respiration, vocalization, and certain types of locomotion. The second type, dynamic pressure, is the result of the inertia of the viscera on the pelvic outlet due to sudden cessation of motion (for example, at the end of a leap). A basic function of the pubocaudalis muscle is to support the pelvic viscera. For this reason pubocaudalis does not greatly decrease in size with tail reduction as does iliocaudalis and ischiocaudalis. This also may explain why pubocaudalis is relatively larger in female *Macaca,* viz. due to the greater need for visceral support.

RELATION OF CALLOSITIES TO BEHAVIOR AND ECOLOGY

Functional Factors Affecting Callosity Formation. There are two func-tional factors that affect the size and the shape of the callosities in *Macaca:* (1) sitting posture; and (2) the nature of the substrate on which sitting occurs.

Although there is slight overlap between callosity form in various species, the main patterns remain fairly consistent and distinct. Each of the four species has a fairly distinguishable callosity shape, and this shape appears to be related to sitting postures. *M. fascicularis* generally sits forward with its weight along the ventral edge of the tuberosity. This results in a consider-able overlap of the callosity on the tuberosity along the ventral edge. It is especially evident in males, in which the callosities form a continuous bar across the midline. The condition is not seen in *M. mulatta* or *M. nemes-trina,* forms that sit farther dorsally on their callosities. The callosities of *M. mulatta* and *M. nemestrina* also show differences which seem to corre-spond to differences in sitting posture. The dorsal aspect of the callosities is narrower in *M. mulatta* than in *M. nemestrina. M. arctoides,* whose posture is not influenced by the tail, may sit with equal ease on any part of the cal-losity, which fact may explain partly why it possesses the largest callosities among available *Macaca.*

Another aspect of sitting behavior that probably affects callosity forma-tion is the amount of time spent sitting. Different species in certain genera (for example *Papio*) have been observed to spend significantly different amounts of time sitting (Crook and Gartlan 1966, Crook and Aldrich-Blake 1968). Further, behaviors such as grooming appear to have different genetical bases in certain species of *Macaca* (Davis et al. 1968). *M. arctoides* is observed spending the greatest amount of time sitting and grooming. Thus its large callosities may be indirectly related to certain social behaviors which are genetically determined.

The substrate on which an animal sits also affects sitting posture. If the substrate is large and flat, the tail may limit sitting postures. However, if the substrate is small (such as a branch), then the tail can hang over the edge and not significantly prevent certain sitting postures. A flat surface gives a large contact area to the callosities, whereas the branch does not. A monkey presumably would not develop callosities larger than those which could make contact with the substrate.

The nature of the substrate is partially a function of the degree to which the monkey is arboreal or terrestrial. A determination of the degree to which each of the species in this study is arboreal or terrestrial is exceedingly difficult to make from available literature. This is largely because of different and sometimes ambiguous ways in which the concepts, arboreal and terrestrial, are used.

The terms arboreality and terrestriality may refer to two fairly separable concepts: (1) the behavioral patterns of the species; and (2) the anatomical adaptations of the form to a particular mode of life. This, however, does not imply that such words have no value, because the majority of forms that are anatomically arboreal would be behaviorally arboreal as well; and the same would be true with terrestrial forms.

Despite the limitations and necessary qualifications of these concepts, it is necessary to examine various macaque species to see whether the degree to which they are arboreal or terrestrial can be correlated with the size and shape of their callosities.

DeVore (1963) lumps all the macaques together as clearly adapted to terrestrial life. Ashton and Oxnard (1964, p. 9) state that *M. radiata, M. silenus,* and *M. nemestrina* are more terrestrial, living on the ground in barren, rocky country or on plains where they move on all fours. Jolly (1965, p. 26) ranks the macaques according to an index of arboreality in which 0 = terrestrial, 100 = arboreal. Jolly gave *M. mulatta, M. sylvana,* and *M. speciosa (M. arctoides)* an index of 55, *M. fuscata, M. radiata,* and *M. fascicularis* an index of 65, and *M. nemestrina* an index of 70. This index was arrived at by "giving each species a score for its use of trees in three basic activities of monkey life—feeding, traveling between food sources, and sleeping" (Jolly 1965, p. 25). Bernstein (1967c) judged that in one Malayan primate community, *M. nemestrina* was more terrestrial than *M. fascicularis.* He noted further that the "pigtail monkey is, nevertheless, still primarily an arboreal animal spending the largest portion of the daylight hours at medium levels of the forest" (1967c, p. 206). The alarm response for these two species is different, *M. nemestrina* fleeing on the ground and *M. fascicularis* through the trees. Tuttle (1969c) notes that there is a differential expression of adaptive shifts to terrestrial habitation in the various species of *Macaca;* that the long-tailed species, including *M. fascicularis,* are primarily arboreal in their foraging, feeding, and locomotive activities; that the intermediate-tailed species, such as *M. nemestrina* and

M. mulatta, are equally at home in trees and on the ground; and that the stump-tailed forms, such as *M. arctoides,* frequently forage on rocky, sparsely forested terrain. Such a division appears to be supported by the terrestrial adaptations of the hands.

Observations on *Macaca sinica* (toque monkey) in the Northwest dry zone of Ceylon indicate that this species is more consistently arboreal than *Presbytis entellus,* a species of langur with which it is sympatric. On the ground toque monkeys are palmigrade and do not engage in facultative digitigrady. In contrast with *M. sinica,* individuals of other species of long-tailed *Macaca,* e.g., *M. radiata* and *M. fascicularis,* are observed to be facultatively digitigrade in quasi-free-ranging conditions (Japan Monkey Center) and in zoos. The intermediate-tailed macaques, which are generally larger than the long-tailed forms, are more consistently digitigrade on the ground, though young animals and adult females may be seen to rest occasionally in palmigrade postures. Similarly, the stump-tailed macaques are among the heaviest species of the genus and are almost wholly digitigrade in terrestrial hand posture (Tuttle 1969c, pp. 193, 194).

In summary, therefore, there appears to exist a general correlation between arboreal behavior patterns and biological adaptations to an arboreal habitat in the macaques.

Also there appears to be a general correlation between the long-tailed macaques and arborealism and the short-tailed macaques and terrestrialism. Callosity size and shape seem to be a consistent aspect of this adaptation when viewed in terms of such functional factors as possible sitting postures and the nature of the substrate.

The Adaptive Advantage of Callosities. That ischial callosities are adaptive characteristics associated with the ability to sit upright has long been recognized (Elftman 1932, R. A. Miller 1945). Washburn (1957, p. 269) observed that since many forms of animals sit without developing callosities, there must be something special about the way they sit that would lead to such an adaptation. This particular way is to sit with the weight on the ischia, with legs extended and feet at higher elevation than the ischia, and with hands near feet. This sitting behavior he notes is "seen only in the primates with well developed ischial callosities, that is, the Old world monkeys and gibbons" (1957, p. 271). Since this type of posture is the normal sleeping posture, the basic adaptation of the callosities is not just for sitting but rather for sleep-sitting.

Summary

In my study of the tail, pelvocaudal musculature, and callosities of *M. fascicularis, M. mulatta, M. nemestrina,* and *M. arctoides,* a number of interspecific differences are observed and various trends are evident. These trends include a radical decrease in tail size and pelvocaudal musculature, a reversal in the relative importance of the pubocaudalis and iliocaudalis muscles, and an increase in the relative size of the ischial callosities and tuberosities, concomitant with changes in their shapes. These trends appear to be interrelated with certain behavioral and ecological differences of these species. One of the ecological differences involves shifts from arboreal to more terrestrial adaptations resulting in differences in posture and locomotor patterns. Terrestrial quadrupedalism differs from arboreal quadrupedalism because terrestrial substrates provide large areas and continuity. Thus, falls are of negligible consequence. Terrestrial quadrupedalism is less dependent upon a highly developed balancing mechanism. Inasmuch as a primary function of the tail in the long-tailed arboreal forms is for balance, the tail of terrestrial forms probably became less essential for balance and more subject to the effects of other selective pressures. And as macaques shifted from tropical, low altitudes to more temperate, higher altitudes, selective pressures probably were in favor of shorter tails. This decrease in tail length, with concordant decreases in the development of osteological and myological components, resulted in tails characterized by less internal stability and more mobility. Sitting behavior was consequently affected in two ways. First, increased mobility of the tail extended the repertoire of possible sitting postures to include sitting on the dorsal portion of the callosities. Second, terrestrial sitting offered a greater contact area for the callosities. This lead to larger callosities and tuberosities. Thus, the factors of habitat, behavior, and anatomy are interrelated in a manner that presents an internally consistent pattern for each of four species of *Macaca*.

These changes and trends—anatomical, ecological, and behavioral—are not to be conceived as occurring independently or in a particular chronological order. There were constant feedback effects in many directions and times.

In conclusion, the methodological perspective presented at the beginning of this paper and the results produced by employing it enables one to construct a reasonable and logically consistent explanation of morphological differences which is based on a functional examination of behavioral and ecological characteristics. The validity of any methodology must ultimately be judged by its capacities to provide bases for the construction of reasonable explanatory models.

12

RUSSELL TUTTLE
THE UNIVERSITY OF CHICAGO

Relative Mass of Cheiridial Muscles in Catarrhine Primates

A prodigious literature, dating from the late seventeenth century, exists on cheiridial muscles in nonhuman primates, particularly in the apes. Nearly all of these studies are descriptive, emphasizing muscle attachments, incidence, and "naked eye" assessments of relative size. Systematic quantitative studies on muscles in nonhuman primates have lagged notably behind mensural and statistical studies on bones which were pioneered and executed most comprehensively by Schultz (1930, 1936, 1956, 1961, 1963). Further, no standardized quantitative techniques of the sort developed for bones and body proportions (Martin 1928, Schultz 1929) have been forthcoming for muscles. The paucity of extensive quantitative studies on muscles may be explained as follows:

First, many anthropologists have elected to concentrate on the bones of extant primates with a view toward interpreting fossil primate specimens, especially hominid remains, the soft parts of which are nonexistent.

Second, there was a dearth of specimens for dissection in most research institutes and universities, so that extensive comparisons of different species

This investigation was supported mainly by NSF Grants nos. GS-834, GS-1888, and GS-3209, and by a Public Health Service Research Career Development Award no. 1-KO4-GM16347-01 from the National Institutes of Health. Supplementary support was provided by USPHS General Research Support Grant no. FR-5367 to The University of Chicago, NIH Grants nos. FR-00165 and RR-00165 to the Yerkes Regional Primate Research Center, NIH Grant no. FR-00164 to Delta Regional Primate Research Center, NIH Grant no. FR-00169-04 to the National Center for Primate Biology, and the Wenner-Gren Foundation for Anthropological Research.

The following persons have generously provided materials, facilities, and technical advice: Drs. C. E. Oxnard, G. Bourne, A. Riopelle, L. Schmidt, C. Rogers, W. C. O. Hill, J. Biegert, F. Ankel, and K. Leim, and Mrs. W. C. O. Hill and Mr. L. Pulchritudoff. Kenneth Glander, Jonathan Lewis, and Marne Hendrickson assisted me in statistical renderings of data. The illustrations were prepared by Miss Helen Cousar. The text was typed by Miss Sharon Kapnick.

were not possible. The collectors of primate osteological holdings for European and American museums and research institutes routinely defleshed specimens in the field. Those specimens that were shipped home in "spirits" were relatively few in number and were scattered widely among numerous laboratories. Thus the researcher was more or less dependent upon the fortuitous death of animals in nearby zoological gardens rather than having opportunities to obtain statistically significant samples of species that might be used strategically to elucidate particular functional and evolutionary problems. Further, there was little opportunity for careful dissection and weighing of muscles by biologists who were fortunate enough to work in the field, due to lack of proper laboratory facilities and equipment. One has only to read Sir Arthur Keith's account of his primatological research while a medical officer in Thailand (Keith 1940) to appreciate at once the poor conditions under which he labored and to be amazed that he could nevertheless extract such a wealth of descriptive information from his specimens.

Third, the principal investigator must devote a considerable period of time to prepare muscles for measurement and description. Technicians with a nominal amount of instruction can macerate and clean bones and arrange them for the principal investigator to describe and to measure. By contrast, it is probably imperative that the principal investigator himself perform all dissections and measurements of muscles. Thus, he must devote extensive amounts of time in order to extract enough data for statistical analysis. For example, it requires approximately two hours to remove carefully all muscles from one forearm and hand (or leg and foot) of a specimen; twenty minutes to arrange the muscles on a desiccation tray; and forty minutes to weigh the desiccated muscles on a torsion balance. Statistical rendering of data on one specimen may be estimated to take one hour. Thus, at least four hours may be spent with one forearm and hand or leg and foot. If, in addition, one notes the areas of attachment, architecture of fibers, development of tendon, relationship to joints, and also measures these parameters for each muscle, the time devoted to one specimen may well exceed twelve hours.

Fourth, and perhaps most important in the context of this symposium, there often are gross deficiencies or a virtual lack of research strategies and theoretical frameworks that would render quantitative data on muscles useful for evolutionary studies. Until recently, most investigators merely noted presence, absence, or "vestigial" condition of selected muscles, summarized this information for intertaxonal comparison, and attempted to use it to support or to deny taxonomic judgments and theories on human evolution. The fallacies intrinsic to the mere counting of traits without proffering how they might be weighted and explained in evolutionary biological contexts have been criticized thoroughly by functionally oriented scientists and probably need not be further belabored at this time. The main point here is that the nonfunctional biases that seem to have pervaded most previous discussions of muscle development in primates probably prevented many scientists from pursuing what might have proved to be more creative lines of inquiry.

Several shifts of emphasis in research have occurred recently which permit and often necessitate the quantification of muscles in comparative and evolutionary studies.

First, many students of comparative primate osteology have come to realize that a detailed understanding of muscle action and development is necessary to interpret features of the bones that have so preoccupied the attention of anthropologists during the past century. It may be possible in the near future to make well documented inferences about the postural and locomotive functions of fossil specimens based on the construction and stressing of models representative of fossil bones (Oxnard 1967, Oxnard and Tuttle 1969). But the reliability of these experimental methods is dependent upon the replication in similar models of conditions observed in living primates. Since muscle contraction and tendon act both to produce and to absorb stresses, we must have as complete a knowledge of the capacities of these structures as possible in order to produce refined models of bone mechanisms. These model systems, like relative mass of muscles, will always have limited value in evolutionary studies, but they promise to be of considerable robusticity if applied judiciously.

Second, live and well preserved specimens of many nonhuman primate species are available for experimentation and dissection due to the recently expanded commercial rape of the primates to provide exotic pets, zoological exhibits, and tools as part of the biomedical research boom. Advanced transportation methods, constituting not only jet transoceanic cargo service but also improved roads in the vicinity of primates, have greatly expedited the primate trade. The successful establishment of breeding colonies of monkeys, apes, and prosimians in zoological parks and primate research centers provides another source for sizable samples of specimens.

Third, improved apparatus for weighing small objects and the development of electronic desk calculators and computers permit relatively speedy recording and processing of data.

Previous Quantitative Studies

Having thus cursorily surveyed and, to some extent, contrasted the differences between past and present opportunities for comparative studies on muscles, let us now look more closely at some of the studies that have included consideration of muscle mass.

Insofar as I have been able to discern from available literature, Haughton, a British anatomist, first investigated muscle mass in certain monkeys (1864, 1865, 1873). Subsequently, several German anatomists determined muscle weights in various anthropoid primates. Notable among the German school were Langer (1879) and R. Fick (1895), who studied the "extremities" of *Pongo;* Michaelis (1903), who described selected muscles in the extremities of *Pongo, Pan troglodytes,* and *"Cynocephalus babuin";* and Schreiber (1936) who noted weights of selected muscles in the forearm of *Pan troglodytes.* In all of these studies, ratios based on muscle weights (or more commonly, merely the weights themselves) were presented as part of the descriptions of specimens. Comparative data were drawn largely from past studies by other observers on single or few specimens of man and apes. Little attempt was made by these authors to explain in biomechanical or historical biological contexts the sometimes remarkable differences between these forms in muscle mass. However, R. Fick (1895, p. 57) judiciously commented on the necessity for caution in using data from different observers and further noted that the extreme variability in the mass of some muscles (viz. gluteus maximus in man) precluded interpretations on the basis of rather scanty available specimens.

It was not until the second half of the present century that studies on relative mass of muscles in the fore- and hindlimbs of nonhuman primates again appeared in the literature with any regularity. This time, several groups of anthropologists and anatomists, including Washburn (1951a) and his students and associates (Tappen 1955, 1961; Ziegler 1964; Tuttle 1965, 1967, 1969a, 1970; Grand, 1967, 1968a, 1968b) in the United States; Preuschoft (1961, 1963, 1965) in Germany; and Ashton and Oxnard (1963) in England, revived the use of relative muscle weights to describe and in some instances to explain differences between the locomotor mechanisms of nonhuman primates and man.

Only a few investigators (Ashton and Oxnard 1963; Oxnard 1963; Tuttle 1965; 1967, 1969a, 1970, Grand 1967, 1968a, 1968b; Grand and Lorenz 1968) dissected sufficient numbers of specimens to present means, variance, standard errors of the mean, fiducial limits, and other quantitative parameters that could be used for intertaxonal comparisons and to test the statistical significance of observed differences. Further, only the studies of Ashton and Oxnard (1963) on muscles of the primate shoulder, Oxnard (1963) on selected forelimb muscles, and Tuttle on hominoid hand (1969a) and foot (1970) muscles actually presented thorough statistical analyses of data before attempting to explain their results in terms of biomechanics and the evolutionary biology of the animals.

Although Grand dissected and weighed muscles of samples that are large enough for refined statistical analysis, he merely presented basic data and mean value for major "functional groups" (or the muscles themselves) without adequately representing the variability that exists in each feature. Further because he weighed all muscles to the nearest decigram (0.1 gm.) his quantitative data are of limited value in studies on *Alouatta* (1968a, 1968b), practically useless in a study on the hindlimb of *Tarsius* (Grand and Lorenz 1968), and of intermediate worth in a study on the hindlimb of *Nycticebus coucang* (1967). This is particularly unfortunate (especially considering the rarity of specimens) in the study of *Tarsius* since weighing to the nearest decigram leads to more than one-half of the muscles in his four specimens having values of 0.1 gm.

The muscles of small animals such as *Tarsius* with total body weights of only 95 to 171 grams probably should be weighed in milligrams. In animals the size of howling monkeys, one also should weigh muscles initially in milligrams (or at least in centigrams) in order to accurately appraise the range of variation among muscles in subsequent intra- and interspecific comparisons. Grand's methodology not only greatly limits the value of his discussions on the meanings of relative muscle mass in the species that he studied, but also precludes or otherwise limits future comparisons of his quantitative data with that on muscles dissected by other observers.

Standing in contrast with Grand's methodology are the univariate statistical methods applied by Ashton and Oxnard (1963) to carefully weighed, wet shoulder muscles of primates. These authors presented the mean and 90 percent fiducial limits for relative mass of major "functional" groups of muscles and calculated the significance of intergroup differences between mean values with tests of "t." Their method of graphically representing data permits the reader to see at a glance the basic scope of variations within a group and the degree to which the groups chosen for comparison are separated from one another.

However, despite its refined statistical methodology, this study by Ashton and Oxnard (1963) has been criticized by Ripley (1967b) on grounds that their comparisons were made between presumed "locomotor groups" which, to a student of primate behavior, have little factual basis. Although the statistical study included 46 specimens, too few individuals of each species were available to perform detailed *intertaxonal* comparisons among lower taxa. In order to conduct statistical analyses, the authors were induced somehow to group their data otherwise. Hence they pooled available specimens into "locomotor groups" that included such diverse primates as *Alouatta, Ateles* and *Colobus* and *Propithecus, Nycticebus,* and *Perodictitus* under terms such as "semibrachiators" and "prosimian hangers" respectively. Consequently this study probably should be viewed as pilot in nature with regard to our understanding of shoulder musculature in primates, and it should be followed by more detailed studies that utilize more specimens of each species.

Tuttle (1969a, 1970) has utilized the statistical approach of Ashton and Oxnard in studies on cheiridial muscles in the Anthropoidea. These studies differ from those of Ashton and Oxnard not only in the anatomical regions and, to a lesser extent, the species chosen for study, but also in the methods by which data were grouped and muscles were treated prior to weighing. Tuttle dissected enough specimens of most species so that they could be grouped prior to statistical analysis more strictly according to taxonomic criteria instead of presumed "locomotor groups." Since locomotor activities are among the factors that one might wish to include in explanations of muscle development, analysis *between lower taxa* allows less circularity to enter into discussions of observed quantitative differences than if the specimens had been pooled initially according to "locomotor groups." Individuals of unique species are probably best represented as points until enough specimens are obtained to determine mean and fiducial limit values.

Dry weights were used by Tuttle (1969a, 1970) instead of wet weights, in an attempt to eliminate possible errors due to differential content of preservative fluids in the muscles. The denseness of fiber packing in different muscles might cause them to hold variable amounts of fluid. Thus, Tuttle desiccated all muscles at 60° C. until successive trial weighings indicated that constant weights were maintained in certain muscles representative of the size range of all muscles in the organ; then they were weighed individually and ratios were computed.

In summary, recent advances in instrumentation, the increased availability of specimens, and the development of statistical approaches to biological materials today permit more extensive and intensive studies of muscle mass and function in primates than could have been performed in the past. But before attempting to illustrate the extent to which the relative masses of cheiridial muscles appear to be correlated with the motile activities of selected anthropoid primates, let us turn briefly to a consideration of some problems that certain physiological factors pose in the use of data on muscle mass.

Some Physiological Factors that Affect Muscle Mass and Function

A wet or desiccated muscle on the pan of a balance is far removed from its active state in the living animal; and concordantly, the weight or volume of a muscle is only remotely representative of its physiological and functional parameters. Thus, great care must be exercised in interpreting possible functional and evolutionary significances of potential functional groups of muscles.

Surprisingly, there have been investigators (e.g. Pfuhl 1926, 1937; Tappen 1961) who have suggested that it is possible to calculate the force exerted by a muscle from its weight and fiber length. The errors of such simplistic approaches have been discussed by Gans and Bock (1965).

The force exerted by a muscle cannot be inferred from any combination of basic morphological measurements such as cross-sectional area, length of fibers, numbers of fibers, or weight (Gans and Bock 1965). Even refined electromyographic recordings only indicate when muscles are active without permitting precise estimates of the actual forces that they exert (Basmajian 1967).

The force exerted by a muscle depends upon many factors, including; (a) motor unit architecture and innervation; (b) contractile properties of its constituent proteins; (c) length and architectural arrangement of fibers (that is, whether predominently parallel-fibered, pennate, fusiform, etc.); (d) extent of attachment to bony surfaces and other muscles; (e) development of tendon versus contractile components; and (f) relationships of muscle attachments to joints and the structure of the joints across which they attach, among others. Thus intertaxonal comparisons of muscles and potential functional groups may suffer from deficiencies in not only morphological comparability but also, in many instances, physiological comparability. Gans and Bock (1965) have discussed features (c), (d), and (e). Feature (f) has been discussed by MacConaill (1946, 1949), Basmajian (1967), and others. Here I will draw upon recent studies and personal anecdotal observations in order to elaborate briefly on features (a) and (b).

A study of Wray (1969) indicates that baboon and man are notably different in mean number of muscle fibers per motor unit in the medial head of the gastrocnemius muscle. Comparing her own calculations from baboon specimens with those of Feinstein et al. (1955) for man, Wray (1969) found that man has in excess of 2½ times more muscle fibers per motor unit in this muscle than baboons have. Clearly, there is a need for more extensive comparative studies on selected muscles *using uniform techniques* to determine differences in motor unit architecture in primates and to relate these to possible differences in function.

A second feature for which we lack good comparative data in primates is the contractile properties of muscle fibers especially as these might be related to the chemical constituents and ultrastructure of muscle tissues. For instance, I have noticed that orangutans seem to be capable of more sustained contraction of their digital flexor muscles than chimpanzees, and that certain muscles in the two species seem to be affected differently by tranquilizers. But these subjective and episodic observations remain to be verified and elucidated by detailed behavioral, pharmacological, and molecular biological studies.

In summary, considering the many factors (only some of which are enumerated above) that may affect muscle weights, it is impossible to determine with any precision the physiological parameters that are represented in the dimension of muscle mass. Thus, the absolute values for individual muscles are generally of very limited worth in comparative studies.

However, intertaxonal comparisons may be performed on values derived from *ratios,* the numerators and denominators of which contain information from the same specimen. These ratios permit the assessment of relative differences between muscles in one organism to be compared with the differences between their counterparts in specimens of other taxa. In such comparisons, the major discrepancies of the measurements are held constant in both elements; hence, the comparative usage renders more valuable inferences drawn from such data. Interpretation of these differences are subsequently based on other sources of information and should be carefully executed in light of the notable limitations of the materials (Tuttle 1969a, p. 317).

Possible Correlations between Relative Muscle Mass and Prehensile Functions: Examples from Catarrhine Primates

The Hominoidea is an excellent superfamily in which to study possible morphological correlates of prehensile, propulsive, and manipulatory functions in hands and feet because of the notable contrasts that exist between species in major locomotor, foraging, and manipulatory modes. Pongid and hylobatid cadavers may be obtained for dissection due to deaths in primates centers and zoos, and live animals are available at these institutions for studies on locomotion, electromyography, joint movements, and other features for which experimental data are required. Further, the recent intensification of research on the naturalistic behavior of apes has greatly augmented the reports of earlier authors and has provided us with fairly complete profiles of locomotor and subsistence activities in gibbons, gorillas, chimpanzees, and to a lesser extent, orangutans.

By contrast with the Hominoidea, the Cercoipthecoidea evidence less obvious diversity in major locomotive, foraging, and feeding modes, although fundamental differences may be noted between certain colobine and cercopithecine species in substrate preferences and the parts of plants that predominate in their daily fares. Except for certain colobine species, adequate and representative samples of cercopithecoid primates may be obtained for dissection. Naturalistic behavioral studies provide sketchy but informative profiles of locomotive and subsistence activities in baboons, geladas, patas monkeys, vervets, guenons, macaques, mangabeys, hanuman langurs, and black-and-white colobus monkeys.

I have elected to discuss here only a few of the more remarkable features of relative muscle mass, emphasizing muscle groups that are involved in manual and pedal prehension, in order to illustrate the comparative strategy and synthetic approach that I have employed in studies on primate cheiridia. More complete presentations of data, methods, and bases for explanation of observed patterns in hominoid cheiridia have been published (Tuttle 1969a, 1970) and fuller expositions on cercopithecoid hands and feet are forthcoming.

Three basic sets of ratios are presented here: (a) ratios that compare selected groups of muscles in the hand with those in the foot; (b) ratios that compare selected groups of muscles within the forearm or hand; and (c) ratios that compare selected groups of muscles within the leg or foot.

Table 12–1. Ratios expressing the relative mass of selected prehensile muscles in catarrhine primates

Ratio	Dimensions
HF–1	Total forearm and hand muscles × 100
	Total leg and foot muscles
HF–2	Total manual lumbrical and interosseous muscles × 100
	Total pedal lumbrical and interosseous muscles
HF–3	Total intrinsic hallucal flexor, abductor and adductor mm. × 100
	Total intrinsic pollical muscles
EH–1	Extrinsic manual digital flexor muscles × 100
	Total forearm muscles
EH–2	Deep extrinsic manual digital flexor muscles × 100
	Flexor digitorum superficialis
IH–1	Total intrinsic thumb muscles × 100
	Total intrinsic hand muscles
IH–2	Adductor pollicis muscle × 100
	Total intrinsic hand muscles
IH–3	Thenar eminence muscles × 100
	Total intrinsic hand muscles
EF–1	Extrinsic pedal digital flexor muscles × 100
	Total leg muscles
EF–2	Peroneus longus muscle × 100
	Peroneus brevis muscle
IF–1	Total intrinsic hallucal flexor, abductor and adductor muscles × 100
	Total intrinsic foot muscles
IF–2	Adductor hallucis muscle × 100
	Total intrinsic foot muscles
IF–3	Intrinsic hallucal flexor and abductor muscles × 100
	Total intrinsic foot muscles

RATIOS COMPARING HANDS TO FEET

In the grossest comparison of hand and foot muscles, viz. a ratio (HF–1) expressing the relative weights of total forearm and hand muscles as a percentage of total leg and foot muscles (Figure 12–1), the lesser apes (*Hylobates, x = 143*) and orangutans (*x = 128*) evidence a clear predominence of the forelimb muscles; gorillas (*x = 108*) possess nearly equal development of total hand and foot muscles; and, chimpanzees (*x = 87*) and the cercopithecoid monkeys (*x = 67 to 91*) have total foot muscles developed relatively larger than total hand muscles.

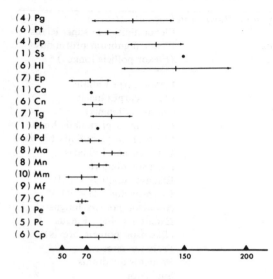

(4) Pg
(6) Pt
(4) Pp
(1) Ss
(6) Hl
(7) Ep
(1) Ca
(6) Cn
(7) Tg
(1) Ph
(6) Pd
(8) Ma
(8) Mn
(10) Mm
(9) Mf
(7) Ct
(1) Pe
(5) Pc
(6) Cp

50 70   150   200

Pg, *Pan gorilla;* Pt, *Pan troglodytes;* Pp, *Pongo pygmaeus;* Ss, *Symphalangus syndactylus;* Hl, *Hylobates lar;* Ep, *Erythrocebus patas;* Ca, *Cercopithecus aethiops;* Cn, *C. nictitans;* Tg, *Theropithecus gelada;* Ph, *Papio hamadryas;* Pd, *Papio doguera;* Ma, *Macaca arctoides;* Mn, *M. nemestrina;* Mm, *M. mulatta;* Mf, *M. fascicularis;* Ct, *Cercocebus atys;* Pe, *Presbytis entellus;* Pc, *P. cristatus;* Cp, *Colobus polykomos;* vertical bars, mean values; double-tipped arrows, 90 percent limits of population; dots, individual specimens. Numbers of specimens are indicated between parentheses beside each species.

Figure 12–1. Ratio HF–1, expressing the relative weight of total forearm and hand muscles as a percentage of total leg and foot musculature in catarrhine primates.

Among the samples of cercopithecoid monkeys, *Theropithecus gelada* ($x = 90$) and *Macaca arctoides* ($x = 91$) resemble most closely the pattern of *Pan troglodytes* among the apes.

The greatly enlarged forearm and hand muscles of the Hylobatidae and *Pongo* may be associated with elongation of the antebrachial and distal segments of the forelimb which presumably is associated with particular patterns of suspensory posturing, arm-swinging, and foraging-feeding behavior in each ape. The apparent emphasis on forelimb elongation in the African apes, geladas, and bear macaques may be considered to indicate less clearly an absolute elongation of the forelimb as opposed to the progressive shortening of the hindlimbs. Chimpanzees and gorillas possess in their forelimbs evidence of a past history that included a notable component of arboreal climbing and suspensory posturing, and they engage in these activities today. By contrast with the African apes, geladas and bear macaques may have achieved their present limb proportions through selection for particular patterns of terrestrial locomotion and foraging on irregular terrain, combined also with unique histories of climbing.

Comparisons of total manual lumbrical and interosseous muscles with total pedal lumbrical and interosseous muscles (HF–2) clearly separate the apes from the cercopithecoid monkeys (Figure 12–2). Gorillas ($x = 192$) have palm muscles nearly twice the total mass of pedal lumbrical and interosseous muscles. Chimpanzees ($x = 150$) and gibbons ($x = 147$) possess palm muscles approximately 1½ times the mass of their counterparts in

Table 12–2. Muscles and muscle groups mentioned in text and figures

	Total Forearm (= Extrinsic Hand; Antebrachial) Muscles	

Extrinsic flexors of the manual digits

Flexor digitorum superficialis
Flexor digitorum profundus
(Flexor pollicis longus)*

Other muscles included in total

Flexor carpi radialis
Flexor carpi ulnaris
(Palmaris longus)
Extensor carpi radialis longus
Extensor carpi radialis brevis
Extensor carpi ulnaris
Extensor digitorum
Extensor digiti II (et III)
Extensor digiti (IV et) V
Abductor pollicis longus
Extensor pollicis longus
(Extensor pollicis brevis)
Pronator teres
Pronator quadratus
Supinator

Total Intrinsic Hand Muscles

Total intrinsic thumb mm.

Thenar eminence mm.

Abductor pollicis brevis
Flexor pollicis brevis
Opponens pollicis

Adductor muscle

Adductor pollicis

Total palm mm.

Dorsal interossei

Dorsal interosseous II
Dorsal interosseous III [radial]
Dorsal interosseous III [ulnar]
Dorsal interosseous IV

Palmar interossei

Palmar interosseous II
Palmar interosseous IV
Palmar interosseous V
(Contrahens II)
(Contrahens IV)
(Contrahens V)

Lumbricals

Lumbrical II
Lumbrical III
Lumbrical IV
Lumbrical V

Other muscles included in total

Abductor digiti minimi
Flexor digiti minimi
Opponens digiti minimi

12–2. Continued

Total Leg Muscles

Extrinsic flexors of the pedal digits	Flexor digitorum tibialis Flexor digitorum fibularis
Other muscles included in total	Peroneus longus Peroneus brevis (Peroneus digiti V) Popliteus (Plantaris) Gastrocnemius Soleus Extensor digitorum longus Extensor hallucis longus Tibialis anterior (Abductor hallucis longus) Tibialis posterior

Total Intrinsic Foot Muscles

Total intrinsic hallucal mm.	Intrinsic hallucal flexor and abductor mm.	Abductor hallucis brevis Flexor hallucis brevis (Opponens hallucis)
	Adductor hallucis	Adductor hallucis—oblique head (Adductor hallucis—transverse head)
Total pedal lumbrical and interosseous mm.	Dorsal interossei	Dorsal interosseous II Dorsal interosseous III [medial] or II [lateral] Dorsal interosseous III [lateral] Dorsal interosseous IV
	Palmar interossei	Palmar interosseous II or III Palmar interosseous IV Palmar interosseous V (Contrahens II) (Contrahens IV) (Contrahens V)
	Lumbricals	Lumbical II Lumbrical III Lumbrical IV Lumbrical V
Other muscles included in total		Abductor digiti V Flexor digiti V (Flexor accessorius) (Extensor hallucis brevis) (Extensor digitorum brevis)

*Muscles and fasciculi enclosed by ( ) are not present in all individuals or species.

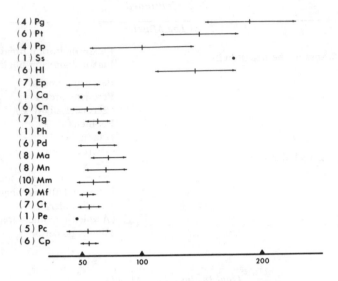

Figure 12–2. Ratio HF–2, expressing the relative weight of total manual lumbrical and interosseous muscles as a percentage of total pedal lumbrical and interosseous muscles in catarrhine primates. (Symbols as in Figure 12–1.)

the foot; the single specimen of siamang (179) is intermediate between gorillas and gibbons in ratio HF–2. Orangutans ($x = 98$) have nearly equal development of total manual and pedal lumbrical and interosseous muscles. In cercopithecoid monkeys, the pedal interosseous and lumbrical muscles are markedly larger than the palm muscles.

The emphasis on palm muscles in excess of pedal lumbrical and interosseous muscles in *Pan troglodytes* and the Hylobatidae may be related to particular usages of manual rays II–V for suspensory posturing and arboreal climbing. Further, in *Pan troglodytes,* manual rays II–V are importantly involved in a particular pattern of weight bearing—knuckle-walking—on the ground and on stout limbs. The pedal interosseous and lumbrical muscles in the Hylobatidae appear to be reduced absolutely concomitant with extreme emphasis on the hallucal ray in the foot. Similarly, the pedal interosseous and lumbrical muscles in *Pan gorilla* are notably small by comparison with those of chimpanzees, probably in association with reduction of pedal rays II–V for more efficient plantigrade terrestrial locomotion. The palm muscles of *Pan gorilla* are well developed probably, as in *Pan troglodytes,* in relation to particular functions of manual rays II–V for knuckle-walking.

A very different pattern emerges among hominoid and cercopithecoid primates when the relative masses of total intrinsic hallucal muscles are compared with total intrinsic pollical muscles (Figure 12–3, HF–3). Not only are species of apes notably separated from one another by values of ratio HF–3, but also certain cercopithecoid species are widely divergent from each other and to a greater extent than are the apes.

In the sample of *Colobus polykomos* ($x = 498$), the intrinsic hallucal muscles are, on the average, approximately five times more massive than the intrinsic pollical muscles, a result that is not surprising considering the greatly reduced thumb and sizable great toe in that species. It is interesting, however, to note that although the mean value of ratio HF–3 for *Presbytis cristatus* ($x = 454$) is somewhat less than that of *Colobus polykomos*, the two species are not significantly different from one another statistically ($p = 0.30–0.50$) in relative development of intrinsic hallucal and pollical muscles. It may be that these results truly represent closely similar developments of the hallucal and pollical muscles in the two colobine species even though the pollex of *Presbytis cristatus*, unlike that of *Colobus polykomos*, contains a full complement of bones. But it is also possible that the hallux of *Presbytis cristatus* is absolutely larger than that of *Colobus polykomos*, thus giving the appearance of equally reduced thumb muscles in the two species.

The position of the single specimen of *Presbytis entellus* (265) is compelling with regard to ratio HF–3, indicating that either its thumb is more heavily muscled or its hallux is less heavily muscled that those of its more strictly arboreal near-relative, *Presbytis cristatus*. But more specimens of *P. entellus* must be investigated before further explanatory statements are proffered.

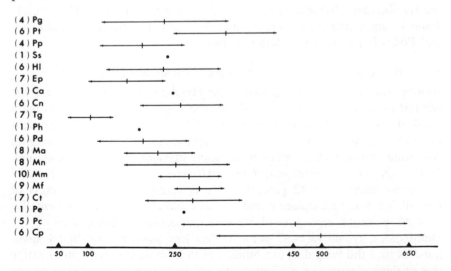

Figure 12–3. Ratio HF–3, expressing the relative weight of total intrinsic hallucal flexor, abductor, and adductor muscles as a percentage of total intrinsic muscles of the thumb in catarrhine primates. (Symbols as in Figure 12–1.)

The position of *Theropithecus gelada* ($x = 105$) in ratio HF–3 is outstanding. *Theropithecus* is the only available species that, on the average, evidences nearly equal development of intrinsic hallucal and pollical muscles. By comparison with macaques, baboons, mangabeys, and most other cercopithecoid species, the external and osseous hallux of geladas is markedly reduced in size. Male geladas do not grasp the hindlegs of the female during copulation as most macaques and baboons do. Reduction of the hallux in *Theropithecus* is probably related to the predominantly terrestrial habits of the species which rarely climb trees to forage or to sleep.

Erythrocebus patas ($x = 167$), a highly terrestrial species, also evidence relatively low values in ratio HF–3, especially by comparison with its arboreal near-relative, *Cercopithecus nictitans* ($x = 260$), and the most arboreal species of *Macaca*. *Papio doguera* ($x = 195$) and, to a lesser extent, *Macaca arctoides* ($x = 221$) fall intermediately between highly terrestrial species such as *Theropithecus* and *Erythrocebus,* and more arboreal species of *Cercopithecus* and *Macaca*. The entire macaque-mangabey-baboon group evidence a continuum of mean values for HF–3, with *Cercocebus atys* ($x = 280$) and *Macaca fascicularis* ($x = 291$) at the high end and *Papio* ($x = 195$) at the low end.

Pan troglodytes ($x = 335$) evidences the highest mean value in ratio HF–3 among the Hominoidea and it surpasses those of all available species of the Cercopithecoidea except *Presbytis cristatus* and *Colobus polykomos*. *Pongo pygmaeus* ($x = 194$), despite its markedly reduced hallucal components, still possesses intrinsic hallucal muscles that are approximately two times the mass of its intrinsic pollical muscles. *Pan gorilla* ($x = 232$) and the Hylobatidae (*Hylobates, x = 230$) exhibit values in ratio HF–3 nearly equal to one another and intermediate between those of *Pan troglodytes* and *Pongo* but closer to the latter species.

RATIOS OF MUSCLES IN FOREARM AND HAND

Among available catarrhine species, the Hylobatidae clearly evidence the greatest development of the extrinsic manual digital flexor muscles (Figure 12–4, EH–1). In both *Hylobates* ($x = 50$ percent) and *Symphalangus* ($x = 49$ percent) the flexor digitorum superficialis and profundus muscles constitute approximately one-half the total mass of forearm musculature. *Pan gorilla* ($x = 44$ percent), *Pan troglodytes* ($x = 42.5$ percent), and *Pongo pygmaeus* ($x = 42$ percent) possess nearly equal relative mass of long digital flexor musculature and compare closely with *Presbytis entellus* ($x = 43$ percent) among available cercopithecoid monkeys. Among the Hominoidea, *Homo* ($x = 37$ percent) has less massive long digital flexor muscles than the Pongidae, exhibiting a pattern most closely approximating that of *Papio doguera* ($x = 37$ percent) among the cercopithecoid monkeys. The catarrhine species in which the long digital flexor muscles are least massive is *Erythrocebus patas* ($x = 35$ percent). In ratio EH–1, *Erythroce-*

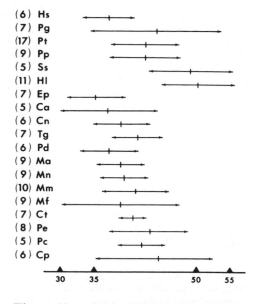

Figure 12–4. Ratio EH–1, expressing the relative weight of extrinsic flexor muscles of the manual digits as a percentage of total forearm muscles in catarrhine primates. (Hs, Homo sapiens; other symbols as in Figure 12–1.)

bus patas is significantly different from *Cercopithecus nictitans* ($p = 0.01$) but not from *C. aethiops*. *Cercopithecus aethiops* is not significantly different *C. nictitans* ($p = 0.30$).

Probably the most striking difference between hominoids and cercopithecoids in relative mass of muscles in the flexor compartment of the forearm is revealed by a ratio (EH–2) expressing the weight of the deep digital flexors as a percentage of the superficial flexor muscles (Figure 12–5).

The Hylobatidae (*Hylobates*, $x = 113$; *Symphalangus*, $x = 101$) exhibit nearly equal development of the flexor digitorum profundus and superficialis muscles. The Hylobatidae are the only catarrhine family in which individual superficial flexor muscles have been found that exceed the weight of the deep digital flexor muscles. The three species of the Pongidae are not significantly different from one another in ratio EH–2. They possess deep digital flexor muscles that are approximately 1.4 to 1.5 times the mass of flexor digitorum superficialis. In *Homo* ($x = 160$), the mean value of ratio EH–2 is somewhat higher than that of other species in the Hominoidea.

In the Cercopithecoidea, the prominent development of the deep long digital flexor muscles is evidenced by mean values of ratio EH–2 ranging from approximately 258 to 367. Comparisons of ratio EH–2 among the

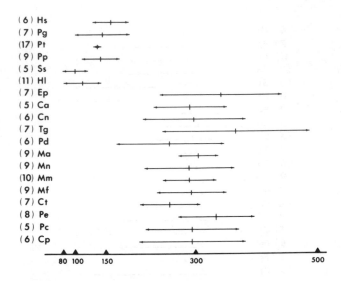

Figure 12–5. Ratio EH–2, expressing the relative weight of the deep extrinsic flexor musculature of the manual digits as a percentage of the flexor digitorum superficialis muscles in catarrhine primates. (Symbols as in Figure 12–4.)

presumed closest relatives of given species suggests that the more highly terrestrial members within each group possess relatively larger flexor digitorum profundus muscles and relatively smaller flexor digitorium superficialis muscles than their more arboreal near relatives. For instance, *Erythrocebus patas* ($x = 343$) has a higher mean value of ratio EH–2 than those of *Cercopithecus aethiops* ($x = 292$) and *Cercopithecus nictitans* ($x = 298$). But *Erythrocebus* is not statistically significantly different from *Cercopithecus* ($p = 0.10$) in ratio EH–2.

Theropithecus gelada ($x = 367$) exhibits a higher mean value for ratio EH–2 than *Papio doguera* ($x = 258$), the four available species of *Macaca* ($x = 290$ to 305) and *Cercocebus atys* ($x = 258$). Tests of "t" reveal that *Theropithecus gelada* is significantly or highly significantly different from *Papio doguera* ($p = 0.01–0.001$), *Macaca nemestrina* ($p = 0.01$), *M. mulatta* ($p = 0.01–0.001$), *M. fascicularis* ($p = 0.001$), and *Cercobebus atys* ($p = 0.001$), but probably not from *Macaca arctoides* ($p = 0.02$).

Papio doguera ($x = 258$) has a lower mean value of ratio EH–2 than would be expected if there were a simple relationship between degree of terrestriality and relative development of the deep and superficial long digital flexors.

Presbytis entellus ($x = 334$), the most terrestrial of available species of the Colobinae, exhibits a higher mean value in ratio EH–2 than those of *Presbytis cristatus* ($x = 294$) and *Colobus polykomos* ($x = 294$), the latter two species presenting nearly identical patterns. Tests of "t," however, do not indicate significant differences between *Presbytis entellus* and the other two populations of the Colobinae ($p = 0.10–0.05$).

The prominence of the flexor digitorum superficialis muscle in the Hominoidea may be related to special emphases on the middle phalanges as part of particular patterns of climbing and suspensory behaviors (*Hylobates, Symphalangus, Pongo, Pan troglodytes*); knuckle-walking (*Pan troglodytes, Pan gorilla*); or a past history of one or more of these activities (*Pan gorilla, Homo sapiens*). But much more knowledge on the biomechanics of middle phalanges during each of these functions must be obtained before fuller explanations can be proffered reliably.

Similarly, since requisite biomechanical studies have not been conducted, one can only speculate that predominance of the flexor digitorum profundus muscle in certain terrestrial cercopithecoids may be related to functions of the distal manual phalanges in propulsion somewhat like the "toe snap" mechanism in the human foot which assists to smooth out the walking stride.

Among catarrhine primates, *Homo sapiens* ($x = 39$ percent) exhibits the most prominent development of total intrinsic thumb muscles when they are expressed as a percentage of total intrinsic hand muscles (Figure 12–6,

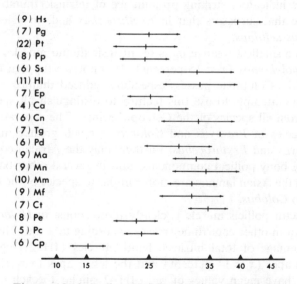

Figure 12–6. Ratio IH-1, expressing the relative weight of total intrinsic thumb muscles as a percentage of total intrinsic hand muscles in catarrhine primates. (Symbols as in Figure 12–4.)

IH–1). The mean value of ratio IH–1 in *Theropithecus gelada* ($x = 37$ percent) closely approximates that of *Homo sapiens* but it is not significantly different from certain cercopithecid species such as *Cercopithecus nictitans*

$(p = 0.10)$, *Macaca arctoides* (0.02), *M. nemestrina* $(p = 0.20)$, *M. fascicularis* $(p = 0.02)$, and *Cercocebus atys* $(p = 0.30)$.

Among apes, the hylobatids (*Hylobates, x* = 33 percent, *Symphalangus, x* = 33 percent) are highly significantly different from the pongids in relative development of intrinsic thumb muscles. The Hylobatidae are the hominoid family closest to *Homo* in mean value of ratio IH–1, although the two groups are highly significantly different statistically. The Hylobatidae evidence a pattern in ratio IH–1 closely similar to *Cercopithecus nictitans* (*x* = 33 percent) and several species of *Macaca*.

The three species of Pongidae (*x* = 24 percent to 25 percent) evidence lower values of ratio IH–1 than do the majority of the Cercopithecinae. The Pongidae evidence patterns intermediate between *Erythrocebus* (*x* = 25 percent) and *Cercopithecus* (*x* = 26 percent), on the one side, and *Presbytis spp.* (*x* = 23 percent to 24 percent) on the other. Eyeball assessments of development of pollical intrinsic muscles in *Erythrocebus* and *Presbytis* clearly reveal an absolute diminution of musculature which correlates with the diminutive appearance of their external thumbs. The Pongidae evidence a notable swelling of pollical muscles in the skinned hand, but the external pollex, expecially in *Pongo,* appears to be short. Nevertheless, superficial examinations indicate a striking prominence of intrinsic thumb muscles in the Pongidae that surpasses that in *Erythrocebus* and, to a lesser extent, *Cercopithecus aethiops.*

Clearly, the smallest relative mass of intrinsic thumb muscles is observed in *Colobus polykomos* (*x* = 15 percent). It is noteworthy that the generalization that the Colobinae possess especially reduced thumbs may be misleading if one attempts to use this feature to distinguish all species of the Colobinae from all species of the Cercopithecinae. The separation between available species of *Presbytis* and *Colobus* is much greater than that between *Presbytis* and *Erythrocebus*. Further, only the African colobines evidence loss of bony pollical components; and in several other basic features of the pollex the Asian langurs are more similar to species of the Cercopithecinae than to *Colobus.*

The adductor pollicis muscle is clearly more robust in *Homo* (*x* = 17.5 percent) than in other catarrhine species according to a ratio that expresses it as a percentage of total intrinsic hand muscles (IH–2, Figure 12–7). The African apes (*x* = 11 percent) and the lesser apes (*x* = 12 percent to 13 percent) have mean values of ratio IH–2 similar to each other and to mean values of several cercopithecine species. *Pongo* exhibits relatively smaller adductor pollicis muscles (*x* = 9 percent) than other species of the Hominoidea, but it is not clearly separated from certain species of cercopithecoid monkeys.

Erythrocebus patas (*x* = 6 percent), *Cercopithecus aethiops* (*x* = 6 percent), and, to a lesser extent, *Cercopithecus nictitans* (*x* = 8 percent), *Presbytis entellus* (*x* = 8 percent), and *Presbytis cristatus* (*x* = 8 percent)

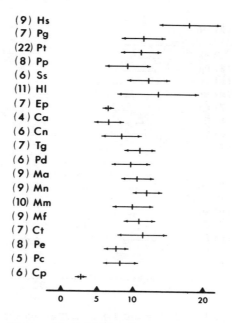

```
 (9) Hs
 (7) Pg
(22) Pt
 (8) Pp
 (6) Ss
(11) Hl
 (7) Ep
 (4) Ca
 (6) Cn
 (7) Tg
 (6) Pd
 (9) Ma
 (9) Mn
(10) Mm
 (9) Mf
 (7) Ct
 (8) Pe
 (5) Pc
 (6) Cp

      0      5     10          20
```

Figure 12–7. Ratio IH–2, expressing the relative weight of the adductor pollicis muscles as a percentage of total intrinsic hand muscles in catarrhine primates. (Symbols as in Figure 12–4.)

exhibit adductor pollicis muscles relatively smaller than those of *Theropithecus* ($x = 11$ percent), *Papio* ($x = 9$ percent), *Macaca* ($x = 10$ percent to 12 percent), and *Cercocebus* ($x = 11$ percent). The smallest adductor muscles among the catarrhine species are possessed by *Colobus polykomos* ($x = 3$ percent).

In a ratio expressing the thenar eminence muscles (abductor pollicis brevis and flexor pollicis bevis) as a percentage of total intrinsic hand muscles (Figure 12–8, IH–3), *Homo sapiens* is not as outstanding among catarrhine primates as it is in relative mass of adductor pollicis muscle. *Homo* ($x = 23$ percent) evidences a pattern of ratio IH–3 similar to those of several cercopithecoid monkeys (Figure 12–8).

The Hylobatidae ($x = 20$ percent to 21 percent) most closely resemble *Homo* among the Hominoidea in ratio IH–3. The Pongidae are markedly distinct from *Homo,* the Hylobatidae, and most species of the Cercopithecinae in ratio IH–3. The Pongidae most closely resemble the Colobinae in relative development of thenar eminence muscles. Both the Pongidae ($x = 14$ percent to 15 percent) and the Colobinae ($x = 12$ percent to 16 percent) possess mean values of ratio IH–3 that are considerably lower than those of most other available species of catarrhine primates.

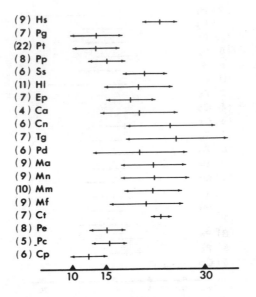

Figure 12–8. Ratio IH–3, expressing the relative weight of thenar eminence muscles as a percentage of total intrinsic hand muscles in catarrhine primates. (Symbols as in Figure 12–4.)

Among the Cercopithecinae, *Erythrocebus* ($x = 19$ percent) and *Cercopithecus aethiops* ($x = 20$ percent) exhibit mean values of ratio IH–3 lower than their arboreal nearest relative, *Cercopithecus nictitans* ($x = 25$ percent). Tests of "t" reveal a highly significant difference between *Erythrocebus* and *C. nictitans* ($p = 0.001$). But the difference between the two species of *Cercopithecus* is not significant ($p = 0.05$).

A simple general correlation between degree of terrestriality and low values of IH–3 is denied by the fact that *Theropithecus* ($x = 26$ percent) evidences the highest mean value of IH–3 among the available species. *Papio doguera* ($x = 20$ percent) compares with *Cercopithecus aethiops* and possesses a somewhat lower, but nonsignificantly different, mean value ($p = 0.20–0.50$) by comparison with available species of *Macaca* ($x = 21$ percent to 22.5 percent).

RATIOS OF MUSCLES IN LEG AND FOOT

Differences between available catarrhine species in muscles effecting grasping actions by the foot are assessed here by five ratios (EF–1, EF–2, IF–1, IF–2, IF–3).

In ratio EF–1, the relative mass of the pedal long digital flexor muscles (flexor digitorum tibialis and flexor digitorum fibularis) is expressed as a percentage of total leg muscles. *Pongo* ($x = 29$ percent) and, to a lesser extent, the Hylobatidae (*Hylobates,* $x = 25.5$ percent) are clearly outstanding among catarrhine primates in possessing large pedal digital flexor muscles (Figure 12–9). The mean value of ratio EF–2 is not significantly different statistically between *Pan gorilla* ($x = 19$ percent) and *Hylobates* because of the extreme range exhibited by the former species. The gorilla sample is comprised of only four individuals, two of which were infants and two of which were large juveniles or young adults. The notable size difference represented in the gorilla sample and the apparent selective onto-

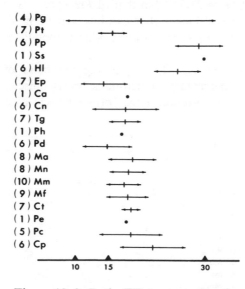

Figure 12–9. Ratio EF–1, expressing the relative weight of the extrinsic digital flexor muscles of the foot as a percentage of total leg musculature in catarrhine primates. (Symbols as in Figure 12–1.)

genetic increase in certain muscles in the leg of gorillas may explain the particular nature of this statistical population estimate.

Among apes, chimpanzees exhibit the least development of pedal long digital flexor muscles. The mean value of *Pan troglodytes* ($x = 15$ percent) in ratio EF–1 compares most closely with that of *Papio doguera* among the Cercopithecoidea.

Erythrocebus ($x = 14$ percent) exhibits, on the average, less development of the long digital flexor muscles than *Cercopithecus nictitans* ($x = 17.5$ percent) but the difference is not highly significant statistically ($p = 0.01–0.02$). Similarly, *Papio doguera* ($x = 15$ percent) exhibits relatively less developed long digital flexor muscles than species of *Macaca* ($x = 17$ percent to 19 percent), the differences between the two genera evidencing

statistical significance ($p = 0.001–0.01$). But *Theropithecus* ($x = 17$ percent) more closely approximates the pattern exhibited by certain populations of *Macaca* than of *Papio* in ratio EF–1. Thus, *Theropithecus* again
precludes comprehensive generalizations on the relationship between extreme terrestriality and reduction of cheiridial muscles related to grasping
functions of the pedal digits.

Among the Cercopithecoidea, *Colobus polykomos* ($x = 22$ percent) evidences the highest value of ratio EF–1, a result not surprising considering
its highly arboreal habitus. But the mean value in *Colobus* is not significantly different ($p = 0.05$) from that of *Presbytis cristatus* ($x = 18$ percent), a highly arboreal langur that exhibits a pattern of ratio EF–1 not
markedly different from those of several other cercopithecoid species (Figure 12–9).

Ratio EF–2 (Figure 12–10), expressing the mass of the peroneous
longus muscle as a percentage of peroneus brevis muscle, is employed to
indicate the relative importance of the former muscle as a flexor-adductor

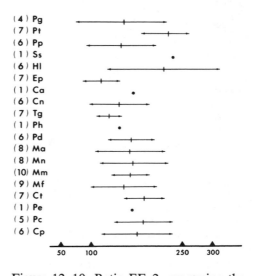

Figure 12–10. Ratio EF–2, expressing the
relative weight of the peroneus longus
muscle as a percentage of the peroneus
brevis muscle in catarrhine primates.
(Symbols as in Figure 12–1.)

of the hallux. According to ratio EH–2, *Pan troglodytes* ($x = 230$) and
the Hylobatidae (*Hylobates*, $x = 220$) possess greatest relative development of the peroneus longus muscle among available catarrhine species. By
contrast with *Pan troglodytes*, *Pan gorilla* ($x = 150$) and *Pongo pygmaeus*
($x = 150$) evidence considerably less predominance of the peroneus longus
muscle. This result is not surprising considering the reduced state of the

hallux in *Pongo* and the notable webbing and reduced grasping capacities of the hallux exhibited by *Pan gorilla.*

Among the Cercopithecoidea, *Erythrocebus* ($x = 120$) and *Theropithecus* ($x = 130$) exhibit the lowest values of ratio EF–2. *Papio doguera* ($x = 170$) exhibits a pattern of EF–2 closely similar to those of several of the species of *Macaca* ($x = 150$ to 170). *Cercocebus* ($x = 190$) is somewhat outstanding by comparison with *Papio* and *Macaca;* its relatively high mean value in ratio EF–2 compares most closely with *Presbytis cristatus* ($x = 180$) among the Cercopithecoidea.

The intrinsic muscles of the foot, like those of the hand, evidence marked contrasts not only between available species of the Hominoidea but also between species of the Cercopithecoidea.

The Hylobatidae (*Hylobates,* $x = 54.5$ percent) exhibit greatest development of intrinsic hallucal muscles according to ratio IF–1 (Figure 12–11) which expresses the weight of intrinsic hallucal muscles as a percentage of total intrinsic foot muscles. *Pan troglodytes* ($x = 47$ percent) exhibits second greatest relative size of intrinsic hallucal muscles among available catarrhine species. The mean value of ratio IF–1 in *Pan gorilla* ($x = 41.5$ percent) is less than that of *Pan troglodytes* and nearer to the mean values of *Macaca nemestrina* ($x = 39$ percent) and *Cercocebus* ($x = 38$ percent). *Pongo* ($x = 28$ percent) evidences markedly smaller intrinsic hallucal muscles by comparison with other species of the Pongidae. *Pongo* compares

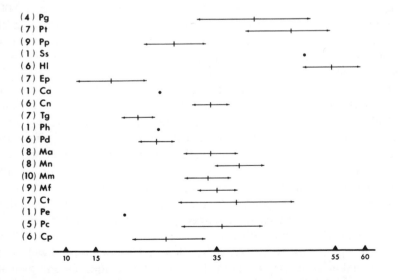

Figure 12–11. Ratio IF–1, expressing the relative weight of total intrinsic hallucal flexor, abductor, and adductor muscles as a percentage of total intrinsic foot musculature in catarrhine primates. (Symbols as in Figure 12–1.)

most closely with *Colobus polykomos* ($x = 27$ percent) among the Ceropithecoidea.

Erythrocebus ($x = 18$ percent) exhibits the lowest mean value of IF–1 among available species of catarrhine primates. Diminution of the intrinsic hallucal muscles is particularly marked in *Erythrocebus* by comparison with *Cercopithecus nictitans* ($x = 34$ percent), its nearest arboreal relative among available species. The single specimen of *Cercopithecus aethiops* (26 percent) its intermediate between *Erythrocebus* and *C. nictitans,* a result that corresponds neatly with the dual arboreal-terrestrial habitus of *C. aethiops.*

Among the Cercopithecinae, *Theropithecus* ($x = 22$ percent) exhibits the second smallest value of ratio IF–1. *Theropithecus* is significantly different ($p = 0.001–0.01$) from *Papio* ($x = 25.5$ percent). But the geladas and baboons are closer to one another than either group is to the macaques and mangabeys (Figure 12–11).

The three available species of the Colobinae are notably separated from one another by ratio IF–1. The single specimen of *Presbytis entellus* (20 percent) is remarkably distinct from its nearest relative, *Presbytis cristatus* ($x = 36$ percent), falling outside the lower 90 percent confidence limit of the *Colobus* sample ($x = 27$ percent) and beyond the lower 90 percent confidence limit and the mean of such highly terrestrial forms as *Papio* and *Theropithecus,* respectively (Figure 12–11). *Presbytis cristatus* compares favorably with *Macaca* in relative mass of intrinsic hallucal musculature. By contrast, *Colobus polykomos* evidence highly significantly ($p = 0.001$) less relative mass of intrinsic hallucal muscles than *P. cristatus* ($p = 0.001$).

Ratio IF–2 (Figure 12–12), expressing the relative weight of the adductor hallucis muscle as a percentage of total intrinsic foot musculature, generally replicates, to a great extent, the pattern exhibited by ratio IF–1 (Figure 12–11). The Hylobatidae, especially *Hylobates lar* ($x = 36$ percent), exhibit greatest relative mass of the adductor hallucis muscle among available species. *Pan gorilla* ($x = 20$ percent) and *Pan troglodytes* ($x = 21$ percent) are not significantly different from each other in ratio IF–2; neither do they present patterns markedly different from certain species of *Macaca* and *Cercocebus.* *Pongo* ($x = 15$ percent) evidences the smallest adductor hallucis muscle among available species of the Hominoidea.

Erythrocebus ($x = 6$ percent) exhibits the smallest adductor hallucis muscle among available catarrhine species; its mean value for ratio IF–2 stands in marked contrast with that of *Cercopithecus nictitans* ($x = 17$ percent). All specimens of *Erythrocebus* lacked the transverse head of the adductor hallucis muscle, a feature in which it is unique among available catarrhine species. The single specimen of *Cercopithecus aethiops* (13 percent) is intermediate between *Erythrocebus* and *C. nictitans,* but somewhat closer to the latter species in ratio IF–2. *Theropithecus* ($x = 9$ percent), *Papio hamadryas* (11 percent), and *Papio doguera* ($x = 11$ percent) are

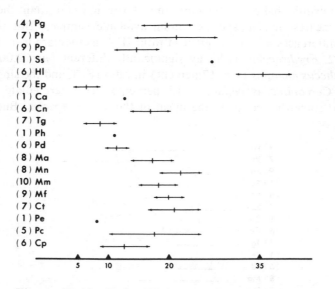

Figure 12–12. Ratio IF–2, expressing the relative weight of the adductor hallucis muscle as a percentage of total intrinsic foot musculature in catarrhine primates. (Symbols as in Figure 12–1.)

clearly separated from *Macaca* ($x = 17$ percent to 22 percent) and *Cercocebus* ($x = 21$ percent) by ratio IF–2.

The single specimen of *Presbytis entellus* (8 percent) falls outside the lower 90 percent confidence limits of the populations of *Presbytis cristatus* ($x = 17.5$ percent) and *Colobus polykomos* ($x = 13$ percent) and within the population estimates of *Erythrocebus* and *Theropithecus. Colobus* exhibits a mean value of ratio IF–2 lower than that of *P. cristatus,* but the difference is not highly significant statistically ($p = 0.02$).

Ratio IF–3 (Figure 12–13), expressing the relative mass of intrinsic hallucal flexor and abductor muscles (flexor hallucis brevis and abductor hallucis brevis) as a percentage of total intrinsic hallucal muscles, presents patterns somewhat different from those of ratios IF–1 and IF–2.

According to ratio IF–3, *Pan troglodytes* ($x = 26.5$ percent) exhibits greatest development of short hallucal flexor and abductor muscles, followed by *Pan gorilla* ($x = 22$ percent) and the single specimen of *Symphalangus syndactylus* (23 percent). *Hylobates* ($x = 19$ percent) exhibits values of IF–3 higher than those of *Pongo* ($x = 14$ percent) and most species of the Cercopithecoidea. The mean value of *Hylobates* in ratio IF–3 is closest to that of *Presbytis cristatus* ($x = 18.5$ percent) among the Cercopithecoidea.

Terrestrial species of the Cercopithecoidea, such as *Erythrocebus* ($x = 11$ percent), *Theropithecus* ($x = 14$ percent), *Papio doguera* ($x = 14$ percent), and *Presbytis entellus* ($x = 12$ percent), evidence a trend toward smaller intrinsic hallucal abductor and flexor muscles than their more arboreal nearest related species among available samples. But the interspecific differences are not as great in ratio IF–3 as they are in ratios IF–1 and IF–2. *Erythrocebus* is highly significantly different ($p = 0.001$) from *Cercopithecus nictitans* ($x = 17$ percent) in ratio IF–3, and the single specimen of *Cercopithecus aethiops* (13 percent) lies intermediate between them and somewhat closer to the mean of the former species. But in ratio

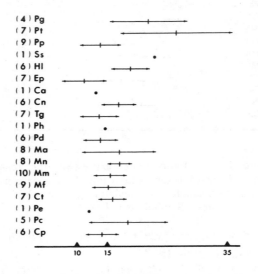

Figure 12–13. Ratio IF-3, expressing the relative weight of intrinsic hallucal flexor and abductor muscles as a percentage of total intrinsic foot musculature in catarrhine primates. (Symbols as in Figure 12–1.)

IF–3, *Theropithecus* and *Papio doguera* are highly significantly different ($p < 0.001$) only from *Macaca nemestrina* ($x = 17$ percent) among available species of *Macaca* and *Cercocebus*.

The single specimen of *Presbytis entellus* falls near the lower 90 percent confidence limits of *Presbytis cristatus* and *Colobus polykomos* ($x = 14$ percent). The latter two colobine species are not highly significantly different from one another ($p = 0.01–0.02$) in relative mass of intrinsic hallucal flexor and abductor muscles.

Overview of Results and Future Studies

The 13 ratios chosen to exhibit relative mass of selected cheiridial muscles, especially those related to grasping functions of hands and feet, evidence imperfect but remarkable trends that may be related to specific aspects of behavioral complexes in certain species.

The Hylobatidae consistently evidence notable prominence of muscles that are related to grasping functions in hands and feet. They exhibit remarkable development of manual and pedal digital flexors, the peroneus longus muscle, and intrinsic muscles that effect pollical and hallucal prehension.

Special features of the hand muscles in the Hylobatidae, such as the extreme development and unique attachments of the flexor digitorum superficialis muscle and palm muscles, are probably related to their particular patterns of ricochetal arm-swinging and suspensory feeding behavior.

At first glance, the extreme development of the forelimbs in the Hylobatidae might draw one's attention away from their hindlimbs. But detailed studies on the hindlimbs reveal their particular importance in the locomotive repertoire of the Hylobatidae. The large heavily muscled halluces are of considerable utility in gibbons and siamangs while climbing branchless tree trunks, vines, and steeply inclined limbs, and while running bipedally along horizontal limbs, a behavior in which they seem to engage more frequently than any other catarrhine primate.

Pongo is unique among the catarrhine primates in possessing remarkable emphasis on flexor muscles and grasping functions of digits II–V in both hands and feet. Concordantly, the extrinsic and, to a lesser extent, certain intrinsic pollical and hallucal muscles are relatively reduced in *Pongo*. In orangutans, as in the lesser apes, superficial morphological examinations of the relative development of forelimbs and hindlimbs might result in underestimates of the importance of the hindlimbs in the repertoire of motile activities. Field and laboratory observations indicate that pedal prehension is a prominent constituent in the suspensory behaviors of orangutans, probably more so than in any other ape. The special functional-morphological complex in the hindlimbs of *Pongo,* emphasizing prehensile functions of pedal digits II–V, may have evolved partly in response to selective pressures in swamp-forest habitats of Southeast Asia (Tuttle 1968, 1970).

The African apes (*Pan troglodytes, Pan gorilla*) do not exhibit the extreme emphasis on total extrinsic and intrinsic hand muscles relative to total extrinsic and intrinsic foot muscles that is evidenced in the Asian apes. This may be correlated with greater employment of the hindlimbs in climbing tree trunks and in terrestrial plantigrade locomotion and the relative infrequency of suspensory posturing in the African apes. Differences between chimpanzees and gorillas are more extreme in relative mass of pedal muscles than in relative mass of hand muscles (Tuttle 1970). *Pan gorilla* exhibits less developed peroneus longus muscles than *Pan troglodytes* does, perhaps in relation to a lower frequency of arboreal climbing and relatively less capacity for hallucal grasping in *Pan gorilla*. The notably smaller pedal lumbrical and interossesous muscles in gorillas, by comparison with those of chimpanzees, correlates well with the more exclusive use of the foot for plantigrade terrestrial locomotion and concomitant shortness of pedal digits II–V in gorillas.

Colobus polykomos is outstanding among catarrhine primates in possessing relatively diminutive pollical muscles, evidence of relative diminution in intrinsic hallucal muscles, and, to a lesser extent, emphasis on extrinsic flexor muscles of the pedal digits and emphasis on long flexor muscles of manual digits II–V. I have observed that hauling movements by the forelimb and pedal grasping are important constituents in the locomotive repertoire of free-ranging *Colobus polykomos* in Arusha National Park, Tanzania. A hook-like manual grasp, employing only digits II–V, may be of special importance to colobus monkeys while running on contiguous leafy substrata and while regaining vantage points atop limbs subsequent to leaps onto pliable substrata. Prominent pollices could be a hindrance during such hauling actions unless they were equipped with special mechanisms for quick release in line with digits II–V and the avoidance of snags. The hallux probably has not been greatly reduced in *Colobus* because selective pressures are operant for gripping less flexible structures while climbing in proximal sections of trees and for securing positions atop branches while sitting-feeding, reclining-feeding, and sitting-sleeping.

Presbytis cristatus and *Presbytis entellus,* like *Colobus,* evidence relative diminution of pollical muscles, but not to the extreme degree shown in the African Colobinae. *Presbytis cristatus* does not demonstrate remarkable development in any other group of manual and pedal prehensive muscles by comparison with other arboreal cercopithecoid species. By contrast, available samples of *Presbytis entellus* evidence trends toward special emphasis on the flexor digitorum profundus muscle relative to the flexor digitorum superficialis muscle and a reduction of the intrinsic hallucal muscles. Both of these features are developed to a greater extent in certain more strictly terrestrial cercopithecine monkeys. Hence they might be related also to terrestrial aspects of the dual terrestrial-arboreal habitus of hanuman langurs.

Erythrocebus patas, Theropithecus gelada, and, to a lesser extent, *Papio doguera* and *Papio hamadryas* are outstanding among the Cercopithecinae in relative development of certain muscles that subserve manual and pedal prehensile functions.

Erythrocebus and *Theropithecus* evidence remarkable reduction of intrinsic hallucal muscles, especially the adductor hallucis muscle. In *Erythrocebus,* both the long pedal digital flexor muscles and the peroneus longus muscle also evidence relatively marked diminution, especially by comparison with arboreal species of *Cercopithecus, Macaca,* and *Cercocebus.*

Theropithecus also exhibits a trend toward less relative mass of the peroneous longus muscle, but virtually no notable difference from *Cercopithecus, Macaca,* and *Cercocebus* in relative mass of the extrinsic pedal digital flexor muscles. Whether this may be correlated with greater cursorial propensities in *Erythrocebus* as compared with *Theropithecus* might be revealed by more detailed biomechanical and behavioral investigations.

Papio doguera, like *Erythrocebus,* exhibits relatively less developed long pedal digital muscles, but the relative development of the peroneus longus muscle in *Papio* is not notably different from that in *Macaca.* It is possible that this complex of features relates to a functional compromise between reduction of pedal digits II–V for effective terrestrial cursorial locomotion and retaining powerful hallucal grasp for climbing trees.

Available species of the Cercopithecinae generally are less separated from one another by features related to grasping functions of the hands than by those of feet; and the nature of existent differences are not readily explained on the basis of locomotive modes. For instance, *Erythrocebus* possesses the most diminutive intrinsic pollical muscles, whereas *Theropithecus* possesses the most robust intrinsic pollical muscles among the Cercopithecinae. If the relative differences between the two species in fact represents absolute differences, they might be explained on the basis of manipulative behaviors instead of locomotive modes.

Data are not available on the pedal muscles of *Homo.* Thus, interspecific comparisons between *Homo* and other catarrhine primates will be confined here to relative development of selected hand muscles.

The clear preeminence of the thumb muscles, especially adductor pollicis, in the hand of *Homo* probably is related to aspects of the wide variety of manipulatory activities to which it is adapted.

It is also interesting to note that although the hand of *Homo* is not used commonly for arm-swinging, knuckle-walking, or other locomotive activities particular to the Pongidae, it exhibits a relationship between relative development of the flexor digitorum superficialis and flexor digitorum profundus muscles that is remarkably closer to that of the Pongidae than to that of other catarrhine Primates. This should direct the attention of experimental scientists interested in human evolution to the middle phalanges as a special focus of research on the hand.

13

JOHN V. BASMAJIAN
EMORY UNIVERSITY

Biomechanics of Human Posture and Locomotion: Perspectives from Electromyography

Today the student of human evolution finds himself tantalized by a host of deducted possibilities. In the special area of posture and locomotion, the clues and guesses which give birth to the hypotheses are subject to new scrutiny, but the simple fact remains that scientists have not extended EMG kinesiology to nonhuman primates. The value of such studies, which we have now begun, speaks for itself.

Electromyographic kinesiology and its sister, *biomechanics,* do not replace observation and visuotactile experiments, both of which have been with us for many years. The deductive methods have yielded many dividends, but these returns have become fewer and smaller with the passing years. Alone, they have become nonproductive, but they provide, through frustration, a stimulus for new growth. Thus they lead to two new types of endeavor which should indicate the direction in which physical anthropology must move in the study of the evolution of man's posture and gait. These two are theoretic biomechanics, and EMG of nonhuman primates and man.

Theoretic Biomechanics

The general laws of kinesiology applicable to all vertebrates are derived from standard physics and engineering, yet they rarely enter into the education (or even the understanding) of many biological scientists, including anthropologists. In the present context, I cannot hope to do more than stimulate the thinking of the host of scientists who should become conversant with this discipline. For considerably greater depth, a recent book (MacConaill and Basmajian 1969) will provide a basis of understanding plus additional leads to new ideas. Here I will glean from that book a sampler that requires neither mathematics nor mathematical thinking.

COMPOSITE MOVEMENTS OF JOINTS

At all joints the most common single movements are *composite movements,* swing and spin together. All hinge joints show this type of motion. For example, the terminal phalanx of an index finger in flexion is also laterally rotated, so that it is brought into a good posture for grasping, say, a needle between itself and the thumb; and the thumb is pronated (rotated ulnarly) in opposition. Correspondingly, this index phalanx rotates medially as it moves into full extension. At the elbow, the ulna is also pronated in full extension and supinated during flexion.

CLOSE-PACKED POSITION OF JOINTS

For every joint there is one position in which there is maximal contact between the male and female surfaces of the mating pair; this is the *close-packed position.* In this position the articular surfaces are pressed firmly together, and the bones to which they belong cannot be separated by traction across the joint. This is because the chief capsular ligaments of the joint are then in the state of maximal tension; they are made taut by a spin that twists the tightened joint capsule and extracapsular ligaments. The close-packed position is the only one in which capsular ligaments may "substitute" for muscles.

HABITUAL MOTIONS

The habitual motions at a joint bring it either toward or away from its close-packed position. For example, the movements of opposition and reposition of the thumb bring it toward and away from close-packing with the trapezium, respectively. Opposition and reposition are the most common movements of a thumb, much more so than extension-flexion or abduction-adduction. The most common movements at the hip joint are flexion with lateral rotation of the femur, and the opposite motion, extension with medial femoral rotation.

These movements are single movements; that is, "locking" and "unlocking" movements. Every locking movement is the equivalent of a swing together with a spin. The swing stretches the main ligaments of the joint, and the spin twists its capsule so that the male and female articular surfaces are screwed together. The unlocking movement is, of course, the same in the reverse direction.

For example, the hip joint is brought into close-pack by a motion that causes extension, abduction, and medial rotation of the femur. Hence the habitual motions of the femur will be extension with medial rotation and flexion with external rotation; these are the two "natural" ways of swinging a thigh, as one can demonstrate easily.

Flexion-extension and abduction-adduction of the thumb are used for modifying the basic grasping movements of opposition in what is called the *precision grip* (Napier 1956). In this grip the first metacarpal is pronated at its carpometacarpal joint, and the first thumb phalanx is also pronated a little at its metacarpophalangeal joint (Napier 1955). The other basic form of grip is the *power grip*. This is used, for example, for holding the handle of a hammer. The four fingers are the principal agents of this grip, the thumb being ancillary to them if it be used at all. In this grip the MP joints are either in or nearly in their close-packed position—full flexion. The interphalangeal joints, however, have moved away from *their* close-packed position—full extension.

EMG Kinesiology of Nonhuman Primates

The informed reader needs no reminder of the dearth of EMG studies in man's closest relatives. If the evolutionary biology of both existing and extinct species is to become clearer, an extensive and intensive attack must be made with modern electronic technologies. I shall attempt to delineate some of the technological approaches that can be most fruitful in the next decade, leaving to the reader's imagination the details that might be investigated.

Multichannel electromyography with intramuscular fine-wire electrodes and concurrent recording of signals on FM tape along with cinematography and/or videotaping is becoming available in well-equipped EMG laboratories for human studies. Transmission of the signals to the recording devices can be mediated through highly flexible lightweight ribbon-type cables or through telemetering. Ultimately, the latter must be widely used if studies are to be carried out in free-ranging animals in the field; yet because it offers certain technical difficulties, telemetering has a limited usefulness in the laboratory today when it is compared with the best cable techniques.

In the decades of the 1950's and 1960's, electromyographic kinesiology in man has come of age. Frequent reports and several substantial books have appeared on the subject. Even the ultimate has happened: an organization known as the International Society of Electromyographic Kinesiology (ISEK) was organized in 1965. With a membership of several hundred scientists from around the world, ISEK is a growing influence in this special branch of science.

APPLICATION TO ANTHROPOLOGY

A host of fascinating problems present themselves in the study of the locomotion and posture of the nonhuman primates *in their own right*. But these should be set aside for the moment in favor of studies that have an evolutionary bias. A diffident way to present this is to suggest that all postural and locomotor muscles already studied in man ought now to be studied in selected species, beginning with the closest relationships and working away. Alas, this response cannot be made in good conscience because the proposal implies a great many lifetimes of work. What must come first are selective studies chosen on the basis of provocative probabilities. Projects are now in their early stages of planning and execution on a cooperative basis between a number of laboratories. At the Yerkes Regional Primate Research Center at Emory University, Dr. Russell Tuttle and I are starting on an ambitious project to electromyograph the hand and forearm muscles of the great apes. We feel that sophisticated functional studies on man's nearest living relatives (gorilla, chimpanzee, and orangutan) provide a broader base for inference on the subtleties of the divergence of hominids from hominoid ancestors. An evaluation of the functional potentialities and actualities of the locomotor systems as they affect the upper limb should throw light on terrestrial locomotion and its modifications through evolution.

Washburn (1951a) held that the single most important factor in the origin of human bipedalism was a change of function of the gluteus maximus from an abductor to an extensor of the hip joint. Although Washburn himself suggested that EMG experiments were needed to test his assumptions, these never were performed. Yet his views have become more and more accepted as dogma, contrary to his own wishes. Recently, however, Sigmon and Robinson (1967) claimed that gluteus maximus of apes and man are not different in function. Where does the truth lie? Obviously new EMG studies should be embarked upon, and plans are being made to do this.

Electromyographic Kinesiology

EMG TECHNIQUES

The Motor Unit. The functional unit of contraction (for the purpose of electromyography) is the motor unit, which is a group of fibers supplied by a single nerve cell. Individual muscles of the body consist of many hundreds of such motor units and it is their asynchronous activity that develops the tension in the whole muscle. The amount of work produced by a single motor unit is rather small. In a living human being it is usually insufficient to show any external movement of a joint spanned by the whole muscle of which it is a part. Even in the case of small joints, such as those of the thumb, at least two or three motor units are needed to give a visible movement.

Motor Unit Potentials. With appropriate equipment one can demonstrate that each twitch of a motor unit is accompanied by a tiny electrical potential which is dissipated into the surrounding tissues. This motor unit potential has a brief duration (with a median of 9 msec) and a total amplitude measured in microvolts or millivolts. When displayed on the cathode-ray oscilloscope, most motor units recorded by conventional techniques are sharp triphasic or biphasic spikes. Generally, the larger the motor unit potential involved, the larger is the motor unit that produced it. However, distance from the electrode, and the type of electrodes and equipment used and many other factors) influence the final size.

Electrodes. There can be a multitude of special types of electrodes (discs, plates, wires, or needles). In every case the electrodes must be relatively harmless, and they must be brought close enough to the motor units to pick up their electrical changes with fidelity. Because kinesiology is often performed by investigators who are not medically qualified, surface electrodes have been used widely. Unfortunately, they have seriously curtailed the scope of many investigations. Indeed, for excellent recordings, the difficulties rising from surface electrodes are prohibitive. Their chief virtue is the ease with which any novice can apply them and obtain—or appear to obtain—reasonable success; this also is the root of their drawbacks.

Perhaps the chief usefulness of surface electrodes in kinesiology is seen with investigation of the interplay between large, widely separated muscles under conditions where palpation is impossible, and especially where the investigators are performing preliminary studies. Thus, substantial progress has been made with skin electrodes in such uncomplicated general investigations; but their routine continued use should be avoided.

Inserted Electrodes. The renaissance of electromyography in this decade is largely due to the improvement of electrodes. For routine multielectrode studies, the best inserted electrode is our fine-wire bipolar electrode. Bipolar fine-wire electrodes are made from a nylon-insulated Karma alloy wire only 25 microns in diameter, but any fine inert insulated wire may be used. They are very simple to make. Their preparation is described in Basmajian and Stecko 1962), Basmajian, Forrest, and Shine (1966), and Basmajian (1967).

These electrodes are (1) extremely fine and, therefore, painless; (2) easily injected and withdrawn; (3) as broad in pick-up from a specific muscle as the best surface electrodes; and (4) give excellent sharp motor-unit spikes with fidelity. With one millimeter of their tip exposed, such electrodes record the voltage from a muscle much better than surface electrodes (Sutton 1962). Bipolar fine-wire electrodes isolate their pick-up either to the whole muscle being studied or to the confines of the compartment within a muscle if it has a multipennate structure. Barriers of fibrous connective tissue within a muscle or around it act as insulation. Thus, one can record all the activity as far as such a barrier without the interferring pick-up from beyond the barrier (such as there always is with surface electrodes). In the case of muscle without internal partitions, the fine-wire electrodes reflect the activity of the whole muscle as broadly as the best surface electrodes.

Apparatus. Electromyographs (either commercial or self-constructed) are high-gain amplifiers with a preference for frequencies from about ten to several thousand cycles per second. An upper limit of 1,000 cycles per second is satisfactory. For kinesiological studies, the best instruments are multichannel and record on FM tape. Direct-writing ultra-violet galvanometers are also quite useful. They require no chemical developing and there is an immediate display. In recent years FM tape recording has provided ready input of data to analogue computers. When combined with analogue-to-digital converters and digital computation, rapid calculations are possible.

Electromyography of Human Posture and Locomotion

Many of the widely held beliefs on human posture—and to a lesser extent, human gait—are based on teleology and metaphysics; many are not borne out by EMG. The most pernicious factor in the creation of the mythology is the idea that man is a bungled patchwork of evolution and that his posture and locomotion are second-rate adaptations. Adaptations they well may be, but electromyography has demonstrated their superb functional efficiencies. In fact, man's upright posture is actually extremely economical. At the same time, there is no clear evidence that man's back and lower limbs are hurt more often than those of quadrupeds.

Man's large "antigravity" muscles are essential to produce the powerful movements necessary for the major changes from lying, to sitting, to standing. Therefore, it is wrong to equate the antigravity muscles of man with those of animals that habitually stand on flexed joints. In the context of this presentation, special attention centers on the varying standing postures and gaits of other primates. Electromyography only now is being exploited in testing the various hypotheses derived from morphological observation and conjecture.

In man, the fatigue of standing is emphatically not due to muscular fatigue, and generally, the muscular activity in standing is slight or moderate. Sometimes it is only intermittent. Walking is generally less fatiguing than immobile standing. Although extreme exertion can produce muscular fatigue, most fatigue in the lower limbs caused by standing is more intimately associated with the inadequacies of the venous and arterial circulation and with the direct pressures and tensions upon inert structures. Thus, a standing person finds relief in constantly fidgeting or shuffling about. These are completely normal unconscious responses. A wise person has learned to employ them consciously to reduce fatigue in his antigravity mechanisms.

FATIGUE AND MUSCLE SPARING

Electromyography (EMG) has been used sporadically for studies of fatigue at its basic level, with confusing results for the general scientist. In the confusion, many onlookers have lost sight of the fact that EMG reveals some striking insights into the fatigue of normal human bodily functions.

Of course, fatigue is a complex phenomenon, but it constantly influences posture in the individual whether he is man or beast. The fatigue of strenous effort is probably quite different from the weariness of prolonged mental effort. Nevertheless, the components of fatigue resulting from maintaining postures and in locomotion derive from factors amenable to study by EMG.

Our investigations have shown postural fatigue is largely due to painful strain not on muscle (which often have been quite quiescent) but on ligaments, capsules, and other inert structures. This has led to the concept that *muscles are spared when ligaments suffice*. This concept is not only important in understanding the sources of postural fatigue in man but also underlies another important principle in man's posture: *there should be a minimal expenditure of energy consistent with the ends to be achieved*. Thus, MacConaill and I (1969) proposed two laws which express the operation of this principle:

(1). Law of Minimal Spurt Action. No more spurt fibers are brought into action than are both necessary and sufficient to stabilize or move a bone against gravity and/or other resistant forces, and none are used insofar as gravity can supply the motive force for movement.

(2). Law of Minimal Shunt Action. Only such muscle fibres are used as are necessary and sufficient to ensure that the transarticular force directed towards a joint is equal to the weight of the stabilized or moving part together with such additional centripetal force as may be required because of the velocity of that part when it is in motion.

EMG has clearly demonstrated these two laws to be valid. Especially dramatic is the biomechanics of the shoulder (glenohumeral) joint. From our studies (Basmajian and Bazant 1959) we concluded that the whole weight of an unloaded arm is counteracted at the glenohumeral joint by the superior part of its capsule alone, no muscles being required. In this area the coracohumeral ligament can be very strong and would appear to reinforce the action of the joint capsule in preventing the head of the humerus from sliding downwards on the inclined plane formed by the glenoid cavity of the scapula. With moderate or heavy loads, the supraspinatus is brought into action to reinforce the original tension in the capsule, and in some instances it was found that the posterior fibers of the deltoid are also called into play, these being more or less parallel to the supraspinatus fibers.

Carrying a heavy weight (say, a suitcase) can become a painful experience. One normally assumes that the reactive component which becomes fatigued and exhibits pain is muscular, but in fact, in the shoulder region it is chiefly ligamentous. This points up the importance of the inert structures. Another example of the Minimal Principle is that of arch-support in the human foot. Generations of surgeons have stressed the importance of the muscular tie-beams in the plantigrade foot. Yet our EMG studies

(Basmajian 1967) have repeatedly pointed to the fundamental importance of plantar ligaments. Indeed, we showed that even very heavy loads do not recruit muscular activity in the leg and foot (Basmajian and Stecko 1963).

COORDINATION, ANTAGONISTS, AND SYNERGY

The central nervous system has an integrating function in the interplay of muscular contractions. But the laws of minimal action enter here as well. Certain simple movements recruit only one or two muscles while all others remain relaxed. This we have shown repeatedly by our various studies. For example, pronation of the forearm is usually produced by one muscle alone —pronator quadratus—unless added resistance is offered to the movement; then, more muscles are called upon (Basmajian and Travill 1961). My colleagues and I have found this to be true in elbow-flexion too, where brachialis alone often suffices, and in other movements. Therefore, it is wrong and misleading to believe that nature always calls upon groups of muscles to produce simple movements. On the other hand, there are complex movements (such as rotation of the scapula on the chest wall during elevation of the limb) which obviously call upon groups of cooperating muscles.

Antagonists, too, have been misrepresented in the normal functioning of muscles. The unfortunate and incorrect impression has been fostered by many physiologists and even more anatomists that during the movement of a joint in one direction, muscles that move it in the opposite direction show some sort of continuous antagonism. The truth of the matter, first proposed by Sherrington as "reciprocal inhibition," is that the so-called antagonist relaxes completely (Travill and Basmajian 1961) except perhaps with one exception—at the end of a rapid motion of a hinge joint. Here, apparently, the short sharp burst of activity in some antagonists occurs to prevent damage to the joint.

One finds that the activity of muscles in the position of antagonists during a movement is a sign of nervous abnormality (for example, the spasticity of paraplegia) or, in the case of fine movements requiring training, a sign of ineptitude. Indeed, the athlete's continued drill to perfect a skilled movement exhibits a large element of progressively more successful repression of undesired contractions.

Many contractions of any one particular muscle may be accompanied by synergistic activity in other muscles to steady the adjacent joints. Gellhorn (1947) thus demonstrated the role of far-removed synergists in movements on the wrist. While flexor carpi radialis was activated in very slight flexion of the wrist, triceps brachii became active with the increasing effort in the prime movers (the extensors of the wrist remaining relaxed meanwhile). Only with very strong static flexion of the wrist would activity —and that only occasionally—appear in the antagonists.

Specific Studies of Human Posture and Locomotion

LEG AND FOOT

The function of the large muscles of the leg has been studied by a considerable number of investigators. As might be expected, any deliberate leaning forward or backward of a standing subject produces compensatory activity in the posterior and anterior muscles to prevent the occurrence of a complete imbalance. A very finely regulated mechanism is in control and the slightest shift is reacted to through the nervous system by reflex postural adjustments; sometimes the motor responses are so fine that they can only be detected electromyographically. In many positions the muscular activity is minimal or absent (Basmajian 1967).

During normal gait on a horizontal surface, Gray and Basmajian (1968) found that the tibiales anterior and posterior, flexor hallucis longus, peroneus longus, abductor hallucis, and flexor digitorum brevis are all concerned with both movements and restraints in the foot during walking. The inversion of the foot seen at heel-strike appears primarily to be related to the activity of tibialis anterior, although tibialis posterior, abductor hallucis, and flexor digitorum brevis show slight activity in flatfooted subjects. All except flexor hallucis may be attempting to maintain inversion in flatfooted subjects during full-foot. However, maintenance of inversion in most "normal" subjects may be due to factors not yet studied.

Tibialis anterior shows a peak of activity at heel-strike and another at toe-off. A peak of EMG activity observed at toe-off of the stance phase is obviously related to dorsiflexion of the ankle, to permit the toes to clear the ground. There is a period of electrical silence at mid-swing (Battye and Joseph 1966, Gray and Basmajian 1968). Our cinematographic records show the foot everting at the end of "acceleration" and remaining everted through mid-swing. Inactivity of the invertor fits the concept of reciprocal inhibition of antagonists.

During deceleration of the swing phase activity builds up to a peak at heel-strike, apparently, because it provides inversion of the foot.

The pattern of activity of tibialis anterior suggests that it does not lend itself to direct support of the arches during walking. At heel-strike, when the muscle shows its greatest activity, the pressure of body weight is negligible (Hicks 1954). Conversely, during maximum weight-bearing at mid-stance when all the body weight is balanced on one foot, the tibialis anterior is silent. When the activity resumes at toe-off, the weight bearing of the involved foot is minimal.

Tibialis posterior during ordinary walking shows activity at mid-stance of the stance phase. Movies show the foot remaining inverted throughout full-foot and turning to a neutral position (between inversion and eversion) just before mid-stance (Gray and Basmajian 1968). First, the fourth and fifth metatarsal heads make contact; then, as the foot everts increasingly toward neutral, more of the ball of the foot makes contact at mid-stance until the entire contact-area of the foot is applied. Although the tibialis posterior is an invertor in nonweight-bearing movements of the foot, its role at mid-stance appears to be a restraining one to prevent the foot from everting past the neutral position. R. L. Jones (1941, 1945) concluded from studies in cadavers and observations on living persons that by inverting the foot the tibialis posterior increases the proportion of body weight borne by the lateral side of the foot.

The plantar flexors, including the tibialis posterior, have a restraining function to control or decelerate medial rotation of the leg and thigh observed at mid-stance; by controlling the eversion of the foot at mid-stance, the tibialis posterior provides an appropriate placement of the foot (Sutherland 1966). The foot must be inverted to accomplish lateral weight-bearing in the early "moments" of the stance phase. This, of course, is because the middle part of the medial border of the foot does not bear body weight in "normal" subjects; the lateral border with its strong plantar ligaments is well-equipped to bear the stresses of body weight in walking (Napier 1957).

Peroneus longus is related to eversion of the foot at mid-stance during level walking. Like tibialis posterior, it is involved in controlling rotatory movements at the ankle and foot (Sutherland 1966). We found that eversion of the foot and medial rotation of the lower limb occur together (Gray and Basmajian 1968). One may conclude that the peroneus longus is in part responsible for returning the foot to, and maintaining it in, a neutral position at mid-stance.

Flexor hallucis longus is relatively inactive in normal feet except during mid-stance, while short intrinsic muscles of the foot show considerable activity from mid-stance to toe-off, the period when the greatest "breaking" force is exerted on the longitudinal arches.

HIP AND THIGH

The muscles of the thigh in man obey the same rules as those of the leg. By and large, the activity during normal, relaxed standing is usually slight. Indeed, it may be absent in most of the muscles for varying periods of time. The reports of many investigators and the work in my laboratory seem to agree. These overlapping and detailed studies (see Basmajian 1967) include most of the large muscles of the gluteal region and thigh.

When subjects carry a load either held in front of the thighs or strapped to the back, Carlsöö (1964) found quadriceps remains completely inactive. Meanwhile, the ischiocrual muscles (hamstrings) show individual variations—from very active to completely inactive—apparently depending on the degree of relaxion of the hip and on whether or not the line of gravity had been shifted anterior to the hip joint.

The iliopsoas, the adductors and the glutei, are particularly interesting. With R. K. Greenlaw, I have studied them extensively during locomotion and posture in a fairly long series of subjects. Many of these studies have been done with simultaneous recording of EMG signals from multiple fine-wire implanted electrodes in as many as eight muscles, along with switches to indicate the exact phase of the walking cycle. Synchronized cinematography and oscillographic recording of the EMG signal has been developed to a very useful stage. Because this extensive study is to provide the material of Greenlaw's Ph.D. thesis, it will not be reported in detail here. Ignoring those aspects that simply confirm the usual teachings (e.g. the functions of the gluteal abductors), let us consider only the most striking departures from the current concepts.

Iliacus and psoas (with independent electrodes) act almost as one muscle. Both are often active during lateral rotation, but contrary to some teachers, not during medial rotation whether the subject is weight-bearing, standing with the limb hanging freely, or lying down. Abduction of the thigh recruits activity in these muscles (which, of course, are powerful flexors, too). During walking the peak of psoas activity occurs as the limb starts its swing phase forward. Iliacus is active continuously through the whole cycle—it is a "dynamic ligament."

The *adductor muscles* reach a peak of activity at the end of the stance phase, fall off slowly during swing, and then rise to a lesser peak at the end of swing and start of stance. Relating this to what is happening to the pelvis and femur, one finds that lateral rotation of the femur (in relation to the pelvis) reaches its maximum at the first and larger peak. Further studies without locomotion (that is, in recumbent postures, etc.) indicate that the adductors are active in rotation. One may suggest with some assurance that the rather mystifying massiveness of the adductor group arises from the very active rotation it provides during walking.

VERTEBRAL COLUMN

While standing erect, most human subjects require very slight activity and sometimes some intermittent reflex of the intrinsic muscles of the back, according to a growing list of authors (see Basmajian 1967). These authors showed that during forward flexion, there is marked activity until flexion is extreme, at which time the ligamentous structures assume the load and the muscles become silent. Floyd and Silver (1955) proved that in the extreme-flexed position of the back, the erector spinae remained relaxed in the initial stages of heavy weight-lifting.

Asmussen and Klausen (1962) concluded that "the force of gravity is counteracted by one set of muscles only, most often the back muscles, but in 20 to 25 percent of the cases the abdominal muscles. The line of gravity passes very close to the axis of movement of vertebra L4 and does not intersect with the curves of the spine as often postulated." Carlsöö (1964) regularly found activity in sacrospinalis in the symmetric, rest position. Klausen (1965) investigating the effect of changes in the curve of the spine, the line of gravity in relation to vertebra L4 and ankle joints, and the activity of the muscles of the trunk, concluded that the short, deep intrinsic muscles of the back must play an important role in stabilizing the individual intervertebral joints. The long intrinsic muscles and the abdominal muscles stabilize the spine as a whole.

Placing a load high on the back automatically causes the trunk to lean slightly forward. The increased pull of gravity is counteracted by an increased activity in the lower back muscles. A load placed low on the back reduces the activity of the back muscles (Carlsöö, 1964). There is increased activity in sacrospinalis with a load held in front of the thighs. Thus, the position of the load—either back or front—either aids the muscles or reflexly calls upon their activity to prevent forward imbalance.

While some investigators believe that the vertebral part of psoas major helps to maintain the posture of the lumbar vertebrae (Nachemson 1966), we found that this muscle shows only some slight activity during standing (Basmajian and Greenlaw 1968). Even strong attempts to increase the natural lumbar lordosis—said to be a function of psoas in man—is not nearly as effective in recruiting psoas as are the movements it produces in the hip joint.

GENERAL STUDIES OF HUMAN GAIT

Multifactorial studies are difficult and time-consuming, and so it is only recently that equipment has improved to the point where electromyography gives especially useful results. Some of this work is now appearing in the clinical literature. A series of large-scaled and complex studies in my laboratory will be the source of a number of papers to be published by our group over the next year or two. In the following section, only a general and incomplete outline is possible.

In walking, there is a very fine sequence of activity in various groups of muscles in the lower limb. As the heel strikes the ground, the hamstrings and pretibial muscles reach their peak of activity. Thereafter the quadriceps increase in activity as the torso is carried forward over the limb, apparently to maintain stability of the knee. At heel-off, the calf group of muscles build up a crescendo of activity which ceases with the toe-off. Before and during toe-off, quadriceps and sometimes the hamstrings reach another (but smaller) peak of activity (Radcliffe 1962).

The following correlation of activity occurs during walking on the level (Liberson 1965).

Contraction of the gluteus maximus is preceded by that of the triceps surae of the opposite side and is simultaneous with that of the iliopsoas on the opposite side. In many cases, two-joint muscles show an increase of tension without EMG potentials because they act as simple ligaments during the contraction of the antagonists.

Gluteus maximus shows activity at the end of the swing and at the beginning of the supporting phase (Battye and Joseph 1966), contrary to the general belief (based on patients with total bilateral paralysis of the muscle) that its activity is not "needed" for ordinary walking. Perhaps gluteus maximus in normal man contracts to prevent or to control flexion at the hip joint.

Envoy

The theories of the evolution of human bipedalism as related to biomechanics and EMG kinesiology dictate a great infusion of these two disciplines into physical anthropology. Despite their support of this idea for two decades, anthropologists have shown a caution that is striking. Further delay is inexcusable; research in this field has begun and should be increased enormously in the coming months and years.

14

CHARLES E. OXNARD
THE UNIVERSITY OF CHICAGO

Functional Morphology of Primates: Some Mathematical and Physical Methods

In a recent review (Oxnard 1968a), some methods used in the analysis of functional morphology in primates were discussed. That paper concentrated initially on a number of relatively older problems (for instance, the assessment of biological variation within closely related groups of organisms, compensation for inadequate material, and the association with morphology of certain aspects of behavior). These considerations then lead to questions of the measurement of shape for eliciting similarities and trends. In particular, it was stressed that sometimes such studies provide new views of interactions between structure and function.

More recently, researches have progressed so that discussion of this type can be taken somewhat further. To more classical anatomical methods, there can be added a number of modern mathematical and physical tools which may aid in the understanding of functional morphology within an evolutionary framework. In particular, these newer techniques (when based upon firm biological groundwork) are providing insights not available from more usual studies. Further, such procedures sometimes supply independent

It is readily apparent that the researches displaying the use of many of these techniques have involved numerous collaborations, and these are recognized within the text. Some of the work is being undertaken with E. H. Ashton, R. M. Flinn, and T. F. Spence, colleagues in the Department of Anatomy, University of Birmingham. Other work is being done in collaboration with P. M. Neely and J. C. Davis, colleagues at the University of Kansas.

The special stimulus and collaboration of Lord Zuckerman, O.M., F.R.S., Professor of Human Biology, University of East Anglia, is gratefully acknowledged.

It is also my pleasure to acknowledge the participation of my research assistant, Miss Joan Hives, at every stage in the production of this paper, and especially in the experimental stress analysis.

The work is supported by U.S.P.H.S. Grant HD 02852, N.S.F. Grant GS 30508, and grants from the Wenner-Gren Foundation for Anthropological Research.

(or at worst, semi-independent) tests of hypotheses that would otherwise remain untestable speculations.

Thus, at an elementary level, we may note the size and shape of bones and bony fragments. We may observe architecture; the positions and orientations of bony buttresses and the changing patterns of trabecular and compact bone are readily apparent to the unassisted eye; the interrelationships of osteones are revealed by the microscope; even at the ultrastructural grade, pictorial information is available about the associations between, for example, collagen and hydroxyapatite. But at all levels comprehension is, in general, confined to major and usually obvious features (for instance, of the main bony buttresses, of principal osteone directions, or of major collagen orientations). This relates inevitably to the pictorial nature of the evidence and the comparisons. The attempt to synthesize *by eye* information contained in sectional or surface views of objects, results in the loss of much subtle information.

From the earliest times therefore, research strategy has included mensurational methods to reveal information contained within pictures but difficult otherwise to elicit. For instance, absolute measurement is a clear aid to assessment of size differences, and the comparison of sizes leads to an understanding of such properties as allometry and correlation, concepts difficult to determine pictorially. Combinations of measurements have resulted in the widespread use of indices (and angles—disguised indices) and, when it is difficult to separate size and shape effects in analysis, such combined dimensions may be reasonable methods of allowing for differences in gross size. For both absolute measurements and indices, univariate and bivariate analyses have long been the chief analytical tactics.

Calculating aids have allowed more detailed investigation of allometric and correlative properties of biological characters by multivariate techniques (for example, Penrose's size and shape statistics). Yet more complicated multivariate analyses (e.g. Mahalanobis' generalized distance) have been available for a considerable time, although it has only been within the last fifteen years that it has become feasible to attempt to solve, with the use of electronic computers, "real" biological problems involving considerable volumes of data.

Again, however, these methods are open to the faults of mensurational methods in general; that is, that they normally involve the analysis of a relatively small number of points. Considerable information in a pictorial representation is thereby lost. It is true that electronic and computer methods are being evolved that make the taking and recording of data easier so that it becomes possible to obtain extensive data sets characterizing, almost *in toto,* pattern and shape. For instance, computer visualization of x-rays or photographs can allow instantaneous identification of coordinate measurements; flying spot scanners can identify and characterize patterns. But often in such cases the data become so extensive that analysis, even utilizing the extremely fast computers of the present day, may be difficult and time-consuming.

These problems are thus leading to a return to the pictorial method, but utilizing modern physical techniques capable of seeing in a picture what is not available to the naked eye. These techniques have been greatly stimulated by research in geology, geography, astronomy, and other physical sciences, and include those research strategies grouped under the heading of "optical data processing." This latter depends upon the fact that pictorial data consist of two independent variables (the x and y coordinates) and may therefore be modeled rather effectively by optical systems which also have two degrees of freedom. Thus for handling pictorial data, such techniques are inherently superior to electrical or electronic analogs. They may allow, for instance, the almost instantaneous computation of Fourier analyses as compared with the hours of computation that may be necessary (depending upon the size of the data set) for Fast Fourier Transformations on a digital computer.

In the foregoing, the problem has been presented as one of shape analysis alone. For evolutionary biology, however, the situation is rendered considerably more complex because there must be tied into the analyses, information relating to the fact that the shapes have functional roles. At all levels (macroscopic, microscopic, ultrastructural) it is readily apparent that, whatever complex hereditary and developmental processes result in structure, impressed mechanical forces are of considerable importance in the formation, maintenance, and change of structure. Recent studies of mechano-electric properties of living tissues reinforce and extend that conclusion. Some definition of impressed mechanical forces due to function is therefore required. This may be obtained through correlation of animal behavior with relationships of bone and soft tissues, especially muscle and connective tissues. There are a number of methodological procedures at this level: (a) biomechanical investigations may involve direct *in vivo* studies (e.g. electromyography); it may be possible (b) to understand such problems indirectly in terms of trends revealed in the analysis of shape and form. We may also, however, test (c) classical anatomical inferences by utilizing *in vitro* techniques of analogy. Thus, in the last case, a biomechanical system may be rendered simpler in order to visualize the contained information, although the simplification should not be of so great a degree as to give misleading results. Modeling may be undertaken, using theoretical computer simulation. Stress analysis may well be of value; here the physical properties of photoelasticity are utilized in experimental simulations.

It is at this point that problems of a different nature are encountered. In the basic researches as just outlined, extant materials are available. For these, there exist clues provided by other facets of the life processes of the individual and of the group, by ability to examine numbers of specimens, and by relative completeness of data sets. They result in conclusions having reasonable degrees of likelihood. Further studies can bring the unique properties of fossil data (time and continuity) to bear on the conclusions. However, there then result new research strategies relating to fitting, into well understood data, information from single specimens. Such information may well be incomplete and may also involve reconstruction. Questions relating to single specimens and incompleteness of data are statistical matters which are reasonably clearly known, if not completely understood. Problems relating to reconstruction are more difficult because there is always the chance, sometimes quite a high chance, of previous conceptions about the fossil unconsciously channeling the nature of reconstruction itself.

Some attempts have been made here to study the incorporation of data from fossil forms. In the main, this has meant working with published dimensions and, *with great circumspection,* working from casts or even photographs of specimens. But extant forms may be treated as "fossils" and this allows investigation of general problems relating to interpolation. Interpolation has utilized both mathematical methods of data analysis and physical techniques of experimental stress analysis.

These then are the purposes of the present paper: to review certain research strategies that are presently available and to assess the extent to which they may provide new insights into problems of characterization of shape and architecture, especially within the framework of evolutionary adaptation of biological function. The applicability of these methods in other fields of primate biology may also be of interest.

Biological Basis for Morphological Investigation

However careful are the techniques of data collection, and however complicated the methods of analysis, new insights into morphology cannot be expected if the biological basis is not understood or is too simplistic. At one extreme, we may think that we require a relatively complete recognition of the detailed behaviors of which an animal is capable. With the present status of research, this large gap in our knowledge is being filled by workers utilizing new hypotheses, techniques, and methods of analysis. However, for the majority of primate genera, it will be some time, if ever, before such data can be garnered. At the other extreme, it is surely inadequate to utilize concepts of overall behavior that relate to such broad (and, especially in primates) almost meaningless terms as "terrestrial" or "arboreal."

What elements, then, of behavior as now known are most relevant in the functional understanding of morphology? The answer is likely to differ according to the nature of the behavior, and the particular animals and anatomical regions involved.

One element of the plastic response of bone is to the vector resultant of forces acting upon it over a period of time. In a behavioral situation, therefore, where one or a few biological functions occasion large forces and most other functions produce relatively small forces, it is clear that the resultant will be more like the former than the latter. (It is also appropriate that account should be taken of the scalar quantity, time. To what extent we can allow for the scalar quantity, adaptive value, is yet to be determined.)

Pending this *caveat,* if we can identify facets of behavior producing large forces, they will be most related to the resultant and will have a discernible relationship to the adaptive shape of bones. This is likely to be the case in, say, the forelimb in animals which utilize that organ for locomotion.

Thus, however frequently a chimpanzee may use its hand for making nests or feeding, by far the greatest forces that impinge upon it are presumably those produced by its involvement as a propulsive strut in knuckle-walking where a large part of the animal's appreciable body weight is supported and impelled by the hand. It is therefore to this locomotor aspect of behavior (and within that cycle to the greatest forces produced) that one may expect to see the most clear-cut morphological adaptation in the hand. (It is nevertheless true that many of the remaining functions of the chimpanzee hand—for example, reaching for and pulling towards its body, food or nesting materials, reaching and pulling its body upwards in climbing—may summate to a function of the hand as a gripping hook; such a summation may also make a considerable contribution to the resultant, and evidence of this may therefore also be expected in its structure.)

In contrast, the hand of man performs a number of different functions, none of which are frequent and powerful enough to contribute excessively more than others to the resultant force acting upon the part. Of these various functions, perhaps the most liable to have an obvious effect upon the resultant is the power grip as it is employed in numerous manual functions. Even here, however, the matter is by no means so obvious as with a non-human primate; occupational, recreational, social, and other differences in modern man may well interfere. Accordingly, it is improbable that one would expect to see any particular adaptation of the human hand to any single function which it is capable of carrying out. In such a case, the adaptation will likely be an overall adaptation to a resultant not particularly similar to any specific subset of forces: an overall adaptation to manual dexterity.

Perhaps the most significant difference between function of the hand of man and the chimpanzee is the virtual absence in the former of functions producing *any* large forces. We can draw an association with this to the distribution of trabecular and compact bone in the phalanges. In man there is little compact bone; most of the architecture consists of finely carved bony spicules. In contrast, in the chimpanzee the phalanges contain fewer trabeculae; the major part of the elements comprise the heavy compact cortex.

Thus, if it is possible to pick out a small number of functions within behavior as producing large forces, then it may be a somewhat easier matter to outline bony adaptations; in general, this is the case for most postcranial parts of nonhuman primates in which all limbs are used in one way or another within a locomotor context. If it is not possible to identify a subset of functions that make major contributions to force patterns, then it may be more difficult to define adaptation of bones; this is in general the case for certain regions (e.g. shoulder, hand, face, jaws) in man. Is it possible to identify the points during the evolution of those regions in man when the changeovers occurred (that is, when a small number of obvious functional parameters ceased to be of importance in appropriate morphologies)?

Such discussion, outlining the complex nature of the relationship between behavior and morphology, may suggest to some workers that we must await new studies of, for instance, detailed movements of bodily parts (e.g. by careful cinematography) and activity of muscles during clearly defined movements and movement cycle (e.g. by telemetric electromyography). However, functional morphological studies cannot be halted pending such full investigations, which may indeed never be made for many anatomical regions. Adequate intermediary studies depend upon detailed and careful inferential analyses of functional components of behavior to the extent that major force patterns may be suggested. These may then be related to myological architecture, and this in turn may be correlated with bony morphology. Such a strategy includes drawing associations that are not only statistical, but that have biomechanical meaning within function. Researches such as these have been carried out by a number of workers (e.g. Ashton and Oxnard 1963, 1964; Tuttle 1967) and they show that an adequate biological basis for functional morphological studies can be provided in the absence of the more technical investigations (these latter must, however, be performed eventually).

Multivariate Morphometric Analysis

With a base in the functional relationships between behavior, and myological and osteological architecture, it is possible to examine in considerable detail single bones and bony fragments. The characterization of bone shape by means of a number of dimensions can be refined by utilizing analyses that make allowance for differing sizes of specimens and that take into account correlation among characters within individuals and groups. Such methods (factor, principal components, generalized distance, and canonical analyses, together with other related techniques—reviewed by Seal 1964) are capable of providing succinct synopses of large volumes of data. Some of the best examples of the use of these techniques in primate evolution are still a number of the earlier studies (for example, Mukherjee, Rao, and Trevor 1955; Ashton, Healy, and Lipton 1957; Howells 1966). These techniques are capable of providing statistically significant discrimination between samples of biological material, together with an understanding of the relationship between the different dimensions (original variables) taken on each specimen.

Thus even when there are only slight differences between samples in terms of individual original variables, multivariate analysis may show where the essential differences in shape actually lie, something that cannot be reliably achieved by univariate methods. For example, the studies of Howells (1966, 1969a) show that samples of data from sets of skulls can define differences between human groups. Further, such analyses indicate important dimensional differences (for instance, that the greatest single range of difference between the groups studied lies in the biauricular cranial breadth). The investigation provides conclusions of distinct importance in historic diagnoses, e.g. in relation to the Fish Hoek and Keilor skulls.

But of further interest is the use of these methods in those very cases where objects are obviously significantly different, and where what is required is characterization of the degree and nature of the differences. In these cases, multivariate transformations may give opportunities for functional interpretation, and possibly provide tests of hypotheses about functional adaptation. This use of the techniques has been carried out to a large extent in sciences other than biology (e.g. sociology: Bock and Haggard 1968) and within biology in areas other than the evolution of primates (e.g. entomology: Blackith and Kevan 1967). Within the primates, they seem to be mainly confined to published studies on the shoulder (e.g. Ashton, Healy, Oxnard, and Spence 1965; Oxnard, 1967, 1968b, 1969a, b; Ashton, Flinn, Oxnard, and Spence 1971) and to studies on the pelvis, hand, and overall bodily proportions currently in progress.

For instance, investigations of the primate shoulder are virtually complete; and they show that of greater interest than the *fact* of separation of different primate groups, is the *nature* of that separation. Initially, that study resolves itself into the examination of 41 genera of primates, when nine dimensions are available from the shoulder girdle of a total of 551 specimens. The particular osteological dimensions are chosen with reference to the muscular anatomy of the shoulder viewed in the light of its presumed function in the locomotion of different primates. Canonical analysis of these data demonstrate that differences among nonhuman primates may be defined by only two canonical variates or axes; these new variates seem to arrange the genera in relation to different locomotor patterns. Thus the first canonical axis arranges nonhuman primates in relation (apparently) to the extent to which the shoulder bears, or is capable of bearing, tensile forces. This seems to be associated with the twisted shape of the scapula and clavicle, a feature related mechanically to the efficiency of the shoulder as a suspensory mechanism. The second variate sorts out these animals according (seemingly) to the extent to which they climb and live in trees. It is associated with a mediolateral decrease in scapular width and increase in clavicular length (that is, with a laterally placed shoulder joint and a dorsally placed scapula). These are features mechanically related to a mobile shoulder that is more efficient within an arboreal environment. Man is uniquely separated from the nonhuman primates by a third canonical variate, and this correlates with the unique functions of the human shoulder girdle.

In other words, it seems as though a biological meaning may be imputed to the mathematical derivatives; and though this is a somewhat speculative suggestion, in this particular case the likelihood that the speculation is correct has been greatly increased by a number of tests.

First, another series of eight measurements is available from the same specimens; analysis of these data suggest that similar information is inherent in the antomical shapes, but to a somewhat less clear extent.[1] This is perhaps to be expected, for although it may be possible to choose certain dimensions as reflecting function in a circumscribed region to the greatest degree, most other dimensions should also be correlated to some extent with the first set, and therefore also with function. The combination of the two sets of measurements has now been carried out and the canonical analysis of the total seventeen dimensions of the shoulder shows a little, but only a little, improvement over the analysis of the best nine (Ashton, Flinn, Oxnard, and Spence 1971). Certainly the addition of the new dimensions has not negated the previous conclusions. In fact, the reverse is the case; it seems remarkable just how few dimensions appear to be able to give an adequate picture (Figure 14-1).

A second test comprises the examination of the original set of dimensions but taken on a series of animals that differ totally, viz., various nonprimate mammals (Oxnard 1968b). Here again, canonical analysis appears to confirm the biological meanings of the parameters. Thus, arboreal mammals lie in the same part of the canonical space as the primates; other mammals do not. The separation of the different arboreal forms by canonical axes one and two seems to hold similar meaning. Animals which hang (and in which, therefore, the shoulder is subject to tensile forces, e.g. sloths) are separated appropriately by the first canonical axis. The most highly arboreal animals are separated by the second axis (e.g. giant tree squirrels). Similar parallels exist in many mammalian orders among those forms which are, like the primates, arboreal.

Yet another test of possible biological meaning attaching to mathematically derived parameters has been made by returning to the original dimensions. Do the particular combinations of original dimensions in the different canonical axes make biological sense? In fact, the combinations, rather than being merely arbitrary, have been shown to reflect obvious overall shape changes in the bones; these changes are not complex and may be demonstrated as sweeping deformations when examined in Cartesian grids (Oxnard 1968b, 1969b). Furthermore, the very nature of the deformations

---

1. In fact, in a previous preliminary study (Oxnard 1967), it had seemed that the functional information was not present to all and that only some information apparently relating to taxonomic groupings was evident. But the more detailed examination of an augmented data set that is strictly comparable confirms the above statement.

is easily referable to the biomechanical situations. This is confirmed by independent experimental stress analyses capable of assessing relative mechanical efficiencies of shapes (Oxnard 1969b and see below).

There can be little reasonable doubt that in this particular study, mathematical variates have truly revealed biological facets of the shapes involved. This may have occurred if the mathematical system is (presumably by chance) modeling in some way what has happened during the macroevolution of the group. Because the mathematical method has picked out a small number of aspects of the shapes (which are mathematically independent), the biological situation may also have been produced by the operation of a small number of biological factors of one kind or another. However, it does not necessarily follow that the biological parameters need to be independent. For many genera there is a considerable association between them, although for other genera it seems to be absent.

Yet other tests of these speculations are also important elements of research strategy in morphology and include the use of independent mathematical and biomechanical methods. (Both confirm the above hypotheses and are described in later sections of this paper.)

A similar analysis has been carried out on a series of dimensions of the pelvis in extant primates. This is an augmented set of the data originally described in a univariate manner (Zuckerman 1966; Oxnard 1966; Zuckerman, Ashton, Oxnard, and Spence 1967). In this case, nine dimensions are available on each of 434 innominate bones in 41 genera of extant primates. Univariate analysis is (as is simple observation) able to define the uniqueness of the human pelvis, but differentiation among nonhuman genera is not clear. The canonical analysis of this data (as noted by Zuckerman 1970 and to be described fully by Zuckerman, Ashton, Oxnard, Flinn, and Spence), while also defining man as distinct, is additionally able to effect marked separations among many nonhuman primates. These seem to relate well to different functions of the pelvis within locomotion. For instance, the slow, clinging lorisines and the fast, hopping galagines are each separated both from themselves and from all other groups. The relatively more quadrupedal New World monkeys and the rather similar lemuroid prosimians are grouped together. The relatively more acrobatic New World monkeys and all Old World forms make up yet another major group. Man is distinct from all others. The major nonhuman separations are achieved primarily by a single canonical axis; a further three axes suffice to separate the subgroups within these major categories. This analysis is yet in the making, so that more detailed inferences about the relationship between the function of the pelvis and its structure are not available.

This information about the shoulder and pelvis in primates is not obtainable from such dissection of muscles or observation of bones as have so far been carried out. For instance, a very extensive and excellent dissectional study of the primate shoulder (Miller 1932) is unable to provide further

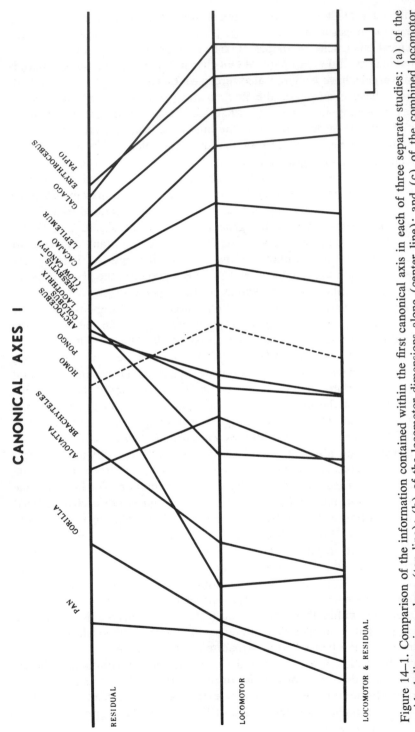

Figure 14–1. Comparison of the information contained within the first canonical axis in each of three separate studies: (a) of the residual dimensions alone (top line); (b) of the locomotor dimensions alone (center line); and (c) of the combined locomotor and residual sets of dimensions (bottom line). The positions of a small number of named genera only are shown. The correspondence among the three studies is most marked.

hypotheses about the evolution of shoulder in primates other than that there is increasing mobility of that joint from prosimians through monkeys and apes; the epitome of mobility is seen in the human shoulder.

Grouping Methods in Morphology

The methods of multivariate analysis are designed for investigation of data which conform (or nearly so) to Gaussian distributions. That is, for one- or two-dimensional sets, the arrangement of the variation within groups approximates to familiar bell-shaped curves and surfaces. Appropriate hyper-bell-shaped distributions are assumed for data in more than two dimensions. It is certainly true that normal or near-normal distributions are common in biology. However, there is no particular reason why they should be ubiquitous and, indeed, there is every indication that they are not.

The nature of the questions that we are asking when we utilize multivariate statistical methods are also restricting factors in their use. Thus, factor analysis attempts to define structure within a single known group (taking account of variation and covariation). Principal component analysis asks for that structure to be viewed in those mutually orthogonal (independent) directions that produce the greatest structural differences. Generalized distance analysis attempts to elucidate the distances between what are known to be a number of groups (taking account of the variation and covariation within and between them). Canonical analysis investigates these distances in mutually orthogonal (independent) planes producing maximum separations. (Other rotated or oblique axes can be superimposed on the data when necessary).

But in many cases the questions that we wish to ask do not relate to given groups. Rather are we interested in knowing *if* there are *any* groups in the data and, if so, what are the structures within and the relationships between them. And we need this answer even if (more especially, if) the groups do not approximate to multivariate normal distributions; groups may be sausage-shaped and even doughnut-shaped in three dimensions (hyper-sausages and hyper-doughnuts in more than three dimensions).

Finally, we may wish to define groups which are only locally stable. For instance, if, in utilizing multivariate normal techniques, it becomes necessary to enter new data, then often the entire data set must be re-examined, every point in the interpoint distance matrix taking a new value. In biology, there may well be no particular reason why the incorporation of new data at one locale should affect other regions. Therefore, methods that have local stability, and that do not involve interpoint distance matrices, have a special place.

For these reasons, biologists have utilized a variety of grouping or clustering methods. Some of these are relatively simple; probably all suffer from technical faults. But some of the most recent attempt to take account of the factors that have just been discussed.

As such clustering analyses have become popular, especially among taxonomists, a large number have been devised and it is difficult to judge their relative merits and demerits (Sokal and Sneath 1963, Gower 1967, Wishart 1969). Some depend upon Gaussian concepts and hence have constraints similar to those of multivariate statistics themselves. Some depend upon starting with most close points and by agglomeration defining groups, ending with the logical group of the entire data set. Others are divisive procedures where the entire data set is gradually broken down. The two general methods do not necessarily lead to the same set of groupings (Gower 1967).

The technique used here is one which attempts to take account of some of these criticisms. This method, known as neighborhood limited classification, is due to P. M. Neely (Oxnard and Neely 1969) and does not depend upon concepts underlying multivariate statistics. It therefore may act as an independent test of results obtained from such techniques. It is an agglomerative procedure that does not scan the entire interpoint distance matrix and accordingly can deal with rather large numbers of objects. It is locally stable; the interpolation of new points (unless of a special nature, see below) affects essentially only the locality where the new points are interjected. It is capable of defining groups that may be any size or shape in a large number of dimensions.

And the fact that the data are examined in "neighborhood space" consisting of indeterminate but finite sets, rather than in Euclidean or other spaces (of infinite sets), may well be more applicable to the examination of certain kinds of biological data, where sampling of groups within groups may differ from taking samples of infinite populations. For the examination of data from, say, a breeding population may well be carried out, utilizing the usual statistical procedures. A sample of 30 animals from a breeding population of 10,000, or in the case of some rare species perhaps only 300, may well be reasonably close to taking a hypothetical small sample from a hypothetical infinite population. But in an analysis where we may be examining many specimens from, say, ten genera which represent a sample of all the genera that may have existed in a given family, this is not the case. Although we do not know, and never can know, how many genera actually exist in the family, it is clear that the number, though indeterminate, is finite and small; indeed, it may well be of the same order as the sample size. In such situations, normal statistics are applicable only with caution. Perhaps a technique such as neighborhood limited classification may have something to offer here.

The method applied to the data on the shoulder girdle (Oxnard 1969b, Oxnard and Neely 1969) produces similar conclusions as flow from the canonical analysis of the data. Thus, neighborhood limited classification suggests that the various specimens of the primates can be grouped into divisions that associate well with the previous notions about the function of the shoulder. The technique is able to differentiate less acrobatic forms from those that are more highly acrobatic in their use of the shoulder in climbing and foraging in trees. It is also able to differentiate those that live primarily on the ground, where the shoulder is more frequently used in craniocaudal movement, as compared to those that are almost totally arboreal, where there is a greater degree of movement of the shoulder in all three dimensions. Man is shown to be different from all other primates.

However, neighborhood limited classification provides further information not available from canonical analysis. This relates to the formation, from individual specimens, of markedly obvious suprageneric groupings or peaks. Not only are these peaks related to data from *specimens* that are closer together at the peaks rather than between them, but the same also applies to the *genera* themselves. Genera are crowded together at the peaks, and those genera closer to the peaks have smaller spreads than do those which lie away from or between peaks.

The study of the function of the shoulder within locomotion does not immediately fit this new result. For instance, although it is possible to suggest that the function of the shoulder in different primates can be described in terms of how the animals are thought to move, it is also recognized that such groupings may be arbitrary. First, this may be a consequence of ignorance about details of locomotion in the rarer primate genera. Second, it results from the attempt to suggest groupings for *genera* which are, after all, collections of species that, within a genus, may well move in different ways. Third, the behavioral plasticity of the primates almost denies attempts at grouping.

It seems, therefore, that the natural history information suggests a spectrum of locomotion; yet the morphological data demonstrate that this has been achieved (in evolution) in terms of what seem to resemble "quantum leaps" from one morphology to another. Does this suggest that in some way or another these morphological peaks may be related to concepts like adaptive peaks? This morphological clustering is most marked; it completely overshadows any generic or other taxonomic grouping that we may expect to discern in the data.

It is fascinating that the preliminary examination of the canonical analysis of the pelvic data also suggests similar results. In this case, canonical analysis produces some fairly obvious suprageneric groupings that make considerable functional sense when viewed in the light of the mechanics of the pelvis in locomotion. At the same time, however, the groupings have created a number of somewhat curious bedfellows (e.g. gibbons and baboons; see above). One explanation of this could reside in the existence of a restricted set of preferred morphological peaks to which many different genera might conform even when, functionally, the animals might be doing somewhat different things. This is certainly confirmed by single linkage cluster analysis of generalized distances. Whether or not neighborhood

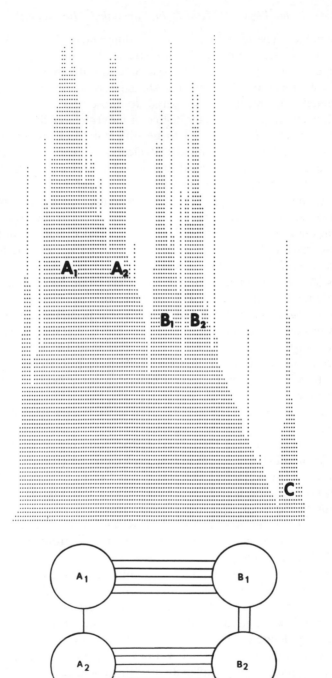

Figure 14–2. Neighborhood limited classification of certain data from Prosimii (*Loris, Nycticebus, Arctocebus, Perodicticus, Propithecus,* and *Indri*) and Anthropoidea (Alouattinae, Atelinae, and Colobinae).

limited classification of that data will confirm the existence of these peaks has yet to be determined.

In other words, although natural selection is undoubtedly selecting morphologies most adaptive to particular locomotor patterns, it is doing so within considerable constraints as to the morphologies that are possible. Thus, though it is likely that selective factors for mechanical efficiency may be partial causative agents, it is also possible that the genetic models of the primate shoulder and pelvis respectively may have been sufficiently fixed at early evolutionary stages as to limit grossly the independent ways in which they may vary.

Finally, the application of neighborhood limited classification to the data from the primate shoulder justifies one of the specific properties of the method: that it is able to perceive groups that may be of shapes other than spherical (or hyper-spherical). In the canonical analysis of that data, one feature of the results is the lack of separate identification of a number of prosimian forms (which move often in such a way that the shoulder bears tensile forces, e.g. potto, angwantibo) from certain New and Old World monkeys (e.g. woolly and howler monkeys, which utilize the shoulder in an acrobatic manner so that it bears tensile forces more frequently than in many other monkeys). Neighborhood limited classification demonstrates a clear separation between these particular prosimians and monkeys. The nature of the groupings is such that (within the multidimensional space) they emerge as two dumbbell-shaped groups lying side by side. Each dumbbell comprises only specimens from its own taxonomic group. The lengths of the bar connecting the heads of the dumbbells is greater than the distance between the homologous heads on each dumbbell. However, the number of long neighbor connections along the bars of the dumbbells (between specimens of like taxonomic groups) is an order of magnitude greater than the number of short neighbor connections (between specimens of different taxonomic groups) from one dumbbell to the other. Neighborhood limited classification reveals this anisotropic nature of the data space and thus separates the groups (Figure 14–2). Canonical analysis reads the situation as equivalent to many closely overlapping spheres and is much less able to distinguish them. Other examples of groups that are not spherical are evident from these analyses.

Another use of clustering methods also presents itself; that is, in combination with multivariate statistics. Multivariate statistical methods are ex-

---

The upper diagram is a photograph of computer output and it demonstrates the groups that exist in the data. Group A comprises both Prosimii and Anthropoidea, though these are separated in each of the two subgroups A_1 and A_2. A similar situation obtains for the group B and its subgroups B_1 and B_2. The lower diagram gives a two-dimensional representation of the relationships among these groups; it demonstrates that in fact the real nature of the groups is as the pairs A_1–B_1 and A_2–B_2. When taken in this manner, the Prosimii are effectively separated from the Anthropoidea. In this case, however, the relationships are defined by the number of links between subgroups rather than by their length.

cellent ways of synopsing large data sets. We are using them in two ways: first the data are rendered by forcing upon them a series of orthogonal (independent) axes which are arranged so as to maximize information—we hope, in a small enough number of axes so as to be discernible to the eye (canonical analysis). However, this method loses the information in later axes that do not make marked separations. Second, we examine the generalized distances within the full dimensionality of the data space; this takes into account all the variation (including the smaller part which is not examined so carefully by canonical analysis). It suffers from the defect that the full information in the generalized distances cannot be visualized because it is multidimensional.

Clustering methods can be used as an aid in understanding the data when already synthetised by generalized distance statistics. Although such methods cannot demonstrate multidimensional figures, they can show the nature of the groupings in the multidimensional space. This means, of course, that whatever clustering method is used immediately takes on the limitations of multivariate methods; given the well understood nature of these methods and their apparent robusticity in practical usage, this is not necessarily an important limitation. It further follows, however, that if we wish to superimpose a clustering method upon multivariate statistics, we do not need to use a method like neighborhood limited classification which is designed to circumvent the concepts of multivariate statistics. Accordingly, we can use the more orthodox techniques; in fact, we have tried single linkage cluster analysis (Sokal and Sneath 1963) and the minimum spanning tree (Gower and Ross 1966). They have been applied to generalized distance statistics in both the studies on the shoulder and pelvis. Particularly for the study of the scapula, they demonstrate clearly the groupings that exist in the data. They are however of special value in the interpolation of unknown data (see below).

Morphological Applications of Optical Data Analysis

One of the principal problems that over many years has vexed those interested in the functional significance of bone form has been the description of the architecture evident within a bone.

Observations of trabeculae have long suggested a relationship to impressed mechanical forces. Much work has been done to elucidate this relationship by study of the alterations that take place during normal growth and following interference in function, both natural and artificial. One difficulty of these studies has related to the problem of how to describe these patterns. Usually such descriptions have relied upon observation of major and obvious elements. The primary problem—more complete characterization of pattern—does not appear to have been tackled. The mere problem of making a set of measurements to characterize such complicated patterns is almost a lifetime work. Analyzing the data is a task that, at the very least, would require the assistance of a computer, and the comparison of many such data sets might take a great deal of time even with a large computer.

The methods of optical data analysis (Goodman 1968) are capable of supplying data collection and analysis for complicated patterns in an almost instantaneous fashion. Our use of these techniques is, as yet, in an early stage; but the clarity of the results and their very obvious value in a whole host of problems relating to functional anatomy of primates within an evolutionary context are so great that a description is worth attempting at this time.

The core of the technique is this: that the many details of any black-and-white transparency can be transformed into its "power spectrum" by appropriate treatment with coherent (laser) light. The power spectrum is a Fourier analysis of the data contained in the original representation. Suppose that we have a picture of a number of small, black, elliptical objects fairly closely packed so that there are a series of clear white channels between them. Suppose next that the orientations and positions of the ellipses are random but that their sizes are Normally distributed around a mean. This information may be sought (a) by making many measurements of the *x* and *y* coordinates of the ellipses; (b) by making a series of angular measurements of the directions in which the ellipses are pointing; and (c) by measuring ellipses. The three sets of data, each in itself a considerable undertaking to acquire and analyze, are all necessary to our understanding of the problem.

The analysis of a black-and-white picture by coherent light performs this almost instantaneously in a photograph of the power spectrum. In the theoretical example quoted, such a photograph consists of a series of light patterns so arranged that they form a thick ring of light around a central point—the reference axis. The fact that the ring is completely circular without any directional parameters demonstrates the random arrangement and directions of the particles within the field. The distance of the most bright center of the thick ring of points from the reference axis gives a (transformed) measure of the mean size of the ellipses. Measures of the dispersion of light away from the central core of the ring give the (transformed) variation of the sizes of the ellipses.

Obviously, data from a nonrandom patterned arrangement such as is presented by trabeculae within a bone produces a rather more involved power spectrum. It can nevertheless be analyzed far more easily than can the original photograph. In any case, it is a relatively simple matter to have even this analysis performed automatically, using a system which can superimpose upon the power spectrum a series of contour lines relating to the brightness and position of different elements. Such analyses can be carried out very quickly and many pictures can be examined. The technique has already been developed for the examination of pattern in rocks. The instrumentation used here is due to J. C. Davis (1970).

Plate 14–1 shows one of the initial analyses that have been carried out in an attempt to see if the technique will indeed work for anatomical material. On the left we have a black-and-white photograph of a cross-section of a fifth lumbar vertebra of man. The pattern displayed here is fairly complex and, while certain major features can be noted (such as the bundles of trabeculae oriented in relation to the position of the pedicles), it is clear that there are also many other details in the picture. The power spectrum of this picture is shown on the right; it is in essence a "fingerprint" of that particular trabecular network. It does not form a simple ring of light (as in the theoretical example described above). Rather, it forms a cloud that is elongated vertically and that has indentations on its lateral aspects.

It may be thought that this picture is every bit as complicated (if not more so) than the original pattern. This is not the case, however, because the information in the original pattern cannot be easily obtained; yet the same information is contained within the power spectrum, and in addition, it can be obtained metrically through a very few measurements (appropriately transformed) that may be made upon the spectrum. For instance, distances from the central ray relate precisely (though transformed) to sizes of elements in the original picture. Angulations within the spectrum (again in relation to the central ray) refer to specific orientational information within the original pattern. The density of the light within the spectrum gives relative amounts of material in the original.

These data are easy to obtain (a) to characterize a single pattern, and (b) to compare two or many similar patterns. The method can be applied not only to sections of the bones of extant animals (as here), but also to data in x-ray form. And it is possible to apply it to sections or radiographs of fossils. More details of the method and further examples of its value within a biological context are given in Chapter 15.

Functional Significance of Bone Form: Experimental
Stress Analysis

In functional assessments of bone form and architecture, anatomical inference, born of intelligent combination of behavioral and anatomical observations, is part of methodology. When, to such compound observations, there can be added methods of increasing the efficiency of observation (e.g. by measurement and analysis), a powerful research strategy is created. However, notwithstanding increased complexity, the links so formed remain correlative; direct investigation of the biomechanics of movement and the function of the bone-joint-muscle structure is essential. Such direct investigation is capable of confirming (which it usually does) or of denying specific anatomical inferences. However, of particular importance is the new information and insight presented by such techniques that cannot be obtained in any way from classical anatomical deductions. The epitome of such methods is the study of movement *in vivo* utilizing (a) telemetric devices that allow relatively unfettered primates to be studied; (b) sophisticated film devices for simultaneous and detailed study of movement; (c) stimulating devices that produce *in vivo* alterations of physiological parameters (e.g. Salmons 1969); and (d) monitoring of physical and physiological properties such as the amount of tension and strain in tissues (Salmons 1969) and, of course, and perhaps most importantly, electromyographical recording of the electrical activity of muscles (reviewed by Basmajian 1967 and in Chapter 13). Many of these techniques are challenges to the technology of our times.

In an attempt to obtain some information relating to mechanical efficiency of skeletal form, but without venturing into these complex technological problems, we have been tempted to utilize some of the analogical methods of experimental stress analysis. Of the techniques available, the photoelastic method seems to provide considerable information (Coker and Filon 1957). However, it must be stressed at the onset that this method works through analogy, involving a series of approximations. One of the chief merits of the method is its visual attractiveness, giving, as it does, an immediate and tangible picture of mechanical efficiency in the structure or detail being investigated.

The technique utilizes the property of reversible birefringence in certain transparent materials when viewed with polarized light under stress. Plastic models of anatomical structures are therefore analyzed. The method involves approximations because models can never be replicas of biological objects. It makes two-dimensional simplifications. Only the more important of the external forces acting on an object can be investigated. Studies in biology have been confined to static systems. These are all serious qualifications which should be understood before attempting to make biological deductions from the physical data.

However, despite these several sources of doubt, the photoelastic method remains an excellent tool for the analysis of stress within structures of complex form. The approximations that are required and the errors arising from them are not usually of sufficient magnitude to invalidate general results. The comparative method of investigation, where models of progressively increasing complexity can be analyzed, provides us with an appreciation for the differences produced by successive approximations. Thus the ultimate leap to the true biological situation is unlikely to provide ideas that are grossly misleading.

In any case, a number of the qualifications to which this technique is subject can be met by more sophisticated researches. For instance, it is possible to allow for three-dimensional situations utilizing frozen stress techniques (e.g. Leven 1955). It is also possible to take into account the

anistropic properties and nonhomogenous nature of bone by using photoelastic skins (e.g. Holister 1961). Dynamic situations can be studied (Flynn
et al. 1962). These improvements are difficult and time-consuming to
apply. Nevertheless, such pilot studies as have been so far carried out
suggest that the new methods, although they give "better" answers, do not
negate the earlier results; rather, they confirm and extend them.

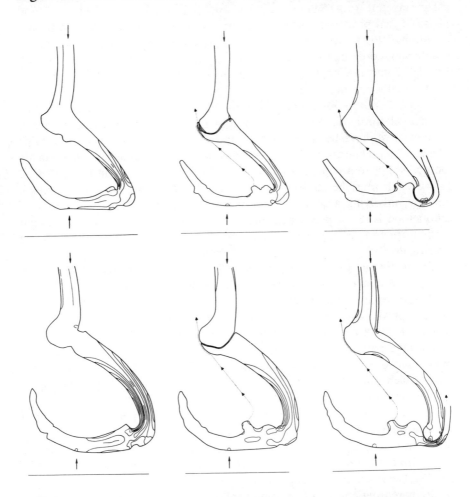

Figure 14–3. Photoelastic analyses of the shapes of the second digital ray in the
chimpanzee and the orangutan, with the elements positioned in a knuckle-
walking simulation.

The left two figures show the comparison (chimpanzee above) when only information about shape is utilized. The number of fringes (indicative of the amount of stress)
is greater in the orangutan, indicating its relative mechanical inefficiency in this postural mode. The second and third vertical pairs confirm that this is also the case when
the simulation is improved by the addition of joints and tendons. It is likely that, if the
investigation could be extended to include the real biological materials, this would still
be the case. Of course, in life the orangutan is not even capable of knuckle-walking.

A number of studies are now progressing; some of these attempt to test ideas that have resulted from knowledge of function and anatomy, and some suggest areas where more classical studies might proceed. One anatomical region now under investigation is the manual digital ray of the great apes and man. Recent investigations (Tuttle 1967, 1969b) outline the mechanism and associated anatomical features of knuckle-walking in African apes. The current study attempts to test and amplify these anatomical inferences. Thus, we are examining, in a knuckle-walking simulation, various architectural features of digital rays (for instance, the relative lengths of metacarpals and phalanges, the nature of their curvatures, the disposition and shape of bony ridges, the orientations of joint surfaces, and the attachments of major tendons).

The studies are being carried out as a series of paired comparisons between individual features of chimpanzees and orangutans so that deductions will be comparable for both forms. In this way, although the mechanical efficiency of neither form can be determined absolutely, relative differences in mechanical efficiency between them can be defined. The study shows (Figure 14–3) that in each comparison the shape of the digital ray in the chimpanzee is much more efficient within a knuckle-walking context than is that of the orangutan (an animal not in fact capable of knuckle-walking).

Another series of comparisons are being carried out with the digital ray in a posture that corresponds to arm-hanging contexts. In this case, though both digital rays are efficient, that of the orangutan is considerably more so (Figure 14–4).

Similar studies are being made for man and the gorilla. The findings for the gorilla more or less replicate those for the chimpanzee; both are highly efficient within the knuckle-walking context, and mildly efficient within the arm-hanging-climbing context. Man is grossly inefficient in both functional roles.

The more detailed results of the above studies supply considerable information as to which aspects of shape contribute to the mechanical differences between the different forms. They thus form an excellent basis for relating function to architecture; and this not only in extant forms but also in fossils. They give clear indication of the facets of shape that may be measured for analytical studies of the type discussed previously; this allows more careful examination of the variation within groups that is not possible, utilizing the photoelastic method.

The technique can thus be seen to be of value in functional morphological studies at different stages. In the case of the studies on the shoulder girdle, it is an independent means of testing the validity of the functional inferences made on the basis of the mathematical treatment of extensive data (Oxnard 1967). In the case of the studies on the digital ray, a situation which in many ways is biomechanically more complex than the shoulder, the technique is being used at an early stage to help clarify those features of the bones which are likely to be relevant to function and which

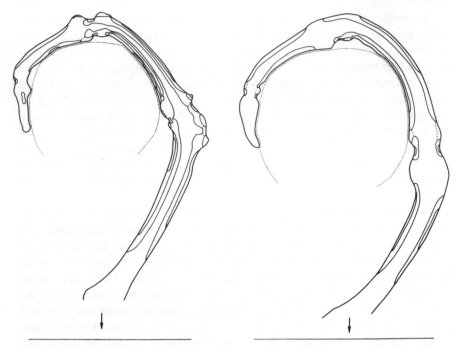

Figure 14–4. Photoelastic analyses of the shapes of the second digital ray in the chimpanzee and orangutan (right) when the digital ray is mounted in a hook-like posture. In this case, only the comparisons of the shaped elements are shown. But the picture demonstrated above (that in this posture it is the orangutan that is mechanically efficient, having fewer fringes) is also confirmed in analyses that include joints and tendons.

may subsequently then be defined and measured for examination by multivariate statistics. A third series of studies on the mechanical efficiency of the pelvis also is currently being carried out. In this case, the analyses are occurring at the same time as the multivariate study of the region mentioned above. However, work to date has not progressed far enough to supply even tentative conclusions. It will be clear, however, that suggestions derived from the canonical analysis of pelvic dimensions (for instance, that the various osteological features of the pelvis appear to congregate in subsets that seem to have functional meaning within locomotion) are exactly capable of being tested by the photoelastic analogy.

Extrapolation to "Unknown" Data

Nearly all of the above techniques can be used in one way or another for the examination of information from unknown or fossil organisms. When data are from a fossil primate, the result represents a speculation that may have greater or lesser likelihood of being correct according to the extent to which it relates to data from extant forms. No further test of such speculation is available unless new data are utilized (new data from the same specimen, similar data from new specimens, or both). Although it is not possible to make direct tests without such new data, it is possible to make tests *in kind,* using extant forms and "pretending" that they are unknown, that they have missing data, that they have been reconstructed, and so forth. Such tests may be useful in that they may indicate the general nature of pitfalls.

A test of this type has now been made for some of the data on the shoulder girdle described by Oxnard (1967), utilizing *Daubentonia* as a test piece.[2] Thus, the canonical analysis of 17 dimensions taken on the primate shoulder have been made (Ashton, Flinn, Oxnard, and Spence, 1971). Groups are analyzed at the superfamily level; this is done simply to keep the number of groups small for the purposes of the test. The analysis is done twice: once with *Daubentonia* excluded from the main study but entered later using the loading factors initially obtained, and a second time with *Daubentonia* included in the main study so that it may make its own contribution to the derivation of loading factors. The first of these techniques is what would normally be done were *Daubentonia* an unknown fossil; the second method is that applied considering the genus as extant.

When *Daubentonia* is entered indirectly after the main analysis, the genus appears to belong to the main group of superfamilies and it lies at its own locus in the same general part of the canonical space (Plate 14–2). To those who know the anatomy of the shoulder of *Daubentonia,* this is a curious result, because, although *Daubentonia* is currently recognized as a primate (e.g. in a taxonomic sense it belongs with the other primates), the shape of the scapula and clavicle differ from those of any other primate. The scapula, for instance, is markedly twisted in a plane at right angles to the general plane of the scapular fossae. This difference from other primates is so great as to suggest that *Daubentonia* is uniquely different (in this respect) from other primates, a conclusion the opposite of that given above by the indirect canonical analysis.

When, however, *Daubentonia* is incorporated directly into the analysis from the beginning, so that the variation and covariation of its own dimensions can make their contributions to the separations, then the results are as follows: *Daubentonia* lies outside the canonical space occupied by the other primates, by a factor of some ten standard deviation units (Plate 14–2); this result is in accord with what one knows about the form of the scapula in this animal (a functional explanation for these features of the shoulder girdle of *Daubentonia* is still wanting; it rests upon the determination of some facts about the use of the shoulder in this rare and curious

2. Ashton, Healy, Oxnard, and Spence (1965) did calculate the position of *Daubentonia* in an earlier study of nine dimensions, but without taking close notice of possible indications that the analysis might be incorrect.

animal). The result shows that it may be a doubtful procedure to investigate an unknown form by merely spotting it indirectly into a previous canonical analysis in this way. This is, for instance, what is attempted by a number of workers examining fossil data using multivariate techniques.

An important question now arises. Given the recognition by the direct study of the misleading results, can there be discerned in the indirect study any information that foretells its illusory nature? Curiously, the hindsight of this investigation suggests that there can. First, careful examination of all the canonical variates may provide evidence. For instance, in the indirect study, the placement of *Daubentonia* in the higher canonical variates (which, once it is determined that they contain little information about the separation of the direct groups, are rarely examined closely) shows that the new form lies at the edge of the separation of genera outlined by each axis. Clearly, although there must always be genera lying at the edges of such separations, one would not expect the same genus to occupy such a position in a number of axes, for if that were the case, then the information relating to that genus could have been brought forward into earlier variates; this is demanded by the nature of canonical analysis—where the maximum separations must occur in the earliest axes. However, for a genus that is entered indirectly afterward, the anomalous result might well occur. This finding therefore suggests that the position for *Daubentonia* in the indirect analysis is indeed "aberrant." The direct analysis confirms this and gives further information as to the large degree of the aberration.

A second way to obtain this information is to utilize the information hidden in the generalized distances by performing a cluster analysis on those values. This technique clearly recognizes when the result obtained from the examination of earlier canonical variates differs from the generalized distances. Using this technique, both the direct and indirect methods give similar (although not identical) results (Figure 14–5).

It must be emphasized that in both of these illustrations the biological situation is deliberately simplified by grouping the data at the superfamily level in order to make the point. However, full examination of the entire data set at the generic level satisfies that similar conclusions apply.

Conclusions such as these allowed the investigation of locomotor adaptations in the shoulder girdles of a number of mammals, and are the basis of the rejection of nonarboreal forms from interpretation (Oxnard 1968b). Of course, in that study these matters are so obvious (the difference between arboreal and nonarboreal mammals is so great) as to require no detailed examination.

Having performed a test of this nature, we may look at an example where a genuine fossil (albeit somewhat reconstructed—the innominate bone from Sterkfontein) is superimposed upon extensive data from extant forms. These data are the nine dimensions of the primate pelvis examined individually and reported by Zuckerman (1966), Oxnard (1966), and

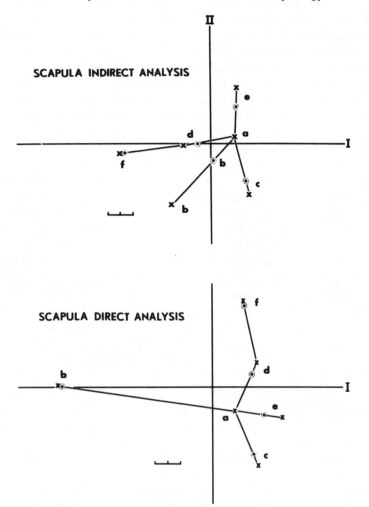

Figure 14–5. These two figures contain the information about the first two canonical variates shown in Plate 14–2, the upper one representing the data with *Daubentonia* (indicated by *b*) added indirectly into the matrix of primate superfamilies (*a,* and *c* to *f*), the lower one representing the data with *Daubentonia* added directly.

Superimposed on the bivariate plots of canonical axes one and two are the appropriate minimum spanning trees for generalized distances. The addition of the minimum spanning tree in the lower figure (shown by x's) adds very little new information to that given by the two canonical variates (shown by 0). The upper figure shows, however, that for *Daubentonia* (*b*) the position given by the first two canonical variates (0) is quite different from that given by the minimum spanning tree in the totality of the 17 dimensional space. This reveals the incorrect data in the indirect canonical analysis.

Zuckerman, Ashton, Oxnard, and Spence (1967). The principal results of the complete canonical analysis have now been reviewed by Zuckerman (1970) and the full details will be published shortly. In this case, canonical analysis demonstrates the unique separation of man from all nonhuman primates—a conclusion readily apparent to the eye. At the same time, how-ever, it reveals very considerable differences among the various nonhuman primates that may be readily related to functional differences (see above).

The interpolation of the Sterkfontein fossil suggests that it lies approxi-mately halfway between man and the great apes. As man, it is unique from nonhuman primates, yet it is far from being identical to man.

Investigations are currently in progress to determine precisely how it differs from man and the great apes. Already it is clear that part of the difference seems to reside in the sizes of the specimens. But it also seems that sub-elements of the innominate bone can be readily defined. For in-stance, in the canonical analysis of those dimensions which can be most closely related to muscle function, the fossil is scarcely distinguishable from the great apes. In the canonical analysis of those parameters which seem most to be measures related to the positions of joint surfaces, the fossil resembles man. *Could it be that in the evolution of upright posture, the changes in the joints in relation to the center of gravity occurred first* (that is, joints were associated in the bipedal configuration early)? *At this stage, muscles still resemble the pattern of ape-like forebears.* However, the full details of this study are yet to be worked out.

In another study, data from fossil remnants became available after com-pletion of the work. These data, from the Sterkfontein scapula and the Olduvai clavicle, are extremely fragmentary and somewhat dubious in nature. Nevertheless, it seems reasonable to attempt interpolation into the extensive shoulder study already performed. As the data are not complete, the technique just described, where a positive position for the fossil is defined, could not be used. We are able, however, to eliminate possibilities rather than suggest a true position for the fossil. This is done (Oxnard 1969a) by allowing the fossil to take up, in turn, for each of the missing dimensions, values equivalent to those already possessed by different extant forms. Such a procedure can never suggest what the missing dimensions may be. But it very definitely can eliminate many sets of dimensions as being outside the realm of likelihood in the sense that the data sets so created lie outside the canonical space normally occupied by primates. (Of course, it is always possible that a fossil may lie completely outside the range of all extant forms; but this is rather unlikely when one is looking at particular fossils that are closely related to extant forms like the great ape and man.)

Finally, unknown data has been inserted into such studies in another manner. Based upon a knowledge of the biomechanics of living forms, an intermediate form is "invented." The interpolation of such an invented form (based upon minimum morphological distance and greatest simplicity in process) is, in general, merely interesting. However, in the particular study, (Oxnard 1969a), the invented form becomes of greater interest because it conforms to the same evolutionary pathway as is already suggested (a) by the examination of extant forms, and (b) by the interpolation of fragmentary fossil data. Such a speculative technique probably only has merit as a confirming mechanism for speculation already obtained from other sources. *Great care is required to prevent circularity of reasoning.*

There are thus a number of ways in which unknown data can be entered into multivariate statistical analyses of data relating to the shape of bones. In a like manner, such data can be interpolated into the various clustering techniques such as neighborhood limited classification. And in this case, further interesting results may obtain because the data for neighborhood limited classification can be in any form or units. Thus, if radiocarbon dates, for instance, are available it is possible to allow time to be one of the dimensions in the grouping process. A dimension in the grouping process that is related to time is obviously unidirectional (thus differing from most other dimensions). This allows the technique the possibility of disconnecting some data points originally connected, and perhaps connecting others initially disconnected, following interpolation of data from fossils. This is of considerable interest because it may indicate convergences and parallelisms. Such analyses have not yet been performed.

Unknown data can also be fitted into experimental stress analysis studies; in this case, a preliminary study has been performed using constructions based upon published photographs of fossils (access to the original material has not yet been obtained). However, the pilot work, insofar as it has gone, gives results which are fairly unequivocal.

What has been done is this: Into the previous stress analysis studies of phalanges described above, additional shapes are analyzed where, for each animal in turn and for each function in turn, the proximal phalanx is removed and the outline of the equivalent member of the Olduvai hand inserted. When the Olduvai proximal phalanx is entered into the chimpanzee hand, then the resulting shape—previously efficient within a knuckle-walking context—becomes markedly inefficient. Within the hanging-climbing context, however, the new shape remains reasonably efficient. When the Olduvai proximal phalanx is placed in the orangutan digital ray, the new shape created remains efficient within the hanging-climbing context but is markedly inefficient within the knuckle-walking context. An analysis of this type cannot tell us what the Olduvai phalanx was actually used for in that organism, but it suggests most strongly that it was unlikely to have been used in knuckle-walking and that it may well have been used in the hanging-climbing situation. *Although we cannot tell whether this applies to what the ancestors of that fossil had been doing (rather than the fossil itself when alive millions of years ago)*, this is nevertheless useful information in the study of the evolution of the free forelimb of man. (Of special interest is the fact that the speculation arising from this data is consonant with that obtained from the totally independent investigation of the shoulder in man and primates, Oxnard 1969a).

Again, let it be emphasized that this study is entirely pilot in nature; and until examination of the actual fossils can be performed, it is only suggestive of the general research tactic that is now available.

Some General Implications

Primate evolution, as seen from the morphological viewpoint, is traditionally an observational, descriptive, and interpretative science. Measurement of variables and quantitative interpretation of data have been undertaken by a relatively small number of workers over many years; but in the main, such studies have not provided the materials out of which our understanding of primate evolution has arisen. Rather, they have supplied confirmation of diagnoses arising from classical anatomical and taxonomic inference.

Sometimes quantitative studies have aimed at the very careful collection of metrical data, with the result that mountains of numbers are analyzed by relatively simple techniques (e.g. Schultz 1930). At other times, concentration is upon complexity and innovation in analysis. This usually involves data that is so meager as to be barely attempting to answer specific evolutionary questions; rather do such studies form examples of the new methods (e.g. Boyce 1964, 1969; Sneath 1967).

Some anatomists and anthropologists look upon quantitative morphological studies of primates as extravagant expenditures of time and manpower, often for results that seem to add little to our knowledge.[3] There is no doubt that criticism of this type is superficial; without these laborious (but pioneering) studies, often rendered the more difficult by lack of techniques and equipment nowadays regarded as indispensable, it would not presently be possible to go beyond the confines provided by interpretation based upon personal observation and description. For it is only the use of currently available formulae to the extreme of their capabilities that confers upon researchers the competence to help propose features and criteria that may be used for the development of yet better techniques.

We are now entering a period in which technical advances in the study of the shape of organisms in functional and evolutionary contexts can provide information and concepts that are of a new order. We are passing the stage of being able to *perform* many of these techniques; rather, we are now able to *utilize* such methods for probing significant problems in primate

3. Needless to say, this is considerably less the case for studies of man himself. Indeed, multivariate techniques were born of studies of human morphology, e.g. Mukherjee, Rao, and Trevor (1955). Even in the field of human morphology, however, there are those who decry the newer methods; for example, see review by Marshall (1969) and rejoinder by Howells (1969b).

morphology, especially those relating to form and architecture, to pattern and texture.

But it seems clear that we are also still in a period in which a number of studies are somewhat repetitive and less rewarding. Some researchers may be criticized for their lack of objectivity and want of conceptual power. Creativity is often lacking. Argument sometimes oscillates without a great deal of forward progress. These factors impinge themselves strongly upon students and intending students who look for new insights and stimulus, and frequently fail to find them.

Yet methods like those discussed in this paper (and, of course, many others used by other workers) can help to give impetus to functional morphological studies of primate evolution. They show the student that exciting progress is possible. Such work cannot be done without instruments such as the computer and the laser. Nor can it be prosecuted without consultation and collaboration with experts in other fields. And the morphologist himself cannot participate unless he is willing to ally mathematical and physical thinking (and therefore training) to his biology. Numerical analysis, computer science, modern physics, all find a place in the education of the morphologist. The new student has the opportunity to become formally educated in these areas; the older researcher must pick them up as he may.

At the same time it should be pointed out that, however effective the computers, however involved the mathematics, however intricate the physical equipment, advances in evolutionary studies of primate morphology cannot be made in the absence of creative, comparative anatomical approaches to the original materials. The biological problems must be clearly recognized; where possible, researchers should attempt experimental or pseudo-experimental strategies; hypotheses should be postulated and tested. In all likelihood, little remains to be discovered by looking at anatomies without specific questions in mind.

A final and most important bonus that is a result of using methods such as these comes from the sense of community induced within scholars in different areas. The very use of these techniques causes, in turn, contact with statisticians, geologists, and engineers, and with methods described by metallurgists, geographers, and physicists. Further, it has not escaped attention that these methods may be used in other areas of primate studies. For instance, multivariate statistics may well be used in a variety of primate researches (for example, in the elucidation of the behavioral structures of these complex animals). Neighborhood limited classification may be used to combine morphological, molecular, and behavioral data in evolutionary studies. The techniques of optical data analysis may well help to disentangle the information resulting from multitudinous interactions of children in playgrounds over time, from patterns of temporal migration of primate troops around waterholes, or from complex sequences of facial expression. A considerable part of the enjoyment of science is such intellectual transplantation among colleagues.

Technical Appendix

So that the reader should be entirely clear about the particular methods that have been described, short technical descriptions are here appended. These descriptions are intended to identify the techniques precisely rather than to provide operational instructions.

Canonical Analysis

If p measurements are taken on q bones, the bones may be considered as q points in a p-dimensional space. The original measurements can be replaced by $k < p$ linear functions, and the values can then be represented in a space of k dimensions. The linear functions are chosen to maintain the greatest possible separation between the populations that are being compared. These functions are the solution of the equation:

$$(\mathbf{B} - \lambda \mathbf{W}) \mathbf{x} = 0$$

when $\mathbf{B}$ and $\mathbf{W}$ are the variance-covariance matrices (dispersion matrices) between and within the populations. The latent roots, λ, are proportional to the between population variances of the corresponding linear functions (See Seal 1964).

In the initial studies, computation depends upon the R technique where it is necessary either to compute the vectors of a nonsymmetric matrix or to perform a special calculation (Ashton, Healy, and Lipton 1957). In the later studies, the considerably simpler computational method (Q technique) is used, which needs only the roots and vectors of a symmetric matrix and standard matrix operations (Gower 1966). Programs due to J. Felsenstein (TXCON) and Biomedical Computer Programs, U.C.L.A., have also been employed.

D^2 and principal component analysis are those associated with these general statistics (e.g. see Seal 1964).

Neighborhood Limited Classification

Again, considering the set of q points in a p dimensional Euclidean space, for each point a set of neighbors can be determined according to some ruling, such as relative closeness. The relatively close rule is that t is a neighbor of s, written $t_\epsilon N (s)$ if

334

$$d\ (s,t) \leqq \max_{x_\epsilon s} [d\ (s,x), d\ (x,t)]\ s \neq t$$

where $d\ (s,t)$ is the Euclidean distance between points s and t. The relatively close rule allows the formation of a connected graph which has no loops and in practice is not simply connected. It determines connected regions in space.

An estimate of density in the neighborhood of a point within the graph is computed as:

$$W\ (s) = [\text{ave}_{n_\epsilon N\ (s)}\ d\ (s,n)]^{-x}$$
$$[W(s)\ \text{is the density};\ x\ \text{has been set to 2}]$$

In order to determine connected regions of continuous high density, the association is computed as:

$$a\ (s,t) = \sqrt{w\ (s) \cdot w\ (t)}/d\ (s,t)$$

for neighbors s and t. This is used in a binary-agglomerative procedure to produce groups (Oxnard and Neely 1969).

The binary-agglomerative grouping procedure is determined by considering, first, all points as single membered groups. The two groups at the highest association are joined, and that association is then removed from the list of allowable associations. Associations to the new group are then recomputed. These steps are then repeated until all the groups are joined in a single group.

The computer program produces as output a list of neighbors and associations; for each step in the binary-agglomerative procedure it gives a list of group members, a list of the interface, and the value of the association between the two groups. If an interior exists, it is listed. Finally, a trunk diagram of the grouping process is drawn. This technique is due to P. M. Neely.

Optical Data Analysis

If an image (for example, a section of a bone) is dichotomized into black and white states, it can be regarded as a two-dimensional square-wave signal which can be expressed by Fourier transformation. One property of a lens forming a real image is that it performs a Fourier transformation on the input signal, and elements of this can be recorded photographically. The Fourier transforms are then enlarged and displayed on a television system. The system can be calibrated, using ruled gratings of known spacing.

In order to evaluate the transforms in a quantitative manner, it is necessary to measure the relative brightness (intensity) of the transforms at different points. This is accomplished by the use of a television-like device, the IDECS processor of the University of Kansas.

This particular technique is due to J. C. Davis (1970) and is more fully described in Oxnard (Chapter 15).

Photoelastic Analysis

Certain transparent materials such as glass, araldite, and other plastics are optically isotropic when unstrained, but when subjected to stress they become anisotropic, and, in the polariscope, exhibit certain features of natural crystals. Thus, a plane polarized light wave entering the stressed plate is split into two components, plane-polarized in the directions of the principal stresses at the point of entry. The two waves are transmitted with different velocities, so that when they emerge from the plate they demonstrate a relative path retardation, the magnitude of which (in a plate of unit thickness) is directly proportional to the difference between the principal stresses.

These two pieces of information, the direction of the polarization of the waves, and the relative path retardation between them, give the primary elements in the solution of a stress problem. They can be obtained from various settings of the polariscope—the machine for producing and operating on the original polarizing ray. Thus, changing the angle of the analyzer maps out the directions of the rotations. The elimination of this information (by using circularly polarized light through the interpolation of quarter-wave plates in the system) allows viewing of the phase differences between the rays. With the equipment as described, other apparatus may be required for the determination of actual stress values. In biological work, this will seldom be necessary because often all that is required is *relative* stress across various parts of the model, especially at the edges.

15

CHARLES E. OXNARD
THE UNIVERSITY OF CHICAGO

The Use of Optical Data Analysis in Functional Morphology: Investigation of Vertebral Trabecular Patterns

Many of the problems relating to evolutionary adaptations of primates may be resolved by the study of skeletal elements. This is especially because temporal effects can be investigated directly by interpolating scattered information from fossils into the matrix of evidence of known forms. Although investigations have often been associated with the characterization and comparison of external aspects of bone, attempts have also been made to include data about internal architecture—especially, about patterns, revealed both in sections and in radiographs, relating to the disposition of trabecular bundles.

That there is an association between such internal structure and stresses imposed by function has always seemed obvious, although there has been considerable argument about the precise nature of the relationship and about possible mechanisms. This was the topic of research at least as early as the 17th century and the numbers of studies rose to a crescendo late last century and early in the present (e.g. Wolff 1870, Roux 1885, Koch 1917, among others). The subject's appeal seemed to wane in the 1940's and 1950's. By then, much evidence seemed to be contradictory and difficult to interpret; different schools of thought had arisen; excellent reviews are available in Murray (1936) and Evans (1957). Recently, interest in these problems has been rekindled with the discovery of piezo-electric and semi-conducting phenomena in different elements of bone tissue (e.g. Fukada and Yasuda 1957, Bassett and Becker 1962), These mechano-electric properties bear a close relationship to the mechanism of adaptation of bone

This preliminary study has been carried out through the kind and helpful collaboration of Dr. John Davis, the University of Kansas.

The work is supported by U.S.P.H.S. Grant HD 02852, N.S.F. Grant GS 30508 and grants from the Wenner-Gren Foundation for Anthropological Research.

to impressed forces. Indeed, numbers of different theories are now being studied (Currey 1968).

But far less attention has been paid to investigating the actual disposition of the trabeculae themselves. In general, the studies in the later decades of last century (when trabecular patterns were investigated in many different planes throughout most of the bones of human and other skeletons: e.g. Wagstaffe 1874) have scarcely been improved upon. Such knowledge, detailed as it is, rests upon descriptions of main trabecular bundles and principal regions of cortical thickness. Many workers, in attempts to support the trajectorial theory of bone architecture, concentrated on mutually orthogonal elements and ignored others. Little is known about more detailed parts of these complex patterns, especially about the smaller features that must have roles in the function of bones.

Very recent studies have been reported (many of them as collaborative work between radiologists and computer scientists) where attempts have been made to reveal all (especially small) details in bone sections or radiographs. These investigations have frequently been pointed towards discerning incipient pathology (for instance, early localized lesions such as metastases, or the onset of generalized bone conditions like osteoporosis). These problems are yet in early stages; they are being pursued by a variety of techniques including (a) the digitizing of the data in a section or x-ray and the mathematical removal (filtering) of unwanted "noise" and other elements by means of matrix manipulations, with the final aim of producing reconstructions that reveal minor (but pathologically important) elements previously hidden (e.g. Nathan and Selzer 1968, Selzer 1968). Other methods that are being utilized include (b) laser techniques whereby noise or other obscuring variables may be removed by the use of physical filters in optical arrangements; again the aim is reconstruction of a sharpened image with the opportunity to visualize features previously scarcely detectable (e.g. Becker, Meyers, and Nice 1969). Nonpictorial techniques are also being utilized: for instance, (c) microdensitometry and radiographic scanning are being employed to render more amenable to investigation such disease processes as osteosarcoma (e.g. Butler 1968).

Most of these techniques have been pioneered in other fields and are undergoing secondary development within biology. Both computational and optical methods for filtering and subsequent reconstruction have evolved as fallout from modern technological advances related, for instance, to space exploration, and the development of instruments such as lasers and computers. The most well known examples are found in the transmission and improvement of pictures taken by artificial satellites and space probes (e.g. Andrews and Pratt 1969) and in pattern recognition studies utilizing powerful computers (Rosenfeld 1969).

A by-product of these methods is the realization that in the procedure of reconstruction, an intermediate stage exists in which the pictorial data is transformed into nonpictorial form. Both computational and optical filtering, for instance, are applied only after the original picture has been transformed in a quantitative manner. These intermediate steps, taken by themselves, may allow the analysis of the pictorial data. With reference to bone, they may provide succinct yet fully comprehensive information about the details contained within complex trabecular patterns.

Again, these analytical applications of the methods have been pioneered in fields other than anatomy. Some of the advances are in the nature of direct analyses of the data in a picture; for instance, the general-purpose computer system of Shelman and Hodges (1970) is capable of directly providing information relating to surfaces (e.g. area, perimeter, and moments) or to shapes (vectorial representations of shape). Other investigations utilize the known properties of particular mathematical transformations to reveal the information in a picture in a particular manner. Thus, computational transformations (such as Walsh/Hadamard functions) have been used to analyze the complex patterns inherent in sound spectograms (Campanella and Robinson 1970); and optical techniques (optical Fourier analysis) have been used to study detailed patterns in different rock formations (e.g. Preston, Green, and Davis 1969; Pincus 1969). Within a relatively short period of time, a considerable literature has sprung up relating to analytical uses of these techniques (e.g. see recent symposia volumes edited by Thomas and Sellers 1969, and by Yau and Garnett 1970). These methods are clearly applicable to biological data, for our purposes, to trabecular patterns within bone. The extent to which they may be of value depends upon matching the specific techniques and appropriate biological questions.

The Basic Technique

Optical data analysis is essentially spectral analysis by diffraction. A coherent monochromatic light source (a ruby laser) is used to produce a Fourier analysis of the raw data. The raw data may be in the form of a transparency of a photomicrograph, an electronmicrograph, an aerial photograph, a contour map, a radiograph, or any other two-dimensional display. In the present study, photographs of cut sections of bones have been used. but further investigations are progressing into the examination of laminographs, whole bone radiographs, and full body x-rays. The analytic result of the procedure is known as the transform or power spectrum of the original picture and is produced by the optical arrangement as shown in Figure 15–1. The transform or power spectrum is the two-dimensional equivalent of the one-dimensional "spectrum" that is produced when a prism is used to split a beam of white light (Figure 15–2).

The diffraction pattern resulting from this process provides information about the distribution of the diffracting elements in the raw data much as an

SYSTEM FOR OPTICAL DATA ANALYSIS

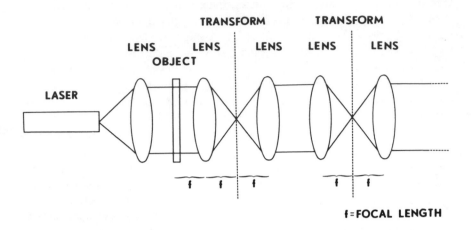

Figure 15–1. Optical arrangement used in Optical Data Analysis.

x-ray diffraction pattern provides information about a crystal lattice. The technique obviates the need to digitize the original data and circumvents the requirement to perform Fast Fourier Transforms on such data with a large computer. At the same time, however, it is clearly an important element of the technique that both computational and optical methods should be applied to the same data set as a test; the precise technique used here has already been confirmed by Davis (1970).

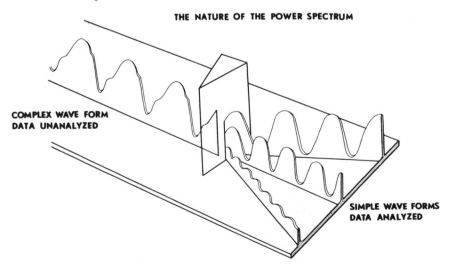

Figure 15–2. The one-dimensional analogy of the process of two-dimensional optical Fourier analysis.

Analysis of Some Geometric Patterns

The information that can result from this method is displayed by the analysis of geometrically simple patterns. Thus the power spectrum of a grid is shown in Figure 15–3. The distance between the lines of the grid is related (albeit, reciprocally and in a transformed manner) to the distance between the central ray and the first harmonic. The angulation of the lines of the grid is given by the angulation (also transformed, in this case by the addition of 90°) of the set of harmonics.

In this regular example, these two items of information can be as readily obtained from the original grating as they can from the power spectrum. But if we suppose that the grid resulted from biological data and possessed, therefore, the irregularities that are often found in such materials, then the first harmonic, instead of being an intense spot of light, would be a rather more diffuse cloud of light. The mean distance apart of the original rulings would then be related to the position of the center of the cloud of light in relation to the reference ray; the dispersion of the cloud of light would provide an immediate measurement (duly transformed) of the variation in the distance apart of the rulings. In this way, the more tedious process of taking many measurements on the grid, followed by statistical manipulation, would be replaced by making only two measurements on the power spectrum.

A further example of the way in which this technique may be applied to a pattern is shown in Figure 15–4. Here there has been analyzed a relatively simple repetitive pattern that has within it a definite orientational

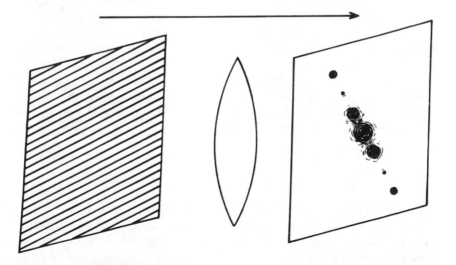

Figure 15–3. The diagrammatic representation of a simple grid and its power spectrum (shown as black on a white ground; the actual transform would be light spots on a dark background).

feature: a set of ovals mounted with regular points of contact and aligned with all the long axes vertical. The power spectrum of this pattern displays clearly: (a) the differing lengths of the major and minor axes of the ovals; and (b) the orientation (transformed by the addition of 90°) of the oval elements. In order to make quantitative assessment, the technique must be calibrated, using a regular grid of known measurement.

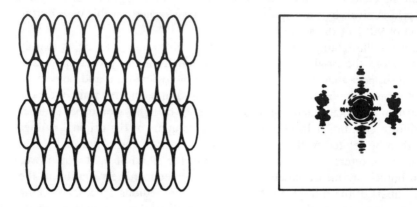

Figure 15–4. A diagrammatic representation of a simple pattern and its power spectrum (shown as black on a white ground; the actual transform would be light spots on a dark ground).

However, periodically regular patterns are not commonly presented in the study of whole organisms, although sometimes such patterns do occur (e.g. growth rings in shells) and may be examined with this technique. Of more frequent occurrence are patterns which exhibit regularity which is so complex that information cannot be culled from it directly, or where the regularity is overlain by other details that we may wish to consider as biological variability, or even as "noise." Such patterns might include the complex branchings found within trees, sponges, and Purkinje cells. As an example of a pattern where most of the regularity (if any) appears to be hidden, the following geological instance is available.

An Example from Geology

Dr. John Davis (1970) of the Kansas Geological Survey of the University of Kansas, has shown how the technique works for the analysis of patterns revealed in thin sections of rocks. In Plate 15–1, just such a thin section is demonstrated; and, apart from the fact that it consists of a large number of particles of varying size, there does not seem to be much other information that can be obtained by eye. Certainly, counting and measurement of the particles for further analysis would be a tedious task.

The power spectrum of such a rock section is shown in Plate 15–2; and although at first sight there appears to be little if any improvement in our ability to see what is contained within the picture, this is in fact not the case. For now, data about the sizes and orientations of the particles in the original picture are contained within the power spectrum in polar coordinate form. Sizes of particles are related to the inverse of the distance of appropriate points from the reference axis (x of the polar coordinates). Angulations of particles are given by the angle plus 90° (ω of the polar coordinates) of angles in the original picture. In order, however, to obtain this information in a manner where it can be observed and rendered quantitative if necessary, the power spectrum must be further examined.

This has also been done at the University of Kansas on the IDECS processor of the Center for Research in Engineering and Science. This system allows the amount of brightness in the power spectrum to be contoured for particular levels of intensity. Plate 15–3 shows how the contoured power spectrum of the rock section appears; the structure within the cloud of light now stands revealed. Most of the contours are circular in outline; and this indicates that for particles of the particular sizes represented by those distances from the central axis, orientations are essentially random. Since these contours are also the largest ones, this means that the smallest particles are those that are randomly directed.

In contrast however, the smallest contour is not circular but oval, and the major axis of the oval is approximately vertical. Thus the particles represented by these contours (the largest particles in the picture—the actual size can be obtained from the calibration experiment) are preferentially oriented in a horizontal manner in the original. If we now return to the original picture (Plate 15–1) with the hindsight of this discussion, we can vaguely see the generally horizontal orientational trend.

In this way it is possible to "fingerprint" thin sections of rocks for geological purposes. However, in addition to this empirical use of the method, it is also being used to test hypotheses about the nature of the patterns present in the rocks. For instance, implicit in the above description is the idea that the patterns in these rocks may represent the realization of some random process in their formation. The extent to which the real situation approximates to this idea can thus be gauged. These two uses of the method can be fairly differentiated, for while the use of the power spectrum as a recognition device is of considerable interest, the hypothesis testing element may be of extra value.

Application to Trabecular Patterns in Bones

A preliminary study has been made of photographs of a number of sections of the bodies of human lumbar vertebrae. Here patterns are clearly different from the geometric examples which are periodic, and different also from the rock patterns which are primarily random. Trabecular patterns are far from random and possess their own regularity, although they are not regular in a periodic sense as are the geometric figures.

Plate 15–4 shows a photograph of one of the bones that has been analyzed: a sagittal section of the body of the second lumbar vertebra. The main elements of the picture are clear; within a generally rectangular outline, trabecular bundles are oriented in a vertical manner, with apparently smaller elements arranged orthogonal to them. There is little else about the pattern that is obvious.

The power spectrum of this section is shown in Plate 15–5, and it is of immediate interest: (a) it does not present a regular pattern of light spots as do the geometric figures; (b) it does not show a generally radially symmetrical cloud of light as does the primarily random rock section. The cloud of light is, in fact, somewhat oval and has irregularities within it; its main dimensions are not vertically and horizontally aligned but appear to be at some slight angle to these axes; within it is a cross of light that is oriented vertically and horizontally. Again, though this blob of light is not particularly impressive, it does contain considerable information, and it may be able to act as a characteristic for the trabecular pattern.

The further analysis of this particular power spectrum by means of the contouring technique previously described produces the picture shown in Plate 15–6. Departures from circularity are immediately and quantitatively obvious; that is, the structure is far from being random. But when we examine each individual contour, information that is of considerable interest comes to our attention. Thus the innermost contour represents data from the largest elements, and as this contour is oriented with its short axis horizontal it represents the trabeculae that are placed with their long axes vertically. Again, the outermost contours, representing the smallest trabeculae, are oriented in a similar manner. This latter fact is not obvious from the original picture, where our eye is captured by the largest elements so that it has difficulty in assessing the smallest.

Of special interest, however, is the information given by the intermediate contours. These represent data from particles of intermediate size; the minor axes are oriented at a small angle to the horizontal; the corresponding trabeculae must then be aligned at some small angle to the vertical, as compared with the other trabeculae.

This information is not obvious from the original picture; indeed, the only other way in which one can readily envisage its discovery entails subtracting larger and smaller elements from the picture to reveal these particular intermediate trabeculae. (This can be done with some of the computational techniques currently available, Seltzer 1968).

Such knowledge may prove important in functional terms. Generally orthogonal elements have long been recognized because of relationships, real or presumed, with the imaginary stress and strain trajectories with which they have been associated in a variety of ways for many years. The existence, therefore, of a specific, nonorthogonally oriented set of trabeculae has implications for biomechanics that require further investigation. (The *caveat* must be introduced: that this information has been obtained from studies that are, after all, only pilot in nature; these findings remain, themselves, to be confirmed and greatly extended before they can be accepted. However the strategy for revealing such information has been outlined and it adds to the value of the method.)

Further Applications of Optical Data Analysis

The preceding discussion has shown in general terms the nature of the technique and the way in which it can be used to examine regular patterns, apparently random arrangements, and structured but nonperiodic designs. When used in this way alone, the method is powerful, being able to provide: (a) fingerprints of complex patterns for purposes of recognition and characterization; (b) detailed metrical parameters relating to quantitative variables of the patterns; and (c) tests of hypotheses about the nature of the patterns, e.g. the extent to which they deviate from random, or from an orthogonal arrangement, etc. But this general method is also capable of further elaboration within the realm of the characterization and comparison of bone form and architecture in the context of functional and evolutionary studies of the primates.

Thus one important element in research strategy relates to the fact that it is possible to examine specific parts of the patterns. For instance an optical filter may be placed at the position of the first transform plane (Figure 15–1), and this may allow us to examine: (a) the reconstituted image showing the nature of the specimen when certain items have been removed; and (b) from the examination of the second transform plane, the power spectrum of the reconstituted image. Since filters may be designed to remove information about sizes, e.g. of all large elements, and about orientations, e.g. of all vertical trabeculae, such a study may give considerable insight into the original trabecular arrangement.

But such comparisons can obviously be taken a useful step further. Instead of using a simple filter in the position of the first transform plane, we may utilize the power spectrum of a first set of trabeculae as a filter in the analysis of a second. This clearly achieves the *subtraction* of one trabecular pattern from another, yielding (at the image position) a picture of the difference between the specimens, and at the next transform position an analysis (as a power spectrum) of the differences between the specimens. *This is comparative anatomy at the flick of a switch.*

Even further examinations can be made, utilizing subtraction techniques.

For instance, information within certain primate fossils relates not only to the trabecular patterns originally present in the bone at the time of fossilization but also to the crystalline and other features that are the result of the nonbiological structures in the replacement material or in the matrix. It is possible to subtract, from a radiograph of the whole fossil, the pattern of the geological elements alone by using, as a filter in the transform plane, the power spectrum of the equivalent contiguous but nonfossiliferous material. (In practice it may sometimes be difficult to obtain the specimens.) Such a filtering process would allow us to reconstruct those elements of the fossil structure that are due to trabeculae existing in the bone, and to make appropriate analyses of those patterns for comparison with related forms. As trabecular patterns may be of particular import in situations where bones may be bearing very different forces (e.g. vertebrae in hanging and quadrupedal apes and in bipedal man), the examination of such trabecular patterns as may exist, for instance, in *Australopithecus* (thought to be bipedal) would be of extreme interest.

Further research strategies of a different nature are possible. These relate to the hypothesis-testing use of the technique. As has already been suggested, the method may be utilized to examine the theoretical idea that trabeculae are orthogonally oriented in some relationship to orthogonal stress or strain trajectories. Recent ideas about such topics suggest that this would not be a very fruitful line of investigation. What might be of considerably greater interest would be the testing of experimentally derived hypotheses about the mechanical efficiency of particular trabecular arrangements. For instance, experimental stress investigations (e.g. the photoelastic studies described in Chapter 14) may be used to arrive at experimental definitions of how given sets of trabeculae function in particular skeletal regions. Again, hypotheses about function in particular anatomical regions may result from computer simulation studies (e.g. as in the investigations of the vertebral column by Aquino 1970). A wide variety of such models may be tested by these general methods.

Conclusions

This paper has not reported a particular methodology that is justified *in terms of the research results that it has generated* (the primary theme of the Burg Wartenstein Symposium). But enough work has now been done utilizing the method in the context of functional and evolutionary studies of skeletal architecture in primates that *its powers and strategies are apparent.*

One question that may be worth considering is: why use Fourier analysis at all? Once we are involved in a field like this, two things immediately become obvious: (a) Fourier transformations are only one of a whole series of mathematically related functions; and (b) all of these are, in principle at any rate, calculable by digital, electrical or optical techniques. Which particular transformation should be used therefore depends upon the precise questions that we wish to ask.

In Fourier analysis the pictorial data is transformed from the spatial domain to the frequency domain. It is therefore of particular value for giving information about particle size and orientation. Walsh/Hadamard functions, a related family of transforms, produce analyses into the *sequency* domain. This technique is useful for analyzing sound spectrographs (Campanella and Robinson 1970) and has been used in morphology to examine the shapes of leaves (Meltzer, Searle, and Brown 1967). Its special value here is that it magnifies the smaller structural features of the leaf edges in comparison with the overall form of the leaf. Yet another such transformation is the Haar. In this analysis, the new domain relates to the comparisons of points in the original picture in pairs, and so it gives information about localities within the picture. I am not aware that this particular transformation has been used in biology; but examples of its use are available in terms of transmitting and sharpening space-craft pictures of the moon (Andrews 1970).

Finally, all of these methods are designed to produce particular results. We must be cognizant of the fact that for some of the problems that we wish to pose, no specific transformation may exist. Accordingly, it may be necessary to utilize the direct methods of Shelman and Hodges (1970) that allow a wide range of *ad hoc* questions to be asked about specific shapes and patterns.

V

Aspects of Behavior
and Ecology

16

BENJAMIN B. BECK AND RUSSELL TUTTLE
CHICAGO ZOOLOGICAL PARK AND THE UNIVERSITY OF CHICAGO

The Behavior of Gray Langurs
at a Ceylonese Waterhole

A variety of strategies and techniques have been used in studies on the behavior of free-ranging primates. In one approach, the observer is mobile, following his subjects either on foot or in a vehicle. This usually maximizes the amount of contact with the species being studied but, especially in the initial stages of the study, the presence of a pursuing human observer may seriously affect the behavioral repertoire. Such an effect is lessened, through the processes of habituation and extinction, as the study progresses, but its overall impact must be considered to be substantial. To avoid these difficulties, the observer may remain stationary and perhaps make some effort at concealment. This strategy minimizes (but does not eliminate) the effect of the observer, but is usually considered to result in a great deal of field time in which no data are being collected. One can resort to provisioning to increase the time that primate subjects are present, but this adds confounding variables. Both of these general strategies were used by van Lawick-Goodall (1968) in her long-term study of chimpan-

This field study was supported principally by a grant to R. Tuttle from the Wenner-Gren Foundation for Anthropological Research, and we gratefully acknowledge the Foundation's commitment to our project and the study of primates in general. The Department of Psychology of The University of Chicago granted B. Beck a stipend during the study.

Analysis of the data was supported by NSF Grant nos. GS–834, GS–1888, and GS–3209 and by a Public Health Service Research Career Development Award no. 1–K04–GM16347–01 from the National Institutes of Health to R. Tuttle. Miss Sharon Levy and Miss Nancy Ping diligently assisted us in summarizing these data.

We are also grateful to the Wild Life Department of Ceylon; its director, Mr. Lyn De Alwis; the warden of Wilpattu National Park, Mr. Percy De Alwis; our tracker, R. P. Somasekaram; and the late John Singho. The Wild Life Protection Society of Ceylon, through its secretary (now president) Mr. Th. W. Hoffman and his wife Mae, offered invaluable aid in planning and conducting the study. We are deeply indebted to the citizens of Ceylon for allowing us to study and share the beauty of their rare and magnificient natural resources. Among those whose hospitality and generosity we had the experience to share personally were Lalith Senanayke and his family, and the Hon. and Mrs. Vernon Jonklaas.

zees, and her work exemplifies the advantages and disadvantages inherent in each. In this chapter we should like to outline a new approach to the study of ecology and behavior of free-ranging primates: specifically, their study at waterholes.

Conservation and wildlife management workers have long studied waterholes, mainly for gathering population census samples (e.g. Simmons 1969). Washburn and DeVore (1961b) recognized the fruitfulness of waterholes as primate study areas, especially in discovering interspecific relationships. However, Tuttle and some associates (Cartmill and Tuttle 1966) were first, to our knowledge, to conduct intensive behavioral observation of primates at waterholes. Their work was done at Wankie National Park in Rhodesia during the summer of 1965, and it was from this study that most of the rationale and techniques of waterhole study originate. We conducted a similar study in Ceylon in the late summer of 1966. Beck and some of his students used the technique in January 1969 to study the Sonoran Desert fauna; and two of the latter, Peter Gelfand and Christopher Engen, returned there in January 1970, to gather additional data. Struhsaker and Gartlan (1970) gained new insight on the behavior of patas monkeys (*Erythrocebus patas*) from a relatively brief study at waterholes in Cameroon.

In all studies, except that of Struhsaker and Gartlan, the observers were stationary, working from blinds or areas of limited visibility to the animals. The Ceylon study was conducted from a tree blind, and many procedures were employed to minimize interference with faunal activity (see below). It is impractical, if not impossible, to insure that the observers will not be detected by any of the subjects; however, in none of our waterhole studies is there evidence for a gradual progressive change in overall activity or behavior in any of the species that were studied. This contrasts sharply with the initial avoidance and slowly increasing toleration shown toward mobile observers during their studies. In our opinion, stationary observers, especially if they take steps to avoid obvious interference, get a much more consistent and typical picture of behavior. Even though primates may eventually cease to avoid pursuing observers, the immediate presence of such an observer probably distorts subtleties of behavior. In addition, the avoidance and escape responses of associated nonprimate species may not habituate or be extinguished during even a long-term study. This further confounds an overall picture.

One of the strengths of the waterhole approach is that it incorporates stationary observation. We have already noted that the principal drawback of stationary observation is the occurrence of long periods where no data are being collected due to the absence of subjects. The Ceylon study included 200 hours, 21.5 minutes of observation on a total of twenty days. One troop of Ceylonese gray langurs (*Presbytis entellus theristes*), whose home range included the waterhole, was observed for 3,889 minutes, 32.4

percent of the total observation time. Other gray langurs, including a lone animal, were observed for 232 minutes; and several toque macaques (*Macaca sinica*) were observed on one occasion for three minutes. In all, primates were observed during approximately one-third of the study time under conditions of stationary observation with no provisioning. This was a very acceptable ratio, especially considering the minimization of interference inherent in our techniques.

More critical, however, is the recognition that useful and productive observations can be made in the absence of primates. Data on the behavior and activity cycles of nonprimate mammals, while being valuable in their own right, are also useful in gaining a total understanding of the primate subjects, especially in the context of interspecific relationships. An example of this, drawn from the Ceylon study, will be presented below. Mammals were absent for only rare moments during the Ceylon study; thus one of the major shortcomings of the stationary strategy was minimized. However, even the absence of subjects becomes very useful data when one considers the larger context. The focus of waterhole studies, then, is an ecological unit: the waterhole area and the total faunal assemblage which uses it, rather than a particular species. The resulting data constitute an ongoing, quantitative summary, continuous over time of all observed behavior at the waterhole.

The Structure of Waterholes

Waterholes are of two general types. The first has various names (e.g. pan, tank) and is an area of low elevation (relative to the surrounding terrain) and has a relatively impermeable substrate. It is simply a passive collector of rainfall and runoff. Its surface area and shape vary with the local contour configuration and with variations in input from rainfall and loss from evaporation and consumption. During dry seasons, when rainfall is scarce and evaporation great, the volume of water will decrease and the surface area will shrink. This leaves "aprons" with few, if any, living trees, and little vegetation with the exception of grass and a few relief features. The breadth of such aprons will vary depending on the contour of the area, but the frequent result is that mammals must cross extensive open areas on the ground in order to use the waterhole. This condition not only greatly facilitates observation per se, but also allows study of the terrestrial behavior of those typically arboreal forms which use the water. Weir (1960) discusses the formation and development of pans and emphasizes the role of faunal populations in these processes.

The second general type of waterhole is one whose input results from an underground source as well as from rainfall. In general, seasonal variations in rainfall and evaporation do not result in such great changes in volume and surface area as is the case with pans, but the same basic principles apply.

Rationale and Method of the Ceylon Study

To exemplify the methodology of waterhole studies and some of the types of data which can best be collected using this approach, we shall summarize and abstract aspects of our Ceylon study which pertain to primates, and indicate the relevance of these data to several ethological, anthropological, and evolutionary issues involving primates.

Southwick, Beg, and Siddiqi (1965) have shown that the behavior of forest-dwelling troops of rhesus monkeys (*Macaca mulatta*) differs considerably from that of temple-dwelling troops. Singh (1968) has shown experimentally that rural and urban rhesus monkeys differ in aggressiveness. This suggests that studies of the behavior of nonhuman primates conducted exclusively in areas of relatively high human population density do not yield representative pictures of behavioral repertoires and natural history. Ripley (1965, 1967a) has studied the Ceylonese gray langur (*P. entellus theristes*). The bulk of her observations were conducted within or near villages. Jay (1963b, 1965) studied the Indian form (*P. entellus entellus*), both in villages and in areas of relatively low population density; and Sugiyama (1964, 1965, 1967), Sugiyama et al. (1965), Yoshiba (1968), and Vogel (1970) studied the Indian gray langur in forests. Forest-dwelling troups of the Ceylonese form had not yet been studied extensively. Partially filling this lacuna was a major objective of our work.

The present study was conducted at a pan called Nelun Wila located in the Wilpattu National Park in northwestern Ceylon (about 8°, 27' N., 79°, 58' E.). The waterhole (Figure 16–1) was of the pan type, roughly triangular, measuring about 555 feet (169 m.) by 450 feet (137 m.) by 685 feet (209 m.) at the time of the study. The forest edge follows the contour of the pan irregularly, leaving almost level grassy aprons of from about 25 feet (8 m.) to about 1500 feet (457 m.) in width as measured from shore line to tree line. The only prominent relief on these aprons were three small copses and the fallen trunk and limbs of a large dead tree.

Wilpattu National Park is composed mainly of mixed thorn scrub jungle and hardwood forest. It receives about 25 to 50 inches (64 to 128 cm.) of rainfall annually, 60 percent of which occurs in the northeastern monsoon from November through January. The dry season occurs from late May to early September (Farmer 1957). Average daily shade temperature at the study site ranged from 78.7° F. (25.9° C.) in the early morning, to 88.1° F. (31.2° C.) at about noon. The lowest shade temperature recorded at the site during the study was 73° F. (22.9° C.) and the highest was 93° F. (33.9° C.). Light to moderate rainfall was recorded on four days during the study. Total rainfall accumulation, while not measured, is

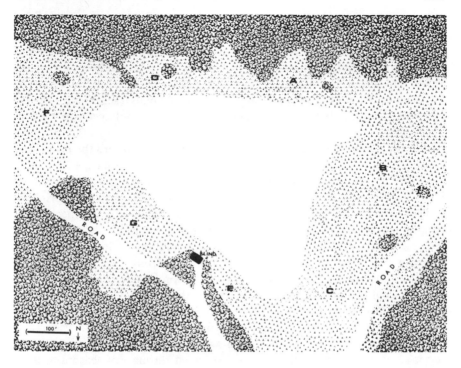

Figure 16–1. Schematic diagram of Nelun Wila, Wilpattu National Park, Ceylon. The letters A–G indicate arbitrarily subdivided sectors of the waterhole area.

estimated to have been very small. During the study, sunrise occurred at about 05:55 and sunset at about 18:25.

Observation was conducted from an elevated tree blind, 22 feet (6.7 m.) above ground and about 25 feet (7.6 m.) from the shore line. We were assisted by a photographer and a Tamil tracker. Observation usually began at about 06:45 and terminated at about 16:45. The 20 observation days included August 10–15, 17–20, and September 3–12, 1966.

Movement to and from the tree blind was always behind the tree containing the blind so as to cause minimal interference with faunal activity. We drove to and from the site and parked the vehicle as inconspicuously as possible. We did not cook at the observation site, and our living areas were more than 3.5 miles (5.6 km.) away. Park visitors, with only a few exceptions, cooperated in the study by avoiding the work site. We entered the study area to collect plant samples, take measurements, etc., only after the behavioral study had been terminated.

Binoculars (7x35, 8x40), pens, notebooks, watches, a thermometer, and still and cinema cameras with a variety of lenses were used.

When an animal was sighted, the time was noted, it was identified, its sex and approximate age noted, and observation of qualitative and quantitative aspects of behavior was begun. Entrance is defined as the time, to the nearest 30 seconds, when the animal is first sighted, and exit as the time when it no longer can be seen. In addition to entrance and exit, onset and cessation of drinking were timed, when possible. The study area was divided into seven sectors, using natural features as markers in order to facilitate notation of the spatial distribution of animals (see Figure 16–1). Most of the time, behavior was being monitored simultaneously by two observers.

Our study was conducted in August and September, toward the end of the local dry season. While behavior at a waterhole is by no means limited to drinking, it is axiomatic that the most distinctive characteristic of water-holes is that they are a source of drinking water. The need for drinking water varies considerably from species to species and within a species, depending on such factors as rainfall, humidity, temperature, wind velocity, and diet (Schmidt-Nielsen 1964). The combination of low rainfall, low humidity, high temperature, and sometimes relatively high wind velocity which characterizes dry seasons increases the probability of drinking. The physiological processes underlying thirst and drinking have been extensively explored (Grossman 1967), but it suffices for present purposes to note that dry season conditions tend to initiate drinking due to increased water loss by evaporation and due to decreased water content of plant and animal food. As a result, waterhole studies conducted during dry seasons will be much more productive in terms of the numbers of animals likely to be present. However, waterhole studies conducted exclusively during dry seasons are very likely to result in heavily biased samples which may result in distorted understanding of intra- and interspecific behavior. For example, Chivers (1969) notes seasonal variation in the daily activity cycles of howler monkeys (*Alouatta palliata*) and cites a personal communication from Aldrich-Blake noting a similar effect in *Cercopithecus mitis*. A complete behavioral picture will emerge only when studies sample all of the seasonal variations in a given locale. This will necessitate some waterhole studies which will be of lower productivity in terms of numbers of animals observed. The 1969 and 1970 Sonoran Desert studies were of this nature.

Results and their Implications for the Waterhole Strategy

The primates observed at the Ceylonese waterhole are, for purposes of discussion, divided into four categories: (1) a "resident" troop of *P. entellus* so designated because of their repeated presence at the waterhole which obviously was included in their home range; (2) an "intruder" troop(s) of *P. entellus,* so designated because they were seen in the area infrequently and their presence usually elicited display jumping and pursuit by the resident troop; (3) one lone, apparently unhealthy *P. entellus;* and (4) four toque macaques (*Macaca sinica*) observed on one occasion only.

The resident langur troop consisted initially of 26 animals and of 27 on the last three days of the study. Following Ripley's (1965) adaptation of Jay's (1963b) criteria and schema, the age/sex composition of the resident troop was 10 adult males, one adult female (an additional old adult female joined the troop during the last three days), seven subadult males, four juvenile males, two infant-2 males, one infant-2 female, and one infant-1, the sex of which we could not confidently establish and which was usually carried ventrally by the adult female. Ripley (1965) notes that the mean size of North Indian and Ceylonese langur troops is 24 to 26 animals. This compares closely with the size of the resident troop. However, the sex distribution is very atypical with only two females initially, the older female which joined the troop near the end of the study and, possibly, the infant-1. The ratio of adult and subadult animals to juvenile and immatures is also unusually high. Jay (1963b), Sugiyama (1964, 1965, 1967), Sugiyama et al. (1965), Vogel (1970), and Yoshiba (1968) report the existence of all-male *P. entellus* troops or groups in India; and Ripley (1965) notes them in Ceylon, but these generally consist of fewer animals (one to 20) and are not very stable. All-male groups have also been reported for other monkey species, e.g. the Nilgiri langur, *P. johnii* (Poirier 1967, 1969), the gelada baboon, *Theropithecus gelada* (Crook 1966, Crook and Aldrich-Blake 1968), the spider monkey, *Ateles geoffroyi*, (Carpenter 1964), and the patas monkey, *Erythrocebus patas* (Hall 1965, Struhsaker and Gartlan 1970). Crook (1970), elaborating on a hypothesis stated earlier by Crook and Gartlan (1966), concludes that all-male groups and solitary males usually occur in conjunction with one-male heterosexual groups (of the same species) which, in areas with minimal food availability and/or low predation pressure, serve to optimize reproductive efficiency. The typical Ceylon langur troop has more than one adult male (Ripley 1965), ample food appears to be available, even in the dry season, and predation on langurs is known to occur (Eisenberg 1969). Thus, if Crook's analysis is correct, the resident troop in the present study is better considered as a heterosexual troop with an atypical age/sex composition than as a male group. However, exceptions to Crook's conclusion have already been reported, e.g. Tanaka (1965) reports no all-male groups of *P. johnii* although heterosexual groups had only one adult male and Ripley (1965) notes all-male groups in conjunction with multi-male heterosexual troops of *P. entellus*. The positon of the present troop in the spectrum of langur social organization cannot be determined until more about the spectrum itself is known and the data derived from this study must be employed cautiously.

One or more members of the resident troop was present on each day of the study and the troop was visible for nearly one-third of the total observation time. The distribution of troop presence (defined as presence of at least one troop member for at least 30 seconds) during 15-minute intervals

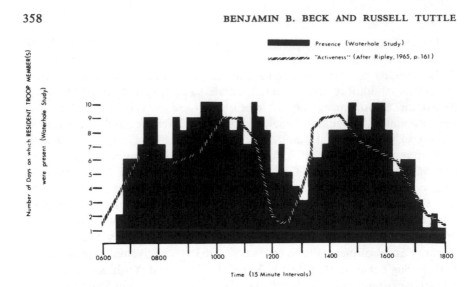

Figure 16–2. The distribution of langur presence as compared to "activeness."

is shown in Figure 16–2. The presence of one troop member, in this graph, has the same ordinate value as the presence of all 26 animals. This method of analysis differs from that used for other species but was adopted because of the basic cohesiveness of langur troops. The observation of any fraction of the troop was taken to indicate the proximity of the troop, but it was only during terrestrial progression that confident counts could be made. The choice of a 15-minute interval as a unit for the abscissa is arbitrary and reflects the nature and the relative lack of fluidity in langur troop movements. Ripley's schematic estimate (1965, Figure 16) of "activeness" of langur troops has been superimposed on Figure 16–2. The similarity between Ripley's schematic function of "activeness" and our plot of presence of langurs indicates that those times when the langurs were being observed at the waterhole are almost precisely those times when the richest and most intense behavior is in progress. It might be noted that an analogous estimate for Nilgiri langurs, *P. johnii*, presented by Poirier (1967, p. 63) is strikingly similar to Ripley's function and our data.

The waterhole strategy offers a unique opportunity to study the terrestrial behavior of the typically arboreal forms. When the animals retreat into the forest they are usually not visible from the blind. For these reasons, and also because the terrestrial behaviors of the langurs are relevant to many interests, we shall begin with discussion of terrestrial activities. Of the 3889 minutes during which the resident troop or part thereof was under observation, one or more members of the troop was on the ground for 1177 minutes, or 30.3 percent of langur observation time. This result agrees well with Sugiyama's (1964, 1967) estimate that 25 percent of the gray langur's "day" is spent on the ground. Our results may be somewhat greater than typical, since most of the study area consisted of treeless space. However, Sugiyama's estimate may be depressed because of the presence of a terrestrial observer. In any event, our data support the conclusion that the niche of the gray langur has a significant terrestrial aspect. In contrast, the Nilgiri langur rarely comes to the ground (Poirier 1967, 1968a, Tanaka 1965) and the silvered leaf-monkey is largely arboreal (Furuya 1961–1962, Bernstein 1968a). The bulk of terrestrial activity can be included under three categories: drinking, progression from sector to sector of the study area, and play behavior.

Drinking

One or more members of the troop drank every day of the study, and the duration of drinking bouts totaled 765 minutes. The bouts ranged from brief drinks by one individual lasting about 30 seconds, to one of 91.5 minutes during which more than 30 observations of drinking were recorded. The mean bout duration was 18.7 minutes. The distribution of 15-minute intervals in which one or more langurs drank shows peaks which generally correspond to the peaks of langur presence (about 09:45, 11:15, and 15:30 in Figure 16–2). In addition, another peak occurs about 13:15–13:30 which, according to Ripley (1965), is the period of increasing activity immediately following the midday resting period. It was not always possible to ascertain the actual number of animals which drank, as some may have drunk several times. We did not usually time individual drinking behavior but, generally, a langur would drink for about 15 seconds, stop and look around for about 5 seconds, and repeat this cycle several times.

Langurs were observed to adopt two general classes of posture during drinking. The most common, adopted during 83.9 percent of the recorded incidences, was one in which the trunk was positioned horizontally over four fully flexed limbs with the ventrum touching the substrate. This was the drinking posture recorded as typical by Ripley (1965, 1967b). Tail posture was variable. At times the tail was fully recurved over the back. On other occasions the tail was almost vertical with a slight posterior bend of the distal one-sixth. On still other occasions it was extended horizontally or nearly so in the air or on the ground behind the animal. Most commonly, however, it was held in the "S" or "question mark" shape typical of Ceylonese langurs. The other drinking posture observed was that typical of baboons, in which the hindlimbs are vertical and almost fully extended and the forelimbs are flexed so that the body is oriented diagonally head downwards with respect to the ground. The variation in tail carriage in this posture was comparable to that noted above. On a few occasions the body and/or tail posture of an animal changed from one type to another during drinking. The infant-1 of the resident troop was usually dropped from the ventral position during drinking, although on one occasion it was held ventrally while the female drank; she used the horizontal, complete crouch posture, unlike the females observed by Ripley (1965, 1967b) who drank in the diagonal baboon-type posture when carrying infants ventrally.

The langurs typically drank with their mouths and lips making direct contact with the water. On only one occasion was a langur (a subadult male) observed to drink by dipping his hand into the water and licking the water from it, and this was interspersed with drinking by direct contact of his mouth with the water. An infant-2 female of a troop observed near our camp consistently drank from a bucket containing provisioned water, using a very stereotyped sequence of dipping her left hand into the water and licking it from the ulnar aspect of her hand and arm. No langurs were observed to swim during the study, although in some cases the hands and even the feet would be submerged during drinking. On other occasions, however, we specifically noted langurs attempting to approach and leave the water and to drink by stepping on raised, dry clods of mud. On one occasion, a male stopped drinking and appeared to grab at something in the water.

Not all members of the troop drank simultaneously. The highest number observed to be drinking together was eleven. One or more animals would come to the water and drink while other troop members fed or rested in nearby trees or played or simply sat on the ground. Frequently, animals would stop to drink during terrestrial progression. During episodes in which many of the troop drank, two to five adult or subadult males were first to drink. Typically, one would emerge from the tree line and sit on the apron for 30 to 90 seconds. He would then go to the water and one or two would take his place on the apron or the first would sit and the second and/or third would pass him and drink first. The net effect was that one animal sat and looked around the area before any drank, and at least one of the first few to drink was sitting and looking while the others drank. Following this period, which lasted two or three minutes, many of the troop members would move more fluidly on the apron between the forest and the shore line. Nonsystematic observation at other waterholes in the area revealed that the drinking bouts of other langur troops and of a troop of toque macaques began and progressed in similar fashion.

There was no correlation between the frequency of drinking at a given point on the shore line, and the distance between that point and the nearest trees. The troop drank in all seven sectors of the study area, although they drank about 32 percent of the time in sector C where the shore line was about 300 feet (91 m.) from the nearest trees; at no other point were the nearest trees farther from the shore line. It should be noted that on one occasion the adult female, two infant-2, and the infant-1 drank as it began to rain; and on another occasion an adult male, an infant-2, and another langur drank in the midst of a light steady rain.

Ripley notes that "the gray langur is able to subsist on the water derived from desiccated leaves during a substantial portion of the year" (1965, p. 26) and concludes that "when water is readily available, on their path so to speak, they will drink but will seldom go far out of their way to obtain water" (1965, p. 119). The same author (1970) concludes that the gray langurs' adaptation to a wide range of habitats is pivotally dependent upon their ability to "eschew" standing water for drinking for long periods. The physiological capacity to survive without drinking water for a long period must, in light of the complex and multifactor causation of thirst and drinking noted above, be considered as a poor predictor of whether given animals will drink under given circumstances at a given time. McCann (1933), in a survey of Indian langurs, reports that gray langurs "drink water regularly about midday and sometimes also in the evening" (p. 624). Yoshiba (1968) notes considerable regional variability in drinking in Indian langurs, but Sugiyama reports that one troop commonly crossed 200m. over an open field to drink in the dry season. Jay (1965) notes the importance of water to Indian gray langurs, but at another point (1963b) notes langurs' ability to live without drinking water. Poirier (1967) reports only five cases of Nilgiri langurs drinking from standing sources during his extensive study.

It would therefore seem that the importance of langurs' ability to survive without drinking is secondary to discovering the actual conditions under which langurs do or do not drink. Ripley, Jay, and Sugiyama all make reference to drinking by gray langurs, but supply few details and de-emphasize its importance in langur activity. In light of our data it would seem that this conclusion must be qualified or otherwise altered. The observation that a troop of langurs whose activity virtually centers around a waterhole, some or all of whose members drink on 20 (of 20) days of observation during a dry season and whose members cross extensive tree-less spaces to drink indicates that water may be more important than was previously conjectured.

Use of the waterhole by primates was not limited to the resident langur troop. On seven occasions, totaling 217.5 minutes, langurs not belonging to the resident troop were observed. It was not possible to ascertain if these animals belonged to one or more troops. On one occasion, more than thirty animals, including seven adult males, eight adult females, and two subadults, eight juveniles, four infant-2s and five infant-1s (carried ventrally) of undetermined sex were observed. On another occasion, three adult males, three adult females, and one subadult, one juvenile, one infant-2, and one infant-1 (carried ventrally) of undetermined sex were seen. Eleven langurs, including eight adult males, one adult female, one subadult male, and one juvenile female were seen on another occasion. The other four observations of nonresident langurs involved two, three, five, and six animals. One individual or a larger group of these "intruder" troops was on the ground for 112 minutes, or 51 percent of the total time that they were observed, and engaged in six drinking bouts totaling 53 minutes. They were always observed in sectors D and/or F. Like the resident troop, the terrestrial activity of the intruders included play but it did not include progression from one sector of the study area to another.

On only four of the seven appearances was the resident troop present. The first time intruders and the resident troop were present simultaneously, the resident troop fed intensively in sector C while two of the intruders drank and at least four more fed in the trees in sector D. There was no apparent interaction, and the resident troop disappeared from observation 10.5 minutes before the intruders left. The next time intruders appeared, this time in sector F, 22 members of the resident troop crossed terrestrially from sector C through sectors B and A to sector D, where they began to display jump in trees. The remaining four members of the resident troop then crossed terrestrially and the intruders disappeared shortly thereafter. The interaction lasted 29 minutes and there was no vocalization until after the intruders had disappeared, when some "whooping" was heard.

On another day, intruders alone were present and drinking in sectors D and F for 21 minutes before the resident troop appeared. Nine adult and three subadult males of the resident troop appeared suddenly from the trees in sector C. This male group rapidly crossed terrestrially about 1,000 feet (305 m.) through sectors B and A to the area occupied by the intruders and disappeared arboreally into the forest as the intruders fled. The interaction lasted only two minutes and there were several whoop vocalizations from the area into which the langurs had gone. One hundred minutes later, four adult and two subadult pursuing males reappeared from the sector into which they had chased the intruders, and were followed by the remainder of the pursuing group 60 minutes later. The animals fed, drank, and crossed terrestrially through sectors F and G to sector E. The remainder of the resident troop was still in sector C. There were several whoop vocalizations and some leaping by the pursuing males as they settled into the trees. They did not rejoin the troop until early the following morning, having spent the night in sector E. On the last occasion that we observed the resident troop and intruders simultaneously, the intruders simply fled into the woods of sector D without vocalization immediately upon the appearance of the resident troop in sector C. "Intruders" were not seen again during the study. It should be noted that in these interactions, the intruders always utilized the sector opposite to and at maximal distance from that sector in which the resident troop appeared.

We mention these interactions mainly to point out that frequent drinking is not limited to the resident troop. Jay (1965) notes that meetings of north Indian langur troops at waterholes are "peaceful," with two or more groups frequently drinking simultaneously. This does not appear to describe the present case, where the resident troop appeared intolerant of the presence of the "intruders" on three of four occasions. Ripley (1967a) concludes that aggressive intertroop encounters among langurs are not directly related to competition for water sources, but such a conclusion is probably limited to the conditions of her study in which water appeared to be of minor importance for langurs.

While overt aggressive behavior was not actually seen in Nelun Wila encounters, one adult male intruder was observed with a lacerated tail on one occasion. Very intense encounters featuring much biting were observed between two langur troops near a bucket of provisioned water in the vicinity of our camp. It would seem that among langurs, any competition over a limited resource (water, space, females, etc.) can result in agonistic intertroop encounters, although most of the aggression is ritualized and few serious injuries result. In addition to the studies noted above and the present study, the observations on south Indian gray langurs (Sugiyama 1964, 1967; Sugiyama et al. 1965; Yoshiba 1968), Nilgiri langurs (Poirier 1967, 1968a, 1968c) and silvered leaf-monkeys (Bernstein 1968a) support this conclusion. Struhsaker and Gartlan (1970) observed a greatly increased frequency of agonistic intertroop encounters among patas monkeys around a waterhole during the dry season in Cameroon.

On August 12, 13, and 14, a lone adult langur, whose sex we could not determine confidently, appeared in the same sector (A) on a total of four occasions for a total presence of 14.5 minutes. On each occasion it drank in the horizontal, fully crouched posture with drinking totaling 7.5 minutes. The fact that it was solitary and displayed slow, labored movements indicated that it was probably sick. No wounds were visible. At 11:31 on August 18, a jackal (*Canis aureus*) emerged from the forest in sector A carrying an apparently dead langur. The jackal chewed on the carcass until 11:37, when an adult male wild pig (*Sus scrofa*) chased him, seized the carcass, and chewed on it until 11:47. Then the jackal, joined by another, approached the pig from opposite sides. The pig dropped the carcass and chased one jackal, and the other picked up the carcass. The pig chased the jackal until 11:50 when they disappeared into the forest with the jackal still in possession of the carcass and the pig in pursuit. Aside from impressing upon us the improbability of finding intact fossilized primate skeletons, this incident, in showing that a lone, sick langur will cross an apron nearly 100 feet (30 m.) wide to drink, further underscores the importance of water in langur ecology.

Terrestrial Crossing

In addition to drinking, crossing the apron from one sector of the study area to another was a common terrestrial activity of the resident troop. Neither "intruders" nor the solitary langur did this. One or more members of the resident troop spent a total of 694.5 minutes on 39 occasions in such terrestrial crossings. The whole troop was never observed to move from one sector to another arboreally when a terrestrial route was more direct. On five occasions however, some (1, 3, 7, 10, and 19) members of the troop took a circuitous arboreal route while the rest of the troop crossed terrestrially. The distribution of 15-minute intervals during which these crossings took place shows peaks which correspond to those of langur presence and drinking (about 09:45, 11:15 and 15:30). The peak at 13:15–13:30 in drinking but not exhibited in presence is also not present in crossing.

Some of the 39 terrestrial crossings involved small fractions of the troop which split off for periods of from 5.5 minutes to more than 24 hours. However, on 23 occasions, 20 or more troop members made terrestrial crossings ranging in duration from 4 to 67 minutes with a mean duration of 19.1 minutes. The progressions were led by adult and subadult males and they proceeded in very much the same manner as the drinking bouts. On several occasions, one of the first males to appear climbed up on the dead tree lying on the apron to visually scan the area. One or more animals drank during 16 crossings, and five crossings took place immediately following drinking bouts, so that the two most common terrestrial activities, drinking and crossing, were closely linked in time. The most commonly used route in terrestrial crosses (crossed 12 times, six in each direction) was between the trees at the border of sectors F and G and sector D, a distance of about 400 feet (122 m.). Four crosses between sectors A and C, a distance of about 500 feet (152 m.) were recorded. Others were about 450 feet (137 m.) used three times, 265 feet (81 m.) used twice, and 100 feet (30 m.) and 105 feet (32 m.), each used once. Leopards (*Panthera pardus*) were observed in the study area during two of the crossings.

While these crossings tend to confirm our impression of the overall cohesiveness of the group, portions of the group did, on occasion, leave the troop for various periods of time. Four of these, involving 1, 7, 14 and 19 animals, were brief forays from the troop. The first began with the animal crossing at 15:12 to 15:12.5 and returning at 15:18 to 15:18.5. The second began with a cross lasting from 12:40 to 12:43.5 and ended with a return from 13:32 to 13:45. Comparable figures for the next were 13:24 to 13:31 and 13:35 to 14:47.5, and for the last 10:44 to 11:03.5 and 11:26 to 12:25. No initiating factor or consistent age/sex composition could be discerned in these forays except that the last mentioned involved ten adult males and seven subadult males. This cross took place on September 5, the same day as the terrestrial pursuit of "intruders" by 12 males of the troop took place. That pursuit by the all-male group can be considered as a fifth case of a foray, although the group did not rejoin the rest of the troop for over 16 hours and the presence of intruders apparently initiated the crossing. On only one occasion was an animal (an adult male), who was seen to leave the troop, not actually observed to rejoin it, although subsequent troop counts indicate that he must have done so.

On September 8, fractionization of the troop became quite frequent with repeated separation and rejoining of troop members. At 11:46, an adult male, a subadult male, the three infant-2s, and the adult female with the ventral infant-1, left the troop and made a terrestrial cross from sector G to sector D which they completed at 11:48. They then moved arboreally on a linear route through sectors D and A until they appeared at the edge of the forest in sector A. At 14:23, they crossed terrestrially to the forest of sector B which, when completed at 14:27, had brought them around two-thirds of the circumference of the study area. They spent the night there, separated from the rest of the troop, and at 06:55 on September 9, whoop vocalizations were heard from the area in which they were last seen.

At 09:14, most of the remainder of the troop was sighted, some moving arboreally and some terrestrially in the direction taken by the group of seven on the previous day. By 10:23, the remainder of the troop had circled to the border of sectors A and D, and at 10:17.5 the group of seven began to recross terrestrially toward the remainder of the troop. The seven completed the cross at 10:19 and the two groups were merged by 10:27. There were no vocalizations during the joining.

The langurs were visible periodically, feeding, drinking, and resting; and at 17:22 they began a terrestrial crossing which brought them back, at 17:38, to sector G which the troop had occupied on September 8 before the group of seven had moved away. However, only 25 of the 26 troop members were counted in this cross. At 07:42 on September 10, the "missing" adult male langur crossed terrestrially from sector D to G and joined the remainder of the troop. He evidently had spent the night separated from the rest of the troop. Whoop vocalizations had been heard at 07:05 from sector G, the area occupied by the bulk of the troop. At 08:34, 21 troop members crossed terrestrially from G to D, completing the cross at 08:40.

At 08:42.5, 18 recrossed (from D to G) terrestrially and were accompanied by an old female who was seen for the first time. She was seen with the troop throughout the remainder of the study and it was at this point that the troop size grew to 27. The 19 langurs completed their crossing at 08:48 but, at 08:49, 16 crossed terrestrially from G to D once again, completing their cross at 08:55. The balance of the troop (eight) apparently crossed by a circuitous arboreal route, as the whole troop was once again seen at 14:44 in sector D. At 15:20, 24 members of the troop recrossed terrestrially, entering sector G at 15:29. One of the remaining langurs (an adult male) was seen at 15:30.5 in sector D and the other two, also adult males, at 15:47 in sector F. At 17:03, another male left the remainder of the troop and crossed terrestrially from G to F by 17:03.5. All four of these animals apparently spent the night separated from the troop.

At 06:55, 06:56, 07:03, and 07:07 on the next morning, September 11, whoop vocalizations were heard from sector G (in which the majority of the troop had last been seen) and at 07:05, one of the four "absentee" males recrossed terrestrially to sector G. The bulk of the troop was visible, resting and feeding, until 09:05. At this point, the remaining three "absentee" langurs appeared and at 09:06.5, six adult, five subadult, and two juvenile males of the troop crossed terrestrially to join them at 09:08 in sector D. There were no vocalizations during the joining and the 16 langurs drank, fed casually, and rested until they went out of sight at 10:05.

At 11:17, the other eleven troop members crossed terrestrially to sector D, completing the cross by 11:22. The troop was seen intermittently during the day and moved arboreally through sectors D and A, the same route taken on September 8 by the group of seven which had split off from the bulk of the troop. At 17:36, 17 langurs, including the adult female with the ventral infant-1 and at least two of the infant-2s, crossed terrestrially to sector B, where the group of seven had slept alone on the night of September 8. The pace of this crossing was rapid, with a duration of only nine minutes. Some of the ten troop members which had not crossed were observed high in the trees in sector A until 17:54.

On September 12 from 06:35 to 07:51.5, repeated whoops came from both sector in which the two groups had last been seen. At 07:29.5, the group of 17 began to return although only 14 crossed terrestrially and they completed the cross at 08:01. The troop was visible intermittently during the morning, feeding and drinking. At 09:25.5, 25 troop members made a terrestrial cross to sector C which had not been used on the preceding four days. They completed the cross at 09:59. The troop was visible intermittently during the day, feeding, drinking, and resting. At 12:42 some soft (low intensity) whoop vocalizations were heard (but an airplane was passing overhead). At 12:58 and 14:22 the two males, who did not cross with the troop at 09:52.5, each made terrestrial crosses of two minutes and joined the troop. Meanwhile, at 12:45, another adult male had crossed terrestrially to sector E, and from 15:23.5 to 15:53 the remaining 26 langurs crossed terrestrially to the same sector. At 16:28, the 27 members of the troop crossed terrestrially beneath the observation blind (one returned but crossed again) and at 16:44.5 the full troop, now consisting of 27 members, entered the trees in sector C where this very fluid sequence had begun five days previously.

We have detailed this crossing behavior because it represents a substantial portion of the langur troop's terrestrial activity. Poirier (1967) notes repeated terrestrial crosses by a male pair of *P. johnii* over a route 500 yards long even though this species is more arboreal than *P. entellus*. Our observations on crossing also reinforce the conclusion that langur troops are cohesive units. While one or more of the members may leave the main body of the troop, sometimes for substantial periods, they may rejoin the troop with no visible disruption. Such fractionalizations of langur troops point up the hazards in sampling troop size and composition in casual roadside surveys. Many "troops" reported in such surveys may simply be portions of the troop separated temporarily from the main body. The purposes of terrestrial crossings are not completely clear. Repelling other langur troops and fully exploiting all available food sources are undoubtedly involved.

Play

A third common class of terrestrial activity was play as defined and classified (e.g. teasing, chasing) by Ripley (1965). The resident troop engaged in 14 observed play bouts totaling 141.5 minutes and the "intruders" in one bout of seven minutes duration. While the trees were used briefly in four of these seven bouts, play was clearly a terrestrial activity and no purely arboreal play bouts were observed in the study area, although several predominantly arboreal play bouts were noted in casual observation near our camp. Again, these results must be qualified by the bias in our study conditions which may support disproportionate observation of terrestrial behavior. The adult female and the infant-1 were never observed to play, but all the other age categories were represented. Membership in play groups changed rapidly and ranged from one to between five and ten troop members. The most common type of play was chasing, sometimes around the tree trunks and, once, among several axis deer (*Axis axis*).

Slapping and wrestling were seen sometimes. On one occasion a subadult jumped from a bough onto the back of another subadult who was passing below carrying a branch. The first snatched the branch and a chase ensued. On another occasion axis deer were teased, and on another an infant-2 slapped the antlers of a grazing axis male. There was only one example of solitary, locomotor play in which a langur hung by one hand from the terminal aspect of a branch and bounced up and down for about one minute before dropping to the ground about five feet (1.5 m.) below. Just as terrestrial crossing and drinking frequently occurred simultaneously or nearly so, play bouts were also observed to occur during drinking bouts (12 times) or just before or after drinking bouts (three times). The distribution of 15-minute intervals during which play took place corresponds to that for presence, crossing, and drinking (including the additional peak at about 13:30 shown by the distribution of drinking). Curiously, Poirier (1967) found essentially the opposite for *P. johnii* with peaks in play occurring during the mid-day rest period (between 11:30 and 13:30) and at 17:00. Ripley (1965) also notes a peak in play during the midday rest period.

Other less frequent terrestrial activities of resident troop members included mounting (seven instances in which adult, subadult, and juvenile males and the infant-1 were "mounters" and adult and subadult males were "mountees" in the dyad), presenting (one instance, in which a subadult male presented to an adult male), displacement (two instances, in one of which an adult male displaced another adult male drinking at the shore line, and another in which an adult male displaced a subadult male under similar circumstances) and "air biting" (one instance by a sitting adult male toward an unidentified animal). Ripley (1965) relates these behaviors to male dominance. The animals involved in all of our observations of these behaviors were on the ground. A subadult male groomed the perineum of the old female. The animals were on the waterhole apron. The female had presented to the male before he groomed her. The infant-1 was also groomed on two occasions, once by the adult female after the infant had withdrawn from her breast. Both of these grooming incidents, however, took place in trees.

Now we will outline arboreal activities of langurs that we studied.

Vocalization

A total of 89 incidents of "whooping" vocalization were noted during the study. Four of these were associated with the presence of an automobile and two were heard as an airplane passed overhead. Of the remaining 83, a total of 53 (63.8 percent) was heard before 09:00 on 12 of 20 days of the study. Both Jay (1963b) and Ripley (1965) note morning whooping by gray langurs, and this has been confirmed by later workers. Morning vocalization has also been reported for howling monkeys, *Alouatta palliata* (e.g. Altmann 1959, Carpenter 1964a, Chivers 1969), titis, *Callicebus moloch* (Mason 1966), black-and-white colobus monkeys, *Colobus guerza* (Ullrich 1961, Marler 1969), Nilgiri langurs, *Presbytis johnii* (Poirier 1967, 1968c; Horwich personal communication), gibbons, *Hylobates lar* (Carpenter 1964c, Ellefson 1968), siamangs, *Symphalangus syndactylus* (McClure 1964), and chimpanzees, *Pan troglodytes* (Reynolds 1963). Such vocalization is commonly thought to be related to the spatial separation of neighboring troops of conspecifics. We could not definitely determine the presence of other langur troops within range of the morning whoops of the resident troop on all occasions, but other troops were seen periodically in the study area.

Morning whooping among the langurs was occasionally perceived simultaneously or successively from two sectors of the study area, but in some of these cases we could determine that members of the resident troop occupied both areas as a result of forays or temporary fractionization. It appears that whooping can serve to establish contact between two separated portions of the same troop as well as to disperse neighboring troops. Of the 30 incidents of whooping which occurred after 09:00, one followed displacement of the "intruders" and seven times were associated with the chasing of the intruders. In the latter class, whoops were heard as the chase began, again 23 minutes later by the portion of the resident troop which had remained behind during the chase, and by the chasers as they returned to the waterhole area after the chase. Aside from these observations and the incidents involving autos and planes, no events were noted which were likely to have caused whooping.

Ripley (1965) notes that morning whooping by gray langurs in Ceylon is not accompanied by physical agitation or excitement, but whooping is always accompanied by agitation when given later in the day. During our study, display jumping and agitated locomotion occurred during one of the incidents of morning whooping. Only five of the 30 incidents which occurred after 09:00 included display jumping; but in some of these cases all of the troop members could not be observed during the vocalization and the display jumping, or some physical agitation may have occurred. Display jumping without audible vocalization was observed in three cases.

The techniques used in this study were not conducive to hearing vocalization of low volume, but "growling" was heard three times in one day, once interspersed with morning whooping and twice in conjunction with display jumping and chasing during an intratroop fight. Growling is probably similar to "snarling" as reported by Jay (1963b). To our knowledge, sound spectrographs of langur vocalizations have not been reported. Precise description, differentiation, and motivational analysis of these vocalizations await such data.

Feeding

Feeding by langurs in the study area was primarily arboreal. One or more of the members of the resident and "intruder" troops fed during prac-

Table 16–1. Common feeding trees of langurs (present study)

Tree	Part Utilized
Capparidae	
Crataeva roxburghii	Leaves
Celastraceae	
Gymnosporia emarginata	Leaves
Combretaceae	
Terminalia glabra	Leaves and fruits (berries)
Ebenaceae	
Diospyros embryopteris	Leaves (new reddish ones preferred)
Maba buxifolia	Fruits
Leguminosae	
Bauhinia racemosa	Leaves
Cassia auriculata	Leaves
Cassia fistula	Leaves (perhaps flowers)
Meliaceae	
Azadirachta indica	Leaves and fruits (berries)
Myrtaceae	
Eugenia jambolana	Leaves and fruits (berries)

tically the entire time the troops were under observation. Table 16–1 presents a list of trees from which langurs were observed to eat during the study. The list is not complete; other trees were undoubtedly used but those on the list were used commonly. The classification is that of Willis and Willis (1911) and identification is based on samples collected after termination of the study. It was not possible to ascertain the proportions of the diet which these trees comprised. Langurs near our camp were additionally observed to eat the leaves of *Toddalia aculeata* (*Rutaceae*), and the leaves and possibly the flowers of *Tephrosia purpurea* (*Leguminosae*).

On five occasions, langurs of the resident troop at the waterhole were observed to eat while on the ground. In one, the leaves from a fallen branch were eaten, and in another a clod of dried soil was broken open and something (perhaps a root) found inside was eaten. On another occasion, a langur was seen sitting on top of a termite mound looking intently into it and chewing something which we could not identify. Near our campsite, langurs spent a great deal of time carefully picking through and eating the nuts of *Aglaia roxburghiana* (*Meliaceae*) which had fallen to the ground.

Sleeping

Troop members frequently drank late in the afternoon, play ceased, and the langurs climbed into trees and adopted sleeping and resting postures (cf Ripley 1967b). Little movement occurred thereafter. While feeding could be seen at this time, distended stomachs indicated that the animals were full and food was picked more slowly and deliberately as dusk approached. Although no night observation was conducted at the waterhole, on the nine occasions on which the langurs were observed to settle into trees in this manner after 16:00, they were always seen in the same sector when first observed before 08:00 on the following morning. On five of these nine nights, trees in the G sector were utilized. On some days, the langurs would drift back into the forest before 16:00 and/or be seen first after 08:00. Even so, including the nine noted above, on 14 of 17 days (82.4 percent) the resident troop was first observed in the same sector in which it was last seen on the preceding day. These observations indicate that movement during the night was limited. Casual observations near our camp tend to confirm this. Additionally, although the troop sometimes penetrates into the forest to feed and sleep, it tends not to move around the waterhole by arboreal routes. As was noted in the previous section on terrestrial crossing, movement from sector to sector is usually accomplished by crossing on the waterhole apron.

Defense Against Predation

Langurs lack the sexual dimorphism and elaborate social organization during progression and response to attack which are commonly postulated to be savanna-dwelling baboons' mechanism of defense against predators. Most observers, e.g. Jay (1965) and Poirier (1967), conclude that langurs' chief mechanism of defense is fleeing to trees. They note that langurs, even when on the ground, are no more than a few seconds from the nearest trees.

Washburn and DeVore (1961b) and DeVore and Hall (1965) postulate an association between savanna baboons and some ungulates, based on the "keen eyesight" of the former and the "keen sense of hearing and smell" of the latter. Proximity of the two groups is said to increase the probability of detection of a predator, and alarm vocalizations by either are reported to stimulate flight by both. No elaboration is supplied on the dynamics of the establishment and maintenance of such associations, though DeVore and Hall (1965) note that "on many occasions [ungulates] seem to actively seek out baboon troops" (p. 48).

Ripley (1965) observed two occasions on which herds of Axis deer (*Axis axis*) were found browsing near a gray langur troop and concluded that "they complement each other's sensory detection of danger in the manner described by Washburn and DeVore for baboons and bushbuck" (p. 131). Schaller (1967) mentions the frequent proximity of axis deer and gray langurs in India and notes at least two occasions on which axis deer responded to langur alarm calls. Jay (1963b) reports several occasions of axis deer grazing near Indian gray langurs, and notes that langurs respond to axis alarm barks by running up the nearest tree.

Poirier (1967) notes that Nilgiri langurs emitted alarm barks after fleeing to the treetops as a hunter stalked a sambar (*Rusa unicolor*) but did not bark as a domestic dog attacked a troop member.

While Ripley implies that the basis of the langur-axis relationship is mutual protection in the fashion of the baboon-ungulate association, she also noted that feeding langurs often drop leaves and suggests that this may contribute to the association. Likewise, Schaller points out that langurs are selective feeders and frequently drop leaves and fruit from the feeding tree, which are eaten by axis deer below. Jay concluded that the "association of langurs and ungulates is coincidence and there is no evidence to suggest that the two species deliberately stay together in the forest" (1963b, p. 287).

Beside some small raptorial birds and crocodiles, the only potential predators on langurs which we observed were jackals (*Canis aureus*) and leopards (*Panthera pardus*). Schaller (1967) found langur remains in 27 percent of a sample of 22 leopard droppings collected in India. Eisenberg (1969) reports leopard predation on langurs in Wilpattu National Park. Schaller (1967) found remains in 0.7 percent of a sample of 138 jackal droppings in India, but this may represent scavenging rather than predation. We saw no distinctive response by langurs to any of these predators! Jackals were present on three occasions when one or more langurs was on the ground. Once, a jackal was lying in a sector adjacent to that where the resident troop was drinking and feeding for 58 minutes. At that point the resident troop began a terrestrial crossing on a route which would have passed close to the jackal. At the approach of the first male langur, the jackal got up and moved off toward the sector where "intruder" langurs had begun to drink. The crossing langurs sat on the apron until the jackal was clear, and then resumed their crossing. The jackal was gone before the resident troop approached and chased the intruders. On another occasion, the resident troop made a terrestrial crossing about 600 feet (183 m.) from two foraging jackals. Jackals were again present when the resident troop made a terrestrial crossing over the same route on another day. The jackals had been sighted 33 minutes before the langurs began this crossing, during which period the jackals captured and killed a Malabar hornbill (*Anthracoceros coronatus*) only 150 feet (46 m.) from the route employed by the langurs.

Leopards were present on two occasions when one or more langurs was on the ground. A leopard crossed an open area about 1,000 feet (305 m.) from the waterhole while the resident troop made a terrestrial crossing on a route roughly parallel to that being taken by the leopard. The langurs were between the shore line and the leopard but they were about 750 feet (229 m.) from the cat and had direct access to trees. The pace of this crossing was leisurely with a duration of 46 minutes (although most of the troop actually crossed in 20 minutes) and many of the animals drank during the crossing. Axis deer were present during two of the jackal observations and during this leopard observation but neither the deer nor the langurs emitted alarm vocalizations.

Axis alarm "barks" were heard on another occasion as an adult male leopard entered and drank in the A sector. The leopard was present for 15 minutes during which he drank for 12 minutes. He appeared to have fed recently and made no attempt to stalk. During the leopard's presence, the langurs completed a terrestrial cross from F to G. The cross took 12.5 minutes, was over the most commonly used route, and no unusual or distinctive response to the leopard or to the axis vocalization by the langurs was observed. About 35 minutes before the leopard was sighted, the langur troop had moved arboreally from sector A to sector F. This arboreal progression featured leaps over a gap about 15 feet (4.6 m.) wide. These were not display jumps but rather were components of direct, linear progression. We had never observed the langurs leaping over this gap before, and the behavior may have been related to the presence of the leopard. About 10 langurs drank shortly before the leopard was seen.

On none of the five simultaneous observations of mammalian predators and langurs in the present study did the langurs show any distinctive avoidance behavior or warning vocalization. Axis deer emitted alarm barks on only one of the four occasions on which they also were present, and the langurs showed no obvious response to this either. One night, near our camp, both axis and langur alarm vocalizations were heard repeatedly during the presence of a leopard, but no visual observations of accompanying behavior could be made. In summary, these observations reveal no distinctive langur responses of any sort to the mere presence of potential predators and provide no direct evidence for an association between langurs and axis deer which is based on mutual protection against predation. However, such a relationship might exist, despite our lack of observation of it in operation. If such were the case, one might predict that more axis deer would be present during the presence of langurs than during the presence of other ungulates. Table 16–2 shows the mean number of axis deer present

Table 16–2. Mean number of axis deer present during the presence of langurs and four species of ungulates

Wild pig (*Sus scrofa*)	13.4
Barking deer (*Muntiacus muntjak*)	12.2
Wild buffalo (*Bubalis bubalis*)	11.7
Gray langur (*Presbytis entellus*)	11.3
Sambar (*Rusa unicolor*)	9.7

during the presence of one or more langurs and during the presence of one or more individuals of a variety of other ungulates. The figures were derived by summing the maximum number of axis deer present during each 3-minute interval during which the species in question was represented, and dividing that figure by the number of 3-minute intervals involved. Members of the other species need not have been present for an entire 3-minute interval in order for the deer count in that interval to be included in the calculation. Examination of Table 16–2 reveals that this line of evidence, again, does not support the hypothesis that axis deer tend to associate systematically with langurs.

However, on 23 of 67 occasions (34 percent) on which one or more members of the resident langur troop was sighted, and on three of seven occasions (43 percent) on which langur "intruders" were sighted, two or more axis deer did move toward and enter the sector occupied by the langurs. By no means did all the deer present move toward them, but on one occasion 38 deer congregated near the langurs. Frequently they would move directly beneath the trees in which langurs were feeding, passing fluidly in and out of sight, making accurate counts difficult. On three occasions, one or more deer was observed to eat leaves and branches dropped by langurs and, on three other occasions, one or more of the deer fed from boughs weighted down by feeding langurs.

It would seem, therefore, that there is some tendency for axis deer to associate with langurs. There is great variability among individual deer in this regard and the precise variables controlling the relationship are unclear. Some deer tend to approach langurs, on some occasions, especially if the monkeys are feeding aboreally. The deer eat browse made available by the langurs. The great variability in this relationship indicates that it is supported by individual learning. We hypothesize that browse made available by the langurs serves as a reinforcer in the learning, by individual deer, of an operant discrimination in which langurs (perhaps, even more precisely, arboreally feeding langurs) serve as a discriminative stimulus. Approach toward the langurs is the learned response. The probability of approach by a given deer will, all else being equal, be a positive function of the number of times it has been so reinforced in the past for approaching the langurs. We do not imply that this operant conditioning phenomenon underlies the baboon-ungulate relationship, but it might bear consideration since it does seem to explain the data on the langur–axis relationship. Eisenberg (personal communication) also feels that an axis–langur relationship is based primarily on food made available to the deer by the monkeys, although the alarm vocalizations of either may serve to direct the attention of the other to potential danger. It is still possible that a more systematic langur-axis relationship exists, based on defense against predation. The demonstration of such a relationship, however, awaits a large sample of detailed descriptions of langur and axis behavior when both are simultaneously in the presence of a predator. If such data reveal a mutual relationship, experiments in controlled environments to discover the precise dynamics (e.g. reinforcement contingencies if the relationship is based on conditioned avoidance) would be warranted. Our observations tend to support the conclusion that the main langur defense against terrestrial predators is flight into trees. Further, the mere proximity of a potential predator is not in itself sufficient to elicit such flight.

Some Conclusions

These data are not presented as a complete account of langur behavior or even of the behavior of this particular troop. Few observations of intra-troop social behavior were recorded; it is difficult to determine whether this was due to the emphases and conditions of the study or because they were, in fact, relatively infrequent, as is the case with Nilgiri langurs (Poirier 1967, 1970; Tanaka 1965). We have emphasized the terrestrial aspect of the daily activity round of the resident troop because it exemplifies the type of data which can best be gathered from the study of primate behavior at waterholes.

Perhaps the most unique feature of waterholes is that they offer a study area which is frequently vast, open, and treeless. This greatly facilitates visual observation per se, and waterholes would therefore appear to be especially desirable for the study of species which inhabit thickly forested areas. In addition, open areas allow detailed observation of the terrestrial behavior of typically arboreal forms.

Of the generally arboreal Colobinae, *Presbytis entellus* is probably the most terrestrial. This makes the langur an ideal model on which to gather data about the pressures on a basically arboreal form whose niche is expanded to include a terrestrial aspect, and about the possible adaptations to such pressures. The widely held hypothesis that hominization was initiated or accelerated by movement of protohominids from a forest to a savanna niche indicates that such data may be especially valuable for drawing inferences about early hominid behavior and ecology. The present study has demonstrated that, at least under some circumstances, drinking water is of far greater importance to langurs than previously expected. Lasker (1969, p. 1483) concluded that "Man cannot drink very much water at one time but can sweat more per hour than any other mammal so far tested. Before man learned how to carry water with him, human occupation of open plains and savanna therefore required behavior that would make it easy to reach drinking water frequently." An obvious solution to this requirement is the inclusion of a waterhole in the home range. Further reinforcement of the inclusion of waterholes in the home range of carnivorous hominids would stem from the large aggregations of potential prey animals found at waterholes. On the basis of paleontological evidence, Howell and Clark (1963) conclude that the bearers of the Late Acheulian industry in sub-Saharan Africa, and probably much earlier human populations, occupied savanna-like areas and camped near water sources. Therefore, to the degree that behavioral studies of contemporary primates are conducted to gain insight into the behavior and ecology of early hominids, waterholes are probably vital study areas.

Once the vast open area of a waterhole is included in the home range, terrestrial activities other than drinking become probable. In the present instance, langurs favored direct terrestrial routes as opposed to circuitous arboreal ones for moving from sector to sector of the study area, and most of the observed play and the few behaviors supporting male dominance took place on the ground.

Frequently mentioned as critical factors in the influence of terrestrial activity on hominization are bipedalism, instrumentation, competition between social units of the same species and between species of the same genus, and resistance to predation. The exact dynamics and nature of the role assumed by each of these factors in hominid evolutionary history are unclear, but it is commonly assumed that each was operative.

In the present study, five brief instances of bipedalism by langurs were observed. On three of these occasions, a langur (not the same animal) stood bipedally on the ground and pulled leaves from an overhanging bough. In one of these cases, the bough was weighted down by other troop members. On another occasion, a langur stood and scanned visually during a terrestrial cross. The function of the fifth occasion was not apparent.

No cases of tool use (as defined by Hall 1963b) were observed, but objects were observed to be carried during terrestrial progression on 10 occasions. The objects were always branches. In six cases they were carried in a hand, in three cases in the mouth, and in another case one branch was carried in the hand and one in the mouth simultaneously. Two of these instances involved a snatch in a play context, and in two cases leaves from the branch were subsequently eaten. Use of the branch in the remaining cases was unclear. One adult male and one subadult male of the male party which pursued "intruders" carried branches when they returned to the vicinity of the troop.

The current study, as detailed above, indicated that intertroop competition at sources of drinking water may be more intense among langurs than previous observations have indicated. Other primates, toque macaques (*Macaca sinica*), were observed at the waterhole only once; an adult female with a ventral infant and two young animals drank briefly. Langurs were not present at the time and, consequently, there was no opportunity to observe interaction between different primate species at the waterhole. It would be fruitful to observe such interactions (especially where species of the same genus occur sympatrically) at waterholes to construct further inferences concerning intergroup competition among sympatric taxa of early hominids.

Terrestrial activity of langurs was frequently accompanied by intent visual observation, but there seemed to be no elaborate behavioral safeguards against predation. The animals were vigilant; on five occasions one or more langurs suddenly interrupted their terrestrial activity and showed startle responses. Four of these cases involved drinking animals; in three cases birds flying by or taking off appeared to be the cause; and in the other, no cause could be identified. In the other case a langur "startled" and jumped back as an axis deer emerged from the tree line. On another occasion, an adult male langur emerged from the tree line and sat on the apron scanning visually. Four more males passed him and began to drink (in the usual manner in which drinking bouts began) but all ran quickly to the trees when the "sentinel" male ran back. There were no vocalizations, and no stimulus for the behavior could be seen. Despite this evidence of vigilance, however, leopards and jackals quite near the langurs did not appear to interrupt their activities and there was no evidence for an interspecific relationship supporting pooling or maximization of sensory capacities to detect predators.

These observations on bipedalism, portation, intergroup competition, and predation do little to cast light on the role of similar activities in hominid evolution. However, observation of these and other activities conducted at waterholes, using a comparative series of primates, would be most useful. Especially valuable would be such observations on a series representing the arboreal-terrestrial spectrum among primates. Comparison would be aided if techniques were similar and observation conditions were equivalent. For reasons outlined at the beginning of this paper, we recommend that such studies be undertaken initially during dry seasons.

Some cautions and reservations concerning waterhole studies must be restated. Specifically, generalizations from the present study must be made in light of the atypical age/sex distribution of the resident troop; the persistent occurrence of drinking and other terrestrial activities may have limited meaning for the understanding of langurs and for inferences about early hominids. We are intrigued by the notion, however, that it may have been just such groups—those with disproportionate numbers of mature or nearly mature males—that were best equipped to cope with the pressures of the forest to savanna transition. More generally, waterhole studies alone should not be expected to yield a representative picture of the total natural history of a species. For example, Spinage (1969) notes that old males in herds of Uganda defassa waterbuck (*Kobus defassa*) tend to inhabit (have territory in) areas far from preferred water sources in Queen Elizabeth Park in Uganda; observation at water sources alone would not have revealed this. We suspect that in the case of primates, waterhole studies yield more meaningful data than studies conducted in areas of high human population density, but studies should be conducted under all conditions represented in a species' range. In addition, study of a particular waterhole would benefit from nighttime and wet season as well as daytime and dry season observation. Another drawback of the waterhole technique is that it is difficult to observe the subtleties of behavior, e.g. direction of gaze, and to identify individuals from the distance required to monitor the whole study area.

With these cautions and reservations, however, we feel that the study of primate behavior at waterholes is a powerful and fruitful approach. The benefits of stationary observation can be exploited. The usual drawback of the stationary strategy—long periods in which no data are collected—can be circumvented by broadening the focus of the study to include animals other than primates. The reliable aggregations of animals attendant at waterholes, especially during dry seasons, provide data relevant to ethology and wildlife management. More important, however, is that a complete understanding of any given primate species must include understanding of the faunal milieu in which it occurs. We should add, parenthetically, that Schaller and Lowther (1969) have argued persuasively that study of the behavior of carnivores (often found near waterholes) is valuable for drawing inferences about early hominids. Observation conditions at waterholes are usually excellent and are especially suited to the study of forms which inhabit dense forests. The terrestrial behavior of typically arboreal mammals making use of the waterhole area can be studied, and this is useful in understanding the evolution of early hominids. Indeed, the waterhole may be more than just a convenient stage in this regard, and may actually represent the site of many types of events crucial in the hominid career.

17

DONALD STONE SADE
CARIBBEAN PRIMATE
RESEARCH CENTER

A Longitudinal Study of Social Behavior of Rhesus Monkeys

A consistent, logical, and generally accepted methodology has not yet emerged in the study of social behavior of primates, in spite of the urgings of Altmann (1967), Plutchik (1964), and others. Much of the work on social organization and behavior of free-ranging primates has therefore been suggestive rather than definitive. Seldom have enough comparable studies of the same or complementary kinds been done on the same problems to provide reasonable proof or rejection of any particular theory or hypothesis. For these reasons, much of the literature on primate social organization remains speculative.

There are as yet no standards or generally accepted criteria on what constitutes reasonable evidence for or against a particular hypothesis, even if the hypothesis only refers to the existence or absence of a particular feature of social organization. For instance, the presence of dominance hierarchies, or their absence, is reported for various taxa. Usually there is no discussion of what kind of study or observation is necessary to determine whether or not a dominance hierarchy exists. Several workers on the Cayo Santiago rhesus monkey colony (Koford 1963a, Kaufmann 1967) reported that a clear-cut dominance hierarchy among adult females could not be

The longitudinal study discussed in this paper has received support from Grant B–2385 from the National Institute of Neurological Diseases and Blindness, Public Health Service to James A. Gavan; a Public Health Service Predoctoral Fellowship from the National Institute of Mental Health; and National Science Foundation grants GS–1777 and GS–2377. Until July 1, 1970 the observations were made within the Laboratory of Perinatal Physiology, NINDS, PHS, HEW, thereafter within the Caribbean Primate Research Center, which is supported by Contract No. NIH–71–2003 from the National Institute of Neurological Diseases and Stroke, PHS, HEW. Since August 1967 the basic observations have been made by James D. Loy, Glenn Hausfater, Judith Breuggeman, and Jay R. Kaplan under the supervision of the author. I am grateful for criticisms of the manuscript by Jack Prost, as well as by the above-mentioned observers.

determined. However, more intensive observation by other workers on the same colony revealed very definite and stable hierarchies among females both in a small group (Sade 1967) and in a large group (Missakian, personal communication). Therefore, when other workers (Hall and DeVore 1965) report the lack of obvious or stable dominance hierarchies among females in groups of another species, one cannot accept the statement without considering whether the kind and duration of the observations were adequate to demonstrate such a hierarchy, if one did exist.

How can aspects of field studies be varied to increase confidence in assertions or inferences about social organization?

Aspects of Field Studies

At least three aspects of field studies come to mind: the focus, the type of measurement, and the time unit. They are diagrammed in Figure 17–1.

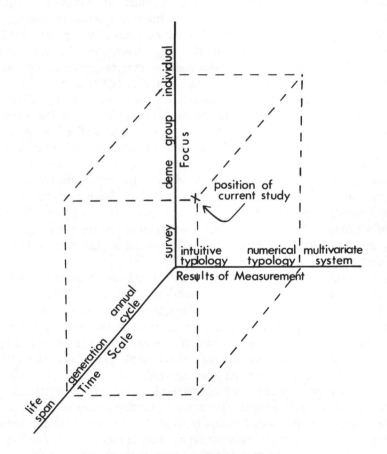

Figure 17–1. Aspects of field studies.

FOCUS

The focus of a study is the level of organization or unit within the species population on which observations are made. The focus ranges in increasing intensity in the following list of studies from a population census of rhesus monkeys over a wide geographic area (Southwick *et al.* 1967), to a study of a system of social groups of howler monkeys (Carpenter 1934), to a study of a single social group of rhesus monkeys (Sade 1966), to studies of social units within a single group (Sade 1965), to studies of life histories of individuals (Sade 1966, 1968). Kummer's (1968) study of hamadryas baboons shifted its focus during its course from broad survey to intense observations on small social units. Many studies, of course, focus on problems other than social organization.

TYPE OF MEASUREMENT

Type of measurement refers to the way in which observations are quantified. Type of measurement varies from intuitive typologies, through numerical typologies, to studies in which the conditionality (Ashby 1962) of quantified variables one upon another are determined by mathematical techniques. Sampling techniques designed to correct for observer bias can increase confidence in the results of these studies (Chalmers 1968).

When an observer states his impression of the usual or normal patterns of interaction as a description of the social organization of the group, he produces an intuitive typology. These impressionistic studies tend to emphasize a few obvious or spectacular aspects of behavior as being somehow more important to group organization than are precisely defined analytic variables. An example of this type of study is the report on baboon social organization by Washburn and DeVore (1961b). Studies of this type produce interesting models which are useful beginnings, but by and large they seem not to have led to further studies designed to test their validity. Rather, they are presented as final conclusions, whereupon the investigator turns to other topics.

A numerical typology results when quantified observations are given as means or other measures of central tendency. These measures sometimes are presumed to represent inherent characteristics of the species. Variance from the central tendency is sometimes assumed to indicate a perturbation, aberration, or other abnormal condition. An example of a numerical typology is the study of group sizes among howler monkeys by Carpenter (1953).

More elaborate quantification leads to studies which attempt to establish conditionality of explicitly defined variables through correlational techniques or other mathematical statements of functions and relations. It is in these studies that the social group is treated as a system in the sense of Ashby (1960) and other theorists. It is these studies which can begin to demonstrate, rather than merely assert, causal relations among the complex of factors which make up social organization. There have as yet been few multivariate studies of social organization. Loy (in press) examined relations among frequency of mating, dominance rank, age, rank of mate, and age of mate among rhesus monkeys.

It is not denied that conditionality has been implied between a variety of ecological, behavioral, and morphological variables even in nonquantitative, highly intuitive studies. However, the point here is to show how an increase of confidence in results and inferences can be gained. It seems to me that quantification of observations is a necessary, although perhaps not a sufficient, way of doing so.

Acquiring a representative sample of behavior in the field poses problems which are not found in laboratory studies. These problems are more like those faced by workers in wildlife management than by laboratory psychologists. Not all members of the free-ranging population are equally eminent. Some individuals or classes of individuals may be observed relatively more frequently than others. If the frequencies with which different individuals perform the same behavior pattern are to be compared, the relative eminence of the individuals must also be measured. The observed frequencies can be adjusted accordingly. Time samples were used by Chalmers (1968) to determine the ratios in which the age/sex classes of mangabeys were seen. Time samples are used in this chapter to rank individuals by relative eminence. The ordinal measure so obtained is used in partial rank correlations to correct for differences in eminence.

TIME UNIT

The final aspect of field studies to be considered is duration. This is a matter of time scale. It can be related to two sets of problems. The first problem is to determine the minimum duration of a study which is able to show conditionality between the social-behavioral variables in question. The second problem is to relate some concepts meaningfully to a time scale. The notion of stability of either group membership or of a pattern of interaction must include an explicit statement as to the duration of stability of the unit in question. In studying animal populations, the meaningful time units are those that are marked by regularly repeated phenomena in the life of the species. These include the daily cycle of activity, the annual reproductive cycle, and stages in the life cycle of individuals of the species. These include also the gestation period, growth stages, duration of generation, life span, and others. Some features, such as the annual reproductive cycle, can be related closely to the human calendar. It is more meaningful, however, to state that a society was observed continuously for two generations than to state its duration in time units which we are accustomed to think of in relation to our own life span. A generation might be a few months among small mammals or a decade among great apes.

Studies whose durations are one annual cycle or less can reveal the struc-

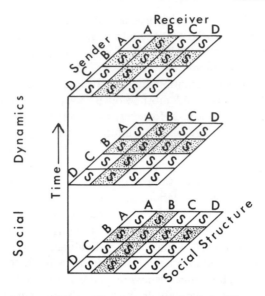

Figure 17–2. The longitudinal study illustrated as a succession of sociometric
matrices.

ture of a social group as a cross-section of activity within the ongoing
social system. These studies are analogous to a single frame excerpted from
a filmed sequence (Figure 17–2). Studies whose duration is considerably
less than an annual cycle cannot provide a convincing description of social
organization, which may change radically between mating and nonmating
seasons. DuMond and Hutchinson (1967) and Baldwin (1969) show that
adult male squirrel monkeys segregate from the adult females and juveniles
during the nonmating season.

Some studies of brief duration, however, have been combined with
censuses of social units over a number of years. Observations on recognizable individuals through one or more generations can reveal the genealogical relations between members of a social group. Investigators can make
use of this information even if their detailed studies of interaction are of
brief duration. Kawai (1958) made use of the genealogical information
which was available on the Koshima group of Japanese macaques in a
study which lasted only ten days. He concludes that the successful monkey
in a paired competition was usually the offspring of the dominant mother.

In true longitudinal studies, however, the same measures of social interaction are repeated at regular intervals on the same individuals in a social
group. If observations are continued for a sufficient length of time, this type
of study provides descriptions of the actual progression of statuses through
which an individual passes during his or her maturation. One might consider
that the minimum unit of time for such a study should be equal to the
period of maturation for individuals of the species. Perhaps the only truly

satisfactory time scale would be one measured in units equal to the life span of individuals of the species. Slobodkin (1967) has estimated that in order to characterize adequately the population dynamics of a species, even under controlled laboratory conditions, populations must be studied through ten generations. If this standard were applied to studies of primate social behavior, then all studies that have been or that are likely to be done would be short-term.

These introductory comments will now be translated into statements which illustrate my point of view, which has developed over the past decade while carrying out an intensive long-term longitudinal study of social behavior within a single group of rhesus monkeys. It is my conclusion that social organization must be studied as a function of life cycle events and therefore must be longitudinal in nature. Since the behavior of adults constrains or conditions the behavioral relations of each new generation through events which occur during the early life of the individual, the minimum time unit of a study must be equivalent to that of the generation span of the individuals of the species. What follows is a brief history of the current study, a review of the techniques and methods employed, a review of conclusions it has produced which differ from conclusions based upon short-term studies, and a concluding discussion of the goals of the longitudinal study of social organization.

History of the Current Study

The study has had a complex history. It was not initially planned as a longitudinal study, nor were its goals and methods well defined at the beginning. It is convenient to describe it as a sequence of two phases: the first, during which its goals and methods were developed; the second, during which these goals and methods are being deliberately applied.

FIRST PHASE

I have described the first phase of the current study in detail elsewhere (Sade 1966). As opportunity allowed between June 1960 and June 1966, I observed Group F of the rhesus monkey colony on Cayo Santiago, Puerto Rico. Other duties took precedence over these observations during most of this period. There were breaks of up to two years in the continuity of the observations. Nevertheless, an adequate set of behavioral categories was developed, the basic factors in the organization of the group became clear, and longitudinal information on the development of social relations among a number of individuals emerged. Some of the findings of the study were obscured by the unavoidable gaps in continuity.

SECOND PHASE

In the fall of 1967 I decided to maintain continuous observations on Group F for as many years as the project could be sustained. It is impossible for one observer to carry out such a longitudinal study of long-lived animals. Therefore, observations have been done by a series of students trained as sociometric technicians. This phase of the study began in August of 1967 and will continue for at least five years, the length of one generation for rhesus monkeys. To date (July 1970), the technicians have been Jim Loy, Judith Breuggeman, Glen Hausfater, and Jay Kaplan. Each technician observes Group F for one year and collects the same basic data on interaction within the group as I collected during previous years. The technician enters into a notebook descriptions of every observed episode of grooming, agonistic behavior, play, mounting and copulation, and other types of behavior. Sociomatrices are tabulated from counts of each kind of interaction. A variety of measures of social status of individuals and subgroups can be computed from the matrices. These measures repeated over a number of years reveal the development of social relations of individuals within the group, and the patterns of stability and change of the social organization of the entire group over time.

INTEROBSERVER CONSISTENCY

Since the longitudinal study currently in progress depends upon a series of observers, confidence in the results is increased if it can be shown that the different observers produce consistent results. Many hours of informal comparison of notes and discussion of observations in the field and in the office gave me every confidence that the various technicians and I observed the monkeys and recorded their behavior in essentially the same way with a minimum of interobserver bias.

We have quantified comparisons of two kinds of observations done by two observers at the same time. These comparisons (Sade 1970) revealed high concordance in judging the dominant and subordinate monkey in simultaneously observed agonistic interactions. The second kind of comparison revealed high correlations between the patterns of social interaction observed by two observers on the same social group during the same period of time as expressed in the row and common marginals of sociometric grooming matrices. These results showed that good agreement between two observers is possible under field conditions.

BASIC DESCRIPTIVE GOALS OF THE STUDY

A succession of sociometric matrices diagrammatically illustrates the project's long range descriptive goals (Figure 17–2). Each matrix represents a time unit of the study. A, B, C, and D are individual monkeys (or subsystems such as genealogies). The symbol S within a cell refers to the set of all statements about communications from sender to receiver during the specified time interval. Entries on the diagonal refer to the physical state or maturational stage of the individual. The succession of stipled cells represents the life cycle of individual B.

Preliminary approximations have been prepared for one genealogy (Sade 1965), for several partial life cycles (Sade 1966), and for dominance relations of the entire group (Sade 1966, 1967, 1969a). Means of stating social dynamics mathematically are being developed, making use of techniques originally applied to small group studies.

Review of Techniques and Methods

CATEGORIES OF BEHAVIOR

The descriptions of behavior recorded in the field notes are of hierarchically organized sequences of motor acts (Table 17–1). The level of description in the field notes is usually that of the bout or of the included component motor acts and postures or variants. Occasionally it may be recorded simply that an interaction, such as grooming, has occurred.

Table 17–1. Hierarchy of behavioral responses.

Level	Examples of Units			
Phase	mating		nonmating	
Relation	consort; adult male/adult male; mother/infant			
Interaction	copulate	groom	fight	play
Bout	present	mount	groom	attack	flee
Component motor acts and postures	tail up curve tail up S tail arched tail down tail aside		hands to hips grasps hips grasps legs insertion pelvic thrusts		slide scrape pick pick-pull lick
Variants	numerous and intergrading			

The categories referring to interactions are polythetic. This means that dissimilar motor sequences may be included in the same category of interaction because they are linked by intermediate forms of the display (see Sade 1967 for an example of this way of categorizing behavior.) Tabulations of the occurrences of these categories are in turn converted into communication matrices for sociometric analysis.

A more complete discussion of the behavioral categories used in the current study was given earlier (Sade 1966).

One further comment on categories of behavior is appropriate here. Within any one level, the named categories are black box categories unless they are analysed into their component parts. Such analysis would be equivalent to describing the combination and sequence of the lower level components of which the higher level categories are composed. For instance, the relation "is the daughter of" is really a black box category. What is significant about the relation is not the name "daughter" but rather the frequency and the kind of interaction, and the position at which these occur in the life cycles of the interacting individuals. We are not, of course, referring here to the genetic relation between mother and daughter. Lower level categories also are black box categories, even those within the level of component motor acts and postures. These categories could (theoretically, if not yet technically) be analyzed into component variables within the nervous systems of the interacting individuals. We would like to approach the state where all our data at any particular level were expressed as interval scale or better measures of the component variables of the next lower level. At the moment, our techniques provide mostly data on ordinal and nominal scales. Perhaps there are some relations which are inherently nonisomorphic with models based on interval scale measurement.

As noted earlier, the descriptive data are tabulated in polythetic classes on interaction matrices. These matrices are or can be analyzed in a number of ways. A variety of mathematical manipulations are available which allow the description of a social group in terms of its subgroups and their membership by individuals. There exist a variety of definitions of cliques and computations which can be performed to reveal them. Another type of information derives from calculations of similarity indices between the interaction patterns of different individuals in the social group. Subgroupings, or cliques, based on similarity measures also can be established. A third type of analysis involves the computation of the status index for an individual which allows him to be ranked among his group members. The best known of these indices is the dominance rank of an individual. However, a variety of other such status indices, based on different types of interactions, can be calculated and used as additional variables in the analysis of group organization. Given the size of the sociomatrices with which we must deal (matrices of approximately a hundred individuals in the case of the present study), it is impossible to compute these various measures and indices without the aid of the electronic computer. Computer science must become an accepted part of the training of behavioral primatologists.

Determination of Conditionality

All the previously discussed analyses are really preparations for the definition of variables and the determination of conditionality between them. The conditionalities we have so far dealt with have been determined by rank order correlation coefficients. Most of our results so far have been based upon bi-variate correlations. We have found that these are often inadequate, if not misleading. It is now clear, for instance, that in any comparison between two variables expressed as a measure of frequency of activity, a third variable must be included: namely, that of observer bias as measured by our time sampling procedures, before the validity of the correlation coefficient as applied to the real population of interactions can be accepted.

Comparison of Conclusions of Short-term
and Longitudinal Studies

GENEALOGICAL RELATION AS AN INDEPENDENT
VARIABLE IN GROUP ORGANIZATION

The two types of studies which include censusing of a group and recognition of individuals over long periods of time, whether they also include continuous monitoring of interactions within the group or whether the interactional studies are done in more restricted periods of time, have revealed a major difference between conclusions derived from short-term studies of macaques and those derived from longitudinal studies. This difference in conclusion involved the identification of an additional variable as one and perhaps the most important predictor of social relations within the group or determinant of social behavior within the group. The identification of genealogical relations over long periods of time allows the social interactions to be ordered according to genealogical relation. Such procedures have provided the well-known results that the initial mother-infant specificity persists and ramifies into behavioral subgroups whose members are usually related to one another (Furuya 1957; Sade 1965, 1966; Yamada 1963). Interestingly, the longitudinal study in the free-ranging condition indicates a slightly different conclusion about the development of social relations when compared to longitudinal studies that have so far been carried out in laboratory colonies. These include the studies of Hinde, Rowell, and Spencer-Booth in England (Hinde and Spencer-Booth 1968) and of various workers in the United States (e.g. Kaufman and Rosenblum 1969). These laboratory studies all agree in showing over a period of a few months, or even a few years, that the frequency of mother-infant interactions gradually decreases as the infant becomes more independent of the mother. The longitudinal study in the free-ranging condition, however, shows that interpreting the decreasing frequency of interaction between mother and young as a waning of the mother-infant bond is an oversimplification. A frequently observed phenomenon in the Cayo Santiago colony is that of a mother responding to attacks directed towards her adult offspring, including her adult sons, on the part of other animals in the social group, even those dominant to herself (Sade 1965, 1966). This indicates that the early interactional bonds have persisted in the form of mutual attitudes between mother and offspring past the time at which interaction occurs with high frequency.

DETERMINATION OF DOMINANCE RANK

Another conclusion which has become well known, and which also is based upon knowledge of genealogical relation in macaque groups, is that dominance relations of the maturing young can be predicted to a greater or lesser degree by knowledge of the dominance relations among their mothers

(Kawamura 1958; Kawai 1958; Koford 1963b; Koyama 1967; Sade 1966, 1967).

Dominance relations in hierarchies revealed by aggressive interactions have been considered important by the students of macaques ever since these animals began to be studied. Another difference between the conclusions of short-term studies and long-term studies can be illustrated by comparing those concerning the relative importance of dominance relations between males, and those between females. Short-term studies have emphasizd that the dominance relations among males are obvious and stable, while those among females are not obvious or are unstable. These assumptions are part of a larger set of assumptions about macaques in which investigators have by and large emphasized the presence of adult males, especially the dominant male, as occupying the central position or performing the most important roles in the organization of the monkey society. These conclusions are in fact reversed by consideration of longitudinal data.

As noted above, intensive observations on both large and small groups of rhesus macaques on the Cayo Santiago colony and observations on several Japanese macaque groups have indicated that, contrary to the usual opinion, linear dominance hierarchies can be observed among the adult females of the group. The longitudinal study carried out on Group F in the Cayo Santiago colony, in which the dominance relations between adults are determined at regular intervals over a number of years, shows that the female hierarchies are highly stable over a ten-year period, whereas male hierarchies are unstable over the same period (Sade 1966, 1967, 1969a, 1969b).

This is not to say that reversals in dominance rank never occur among females. However, in contrast to the unpredictable reversals which occur occasionally in the adult male dominance hierarchy, the longitudinal study shows that the changes which occur between females are regular and predictable events during the social maturation of the females born into the group (Figure 17–3). These predictable and regular changes consist of young females, who as juveniles are subordinate to most or all of the adult females in the group, regularly rising in rank above those females who rank below their mother, until they come to occupy a rank in the adult hierarchy of females at a position just below that of their mother. The observation that younger sisters regularly rise in rank above their older sisters, so that by the time sisters have reached adulthood they rank among themselves in inverse order of age, is possibly a subclass of this phenomenon.

This latter phenomenon has also been observed by various Japanese workers (Kawamura 1958, Kawai 1958, Koyama 1967). The studies by these authors were on groups in which genealogical relations were known by long-term censuses. However, the actual interactional studies on which the conclusions were based were carried out during fairly brief periods of time. Therefore, these workers do not have—or at least do not present— before-and-after observations on the rank of sisters.

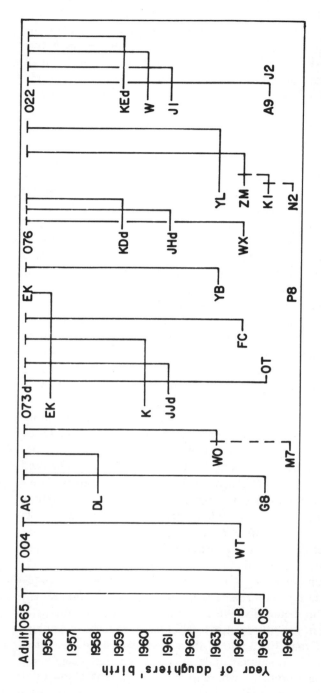

Figure 17–3. Origin of dominance hierarchy of adult females in Cayo Santiago, Group F, by October 1969.

Females who were adult in 1960 are named in the top row from left to right in order of decreasing dominance. A lower-case "d" following a female's name means the female died before October 1969. The daughters of the females in the top row are listed in the column under their mother's names, and in the row of their respective year of birth. Life lines lead from the names of daughters to the dominance position (decreasing from left to right) they held by the time they were adult. Daughters of females who were born into the group are listed in the column under the termination of their mother's life line in the top row. For instance, YL, ZM, K1, and N2 are daughters of KD, who is dead and who was the daughter of 076, who still lives. Dashed life lines are of daughters whose positions were unstable at the time the chart was drawn. A life line that crosses another without breaking indicates that a reversal of dominance occurred between the two individuals. A life line that breaks when crossing the line of a second female indicates that the second female died before the predicted reversal in dominance occurred.

The Japanese workers have suggested that the closer affectional bonds between the mother and her youngest daughter conditions the rise in rank of the daughter. Another possibility is that physiological events during the younger sister's first estrous period conditions the rise in rank by increasing her general level of activity. The longitudinal observations on a number of females in different genealogies in Cayo Santiago Group F shows that these two hypotheses are not sufficient. The age at which younger sisters begin to dominate their older sisters is variable, ranging between one year to about four and one half years for the younger sister at the time of her rise in rank over an elder sister. There have been no obvious associations between a younger sister's rise in rank and estrus. At least four younger sisters rose in rank before their first mating season. At least three rose in rank during the nonmating season. One of these had come into estrus several times during the preceding mating season, but did not rise in rank over her older sisters at that time. Unfortunately, many of the cases so far observed in Group F occurred during the first phase of the study when it was not possible to maintain complete continuity of observations. A number of further cases are accumulating, and over the next few years we hope to be able to state the statistical characteristics of this phenomenon more precisely.

The most striking cases illustrating the value of the longitudinal study, the type of analytic control which becomes possible when consistent observations are made over a number of years, and the advantage which can be taken of natural experiments, occurred in two different genealogies in Group F. Several years after their own mothers died, younger sisters rose in rank predictably above their older sisters and other older females who had ranked below their own mothers. In one case, the younger sister was nine months old at the time her mother died. Her interactional relations focused upon one of her older sisters who was adult at that time (Sade 1965). The older sister therefore occupied the interactional node that her mother had formerly occupied before her death. One would have supposed that the adult dominance rank of this orphaned infant would then be determined not by her own mother's position but by her elder sister's position, if the social dynamic or interactional explanation were sufficient to predict adult dominance rank in females. In fact, however, the orphaned infant rose in rank above this older sister, who was essentially her foster mother, just as would have been predicted had the mother survived.

The regular rise in rank of younger sisters over older sisters is not duplicated among brothers, as a general rule. Rather, older brothers often remain dominant to younger brothers. However, during the last few months of observation, Breuggeman has observed several cases of the rise of younger brothers over older in Group F. In one of these cases the oldest brother of a set of three had been the dominant male in the group. He recently lost rank to two of his younger brothers, whereupon he fell in rank below other males from a genealogy lower ranking than his own. During the following mating season, he left the group and became solitary. These again are features not observed in the changing dominance relations between sisters.

STABILITY OF GROUP MEMBERSHIP

When the stability of membership of males and females is compared over a period of time exceeding two or three years (ten years in the present study), it is seen that the membership of adult females forms the stable core of the social group, whereas the adult males form an unstable adjunct to that core (Figure 17–4). Therefore, the continuity of organization within the group cannot depend upon the presence of a single dominant male nor of a stable core of adult males. Interestingly, Figure 17–4 illustrates that if a study were conducted over a period of time of even two or perhaps three years, this particular conclusion might not be obvious, because some males remain in the group for that period of time. In spite of the annual departure of males at the beginning of the mating season, and occasionally at other times, one might conclude that some males nevertheless remain stable members and form the core of the group. However, the data over a ten-year period of time indicate that no males have as yet remained members of Group F for more than four years (with one exception, which is a special case to be discussed below). The longitudinal study therefore again reverses the conclusion of the short-term study.

Furthermore, the longitudinal study provides a rationale for interpreting

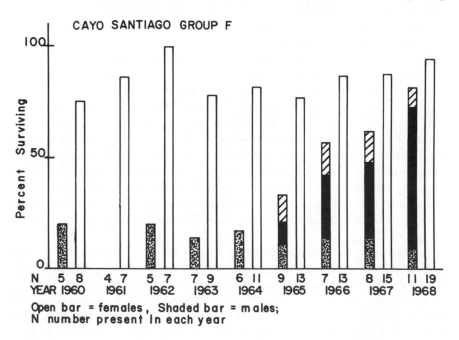

Figure 17–4. Survival of adults from June of each year through June 1969.

observations, such as that of the Minoo-B group of Japanese macaques (Kawamura 1958) which was observed to maintain a stable organization during at least one summer in spite of the absence of adult males, or the group observed on eleven days over a period of six months by Neville (1968) in India, which was alleged also to lack adult males.

INHIBITION OF SON-MOTHER MATING

A particularly interesting illustration of the importance of longitudinal data when compared to cross-sectional, short-term observations can be found within the context of the present study. Observations during the mating season of 1965 resulted in a hypothesis (Sade 1968) that an adult male who remains with his mother's group was inhibited from copulating with her by her higher dominance rank, although this inhibition did not apply to those unrelated females who were dominant to him. The single well documented case of mother-son copulation which, also, followed the son's rise in dominance rank above his mother, was a unique observation during that period. These observations, although part of study which had continued over a number of years, were essentially cross-sectional in nature. This was because of the lack of complete continuity of observation and because in the observed case of mother-son copulation, the son had just reached his reproductive maturity at the age of four and one-half years. There were no observations on Group F during the following mating season. It was expected, however, that following his rise in rank above his mother and his copulation with her during the 1965 mating season, they would continue to mate during succeeding mating seasons.

The second phase of the project, based upon continuous observation initiated in 1967, has provided additional information which requires a revised hypothesis. During the mating season of 1967 there were ten potential mother-son mating pairs in Group F. The four-year-old male that copulated with his mother in 1965 was still dominant to her in 1967 but was not seen mating with her. He was inactive during the later part of the mating season, for unknown reasons. Another male who as a five-year-old in 1966 was clearly subordinate to his mother, was clearly dominant to her in 1967. Exactly when he rose in rank is not known because there were no observations made between June 1966 and August 1967; this period included the critical mating season of 1966. He was not observed mating with his mother during 1967.

An eleven-year-old male castrate was first seen to be dominant to his mother in 1967. Although he was active in mating (including with a female who outranked him), he was not seen copulating with his mother.

A five-year-old male was clearly subordinate to his mother at the beginning of the mating season of 1967. Late in the season Loy saw him mount his mother about 30 times within one hour, before Loy had to discontinue observations at dark (Loy in press). Loy did not see ejaculation occur. The mounting series was unusual in the large number of mountings and in the frequent interruptions during which the male mounted and wrestled other monkeys. Several weeks after this mounting series, he was beaten by his mother in a fight. However, he won subsequent fights with her, indicating that he rose in rank above her within two months following the copulation.

Sons of the other six potential pairs of mother-son mates remained subordinate to their mothers and did not copulate with them in 1967.

During the mating season of 1968 both Hausfater and I observed that another male rose in rank above his mother and later copulated with her several times. However, neither of the two males who had copulated with their mother in previous years did so in 1968, although each mated with other females of the group. These three males were brothers. During the mating season of 1969 Breuggeman did not observe copulations between these three males and their mother, although all remained dominant to her and all of them were active in mating. However, the male, who had mated with his mother the previous year, was observed mounting her three times in a series without ejaculating. This occurred at the beginning of the mating season during an intergroup squabble. There were no other cases of mother-son copulation observed nor were there other cases of rises in rank of sons over their mothers that had not been reported previously.

Viewing the observations from all years we find: no cases of three-year-old males (adolescents) mating with their mothers; one case of a four-year-old (late adolescent) male mating with his mother within two months of rising in dominance rank above her; one case of a five-year-old male (young adult) mating with his mother after rising in rank above her; one case of an atypical mating between a five-year-old male and his mother, followed within two months by his rise in dominance rank above her; one case of a six-year-old male mounting his mother in series without ejaculating, during a group disturbance; no cases of six-year-old or older males (full adults) copulating with their mothers, despite the higher dominance rank of the sons and despite the fact that each mated with his mother in a previous mating season.

The occurrence of mother-son mating still appears to be related to relative dominance within the pair, but also to the late adolescent/early adult stage in the son's life cycle. Only by continuing observations on these males and by adding more males to the sample (by continuing observations on males who are at present juveniles) can the relations between the various factors which result in the occurrence or inhibition of mother-son mating be clarified.

DISCLAIMER

I would like to enter a disclaimer at this point. I am not suggesting that the adult male has no role in the organization of the social group of macaques.

In the writings of Bernstein and Sharpe (1966), C. R. Carpenter (1942a, 1942b), and Southwick and Siddiqi (1967), attempts have been made to analyze the role of the adult males in the social group. One problem which will receive high priority in the current study will be the comparative sociometry of the four different males who have been at one time or another dominant in Group F in order to isolate the common factors or similarities in their role behavior. Much of what has been written about the role of the dominant male is in fact based upon rather short-term observations. We may expect surprises when longitudinal data are analyzed.

One criticism sometimes voiced against the longitudinal study is that it simply produces a larger amount of data. The problem in analyzing the quantity of data which can be collected even during brief periods of time is, according to this criticism, simply compounded in the longitudinal study. In other words, it is sometimes argued that the longitudinal study rapidly produces a diminishing return for the effort involved. The above comparison of the conclusions of the longitudinal study with the short-term study shows that this is not the case. In fact, the longitudinal study, by presenting a sequence of statuses for individuals over time, produces new kinds of data. The information which in the short-term study becomes an end in itself is converted into a tool in the longitudinal study and is then used to produce conclusions which go beyond that of the short-term study.

Theoretical Goals of the Longitudinal Study

The theoretical goal which has been developed over the years in this longitudinal study, and which is now explicitly recognized in the planning and execution of the study, is the discovery of the relation of the life cycle of individuals to the regularities of behavior which we observe as social organization. The elements of social behavior, the displays and signals of communication, are considered the behavior of other individuals over short periods of time, as Altmann (1962, 1967) has argued. However, in addition to the moment-to-moment constraints imposed by the exchange of signals between individuals, there are long-term constraints imposed through interactions of individuals during critical or sensitive periods, or through the transfer of behavioral patterns by imitation (traditions) which persist over many years. The operation of these constraints results in regularities in spacing, orientation, and activity which we observe and measure and which we call social organization.

Discovery of these constraints requires identifying the conditionality of events separated by long periods of time, and the identification and characterization of critical or sensitive periods during maturation, and those alterations in typical mood of individuals that occur through such maturational changes as puberty. Of constraints which operate during sensitive or critical periods, the more intense operate from adult to young. However, mothers "imprint" on their young. This phenomenon is documented experimentally among the Artiodactyla (Klopfer et al. 1964), and is implied by many observations of specificity of reaction of monkey mothers to the distress of their young (even their older young, as noted previously). Also, monkey mothers orient differentially to their own offspring among the juveniles of the group, even at times when their young attempt to avoid their approach. These phenomena imply the presence of sensitive or critical periods during the adult life of females, perhaps related to a particular lactational stage or a particular post- or peripartum period.

In addition, there seem to be significant transitional periods during maturation in both males and females. Critical and sensitive periods may in fact be a subclass of transitional periods. Examples are the perinatal transition from fetus to neonate, the transition from a more-or-less helpless Stage I to a Stage II infant with well-developed motor coordination, the interactional transition which occurs at the birth of a new infant whereupon the yearling is weaned, the hormonal changes which occur at puberty and which produce alteration in typical moods, the apparent role conflict which reaches a crisis point among young adult males in their relation with their mothers, the period in the maturation of the females in which younger sisters rise above older sisters and take their place in the adult female hierarchy, and perhaps others. The boundaries and characteristics of these transitional periods are not as yet precisely defined.

INFERENCES ABOUT MECHANISMS

The analytical power of the longitudinal study lies in the regularly repeated alteration of conditions which occur in the annual cycle of reproductive and nonreproductive phases, the before-and-after conditions within which the same animal is observed during its maturation on either side of the various transitional periods, the irregularly occurring contrast of conditions, such as between orphan and nonorphan, and the regularly repeated measurements of social interaction which allow a repeated confirmation of patterns which show regularities about which the short-term study can only provide speculation. In these aspects the longitudinal studies show some of the features of more experimental investigations. One of the results is that speculation and inferences about mechanisms become possible.

Another major goal of this study has been to develop sets of inferences about presumed mechanisms. Ultimately the success of the study will be judged on whether those inferences can be translated into hypotheses that lead to experimental analysis of the biological processes which produce the observed patterns of behavior.

The remarkable differences in the life cycle of males and females in relation to the social organization of the rhesus group, as revealed by the current study, can be used to illustrate the kinds of inferences which we have so far tentatively developed. In some cases, the inferences only relate to the likelihood that some mechanisms must exist, although their natures remain completely unknown at the moment. For example, the regularity in rise in rank of younger over older sisters and unrelated adult females from low ranking genealogies suggests that two sets of mechanisms must exist. The first set includes those factors which initiate the rise in rank of the young female. The second set includes those factors which limit that rise in rank to the position just below her mother or to the position which her mother would have occupied had she lived. The contrast between condition of orphan versus nonorphan, also described above, allows us further to infer that the social dynamical explanation offered by the Japanese workers cannot be considered to be complete, whatever merits it may have as a partial explanation. The observation that a female orphaned at an early age was nevertheless predictable in her future dominance relations helps us to bracket the period of time at which the critical constraints may have first been imposed, in this case during the first nine months of life.

The dispersal of males from the natal group provides another set of observations which allow speculation about mechanisms. Not all the data on which these speculations are based come from this particular study. The current study of Group F, as well as observations by Koford (1966) on the entire population over a number of years, have indicated that the dispersal of males is most frequent at (1) the onset of the mating season; and (2) during the early years of the reproductive stage of the males. These two observations imply obviously that the activation of some CNS mechanisms by the onset of testosterone production is at least a part of the mechanism involved in male dispersal. A further set of observations on one male born into Group F, and castrated as a four-year-old (Conaway and Sade 1965), lends support to this hypothesis. Following castration, this male rejoined his natal group from a peripheral male group, which he had joined prior to the operation. This occurred during the fall of 1961. It can be seen (Figure 17–4) that one male, indicated by the stipled section of the bar, persisted as a member of Group F from 1961 to present. This male was the castrate.

These speculations should surprise no one, since the dispersal of males from their natal area at about the onset of puberty is a widespread phenomenon among mammalian societies and, indeed, among vertebrate societies in general (Count 1958). From the point of view of the general vertebrate biogram, rather than from the specialized one of the single primate species, we see that what needs to be explained among macaques is not the so-called process of peripheralization, but rather why some males fail to disperse.

To carry this line of reasoning one step further, we can suggest that the differences in social organization as observed between some primate species, at least, may be explained in terms of this common process of dispersal of males. The difference in the complete withdrawal of the adult male from the heterosexual group, as in squirrel monkeys (Baldwin 1969), and the incomplete dispersal of macaque males, could be described as resulting from the differing intensity of operation of a homologous process, rather than a different type of social organization. This interpretation is supported by the observation that other aspects of the male reproductive cycle (for instance, regression of the testes) are more pronounced in squirrel monkeys (DuMond and Hutchinson 1967) than in rhesus monkeys (Conaway and Sade 1965).

A counter-argument would be that it is at the end of the mating season that squirrel monkeys withdraw from the heterosexual group. A counter-counter-argument would be that the homologous phenomenon in rhesus society is the all-male group, which is more distinctly structured and less intermingled with the heterosexual group during the nonmating season than during the mating season. These arguments and counter-arguments will persist until precise, quantitative, interactional data replace the typological statements in which the arguments have been framed.

Evolutionary Implications

The findings of the study must ultimately be related to evolutionary theory. It might be of interest here to note some of the speculative implications of some of our findings.

GENETIC IMPLICATIONS

The first findings imply that whatever mechanisms select particular males for emigration from their natal group, our observations suggest that they may operate to select genotypes at random. This would make sense if the ultimate function of the dispersal were gene flow within the intergroup system, rather than competition between males for the production of offspring within the group. At least it seems that emigration is random with regard to socially acquired status, since males from high, middle, and low ranking genealogies have been known to depart from the group. Since the members of the same genealogy share genetic descent (through the female line, at least), it may be that genetic differences are unimportant in emigration. There may, of course, be differential mortality among the emigres or differential success in establishing themselves in other social groups. One of the limitations of the current study is that through necessity it has been confined to a single group. A recent view of social organization which is developing among primatologists is that the basic unit of social organization is not the single social group itself, but rather a system of social groups. The system of social groups interacting in a restricted area would correspond to the deme of the population geneticists.

Again, the point here is that better understanding of these matters will require the type of life history data on individuals which only longitudinal studies can provide.

IMPLICATIONS FOR EVOLUTION OF SOCIAL ORGANIZATION

The brief discussion of squirrel and rhesus monkeys, in which different types of social organization were considered as resulting from differential emphasis of a homologous process, was an example of a model which was applied earlier to differences of social organization within the cynomorph Cercopithecidae (Sade 1966). This attempt was to describe the differences in social organization of woodland dwelling rhesus and Japanese macaques, of savanna baboons (*Papio anubis* and *P. chacma*), and of hamadryas baboons as representing differential emphasis on three aspects of social organization which were present in all the species. These aspects were genealogical relations, intramale dominance, and male-focal relations. Needless to say, these aspects, with the exception of the genealogical among macaques, have yet to be analyzed into their component, interactional variables, to say nothing of their underlying neurological and endocrinological bases. Nevertheless, this approach may lead us out of the rather sterile discussion of differences in type of social organization that seems to be current in the literature today.

18

IRWIN S. BERNSTEIN
UNIVERSITY OF GEORGIA AND EMORY UNIVERSITY

The Organization of Primate Societies: Longitudinal Studies of Captive Groups

Implicit in any comparative study of nonhuman primate societies is the presumed relationship between the organization of nonhuman and human societies. Being the animals biologically most similar to man, it is expected that the nonhuman primates will also be behaviorally and sociologically most similar to man. The biological and evolutionary basis of the complex social structure seen in human societies may be related phylogentically to the organization of nonhuman primate societies. The basic social mechanisms, however, produce not a single type of social organization varying only from primitive to more complex, but instead, a variety of social formats, even within the single species *Homo sapiens*. Although the social organization of any single nonhuman primate species may appear relatively invariant, examination of a broad range of nonhuman primates reveals a wealth of variation in the organization of societies.

There are several basic principles of organization which may vary, and primate societies must be regarded as alternative forms of social organization rather than as merely more or less primitive or complex.

The societal structure of each primate population may be assumed to be suited to the ecological requirements of the niche occupied, and the morphological adaptations of that species to its environment. Each type of organized society, however, will retain flexibility with regard to specific aspects of organization as a function of the extent to which the Order Primates appears to have specialized in generality and plasticity of behavior. In much the same way that individual learning functions within the genetic limitations of behavioral development, so too may the basic structure of a species' social organization be modified. Great variation in the final response patterns is possible for an individual, but always within the limits of the framework of species typical responses. So too may a primate society vary.

This research was supported by National Institute of Mental Health Grant 1 RO1 MH 13864 and in part by National Institutes of Health Grant FR–00165.

As such, many primates are capable of living under a variety of changing conditions and are capable of adjusting to a range of social circumstances. Various existing types of social organization must therefore be regarded only as alternative solutions to the problems of survival in the available habitats. No one type of society may be assumed to be superior to another inasmuch as all have passed the crucial biological test, survival.

As such, we should not be surprised if we failed to find a single universal principle upon which all primate societies are based. Only given the absence of a single unifying principle can we account for the diversity of organizational types to be seen in the Order. Notions of sexual motivation or status hierarchies cannot of themselves account for such diversity. These motivating and organizing principles may be present in a wide variety of primate taxa, but their importance and prevalence may range from primary to near negligible, whereas in other primate taxa they may in fact be absent. We should instead expect that the multiple functions which a society can and does serve will be of differential importance in the diverse primate taxa and that the multiple effects might account for the variety of specific social organizations to be found.

The first problem in the study of nonhuman primate societies is to identify and describe the types of existing social organizations. These descriptions must allow for comparison between various primate taxa so that we may later try to understand the relevant selective pressures and their effects upon each of the groups under study. In this way we can hope to appreciate the interrelationships between social organization and biological requirements.

If a society consists of an organization of interrelated roles (Sarbin 1954), then the description of a society will require identification of the roles and interrelationships characteristic of that society. The economics of evolution suggest that the differentiation of roles and division of labor in a society must serve some function in the preservation of the group. The differential roles in a primate society, however, mean that not only may several individuals serve the same role functions, but a single individual may satisfy multiple role requirements, either in succession or simultaneously. Thus, an individual adult male may show the same role pattern as do other adult males with respect to the formation of consort relationships with females, and at the same time he may function in the special role of control animal during episodes of group disruption due to either intragroup or external sources of disturbance (Bernstein and Sharpe, 1965, Bernstein 1966). The same male may also ally with other selected group members in forming coalitions, and may demonstrate paternal role behavior, depending both on his species membership and individual behavioral traits (personality). In fact, "personalities" (or the animals' individual response patterns) influence not only whether an animal participates in particular role functions, but also the exact expression of the role. As a corollary, during the time an animal is functioning in certain role capacities, its own idiosyncratic behavior patterns will be modified by the requirements of the role. The role expressions may be modified, but the role functions in the society must be served.

One problem common to all societies is the integration of newcomers into the society, be they new recruits or infants. A number of socialization mechanisms may be involved, and these may vary according to specific situations, group characteristics, and the taxonomic identity of the group. The basic requirement, however, is that newcomers must participate in group functions and be responsive to the pressures, requirements, and demands of the social group. The group itself, as a coherent entity, will utilize the resources available in the habitat and coordinate the activities of the individual members. As such, one approach to the study of social organization is the study of the utilization of resources by the group; that is, the investigator will study the selective use of available resources and the coordination of individual utilization of these resources as a function of group membership. The biological relevance of a particular group structure may lie in the controlled use of available resources through the coordinated activities of group members.

The practical application of these theoretical formulations is revealed in the study of social units. The theoretical framework defines the social unit and the level of organization which serves as a model of a particular primate society. The study itself must take place under defined conditions, and it would seem that the most relevant studies would be of undisturbed groups living in a habitat which most closely resembles that in which the species evolved. In such circumstances, the relevance of particular social mechanisms and the value of particular social structures may be most readily appreciated. Finding such an ideal natural habitat in today's world, however, may be impossible. The definition of a natural habitat must therefore allow for some compromises from the ideal situation (see Bernstein 1967a). In reality, wild primates live under a great range of disturbed conditions: nonhuman primates are subject to varying degrees of human predation, and are also subject to displacement from their original habitat through the destruction or modification of such habitats by man. Although it is true that man has been an element in the natural habitat of many nonhuman primates for many millennia, and it is quite possible that nonhuman primate societies reflect adjustments to human predation and disruption, it is not conceivable that the level of disturbance found in the natural habitat today, as a result of rapid technological advances in *Homo sapiens* culture, have allowed nonhuman primate societies to make any kind of genetically controlled adaptive adjustment. The elimination of nonhuman primates from many areas of former habitat and the vast reduction of numbers in many species testifies to the lack of adaptation to the present heavy human predation and disruption. In only a very few cases have the nonhuman primates been able to exploit the new opportunities presented by human habituation, such as cultivated areas, construction features, and garbage concentrations. Where such exploitation has occurred, it must be remembered that the present habitat is not the one in which the animal evolved, nor the one for which its adaptive mechanisms were originally selected. This secondary adaptation results from the fortuitous suitability of mechanisms developed in the original habitat, in the newly available habitat.

To say that all natural habitats are disturbed is not to say that they are all disturbed equally, and the study of natural troops still allows considerable leeway in the selection of study sites. The kind and degree of disturbance acceptable to the aims of a particular study must be considered. Some areas may appear deceptively undisturbed. The elimination of natural predators or other sympatric competitive species may be overlooked, as may also be the case with the effects of selective logging operations. The discovery of the nature of habitat disturbance may require considerable effort.

The selection of a suitable study site cannot ignore factors related to comfort and convenience. In fact, the most undisturbed habitats may be the most inaccessible to humans and therefore rarely subject to investigation. The more disturbed habitats, however, may be the most convenient for study and may be those habitats which have received the most study in field investigation. Natural troops which have become habituated to a human population and which have learned to exploit cultivated fields, gardens, market places, and garbage dumps are often those troops most easily studied in the wild. Such studies, of course, cannot hope to determine which selective pressures have been responsible for the development of various features in the social organization, but they may nonetheless describe the organization of wild troops and the social mechanisms seen in these troops. When investigators attempt to study wild troops which have not been previously habituated to humans, the habituation process itself must influence the animals. Recognizing this, some investigators have established a system of provisioning troops. This procedure admittedly influences the range of the troop and many of its foraging and feeding patterns, but the social organization itself appears unaffected. Indeed, it appears as if the communication responses in a social organization are very resistant to change, and the same forms of behavior have been described in a great variety of situations (Bernstein 1967b, Eisenberg and Kuehn 1966, Gartlan and Brain 1968, Kummer and Kurt 1965, Rowell 1967). There is, of course, still the danger that particular situational factors will distort the social organization and modify behavioral patterns of a wild troop. This argues very strongly for multiple studies of the same nonhuman primate taxa living under the widest possible range of conditions (see DeVore and Washburn 1963, Gartlan and Brain 1968, Hall and DeVore 1965, Rowell 1966, Struhsaker 1967b; Yoshiba 1968).

From the consistently heavily provisioned troop, maintained for many years in a single area, to the captured and transplanted troop, is but a small step. Transplanted free-ranging animals, such as the African green monkeys found on St. Kitts Island in the Carribean, represent a situation intermediate between captive and wild. These animals survive and maintain their numbers in a habitat very unlike that in which they originally evolved; an Old World monkey has adapted to a New World ecology. Comparative studies of green monkeys on St. Kitts Island and in their West African habitat would be most instructive.

If one were to continue on a scale from the most natural to the most artificial captive situation, the next step from the transplanted colony on St. Kitts would be to the transplanted and provisioned colonies such as those maintained at Cayo Santiago and La Cueva near the island of Puerto Rico These animals are restricted in space and live in a New World ecology or habitat where there is insufficient food to maintain the population; they are dependent on human intervention, in the form of provisioning, for survival. Although restricted in space by a water barrier or some other type of artificial barrier, these populations usually enjoy a living space of many acres complete with vegetation. As such, they may be said to be "free-ranging in a naturalistic setting."

By reducing the size of the enclosed area, we reach the next recognizable step in captive living, that of the zoo island, or outdoor enclosure, or compound. Zoo islands and enclosures are usually too small to support significant vegetation or accommodate more than a single social unit, whereas the larger island facilities often provide sufficient resources to support the social requirements of multiple social units. In addition, zoo islands or enclosures usually include an artificial shelter of some sort inasmuch as such facilities are often constructed in areas where the weather is more extreme than the animals can normally tolerate. Roofing of an enclosure produces a very large cage. Reduction in the size of the cage eventually eliminates the possibility of maintaining a social group.

Despite the variation in physical facilities, we must still recognize a continuum from the large social cage to the wild troop living under idyllic conditions. The dimension of variation will include both negative and positive human intervention, in the sense of interference which decreases or enhances the probability of continued survival and/or the expression of the behavior we wish to study. Along this continuum we can recognize only a few discrete conditions: wild troops living under conditions of minimal disturbance, wild troops living under very badly disturbed conditions, wild troops which are provisioned and even completely dependent upon provisioning for continued survival, transplanted troops which are not dependent upon man for their continued survival, transplanted troops which are dependent upon man for their survival, and troops which are bound in space as a result of human intervention. Inasmuch as the spatial requirements of many primate taxa have not been fully studied, it is not possible at present to differentiate enclosures or islands which are sufficient to permit full expression of day and seasonal ranging patterns from the extreme situations which we know are so restrictive that the daily and/or seasonal ranging patterns of a troop are thwarted.

All of the above described living conditions are suitable to the study of selected aspects of primate societies. Each has its advantages and limitations. In order to select the most suitable study condition, an investigator must first specify the problem area he wishes to examine. He must next consider the variables which influence this behavior, and then consider the possible alternatives as they exist.

When considering the study of wild troops living in the natural habitat, one must consider the degree of disruption of the habitat as well as the characteristics of the habitat of the particular species to be studied. The habitat exerts a profound influence on possible field techniques and there are a diversity of primate habitats (e.g. semiarid plateaus, open savannah, woodland forest, gallery forest, tropical forest, mangrove swamp, and various montane ecological niches). On the open savannah it is possible for an investigator to observe every animal in the troop simultaneously, provided the grass is not too tall and the number of obstructions to visibility is not too great. If the investigator can achieve the ideal of neutrality with reference to the troop, then under savannah conditions he may choose when to observe, which animals to observe, and under what conditions he will observe. In contrast, in the tropical rain forest, or swamp forest, the investigator seldom, if ever, can see all the animals in the group under study simultaneously, and the investigator must satisfy himself with observations of animals as they become visible (Chalmers 1968). Under these conditions, it is the animals that decide when and for how long they will be observed; who, and even what, will be observed (Booth 1962). Under such conditions, even the best field workers may achieve very little, despite prolonged periods in the field and great personal sacrifice. Field workers such as Struhsaker and Gartlan have demonstrated their abilities in the study of savannah troops of vervets (Struhsaker 1967a, Gartlan and Brain 1968), and yet each of them has run into near-insurmountable problems attempting to study animals living in tropical rain forest conditions (Gartlan 1970, Struhsaker 1970). Other investigators attempting to work with rain forest primates have attained some measure of success by studying troops living under more disturbed conditions outside of the rain forest proper, or by instituting provisioning procedures. Both the value and limitation of provisioning techniques can be appreciated by examining the detailed data collected by the workers of the Japan Monkey Center on wild troops living in the temperate forests in Japan.

Wild troops living in and around areas of human habitation are often far more visible than animals living in more remote regions, mainly because they are somewhat habituated to observers and recover rapidly from the types of disturbance associated with living in close proximity to humans.

After considering the many advantages to field work apropos the various types of problems which can probably be answered under no other conditions, many investigators choose to work in the tropical rain forest, sacrificing visibility in exchange for the opportunity to acquire information relating to problems of ecological relationships, the presence or absence of territoriality, home range, both day ranging and seasonal ranging patterns, spatial and social relations with other animals, migration if any, immigration and emigration mechanisms, and the utilization of available resources. Certain types of field experiments may also be performed in the natural habitat, and some problems involving long-term kinship can be studied either in the natural habitat or using specially maintained captive groups. Captive groups have the advantage of easy replication of specified conditions, and the investigator has control over a number of important variables such as food supply, group size, and available spatial dimensions and attributes.

Field and captive group studies will attain maximum power when designed to supplement each other. Captive groups allow for detailed analysis of many hypotheses generated as a result of field studies and experiments. Captive groups should also be considered as a resource in training and preparing investigators for field work. By observing captive troops, an investigator can become familiar with the forms of social expression typical of the taxa, typical social mechanisms and patterns of social organization, and individual markings and behavioral patterns useful in identifying animals at greater distances. Upon completion of his field study, the investigator can return to the captive situation to replicate certain rare situations observed in the field so that he may better understand the responses of the animals to these situations. Where problems of habituation may have prevented certain field experiments, in the captive situation such habituation problems are more readily solved. Captive and field studies must therefore be regarded as complementary and supplementary; facilities and abilities permitting, the best results should be attained when an investigator or a team of investigators studies a selected species under both conditions.

The ease of access to a captive population makes longitudinal study strategies readily attainable. Such studies, of course, are only part of a totality which includes field and cross-sectional laboratory studies. Much as the field study provides certain research opportunities not readily attained with captive groups, the longitudinal study affords opportunities only approximated by cross-sectional approaches. For example, whereas a cross-sectional study may allow an investigator to estimate the degree of relationship between animals, only a long-term study can establish the genealogy of a social troop or group. These data can then be used to determine the influences of genealogical membership upon participation in particular activity patterns and social interactions. For example, a longitudinal study of a captive group of pigtail monkeys (*Macaca nemestrina*), maintained at the

Yerkes Regional Primate Research Center Field Station, has suggested that the long-term associations of mothers and young and siblings not only influences social rank but also the expression of "aunt" behavior. The role of "aunt" has been played by a number of nulliparous, primiparous, and multiparous females and seems to be elicited especially at the time of the birth of a sibling. In one instance, a female carrying her own infant kidnapped her mother's newborn infant as well. In this particular case, the daughter outranked the mother in the status hierarchy, as revealed through independent measures of status. The true mother was, however, able subsequently to retrieve her own infant, and although her daughter has several times stolen the infant from her, she has been able to retrieve it each time. In another case of kidnapping, a female lost her baby to a nonrelated animal of higher social rank and was unable to retrieve her infant.

On the matter of social rank itself, we have witnessed (1) the rise in rank of a one-year-old male adopted by a female of higher rank than his mother; (2) the fall in rank of the same young male after the death of his adoptive mother; and (3) the return of his status to that reflecting the status of his biological mother. At no time in this process did he ever dominate his own mother, thus violating the normal principle of transitive linear hierarchies in the group.

Other observations demonstrate long-term associations between siblings. The death of two of the senior matriarchs in the group has not disrupted the relationships among their offspring. The two sets of siblings remain cohesive units, associating and interacting with one another with high relative frequency, and also maintaining close hierarchial status. The persistence of these bonds, in both orphaned siblings and in animals whose parents continue to survive, is reflected in the organization of the group and its activities. Even the spatial arrangement of the unit, so typical of pigtails, is modified by these genealogical units (Rosenblum, Kaufman, and Stynes 1964) and these subunits cluster in close proximity.

Other observations also demonstrate long-term associations between siblings. One nulliparous female began carrying and nursing her juvenile brother after a new sibling was born. In another case, an adolescent male has been seen to carry and hold his younger brother. Such observations, although still barely more than anecdotes, can be used to generate hypotheses testable in a longitudinal program. The knowledge of relationships in several matriarchial lineages presently encompassing three generations will allow testing of these hypotheses.

Longitudinal study programs also allow for examination of seasonal, weather, or other periodic variables and their influences on group activities. Events such as episodes of predation and changes in leadership may be rare (Bernstein 1969), but are extremely significant with respect to the survival and adaptation of a troop. Such events, as well as more gradual processes, such as aging, require longitudinal studies.

Although a cross-sectional study may reveal the age and sex composition of the troop or group, and allow one to estimate the birth intervals in a troop, longitudinal study is required to provide precise answers to such questions as infant mortality, birth intervals, and possible interrelationships. The composition of a troop results from mortality, natality, and emigration and immigration, where possible. In captive situations, however, emigration and immigration are usually under the control of the investigator. Nonetheless, data can be collected on infant and other mortalities, unsuccessful pregnancies, and intervals between pregnancies following both live and dead deliveries. When summarized for both individuals and total group, this data can be used to search for lawful relationships with data on social rank, group size, and age/sex composition.

The study of social status requires a longitudinal study, inasmuch as a status hierarchy implies long-term relationships between the animals rather than a transitory one decided at the time of each interaction. If one says that a social status hierarchy exists, this implies that knowledge of the directionality of previous interactions permits prediction of directionality of future interactions (e.g. if animal A defeats animal B in a series of fights, then in future agonistic interactions one would expect A to defeat B). Further, if the social hierarchy appears to be relatively stable, relatively little fighting should be required to maintain the social status or dominance hierarchy. Agonistic relationships and agonistic interactions may take the form of threats, chases, and token aggressive responses rather than full-fledged fighting or other severe forms of aggression. Data collected on groups representing five of six taxa selected for study at the Yerkes Field Station supports these hypotheses (a consistent pattern of agonistic interactions could not be demonstrated in the sixth, Bernstein 1970). On the other hand, the hypothetical pervasive influence of the status hierarchy, with reference to the directionality of other social patterns in these groups, could not be demonstrated.

The longitudinal strategy permits the study of each individual through all developmental stages from birth through death. The individual animal serves many roles in a primate society, and these role patterns change with age. If early experiences and socialization processes exert a pervasive influence on individuals, these early experiences must be recorded and correlated with activity and role patterns throughout life. The study of postprime animals and their changing functions and relationship to other animals in the troop must be examined to determine the nature of the differences between roles of postprime animals and young adults of equivalent social rank. In the pigtail monkey group previously referred to, the control animal role (Bernstein 1966) was selected for intensive study. The control male was carefully observed as his physical abilities began to deteriorate and as other males in the troop became prime. "Prestige" factors appeared to allow the original troop leader to retain his function as control male long past the time that he was physically capable of enforcing his dominance over young males who were his physical superiors. The biological merits of this phenomenon can be understood when one considers that at one time this animal had indeed proved himself to be of superior physical and social abilities within the troop, and that despite his advanced age he remained a superior animal to the extent that these qualities were genetically controlled. It must be admitted, however, that conclusive data is lacking to support the hypothesis that the control male has the greatest reproductive success in the group.

The longitudinal study also allows us to investigate the influences of various "personalities" upon critical roles in the troop. As different animals function within key roles in the same group, we can note how each animal influences the expression of the role pattern. For example, in the control role pattern, an individual may show certain behaviors which appear to be extremely aggressive. In order to know whether such behaviors are part of a role pattern or are an individual characteristic of this individual, it is necessary to observe several different individuals functioning in the same role in the same troop. Similarly, the role pattern may be assumed to influence the animal's personality, and if one has information concerning the previous behavior of an individual, this information can be used to compare the animal's behavior as he now functions in a new role within the social group.

One example of the value of longitudinal studies is our study of the bachelor male role in a captive group of geladas (*Theropithecus gelada*). In a group consisting of two one-male units and a bachelor male, it was noted that the bachelor male carried infants and juveniles of both sexes, was frequently the center of a play group including immature animals from both one-male units, and was active in play fighting, mounting of juveniles, and in the protection of immature animals any time violent activity broke out in the group. Such behavior might be due to the individual characteristics of this male; or it might be the normal role function of the bachelor male. Two years after the group had been established, it was possible to resolve these alternatives when the former bachelor male deposed one of the one-male unit leaders and took over the functions of the one-male unit leader. His former association with immature animals ceased, his interactions with adult females went from zero to the high levels associated with his new role, and, moreover, the former unit leader began associating with immature animals and engaged in the same patterns as had been typical of the former bachelor male. Thus, the activities of the bachelor male were demonstrated to be role-specific and not idiosyncratic. Furthermore, the activity patterns of the oldest immature males born into the group are beginning to take on the characteristics of the bachelor male patterns. What might have been hypothesized as a tradition in the group thus is interpreted as normal bachelor male role behavior.

My own approach to the study of nonhuman primate social organization is through the use of comparative studies utilizing a longitudinal strategy and several different selected taxa of nonhuman primates. Comparisons are made of the differential influences of selected variables, and the activities of animals living in the same troop as a function of age. Comparisons between animals of different primate taxa, each living in its own troop with situational factors specified, allow us to identify species typical patterns and situational influences (Bernstein 1971). Through the manipulation of both situational variables and the primate taxa selected for study, one can hope to attempt to generalize on the biological and social adaptations common to the animals within the Order Primates. The generality of conclusions regarding social behavior, social organization, and individual abilities must be demonstrated before one can extrapolate from the nonhuman primates to the human situation. Even so, such extrapolation is dangerous in that it is always possible that a quantum step has occurred with the evolution of man within the Order Primates.

Despite systematic control of the habitat of captive animals, it is always possible that the situation itself has influenced the behavior more than has the phylogenetic membership of the animals under study. As such, it is necessary to validate studies conducted in captive situations by studies conducted in natural habitats. In my own work I have attempted to conduct field and captive studies on both the same and different primate taxa (Bernstein 1964, 1967b, 1967c). Various parameters in captive situations have also been manipulated to demonstrate their influences on activity patterns (Bernstein and Mason 1963). Quantitative measures have indicated that many factors influence the frequencies and durations of activities, but that the social expressions remain unchanged.

In the comparative approach, few differences are of such an order of magnitude that a qualitative difference can be observed. Some of the subtle influences may be of the greatest interest to us, and these may be completely obscured if quantification is not possible. As Hans Kummer (1970) has so ably summarized, the study of social organizations begins with the study of social response elements. These elements must be identified and defined. They must then be measured and quantified; such measurement may consist of frequency and duration measures of the responses of each of the individuals living in a social unit. The next step involves the recognition of sequences and the determination of the probability of certain sequences of social responses. This sequential analysis will produce data which will allow one to verify subjective interpretations of response significance. Such data will also identify differential probabilities of various response patterns for each of the individuals in the troop; the differential probabilities of response sequences can be used to identify various role patterns in the social unit. Following the identification of social mechanisms and roles, and the interrelationships of the roles within the social unit, the type of social organization may be specified. Specification of social organization consists not only in identifying the numerical age and sex composition of the social unit, but also requires an explanation of the relationships between animals, thereby explaining the significance of the numerical age and sex composition of the social unit.

Social response measures may be obtained either from continuous recording or time sampling techniques. Many investigators choose to use some convenient unit of time, and simply check the occurrence or nonoccurrence of the response pattern in the individual under observation. The use of a time unit in a time sampling technique does not preclude estimation of true frequency rates if the time units are longer than the average response duration or if only the response onset is marked. Altmann and Wagner (1970) have demonstrated how time unit data may be used to estimate the true rate of occurrence. This frequency rate is somewhat different from estimation of the percent of time which an individual spends occupied in various selected response categories. The latter requires selection of observation units shorter than the typical duration of the responses. It is a relatively simple matter to convert such data to duration estimates. Thus, it is possible to use a time sampling technique (employing a selected time unit) to obtain information for each individual concerning the frequency rate, average duration, and the percent of time occupied in selected responses. This last may also be expressed in terms of the probability that the response will occur in a given period of time.

These procedures have been used to obtain activity profiles for selected nonhuman primate groups (Bernstein 1971). Captive groups were maintained in specified situations, and time sampling techniques were used to measure the influence of such variables as taxonomic affiliation, age and sex composition, weather, season, and time of day. Nine different primate taxa have been sampled to date, but the tyranny of numbers has limited the number of groups representing each taxon. As a result, there is inadequate data to specify the range or variation between groups of the same primate taxon. Without such information, attributing differences between groups to taxonomic affiliation alone must be suspect. In a longitudinal study, however, test retest techniques not only indicate the reliability of the measures, but also suggest the validity of the measure for the entire taxon inasmuch as natality and mortality slowly changes group membership and the group itself. When the phenomena being compared are simply present or absent in a taxonomic group, then intertaxa comparisons present few difficulties. Most comparisons, however, reveal differences in relative frequencies rather than the quality of response expression. In such cases, it would be useful if measures taken at specified intervals in a longitudinal study could be regarded as measures of different groups. There is no ready rule of thumb for such determination, at persent.

The activity profiles obtained have revealed good consistency between repeated measures of the same group and also between measures of different groups of the same species. Even the influences of situational variables do not obscure the intertaxa differences in activity profiles. Furthermore, cogeneric species can be demonstrated to have more similar activity profiles than species of more widely divergent taxonomic affiliation (Bernstein 1971).

The quantitative methodology evolved in the study of captive groups assumes complete visibility of every subject in a social unit at all times, the ability of the investigator to control the sampling procedure and the time of observations, and the ability of the investigator to specify the conditions under which the data are to be collected. The extension of this technique to field situations is readily achieved in only a few habitat situations. In most situations, poor visibility requires some modification of procedure. Where the assumption of a Poisson distribution of responses independent of visibility can be justified, the data can be collected whenever subjects are available. Even when this assumption cannot be justified, more information can be obtained with this procedure than by the qualitative descriptions and impressions characteristic of initial field observations. The transition from qualitative description to frequency counts requires definition of responses and recognition of individuals. This will allow for matrix analyses according to relative response frequencies and individual participation. Even with only an ordinal scale, it is possible to perform certain sociometric analyses of a social unit with reference to theoretically significant dimensions.

The significance of comparative nonhuman primate social studies for the study of human social problems may not be immediately apparent. Certainly, human beings are organized in a vast multitude of societies with many forms of social organization. The nonhuman primates, on the other hand, each appear to have a species-typical social organization and a species-specific social response repertoire. The variation found in human societies, however, may be approximated by the variation in societies of different nonhuman primate taxa. For example, an animal model of various human childrearing practices can be attained by selecting several different nonhuman primate taxa. In so doing, we can find groups wherein the biological mother of an infant appears to be solely responsible for the care of her infant and will preclude maternal responses shown to her infant by other members of the social unit, and other groups wherein the infants are reared by the adult females at large, or even groups wherein paternal care is as important as maternal care. Of course, individual variation exists within a taxon, and more or less latitude may be shown by different females with reference to the amount of contact that other animals in the troop may have with her infant; but the most dramatic differences involve different taxonomic groups. For example, in the marmosets, the male typically carries and protects the infants and gives them to the mother only during periods of nursing. Among some of the leaf monkeys, infants are passed from one female to another (Bernstein 1968a, Jay 1963a, Poirier 1968b, Sugiyama 1965). Among some macaques such as the bonnet (*Macaca radiata*), mothers are relatively permissive about other females contacting their infants (Rosenblum and Kaufman 1967), whereas in other macaques, the female may allow contact only between her infant and its siblings, or she may preclude contact between her newborn infant and any other animal in the group.

There is some variation in maternal care patterns even within a given species. In the longitudinal study of the pigtail monkey group previously referred to, we have noted not only individual variation among the mothers in the degree of protectiveness shown, but we have also noted changes in both the responses of individual females to successive infants, and changes in the care infants receive as genealogical relationships have proliferated with the second generation being born into the group. This latter point assumes particular importance when one realizes that the "normal" pigtail pattern has been described as mothers who resist all efforts of other females to carry their infants, whereas with the second generation being born into this group we note that infants are increasingly to be found being carried by grandmothers, siblings, and female peers of the young mothers born into the group. Such patterns are nowhere near those seen in langurs or reported for bonnets, but are nonetheless a significant departure from the original patterns reported prior to the development of the genealogical complex and past history of play group association now found in the group.

The organization of the social unit itself is relatively constant within a single primate species, but profound differences in the social organization of different species do exist. Among the gibbons and titi monkeys we find the nuclear family consisting of a single mated pair and their offspring (Carpenter 1940, Ellefson 1968, Mason 1966). In the hamadryas baboon and the gelada, we find a single adult male with multiple adult females and young constituting the basic social unit (Crook 1966, 1967; Kummer 1968). In both cases, however, the basic social unit is combined in a higher order social organization consisting of multiple such social units combined into another social organization. Among the patas monkeys (*Erythrocebus*) and lutong (*Presbytis cristatus*), however, single social units of an adult male and several adult females and young do not appear to allow for the creation of larger units combining several such basic social units (Bernstein 1968a, Hall 1967). It is problematic whether the *Cercopithecus* monkeys are typically organized into social units with only a single adult male, or whether (at least under some circumstances) multiple adult males may be found in a single social unit (Struhsaker 1969). Within the *Cercopithecus aethiops* group, however, multiple adult males have been reported to be in a single social unit (Struhsaker 1967b). In the genus *Mandrillus,* again it has been hypothesized that the basic social unit consists of a single adult male, multiple adult females, and young, but that these basic social units may combine from time to time in temporary associations (Gartlan 1970).

This both contrasts with and parallels the case found in the chimpanzee, where individuals may combine and recombine to form social units which apparently are fluid within the context of a larger community of individuals (Azuma and Toyoshima 1961, Goodall 1967, Itani and Suzuki 1967, Reynolds 1963). Among the gorillas it appears that multiple males, females, and young form a basic social unit, but that these social units may come together and combine temporarily (Emlen and Schaller 1960, Schaller 1963). In the savannah baboon, multiple males, females, and young form a basic social unit, and these units may aggregate at resources such as water holes only temporarily (DeVore and Hall 1965, Hall and DeVore 1965, Washburn and DeVore 1961a). The same condition appears to apply to the macaques and perhaps also to the mangabeys and several other lesser known primate species. The ratio of adult males to adult females varies enormously from one primate taxa to another (Carpenter 1954). In the case of the macaques and baboons, where multiple males and females associate together in a single social unit, there is usually a surplus of females to males; that is, the "socionomic sex ratio" (Carpenter 1954) frequently shows a high preponderance of females. Exceptions have been reported in the case of forest living baboons (Rowell 1966) and bonnet macaques, (Simonds 1965, Rahaman and Parthasarathy 1967). Among the capuchins, the socionomic sex ratio may typically be balanced (Bernstein 1964), although other New World monkeys, such as the howler monkeys, (*Alouatta*) and spider monkeys (*Ateles*), typically have a preponderance of adult females (Carpenter 1939, Collias and Southwick 1952, Eisenberg and Kuehn 1966, Chivers 1969). Other studies of primate societies have suggested groups in which the females and young form a permanent social unit, and clusters of associated males are associated but only join the unit during special times of the year, such as the breeding season, e.g., among the squirrel monkeys (Baldwin 1968, Thorington 1967).

This range of social organization certainly appears to include all the conditions found in human societies, with the possible exception of polyandry, and also appears to include some conditions which are rarely, if ever, found in human social organizations. To select any one of these as a model for the precursor to present human societies seems premature, and we must consider that the precursors to human societies may have included any of these conditions either singly or in combination. The net result, however, seems to suggest a greater modifiability of human social organization than that seen in any of the nonhuman primates. Thus, along with the greater plasticity of behaviors seen in individual learning response patterns, we also see a greater plasticity with respect to the social organizations possible in human society. This plasticity may be characteristic of the Order Primates, and appears to have found its greatest expression in the species *Homo sapiens*. Different degrees of plasticity, however, may be found in all of the nonhuman primates.

Questions concerning the territorial nature of man may or may not be relevant; however, the nature of territoriality in nonhuman primate societies and the correlation of various response patterns to the presence or absence of territoriality may suggest social influences and social functions of territorial versus home range distributions and organization of space. In fact, in the same superspecies or species *Cercopithecus aethiops,* both territorial and nonterritorial troops have been described (Gartlan and Brian 1968, Struhsaker 1967b). In these cases, the ecology seems to determine the sociology. Ecological influences on social units have also been reported in groups of baboons (DeVore and Washburn 1963, Hall 1963a, Hall and DeVore 1965, Rowell 1966) vervets (Gartlan and Brain 1968, Struhsaker 1969), and langurs (Yoshiba 1968).

Variations in human childrearing practices have been considered by many as the determinants for various types of later adult social behavior. Among the nonhuman primates we do see a great range of infant care practices and a great range in the final social organization found in the species. The available data do not suggest that the one necessarily determines the other. In fact, using nonhuman primate social organizations, it would be necessary to cross foster animals from a primate taxa showing one form of childrearing practice and social organization, to another group showing different maternal rearing procedures and social organization, in order to determine the extent to which adult social patterns were influenced by early experience and maternal rearing practices. We could not, of course, control for the overwhelming influence that the group social structure might have on the individual animal, nor could we hope to control the influences of taxonomy and differential morphology.

All societies face certain problems in common, such as the control of aggression, the integration of new members into a society, and the control of deviant behavior or deviant individuals either through socialization or through exclusion. The captive nonhuman primate society allows us to ethically explore the consequences of such practices and mechanisms as might influence the control of aggression in an individual and in a society, or the control and expression of deviant behavior and deviant individuals in a society. Such studies should not be regarded as directly applicable to human situations, but they will allow us to consider alternative means of dealing with the various problems of human society as they are already dealt with in nonhuman societies. The extent to which such answers are applicable to human societies must be subjected to further comparative study to validate attempted extrapolation to human societies. If nothing else, the comparative study of multiple nonhuman primate societies will allow us to appreciate the range of solutions possible to the same types of problems, and the fact that multiple solutions to the same problem can exist without one necessarily being superior to all others.

19

JOEL E. COHEN
HARVARD UNIVERSITY

Aping Monkeys with Mathematics

This chapter offers evidence that as a means of understanding the functional and evolutionary biology of the primates, mathematical models are at least worth their weight in paper. The basic idea of mathematical models and the possibilities and the limits of their usefulness are outlined in Section 1, "Mathematical Models." As morphological arguments for evolutionary conclusions have become better defined, the use of mathematical models in morphology has increased. This trend is not without difficulties (as examined in Section 2, "Morphology in Evolutionary Arguments"). As behavioral and social characteristics of animals have appeared in evolutionary arguments, mathematical models are needed even more to help keep ideas dynamic (rather than typological), clear, and consistent (as explained in Section 3, "Behavioral Taxonomy"). Recent studies of casual social groups among human and nonhuman primates exemplify an explicit mathematical characterization of one aspect of social life. This characterization may provide a useful base for comparative evolutionary inferences (detailed in Section 4, "Monkeys en Masse").

The data used in quantitative studies of social groups can be related to frequencies of other aspects of behavior (asymmetric relations such as dominance, symmetric pairwise relations such as play partnership, and individual behaviors) by representing data in multidimensional contingency tables (in Section 5, "Data Structures and Models for Social Behavior"). The frequencies of behaviors obtained from multidimensional contingency tables, and the frequency distributions of choice of habitat, may be interpreted functionally on the basis of recent mathematically formulated discoveries in the psychology of operant conditioning (discussed in Section 6, "Frequencies of Behavior"). But most of the use of mathematical models in

I thank Scott A. Boorman, Stephen E. Fienberg, Richard J. Herrnstein, William W. Howells, and Russell Tuttle for helpful criticisms of the manuscript.

studying the natural behavior, sociology, and ecology of primates (human and nonhuman) remains to be done (see Section 7, "Math and Aftermath").

Section 1. Mathematical Models

Like Moliere's gentleman who learned at an advanced age that he had been speaking prose all his life, those who claim ignorance of mathematical models may be surprised to discover that they have used them ever since elementary school (even without benefit of the "new math"). This recognition is important because insight into simple mathematical modeling can serve as a bridgehead to the understanding of much more complicated modeling.

By way of introduction to simple mathematical modeling (Cohen 1970), consider the two statements, "One plus one is two," and "One apple plus one apple makes two apples."

The second statement is an empirical statement about the world of experience. Once we agree on a procedure for counting, if we place one apple next to another apple we will both count two apples. The first statement is an arithmetical theorem which can be proved from simpler axioms along with many interesting results such as "Two times two is four."

Mathematical modeling is the act of imagination which proposes a connection between mathematical statements embedded in a mathematical structure, and some empirical observations. The difference between mathematics and mathematical modeling is like the difference in points of view one can have toward a microscope, which can be intrinsically interesting, or which can be a tool with which one extends one's perceived world.

In full flower, a mathematical model exemplified by "the laws of elementary arithmetic apply to apples and the way we count them" leads to such nonobvious predictions as "32,479 apples plus 90,503 apples makes 122,982 apples." Although this prediction has probably never been checked empirically, the behavior of small numbers of apples conforms so well to the elementary assumptions of arithmetic that few would doubt it. But the conscientious mathematical modeler cannot rest with proposing, generalizing, and verifying his model in special cases. He must also search for the limits of its validity. After observing apples in large quantities (or making calculations about the strength of materials), the modeler should point out that when more than one hundred thousand apples are gathered together, some of them become applesauce. Ordinary arithmetic does not then apply.

Nor does the truth of a model (arithmetic or any other) guarantee that the model is of interest. If instead of apples we put twenty monkeys together, the arithmetical details may be of less interest than the formation of a troop. And if we put two troops together, the number of troops (though arithmetically predictable) may be of less interest than the interactions between them.

Whether a particular mathematical model interests a reader depends in part on the temperament of the reader, in part on the skill and insight of the modeler, and in part on the cooperation of nature.

Mathematical models are important in the study of primate evolution when they assist scientists to clarify what they believe about the phenomena they study. By thus sharpening the beliefs and expectations of scientists, mathematical models help to confirm or reject their assumptions and hypotheses. Mathematical models may facilitate finding explanations because they permit "what-if" experiments. If certain simplifying notions that are embodied in a mathematical model lead to qualitatively wrong predictions, then the scientist can abandon those bad notions. And even if they never lead to full explanations, mathematical models can clarify how much of the phenomena has been comprehended and expressed, and how good the resulting approximations are.

But mathematical models are not always necessary. Where ordinary concepts and everyday chains of logic suffice to give intellectual comfort with familiar kinds of materials, there is no scientific virtue in mathematical formalism. Three kinds of events can upset this amiable state of affairs: (1) a rise in the standard of intellectual comfort (as exemplified by other chapters in this book), so that what was once acceptable as an explanation is no longer adequate; (2) an extension of study to new kinds of materials where ordinary concepts and everyday logic are no longer competent (as in the study of "social evolution"); and (3) initially unrelated conceptual inventions, often new mathematics, which change the setting and illumination of familiar materials (see sections 5 and 7 to follow).

Section 2. Morphology in Evolutionary Arguments

The very progress of the study of primate evolution has created a need for mathematical models.

A fair caricature of classical morphological arguments in primate evolution may be provided by arranging four mandibles, one each from *Parapithecus, Propliopithecus, Pliopithecus,* and a modern gibbon, in a row (Clark 1957, p. 91). By looking at this arrangement, the reader is supposed to conclude that the modern gibbon mandible evolved from the smaller but similarly shaped *Pliopithecus* mandible, which in turn evolved from the still smaller *Propliopithecus* mandible, which originated from the smallest mandible, that of *Parapithecus*. But the graphic presentation of information does not make explicit at least two important steps which precede this conclusion.

First, the reader and author of the figure must assume that mandibles are of evolutionary significance and that, unlike certain other morphological complexes in the organisms, the mandible is reliably and importantly related to the life, form, and evolution of the whole animal. Then both reader and author must characterize each mandible in a way that permits comparison among them, perhaps according to size, particular shapes and relative proportions of mandibular components, and specifications of the distribution of bone density. This characterization of the mandibles in a way that makes them comparable is actually the process of model construction; it is the abstraction from the complexity of real objects to an intellectually tractable description of what matters.

Second, the reader and author construct transformations of their characterizations of each mandible into those of the next mandible. The reader's sequence of transformations must be internally consistent (so that the step from *Parapithecus* to *Propliopithecus* is not too different by some measure from the step from *Pliopithecus* to *Hylobates*). Further, these transformations must be consistent with additional knowledge that the author and reader possess about the animals and evolutionary principles. Boorman (1970) offers an enlightening discussion on measures of distances between complicated objects.

Because, in classical morphological arguments, so much of the process from evidence to conclusion is private, there is enormous room for the author and his readers to arrive at different conclusions. If I think that only the shape and size of the canines significantly affect diet and evolution, while you attach much more importance to molars and mandibular size, and a third person believes that the scapula really matters and the mandible is not worth bothering with, then each of us will characterize the evidence differently. If I have one belief about phylogenetic allometry and you have another, even starting from the same characterizations we may differ in which evolutionary transformations we employ as standards of distance.

In spite of the great room for disagreements in the abstract, in fact substantial agreement has obtained in classical morphology. The whole mandible is one important determinant and consequence of diet and it is a morphological concomitant of other evolutionary changes. Progressive increases in its size without drastic changes in its shape constitute a plausible evolutionary sequence.

A great part of the rise in the intellectual standard of living among primate morphologists has been devoted to making explicit and otherwise improving procedures for characterizing (modeling) the objects of their studies. D'Arcy Wentworth Thompson superimposed square grids (Cartesian coordinates) on drawings of forms, and showed how they could be simply distorted to produce drawings of apparently related forms.

Oxnard (1969c, pp. 75–76) observed:

The shape of a biological specimen may be represented by a series of measurements of different kinds taken on the specimen. One way, therefore, of representing the specimen is as a single point located in a multidimensional space. The coordinates of this point are the actual values of each of the measurements, and the many dimensions are the many different measurements. . . . [See also Howells, Chapter 5.] But each original measurement does not necessarily give completely new information about a particular specimen; for instance, a second measurement of the radius of a circle tells us nothing new at all about the circle.

A mathematical technique called canonical analysis provides a way of combining measurements into a new set of dimensions such that each additional dimension in this new set provides the maximum increase in information (by some measure of information) about the shape of the specimen. Canonical analysis may show that a much smaller number of dimensions suffices to characterize specimens, though these dimensions would not have occurred to a naive observer. Given improved characterizations, improved speculations about how evolution transforms one to another are possible (Oxnard 1969c, pp. 92–95).

Multivariate statistical methods have also been used to reveal morphological relations among populations of men. Howells (1970a) found discriminant functions which would characterize efficiently the differences among 17 populations of contemporary human skulls. He used these functions (1970b) to assist in the interpretation of skulls of Mount Carmel man.

These multivariate statistical methods improve earlier eyeball "guesstimations" of form. But these methods can cause problems which call for more refined and different mathematical modeling. Here is an example:

Suppose we are investigating a polygenic character. Suppose also that the number of genes (additive polygenes, not loci) for this character is proportional to the length of time that some spherical morphological feature grows. But the measured aspect of that feature is its volume. In this case, the number of genes for the feature will be related to the measurement of the feature as x is to x^3, and in general as x is to $f(x)$.

Suppose that for five known populations, numbered 1, 2, 3, 4, and 5, the numbers of genes for this character are, respectively and exactly, 1, 2, 3, 5, and 6, and hence that the phenotypic measurements (which are the only direct measurements we can make) are 1, 8, 27, 125, and 216. We observe now a new sixth population with phenotypic measurement 343. The linear phenotypic distance between population 6 and population 5 (namely, $343 - 216 = 127$) is greater than the linear phenotypic distance $125 - 1 = 124$ between population 4 and population 1. We therefore conclude from linear analysis of phenotypic measurements that populations 1 and 4 are more closely related than are populations 5 and 6.

But if our concept of relatedness is to be based on genetic similarity, it is clear that two populations (5 and 6) which differ in only one polygene are more closely related than populations (1 and 4) which differ by three polygenes. Further, on the genotypic level, populations 1, 2, and 3 are about as closely related as are populations 4, 5, and 6; but on the pheno-typic level, the discovery of population 6 appears to make populations 1, 2, and 3 more closely related, relative to the whole scale of variation.

If, as Norbert Wiener purportedly said, what we want are measurements that will stay put when our backs are turned, the accessible phenotypic measurements are clearly less satisfactory than the inaccessible genotypic ones. One role for theory is to transform measurements, e.g. $f(x)$, into evidence, x, which will not require a new view of the world to accommodate every new experience. The mechanized algorithms of linear statistical analysis may be very useful aids in preliminary probes designed to discover what theory must accomplish, but they are not adequate theories.

Section 3. Behavioral Taxonomy

In the study of behavior, the temptation to affirm that seeing is believing is no weaker than in the study of morphology, and the need for theory to relate observations to some underlying invariants is no less urgent.

In 1898, Charles Otis Whitman emphasized behavioral approaches to evolution: "Instincts and organs are to be studied from the common view-point of phyletic descent." His successor Konrad Lorenz (1958) put it thus:

> As phylogenists, Whitman and Heinroth both sought to develop in detail the relationship between families and species of birds. To define a given group they had to find its "homologous" traits; the resemblances between species which bespeak a common origin. . . . Behavior, as well as body form and structure, displays homologous traits.

Marrying behavior and morphology in the study of primate evolution, Tuttle (1969b) studied the hands of the great apes. Although using no mathematical models, he characterized the role of the hands in locomotion by the bone structures in modern and fossil apes, and by the behaviors of modern apes in knuckle-walking, fist-walking, and modified palmigrade walking. He considered plausible transformations of these descriptions and showed them to render unlikely some suggested evolutionary sequences leading to man. Remarkably, Tuttle (1969b, p. 957) declined to claim that knuckle-walking devolved from fist-walking and not modified palmigrade locomotion, or vice versa. Even without a mathematical model, Tuttle obtained what is often one of the chief benefits of a mathematical model: a demonstration that present data are insufficient to choose among different possible transformations leading to a given result.

Detailed studies of one aspect of primate morphology or behavior, such as those of Oxnard and Tuttle, have the virtue of specifying the procedure for measuring the dimensions of the object of study with sufficient public detail that another worker can check the measurements on his own material or adapt the procedure to new kinds of materials.

Studies in similar detail of the behavior of human individuals in a social setting led Bales (1955) to propose four broad categories of talk: positive emotional reactions, attempts to solve a problem, questions, and negative emotional reactions. In observing primates whose language, if it exists, is not understood or difficult to observe, only the first and last categories of acts seem applicable. These categories reappear in the behavior of rhesus monkeys. Altmann (1968, pp. 62–63) first defined a detailed behavioral repertoire for rhesus monkeys, and then looked inductively for groupings in the sequences of those behavioral elements. He found these groups: primarily affinitive and mild tactile signals; primarily agonistic signals; primarily sexual signals of males; primarily sexual signals of females; play; suckling and weaning; aggressive chases; carrying of infants; and two miscellaneous agonistic patterns. Applied to other species, Altmann's detailed procedure might lead to comparable groupings of behavior patterns.

Section 4. Monkeys en Masse

Even some of the most systematic attempts to advance from individual behavior to social systems return to a level of discourse comparable to that of morphology before D'Arcy Thompson, because they follow the strategy of arranging homologous bits of (social) anatomy in suggestive sequences as Clark (1957, p. 91) arranged four mandibles to represent a hylobatid lineage.

Crook and Gartlan (1966) named five "adaptive grades" of primates, moving from forest dwelling through forest fringe and tree savannah to grassland or arid savannah, with concomitant changes in diet, diurnal activity, size of troops, reproductive units, male mobility, sexual dimorphism, and population dispersion. In this typology, Crook and Gartlan specified "size" of troops (which they called "groups") in successive grades as "usually solitary, very small groups, small to occasionally large parties, medium to large groups, . . . [and] large groups. . . ." Here any actual measurement of the size of troops is implicit. The introduction of a Cartesian grid for all visible dimensions or aspects of the primate adaptations is clearly not even contemplated.

Kummer (1968) and Altmann and Altmann (1970) provide explicit measurements of many different aspects of social patterning in baboons. These cross-sectional studies of single species run no risk of overlooking aspects of behavior or ecology which lack apparent homologues in other species. Perhaps in reaction to global and glib comparative essays less scrupulous than that of Crook and Gartlan (1966), these studies indulge sparingly in comparative or evolutionary speculation. The detail with which their observations are made and reported offers the possibility of making similar observations on other primate species, and then making potentially credible comparative speculations involving explicit mathematical models.

Rather than studying in detail the social behavior of one entire species, and leaving comparative work for later, another approach is to select one aspect of social behavior and study it comparatively in great detail, leaving other aspects of social behavior for later. Such a narrow but longitudinal approach may provide suggestions and indications, early in the course of study, about those aspects of social behavior which have evolutionary interest or which correlate in interesting ways with other characters of evolutionary interest. If both the longitudinal and cross-sectional approaches are well based in careful observation, and if the principal elements of each species' sociobiology have evolutionary meaning and therefore comparative homologues, the two approaches should lead to similar evolutionary inferences.

One narrow aspect of primate social behavior which has been studied comparatively by means of mathematical models is the formation of casual or spontaneous social groups (Cohen 1971; and in press).

In the early 1950s, John James observed the sizes of freely forming small groups of human beings in a variety of situations. James defined freely forming groups as "those whose members are relatively free to maintain or break off contact with one another; that is, they are ones where informal controls on behavior are at work and spontaneity is at a maximum." He included in his observations only those "groups in which the members were in face-to-face interaction as evidenced by the criteria of gesticulation, laughter, smiles, talk, play, or work. Individuals who merely occupied contiguous space were not counted as members of a group." Goffman's (1963) far subtler analysis of social interactions in public places makes it clear that James' definitions and approach overlook a great deal, but does not render them less useful for present purposes.

The full frequency distributions of the nearly 18,000 groups James observed in 21 different situations were published in Coleman (1964b, pp. 368–373). Situations observed included pedestrians in Eugene, Oregon on a spring morning; shopping groups in two Portland, Oregon department stores; and play groups in Eugene in the spring on the playgrounds of 14 elementary schools (directed or organized play was not included in the observations).

In 1961 Coleman and James observed that a Poisson distribution with the zero value truncated described nearly all of the observed frequency distributions very well. Coleman proposed a set of assumptions about a group's probabilities of transition from one size to another which led to a prediction at equilibrium of a truncated Poisson distribution of group size.

The next year Harrison White pointed out that Coleman's model treated each group as if its behavior were independent of the sizes of all the other groups in the system of groups (in the playground or in the department store). White proposed several "sociological" models in which the behavior of particular groups depended on the number of groups of other sizes in the system. These models differed in assumptions about the flow of individuals between the system and its environment, the flow of members among the groups, and the dependence of rates of arrival to and departure from groups on sizes and numbers of groups. In spite of substantial differences in assumptions, White claimed to show that all the models led to an equilibrium distribution of group size given by the truncated Poisson. As did Tuttle in considering alternate pathways for the evolution of knuckle-walking, White concluded that "more elaborate empirical investigations . . . will be needed to test the applicability of and to discriminate among simple stochastic models for the circulation of members among casual groups."

In 1964 Goodman criticized the mathematics of Coleman and White, but concurred in the conclusion that a variety of models could lead to the same equilibrium distribution. He proposed yet another model.

Thus James discovered a striking regularity in the social aggregations of large collections of people, although only 18 of James's 21 observed frequency distributions were well described by the truncated Poisson distribution. Coleman, White, and Goodman provided several different explanations, each accompanied with a plea for further empirical research.

These stochastic models for systems of social groups appeared at the same time as the first of the contemporary spate of primate field studies. The coincidence raised the hope of finding primate field data on casual social groups which, first, would show whether the Poisson regularity held for primate species other than man; and second, would discriminate among the existing models leading to the Poisson distribution.

In 1967, Stuart Altmann directed me to the last table in Thomas T. Struhsaker's doctoral dissertation on vervet monkeys in East Africa. For each night of observation in this table, Struhsaker recorded which individual monkeys of a particular troop were sleeping together in one tree top, which monkeys if any in another, which if any in a third, and so on.

These sleeping groups are not casual groups within James' definition because the vervets cannot leave in the middle of the night, and during sleep there may be hardly any interactions. Nevertheless I assumed that the groups represented a snapshot of the state of the system of groups within the troop at the time that the monkeys ascended into the trees for the night. The frequency distribution of group size was clearly not Poisson. But it approximated a truncated negative binomial.

Now, either the process of forming human casual social groups and the process of forming vervet sleeping groups are different, or both are manifestations of some common underlying process (at least as far as their size is concerned). I preferred the latter alternative, and invented a family of stochastic models, called linear one-step transition (LOST) models, which depend on two ratios, a/d and b/d, of three parameters, a, b, and d. Parameter a is a proportionality constant which measures the rate at which individuals join groups, independently of the size or identity of the individuals in the groups. Parameter b is a proportionality constant which measures the rate at which individuals join groups in proportion to the numbers of individuals in those groups (but independently of the identities of the individuals in the groups). Parameter d is a proportionality constant which measures the rate at which individuals already in groups make independent decisions to leave their groups and become isolates. For positive values of b, these models predict a truncated negative binomial distribution of group size; and for $b = 0$ they predict a truncated Poisson distribution of group size.

Thus, these models encompass Struhsaker's data on vervet sleeping groups and those data of James previously described by the truncated Poisson distribution. Reexamination of James' three distributions which did not fit the Poisson showed that two of them also could be fitted by the truncated negative binomial.

These data were not sufficient to test whether the detailed assumptions of the LOST models concerning arrivals and departures were true, but the detail in Struhsaker's observations was sufficient to rule out a variety of other models that had been proposed as mechanisms leading to the negative binomial distribution.

To test directly the detailed mechanisms of the LOST models, I observed the free play of four-year-old humans in the Cambridge Nursery School. There appeared to be no large differences between what the LOST models assume about the dynamics of group formation, and how the actual sizes of play groups changed over 30-second intervals. This surprising result did not show that the same dynamics were at work in any species of non-human primate; it merely failed to rule out the possibility that the LOST dynamics were at work among humans.

I then wanted to see whether monkeys' undirected socializing could be described by the same models as human socializing. In 1969 Stuart Altmann provided me with an opportunity to answer my questions by joining his field project on baboons in Kenya. The setting of this study, as of 1964, is described by Altmann and Altmann (1970). Details on my observations appear in Cohen (in press). In brief, of the seven troops I observed for sufficient periods, only one had a group size distribution which fit the predicted distribution poorly (using less than fully efficient methods of estimation and a test of goodness of fit with artificially increased power).

The generally acceptable agreement between the observed frequency distributions of group sizes in men and monkeys and the fitted distributions shows that a single family of models may suffice for present purposes. The dynamic assumptions of the models have been tested only on nursery school observations. Since the models characterize those situations they describe by the two numbers a/d and b/d, the net result of the studies to date can be summarized by a list of the primates observed and their associated values of a/d and b/d.

Table 19–1. Median values of the parameters of the linear one-step transition (LOST) models for various primates

	a/d	b/d
Vervets (Struhsaker)	1.15	0.66
Yellow baboons	0.12	0.16
Human four-year-olds	0.33	0.10
Mixed humans (James)	0.86	0

The median values of the parameter ratios for baboons in Amboseli were chosen as typical. The vervet parameter values are based on Struhsaker's observations during nights when all members of the troop were noted. The nursery school figures are an average of the estimates obtained on different days of observation weighted by the number of observations on each day. Finally, the figures based on James's observations are the medians of Coleman's estimates for the 18 cases where the truncated Poisson distribution was descriptive and my estimates for the two other cases where the negative binomial distribution was descriptive.

The systems of casual social groups characterized by the LOST models bear the same relation to the whole social behavior of a primate species that the hand or perhaps a single digit do to the total morphology of an organism: they are important, and not the whole picture. The models show one way to characterize, perhaps usefully, a complicated kind of social behavior by a small number of parameters. Characterizing a primate social system by parameters embedded in a predictive model is self-validating: one need take the characterization and the parameter values seriously only when the predictions of the model are confirmed. If the equilibrium distribution of group sizes in a system of casual social groups is neither a truncated negative binomial nor a truncated Poisson, then the parameter values—no matter how operationally defined—are no longer providing useful information.

The characterization of social systems by operationally defined parameters embedded in models gives concrete meaning to the phrase, "the evolution of social systems." Such evolution is simply the trajectory in time of the parameter values typical of a taxon and situation, within the span of evolutionary time that the models are useful.

The estimates of b/d in the rightmost column of Table 19–1 decline as one reads down the table. Though the difficulties of evolutionary inference from the behavior of presently existing species are manifold, I wonder if there is any evolutionary meaning to the decrease in the role of individual attraction in relation to individual departures (b/d) and the ascendancy of group attraction in relation to individual departures (a/d). A theory of this phenomenon, if it exists, is evolutionary theory, and can be distinguished from the theory of the social system itself.

Section 5. Data Structures and Models for Social Behavior

The studies of casual social groups just described report observations on the frequency distribution of the size of groups engaged in a particular class of activities (informal socializing).

Transaction flows are another common form of reporting quantitative observations of social behavior. In transaction flows, an actor and an object are designated; for example, a speaker and a listener (Bales 1968), a dominating chicken and a dominated chicken (Guhl 1956), or a threatener and a threatened subject. Transaction flows are usually reported as a matrix in which the rows correspond to actors and the columns correspond to objects. The number in the ith row and jth column reports the frequency with which individual i acted upon individual j. In such matrices, the diagonal is zero or undefined. The statistical analysis of such matrices, assuming certain probability models, has been treated by Bishop and Fienberg (1969), Fienberg (1969), Goodman (1968), and Wagner (1970). When the individuals are lumped into classes such that one member of a class may act upon another member of the same class, the diagonals need not be zero (e.g. Altmann 1968).

A third form of recording quantitative observations of social behavior consists of matrices in which rows and columns are identified as before but in which the individual entries record the frequency of some conjoint action, such as the frequency with which monkey i and monkey j slept together in the tree tops (Struhsaker 1967b, p. 112). Clearly this relation is symmetric: the entry in row i and column j equals the entry in row j and column i. When columns and rows are identified with individuals, the diagonal may be left undefined or may record the frequency of the action (if meaningful) by the isolated individual. The half-matrix above the diagonal suffices to record the pairwise frequencies when rows and columns are identified with individuals. When individuals are pooled into classes, the diagonal is required as well.

A fourth form of recording quantitative observations of social behavior (to which I will return in Section 6) constitutes a frequency distribution on patterns or elements of behavior. This frequency distribution tells, for example, how often all individuals chased, fled, slept, or played. Altmann (1965) gives a frequency distribution on a behavioral repertoire of 120 elements.

A fifth method that could be employed to record quantitative observations of social behavior is the multidimensional contingency table. Each dimension is associated with an individual in the social system. Each category along the dimensions is associated with an element in a behavioral repertoire. This repertoire must be defined so that its categories are mutually exclusive. This requirement excludes a repertoire which permits an animal, for example, to be groomed and to be grooming at the same time if these are considered two different elements of the repertoire. Multidimensional contingency tables have recently been used to analyze lizard feeding behavior and interspecific interactions in ecology (Schoener 1970). A simple guide to their statistical analysis in this context is available (Fienberg 1970). To my knowledge, these tables have not been used to record social behavior.

The information contained in the first four forms for recording observations of behavior can be derived by arithmetic from this fifth form. Further, a model for the probabilies of observations in the cells of the multidimensional contingency table implies models for each of the other four arrays of observations, by the same arithmetic operations performed on the predicted probabilities. In this sense, a model for the multidimensional contingency table is an underlying model for the other kinds of observations. In particular, a model of independence in conjoint action as recorded in symmetric matrices (the third form above) and the Poisson or negative binomial theoretical distributions of group size (as predicted by the LOST models of the previous section) can both be derived from a single, simple partial model for the multidimensional contingency table. (The much more difficult problem of the effects of collapsing multidimensional contingency tables when a model of independent dimensions is not assumed is being studied by Bishop [1971].)

To keep the example simple, suppose we have a primate troop of only three individuals that do only four things: chasing, fleeing, sleeping in the troop's roost, and playing in the troop's playpen. All individuals in the troop are presumed to be in view with each observation of the troop. (This assumption avoids all the difficult problems of sampling.) At any time, each of the animals is engaged in precisely one of these activities. An animal can be engaged in chasing if and only if there are one or more animals fleeing from it, and an animal can be fleeing if and only if there are one or more animals chasing it, and no animal can be both chasing and fleeing at the same time. (In the common sense of the terms "chasing" and "fleeing," an animal may be both chasing and fleeing at the same time; but we require the categories to be defined as disjoint.) Any binary social relation, such as talking and listening, which satisfies the same properties may be substituted for chasing and fleeing. Arbitrary n-ary relations may be recorded in the same contingency table format so long as there are at least n individuals in the troop.

Then observations of the entire troop's social activities may be entered in the three-dimensional array with four categories in each dimension, as shown in Table 19–2. Here the letters $a, b, c, \ldots x, y, z, A$ are numbers of observations. Roman numerals I, II, III identify the three individuals, and (1), (2), (3), (4) identify chasing, fleeing, sleeping, and playing, respectively.

The zeros are logical zeros. For example, the entry in the third row and third column of the first panel is zero because if animal I is chasing (1), and animal II is sleeping (3), by definition of "chasing" animal III must

Table 19–2. A multidimensional contingency table to record social
behavior in a troop of three individuals (I, II, III), each having
four patterns of behavior [(1), (2), (3), (4)], entries
a, b, c, ... y, z, A being observed frequencies

		II(1)	(2)	(3)	(4)
	III(1)	0	a	0	0
I(1)	(2)	b	c	d	e
	(3)	0	f	0	0
	(4)	0	g	0	0
		II(1)	(2)	(3)	(4)
	III(1)	h	i	j	k
I(2)	(2)	1	0	0	0
	(3)	m	0	0	0
	(4)	n	0	0	0
		II(1)	(2)	(3)	(4)
	III(1)	0	p	0	0
I(3)	(2)	q	0	0	0
	(3)	0	0	r	s
	(4)	0	0	t	u
		II(1)	(2)	(3)	(4)
	III(1)	0	v	0	0
I(4)	(2)	w	0	0	0
	(3)	0	0	x	y
	(4)	0	0	z	A

be fleeing (with frequency d, row (2)), and cannot also be sleeping. All the other zeros are argued similarly.

Table 19–3 illustrates the four derived data formats, in the order in which they were first described. The entries in these arrays show how they were derived from the counts in the multidimensional contingency table (Table 19–2). Part I of Table 19–3 gives the frequency distribution of sleeping group sizes and the frequency distribution of play group sizes. For example, the frequency with which exactly two animals were observed playing together is $u + y + z$. The frequency with which all three animals slept in one group is r.

The frequencies with which particular pairs of animals engaged in chasing and flight are given in part 2 of Table 19–3. Thus animal I chased animal III with frequency $b + c + d + e$, whether or not animal I also chased animal II or was joined in chasing III by animal II.

The frequencies with which each pair of animals slept together and played together are given in part 3 of Table 19–3 regardless of the presence or absence of the remaining animal in the troop. Thus animal II and animal III slept together with frequency $r + x$.

Finally, the frequencies with which the acts of chasing, fleeing, sleeping, and playing were observed are given in part 4 of Table 19–3. Since A is the frequency with which all three animals were observed playing at the same

Table 19–3. Four frequency distributions derived from the multidimensional contingency table (Table 19–2): 3.1, group sizes; 3.2, transaction flows; 3.3, pairwise conjoint actions; 3.4, behavior patterns

3.1

Size of Groups	Frequency of Sleeping (3) Groups	Frequency of Play (4) Groups
0	$a+b+c+e+g+h+i+k$ $+1+n+p+q+v+w+A$	$a+b+c+d+f+h+i+j$ $+1+m+p+q+r+v+w$
1	$d+f+j+m+u+y+z$	$e+g+k+n+s+t+x$
2	$s+t+x$	$u+y+z$
3	r	A

3.2

		Fleeing (2)		
		I	II	III
	I	0	$a+c+f+g$	$b+c+d+e$
Chasing(1)	II	$h+1+m+n$	0	$b+1+q+w$
	III	$h+i+j+k$	$a+i+p+v$	0

3.3

Sleeping (3) pairs

	I	II	III
I	0	$r+t$	$r+s$
II	$r+t$	0	$r+x$
III	$r+s$	$r+x$	0

Playing (4) pairs

	I	II	III
I	0	$y+A$	$z+A$
II	$y+A$	0	$u+A$
III	$z+A$	$u+A$	0

3.4

Behavior	Frequency
(1)	$2a+2b+c+d+e+f+g+2h+i+j+k+1+m$ $+n+p+q+v+w$
(2)	$a+b+2c+d+e+f+g+h+2i+j+k+21$ $+m+n+p+q+v+w$
(3)	$d+f+j+m+y+z+u+2(s+t+x)+3r$
(4)	$e+g+k+n+s+t+x+2(u+y+z)+3A$

time, each such observation contributes 3 to the number of times which the act of playing was observed; similarly for the other entries multiplied by numbers.

The point to be made is that if the contingency table in Table 19–2 is available, all the other forms in Table 19–3 of describing social behaviors follow by arithmetic.

A model of quasi-independence which is widely used for the analysis of transaction flows (Bishop and Fienberg 1969, Fienberg 1969, Goodman 1968, Wagner 1970) attributes to each individual $i = $ I, II, III a chasing-strength $\alpha_{i(1)}$ and a fleeing-strength $\alpha_{i(2)}$, both of which are dimensionless probabilities between 0 and 1. The model predicts that the frequency with which individual i will be observed chasing individual j will equal $N\alpha_{i(1)}\alpha_{j(2)}$, where N is the number of observations of the whole troop. For symmetric relations such as sleeping or playing, the matrices of expectations $N\alpha_{i(3)}\alpha_{j(3)} = N\alpha_{j(3)}\alpha_{i(3)}$ and $N\alpha_{i(4)}\alpha_{j(4)} = N\alpha_{j(4)}\alpha_{i(4)}$ are symmetric. That there is no statistical or logical necessity for observations of pairwise interactions to be described by a model of quasi-independence is most convincingly demonstrated by the failure of real data to be so described (Altmann 1968).

To relate the symmetric model for a conjoint action, e.g. playing (4), to a Poisson or negative binomial distribution for the size of play groups, we construct a model for the $2 \times 2 \times 2$ contingency table in part 1 of Table 19–4, which is obtained from Table 19–2 by collapsing the four elements of the behavioral repertoire into two elements, namely, "playing" and "not playing." (When there are n animals in the troop, this censorship yields a 2^n contingency table for the frequencies of conjoint playing.) The model of independence, which gives each animal i a probability $\alpha_{i(4)}$, $i =$ I, II, III, of joining in play, independently of all other animals, assigns to each combination of animals the probabilities in part 2 of Table 19–4. When these are transformed to a symmetric matrix part 3 of Table 19–4 analogous to part 3 of Table 19–3, it appears that the probability that any animal is paired with any other animal in play equals the product of their respective probabilities of playing; hence that the model of quasi-independence is satisfied.

When the probabilities in part 2 of Table 19–4 are transformed to a frequency distribution of group size as in part 4 of Table 19–4, it appears that if all the animals' probabilities of play $\alpha_{i(4)}$ were equal to $\alpha_{(4)}$, then group sizes would be binomially distributed, with probability parameter $\alpha_{(4)}$ and with $n = 3$. It is well known that if n becomes large and $\alpha_{(4)}$ remains small so that $n\alpha_{(4)}$ is moderate, the binomial distribution approximates the Poisson distribution with parameter $\lambda = n\alpha_{(4)}$. If the animals in a large troop were divided into classes, the members of which played only with other members of the same class, and the probabilities $\alpha_{(4)}$ were distributed over the classes in an approximately gamma distribution, then the group size distribution for the whole troop would be approximately negative binomial.

This derivation of the Poisson and negative binomial distributions from the underlying model of independence presented in part 2 of Table 19–4 rests on the assumption that all $\alpha_{i(4)}$ are equal. This assumption may be relaxed. If the probabilities $\alpha_{i(4)}$ are distributed over the animals i in a beta distribution, then it may be proved that the distribution of group sizes will be hyperbinomial. As n gets large, for small group sizes and subject to certain reasonable constraints on the beta distribution (small variance), the hyperbinomial distribution approximates a binomial distribution. The details of this argument and the explicit form of the restrictions on parameters which make this limiting process valid may be found in Pratt, Raiffa, and Schlaifer (1965, ch. 9). This binomial distribution of group sizes yields

Table 19–4. Modeling the multidimensional contingency table in order to relate models of pairwise conjoint actions and group size distributions: 4.1, Table 19.2 censored to show only playing (4) or not playing (−4); 4.2, model of independence for 4.1; 4.3, model for pairwise conjoint actions implied by 4.2; 4.4, model for group sizes implied by 4.2*

4.1

		II(4)	II(−4)		II(4)	II(−4)
I(4)	III(4)	A	z	I(−4)	u	g + n + t
	III(−4)	y	v + w + x		e + k + s	a + b + c + d + f + h + i + j + l + m + p + q + r

4.2

		II(4)	II(−4)
I(4)	III(4)	$\alpha_1\alpha_2\alpha_3$	$\alpha_1(1-\alpha_2)\alpha_3$
	III(−4)	$\alpha_1\alpha_2(1-\alpha_3)$	$\alpha_1(1-\alpha_2)(1-\alpha_3)$
I(−4)	III(4)	$(1-\alpha_1)\alpha_2\alpha_3$	$(1-\alpha_1)(1-\alpha_2)\alpha_3$
	III(−4)	$(1-\alpha_1)\alpha_2(1-\alpha_3)$	$(1-\alpha_1)(1-\alpha_2)(1-\alpha_3)$

4.3

	I	II	III
I	0	$\alpha_1\alpha_2$	$\alpha_1\alpha_3$
II	$\alpha_1\alpha_2$	0	$\alpha_2\alpha_3$
III	$\alpha_1\alpha_3$	$\alpha_2\alpha_3$	0

4.4

Size of groups	Probability that a play (4) group will have the size given
0	$(1-\alpha_1)(1-\alpha_2)(1-\alpha_3)$
1	$\alpha_1(1-\alpha_2)(1-\alpha_3) + (1-\alpha_1)\alpha_2(1-\alpha_3) + (1-\alpha_1)(1-\alpha_2)\alpha_3$
2	$(1-\alpha_1)\alpha_1\alpha_2 + \alpha_1(1-\alpha_2)\alpha_3 + \alpha_1\alpha_2(1-\alpha_3)$
3	$\alpha_1\alpha_2\alpha_3$

*In this table α_i means $\alpha_{i(4)}$, i's probability of playing.

the Poisson and the negative binomial distributions by the same processes as before.

Hence it is possible to go from a single underlying model as represented in part 2 of Table 19–4 to both the model of quasi-independence in pairwise interactions and the same equilibrium distributions of group sizes (Poisson and negative binomial) which are predicted from other (LOST) models. The multidimensional contingency table (Table 19–2) offers a way to unify apparently different quantitative approaches to the analysis of social behavior, and raises an obvious further question (as Stephen E. Fienberg has pointed out privately): What model of the full Table 19–2 will predict quasi-independence for part 2 of Table 19–3 and symmetric quasi-independence for symmetric pairwise activities such as part 3 of Table 19–3? The answer is instantly obvious to neither him nor me.

Section 6. Frequencies of Behavior

Why are the frequencies of behavior what they are? If the frequencies are presented as a summed distribution over the repertoire (e.g. part 4 of Table 19–3), then an initial procedure is to explain the frequencies by deriving them from some probability model of the underlying contingency table (Table 19–2). Thus if one is interested only in the frequency of playing, one could relate the frequency to the underlying probabilities $\alpha_{i(4)}$, $i = $ I, II, III of joining in play. This is only temporarily satisfying, and merely changes the question to: why are the probabilities $\alpha_{i(l)}$, $(l) = (1)$, (2), (3), (4) what they are?

One answer to this question may be based on physiological mechanisms in the animal which are triggered by its environment with a frequency corresponding to the frequency of the behavior. When a clear-cut trigger exists, as in many mating rituals, this answer can be illuminating. It is less satisfactory when animals behave in similar ways in a variety of circumstances and when, under similar circumstances, different individuals and species behave variously.

A response of evolutionary biologists is that when individuals of a species achieve behavioral equilibrium in an environment, the frequency distribution of observed behavioral patterns is that which is most adaptive, in the sense of maximizing Fisher's Malthusian parameter. For example, Altmann and Altmann (1970, pp. 198–201) attempt to explain a baboon troop's allocation of time to different quadrats of a habitat as one which maximizes the sum of the differences between benefits and costs (measured in terms of reproductive success) associated with each quadrat. Testing this approach requires a direct measurement of the benefits and costs associated with each quadrat occupied or behavior pattern observed. Field studies of primates are only beginning to provide such measurements.

However, direct quantitative evidence is available that pigeons, rats, and people in psychological laboratories match their behavior to the rewards for different acts provided by the environment. The following description is based on Herrnstein (1970; and in press).

Suppose a pigeon in a Skinner box has a choice of two disks to peck at. Suppose that for each disk there is a minimum interval of time (possibly different for each disk) such that as soon as the pigeon pecks after that interval it is reinforced with food, but if it pecks before the end of the interval it gains nothing; and suppose this interval fluctuates around some mean. This schedule of reinforcement is called a "variable interval" schedule and usually produces in pigeons a rate of pecking far greater than the rate of reinforcement.

Herrnstein showed that the pigeons matched their pecking to the rewards from each disk. For example, when the average rate of reinforcement from the left disk was 30 reinforcements per hour and the average rate from the right disk was 10 per hour, then each pigeon delivered three-quarters $= 30/(30 + 10)$ of its pecks to the left disk and one-quarter to the right. An extensive series of experiments led to the conclusion (Herrnstein in press) that "responding and reinforcement are related by a constant of proportionality, k. Reinforcement, however, is not measured absolutely, but as a ratio between the reinforcement conditional upon the response and total reinforcement." If $P_{i(l)}$ is animal i's rate of performing acts of category (l), and $R_{(l)}$ is the reinforcement to acts of type (l), then Herrnstein found:

$$P_{i(l)} = \frac{k_i R_{(l)}}{R_{(0)} + R_{(1)} + \ldots + R_{(m)}}, l = 1, \ldots m. \qquad (1)$$

The number of different acts visibly reinforced is m. $R_{(0)}$ is a residual reinforcement for acts not listed in the repertoire, akin to the animal's proclivity for self-generated entertainment without external reinforcement. It follows that for each animal i the probability $\alpha_{i(l)}$ of acts of type (l), $l = 1, 2, \ldots m,$ among all m overtly reinforced acts in the repertoire, is:

$$\alpha_{i(l)} = \frac{P_{i(l)}}{P_{i(1)} + \ldots + P_{i(m)}} = \frac{R_{(l)}}{R_{(1)} + \ldots + R_{(m)}}.$$

Hence there is a matching between relative rates of response and relative reinforcement.

An important implication of expression (1) noted by Herrnstein is:

Contrary to intuition, responding may therefore be more or less indifferent to the reinforcement it produces. If the response's reinforcement is a large part of the total reinforcement, which is to say that the reinforcement ratio in equation [1] is close to 1.0, then responding will stay close to the value of k. And conversely, if the response's reinforcement is only a small fraction of the total reinforcement, then the response will be quite sensitive to variations in its reinforcement, in the limiting case being directly proportional to it.

The confirmation of the generalization summarized by equation (1) in more than 50 experiments suggests that it also might be a valuable aid in interpreting frequencies of behavior outside of Skinner boxes as attempts to match environmental reinforcements.

The baboons' occupancy of different quadrats is not a behavior with a rate, but a choice with duration. An experiment which shows that the same regularity, Equation (1), describes such choices is described by Herrnstein:

Pigeons were given the chance to choose between blue or amber illumination of the experimental chamber. One peck at a disk changed it to amber if it was blue, and vice versa. Every now and then, the pigeon (who was hungry) was given a bit of food irrespective of its responding. The rate of feedings depended on the color of illumination. At any moment, the pigeon could switch from the prevailing color to the other. Except for the lack of a response requirement, the two schedules were variable intervals running concurrently, so that it was advantageous for the pigeons to switch back and forth to collect the feedings that were coming due. . . . the pigeons kept the proportion of time spent in a given color of illumination equal to the proportion of reinforcements obtained therein.

If the various quadrats of a baboon's habitat provide resources in an approximation to a variable-interval schedule, then the observed durations of occupancy of the different quadrats may indeed be matching the net gains from the quadrats, as the Altmanns have suggested. The finding of the experimental psychologists shows that such matching occurs in the laboratory and defines the precise form of the distribution of net benefits which should be looked for in the field. That the net gain from different quadrats, or from different patterns of behavior in the field, in fact matches the duration of occupancy or frequency of performance, remains unproved.

Section 7. Math and Aftermath

In addition to those already mentioned, other mathematical techniques are ripe for a fruitful union with empirical techniques and observations within the next few years. Moreover, the growth of mathematical models peripheral, but nonetheless related, to primatology may offer suggestive leads.

One area of primatology ready for a union of existing models and data is the study of dominance. Many data of varying degrees of detail (Kawai 1958; Itani et al. 1963; Mizuhara 1964; Bernstein and Draper 1964; Bernstein 1968b, 1969, 1970; Uyeno 1967; Delgado 1967; Vandenbergh 1967; Sade 1969a; and others) have never been compared systematically with theory (e.g. Landau 1968; Boorman 1970, and in preparation). Mathematical models recently developed in the theories of preference and measurement (Roberts 1969) seem readily translatable into sociological situations where transitivity is absent. But their relevance remains unexplored. Structural models of roles in human societies may well offer insight into primate social relations when simple dominance models fail (Lorrain and White 1971).

Second, the population genetics of primate troops is now becoming accessible to study through the conjoint development of improved methods of assessing and understanding the genetic status of individuals (Court-Brown 1967, Harris 1970), improved models of small, nonrandomly breeding populations (Karlin 1969), and long-term studies of primate populations including observations on obvious genetic malformations (Itani et al. 1963, pp. 29–36) and genealogy (Sade, Chapter 17).

Third, the study of primate diseases (Cockburn 1963, Fiennes 1967, Bray 1968) offers an excellent testing ground for theories of parasitism, prudent predators (parasites), and the mutual adaptation of host and parasite. Mathematical models of the transmission of malaria, for example (Macdonald 1957), require primate studies when monkeys as well as men belong to the reservoir of potential hosts (Contacos and Collins 1969). The same conclusion applies to mathematical models of schistosomiasis (Goffman and Warren 1970) for the same reason (Miller 1960).

Several potentially applicable developments in peripheral areas related to primatology have already been mentioned. Others have been reviewed in the fields of social psychology (Abelson 1967, Whitla 1968), sociology (Coleman 1964a, 1964b), ecology (Watt 1968, Pielou 1969), political science (Bernd 1966), and stochastic models (Bartlett 1960, Bartholomew 1967). Recent flowering of the analysis of incomplete demographic data has immediate utility in primate studies, a field in which observations are often fragmentary (United Nations 1967).

What is required of data collectors and of data analyzers in order to promote full and rapid employment of available mathematical models, and development of new mathematical models, in primate behavior, sociology, and ecology?

As for the data generator, I subscribe to the dictum which E. O. Wilson enunciated at an informal Harvard seminar: the state of biology is such that a person at any level of mathematical sophistication can make substantial contributions. Nevertheless, the magnitude and effectiveness of those contributions can be improved by sufficient mathematical self-confidence to scan theoretical literature critically and to search for conclusions or predictions relevant to field work.

A simple example illustrates how an appreciation of the models which underlie field techniques can improve the use of field data. A familiar method of estimating the density of points in a plane is to choose random points and measure the distance from each to its nearest neighbor. The density of points may be estimated from the square of the mean of the distances or from the mean of the squares of the distances. Kendall and Moran (1963, p. 38) have shown that it is more efficient (i.e., gives an estimate with smaller variance) to use the latter estimate than the former. Hence following the inattentive routine of just presenting the average of nearest-neighbor measurements would not be taking full advantage of available information. Such mishandling of field data can destroy valuable information in them.

I would urge field primatologists to publish as full and detailed accounts of systematically collected data, along with the methods of collection, as possible. And I would urge them to encourage their students to have calculus through ordinary differential equations, probability theory, and some modern algebra. Properly taught, such courses can lay the foundation for confident consultation with active modelers.

The mathematical modeler also needs to gain an appreciation of what is important and what is feasible in primate field studies. He can gain such appreciation by participating actively in primate field studies, or by consulting frequently and for a long time with field workers, or by trusting his armchair insight and luck. The first option is best. The armchair is an ideal vantage point from which to overlook the obvious and make impossible demands: when estimating density by choosing random points and measuring the distance to the nearest neighbor, the field worker must have alternative procedures if the random point is in a patch of six-foot-tall grass occupied by an elephant.

A warning issued to physicists by Bridgman (1927, p. 209) has equal relevance to those who construct and take seriously mathematical models in the life sciences:

> There is an aspect here of our physical research that is often lost sight of, namely, the small proportion of successful discoveries compared with the number of investigators. Certainly the number of unsuccessful attempts, even in the case of those fortunate individuals who make the great discoveries, is very much greater than the number of their successful attempts. (Faraday's reputed satisfaction with a 1/10% return comes to mind.) This must always be taken into account in estimating the probable chances of correctness of any new theory. With so many physicists working to devise new theories, the chances are high that many false theories will be found, in which a number of phenomena may apparently fit together into a new relation, but which eventually prove to be inconsistent with other phenomena, so that the proposed theory has to be abandoned. As physics advances and the number of investigators and the amount of physical material increases, one has to be more and more exacting in one's requirements of a new theory.

References

Abelson, R. P. 1967. Mathematical models in social psychology. *Adv. in Exper. Soc. Psych.* 2:1–54.

Allee, W. C. 1926. Distribution of animals in a tropical rain-forest with relation to environmental factors. *Ecology* 7:445–468.

Allen, G. M. 1938. *The mammals of China and Mongolia.* Part 1, pp. 279–293. New York: American Museum of Natural History.

Altmann, S. A. 1959. Field observations on a howling monkey society. *Jour. of Mammal.* 40:317–330.

————1962. A field study of the sociobiology of rhesus monkeys, *Macaca mulatta. Ann. N.Y. Acad. Sci.* 102 (art. 2):338–435.

————1965. Sociobiology of rhesus monkeys. Part 2. *Jour. of Theor. Biol.* 8:490–522.

————1967. The structure of primate social communication. In *Social communication among primates.* Altmann, S. A., ed. Chicago: University of Chicago Press.

————1968. Sociobiology of rhesus monkeys. Parts 3, 4. *Behav.* 22:17–32, 49–69.

Altmann, S. A. and Altmann, J. 1970. *Baboon ecology: African field research.* Basel: S. Karger.

Altmann, S. A. and Wagner, S. S. 1970. Estimating rates of behavior from Hansen frequencies, *Primates:* 11:181–183.

Andrew, R. J. 1964. Displays of the primates. In *Evolutionary and genetic biology of primates* vol. 2, pp. 227–309. Buettner-Janusch, J. ed. New York: Academic Press.

Andrews, C. W. 1906. *A descriptive catalogue of the Tertiary vertebrata of the Fayûm, Egypt.* London: British Museum.

Andrews, H. C. 1970. Multidimensional rotations in feature selection. In *IEEE Conf. Rec. of the symp. on feature recognition and selection in pattern recognition,* pp. 9–19. Yau, S. S. and Garnett, J. M., eds. New York: Inst. of Elect. and Electr. Eng.

Andrews, H. C. and Pratt, W. K. 1969. Transform image coding. In *Symposium on computer processing in communications,* pp. 63–84. Brooklyn: Polytechnic Inst.

Andrews, H. N., Jr. 1961. *Studies in paleobotany.* New York: John Wiley & Sons.

Andrews, P. 1970. Two new fossil primates from the Lower Miocene of Kenya. *Nature* 228:537.

Andy, O. J. and Stephan, H. 1968. The septum in the human brain. *Jour. Comp. Neurol.* 133:383–409.

Ankel, F. 1962. Vergleichende Untersuchungen über die Skelettmorphologie des Greifschwanzes südamerikanischer Affen. *Z. Morph. Oekol. Tiere* 52:131–170.

————1965. Der Canalis sacralis als Indikator für die Länge der Caudalregion der Primaten. *Folia primat.* 3:263–276.

————1967. Morphologie von Wirbelsäule und Brustkorb. *Primatologia* vol. 4(4):1–120.

————1966/67. Morphologische Spezialisationen der menschlichen Wirbelsäule. *Bull. Schweiz. Ges. Anthrop. Ethnol.* 43:70–81.

Anthony, J. 1946. Morphologie externe du cerveau des Singes Platyrhiniens. *Ann. Sci. Nat. (Zool.)* 8:1–150.

Aquino, C. F. 1970. A dynamic model of the lumbar spine. *Jour. Biomech.* 3:473–486.

Arambourg, C. 1945. Au sujet de l'*Hippopotamus hipponensis* Gaudry. *Bull. Soc. géol. Fr.* (Paris) 5(14):147.

————1959. Vertebres continentaux du Miocène Supérioure de l'Afrique du Nord. *Publ. Surv. Géol. Algerie* Pal. 4:1.

Arao, T. and Perkins, E. 1969. The skin of primates 43. Further observations on the Philippine tarsier (*Tarsius syrichta*). *Amer. Jour. Phys. Anthrop.* 31:93–96.

Ashby, R. 1960. *Design for a brain.* 2nd ed. New York: John Wiley & Sons.

————1962. Principles of the self-organizing system. In *Principles of self-organization.* Foerster, H. von and Zoph, G. W., eds. New York: Pergamon Press.

Ashton, E. H., Flinn, R. M., Oxnard, C. E., and Spence, T. F. 1971. The functional and classificatory significance of combined metrical features of the primate shoulder girdle. *Jour. Zool.* 163:319–350.

Ashton, E. H., Healy, M. J. R., and Lipton, S. 1957. The descriptive use of discriminant functions in physical anthropology. *Proc. Roy. Soc. B.* 146:552–572.

Ashton, E. H., Healy, M. J. R., Oxnard, C. E., and Spence, T. F. 1965. The combination of locomotor features of the primate shoulder girdle by canonical analysis. *Jour. Zool.* 147:406–429.

Ashton, E. H. and Oxnard, C. E. 1963. The musculature of the primate shoulder. *Trans. Zool. Soc. Lond.* 29:553–650.

————1964. Functional adaptations in the primate shoulder girdle. *Proc. Zool. Soc. Lond.* 142:49–66.

Ashton, E. H. and Spence, T. F. 1958. Age changes in the cranial capacity and foramen magnum of hominoids. *Proc. Zool. Soc. Lond.* 130:169–181.

Asmussen, E. and Klausen, K. 1962. Form and function of the erect human spine. *Clin. Orthop.* 25:55-63.

Avis, V. 1962. Brachiation: the crucial issue for man's ancestry. *Southw. Jour. Anthrop.* 18:119–148.

Azuma, S. and Toyoshima, A. 1961–62. Progress report of the survey of chimpanzees in their natural habitat, Kabogo Point Area, Tanzanyika. *Primates* 3(2):61–70.

Baldwin, J. D. 1968. The social behavior of adult male squirrel monkeys (*Saimiri sciureus*) in a seminatural environment. *Folia primat.* 9:281–314.

————1969. The ontogeny of social behavior of squirrel monkeys (*Saimiri sciureus*) in a seminatural environment. *Folia primat.* 11:35–79.

Bales, R. F. 1955. How people interact in conferences. *Scient. Amer.* 192:31–35.

————1968. Interaction process analysis. *Inter. Encyc. of the Soc. Sci.* 7:465–471. New York: Macmillan.

Banks, E. 1931. A popular account of the mammals of Borneo. *Jour. Malay. Branch Roy. Asiatic Soc. (Singapore)* 9(2):1–139.

Bartholomew, D. J. 1967. *Stochastic models for social processes.* New York: John Wiley & Sons.

Bartlett, M. S. 1960. *Stochastic population models in ecology and epidemiology.* New York: John Wiley & Sons.

Basmajian, J. V. 1965. *Muscles alive: their functions revealed by electromyography.* Baltimore: Williams & Wilkins.

————1967. *Muscles alive: their functions revealed by electromyography.* 2nd ed. Baltimore: Williams & Wilkins.

Basmajian, J. V. and Bazant, F. J. 1959. Factors preventing downward dislocation of the adducted shoulder joint: an electromyographic and morphological study. *Jour. Bone and Joint Surg.* 41-A:1182–1186.

Basmajian, J. V., Forrest, W. J., and Shine, G. 1966. A simple connector for fine-wire electrodes. *Jour. Appl. Physiol.* 21:1980.

Basmajian, J. V. and Greenlaw, R. K. 1968. Electromyography of iliacus and psoas with inserted fine-wire electrodes. *Anat. Rec.* 160:310–311.

Basmajian, J. V. and Stecko, G. 1962. A new bipolar indwelling electrode for electromyography. *Jour. Appl. Physiol.* 17:849.

————1963. Role of muscles in arch support of the foot. *Jour. Bone and Joint Surg.* 45-A:1184–1190.

Basmajian, J. V. and Travill, A. 1961. Electromyography of the pronator muscles in the forearm. *Anat. Rec.* 139:45–49.

Bassett, C. A. L. and Becker, R. O. 1962. Generation of electric potentials by bone in response to mechanical stress. *Science* 137:1063–1064.

Battye, C. K. and Joseph, J. 1966. An investigation by telemetering of the activity of some muscles in walking. *Med. and Biol. Eng.* 4:125–135.

Bauchot, R. and Stephan, H. 1964. Le poids encéphalique chez les Insectivores Malgaches. *Acta Zool.* 45:63–75.

————1966. Données nouvelles sur l'encéphalisation des Insectivores et des Prosimiens. *Mammalia* 30:160–196.

————1967. Encéphales et moulages endocraniens de quelques insectivores et primates actuels. Problèmes actuels de paléontologie. *Colloq. inter. Centre Nat. Rech. Sci.* 163:575–586.

————1968. Etude des modifications encéphaliques observées chez les Insectivores adaptés à la recherche de nourriture en milieu aquatique. *Mammalia* 32:228–275.

————1969. Encéphalisation et niveau évolutif chez les simiens. *Mammalia* 33:225–275.

Beadnell, H. J. L. 1905. *The topology and geology of the Fayûm province of Egypt.* Cairo: Surv. Dept.

Becker, H. C., Meyers, P. H., and Nice, C. M., Jr. 1969. Laser light diffraction, spatial filtering, and reconstruction of medical radiographic images: preliminary results. In *Data extraction and processing of optical images in the medical and biological sciences,* pp. 465–486. Tolles, W. E., ed. *Ann. N.Y. Acad. Sci.* 157.

Benjamin, R. M. and Welker, W. I. 1957. Somatic receiving areas of cerebral cortex of squirrel monkey (*Saimiri sciureus*). *Jour. Neurophysiol.* 20:286–299.

Bennett, K. A. 1965. The etiology and genetics of wormian bones. *Amer. Jour. Phys. Anthrop.* 23:255–260.

Bernd, J. L., ed. 1966. *Mathematical applications in political science.* Arnold Foundation Monograph 16. Dallas: Arnold Foundation.

Bernstein, I. S. 1964. A field study of the activities of howler monkeys. *Anim. Behav.* 12:84–91.

————1966. Analysis of a key role in a capuchin (*Cebus albifrons*) group. *Tulane Studies in Zool.* 13(2):49–54.

————1967a. A field study of the pigtail monkey (*Macaca nemestrina*). *Primates* 8:217–228.

————1967b. Defining the natural habitat. In *Progress in primatology,* pp. 176–179. Starck, D., Schneider, R., and Kuhn, H.-J., eds. Stuttgart: Gustav Fischer Verlag.

————1967c. Intertaxa interactions in a primate community. *Folia primat.* 7:198–207.

————1968a. The lutong of Kuala Selangor. *Behav.* 32:1–16.

————1968b. Social status of two hybrids in a wild troop of *Macaca irus. Folia primat.* 8:121–131.

————1969. Stability of the status hierarchy in a pigtail monkey (*Macaca nemestrina*) group. *Anim. Behav.* 17:452–458.

————1970. Primate status hierarchies. In *Primate behavior (developments in field and laboratory research)* vol. 1, pp. 71–109. Rosenblum, L. A., ed. New York: Academic Press.

————1971. Activity profiles of primate groups. In *Behavior of nonhuman primates* vol. 3. pp. 69–106. Schrier, A. M., and F. Stollhitz eds. New York: Academic Press.

Bernstein, I. S. and Draper, W. A. 1964. Behaviour of juvenile rhesus monkeys in groups. *Anim. Behav.* 12(1):84–91.

Bernstein, I. S. and Mason, W. A. 1963. Activity patterns of rhesus monkeys in a social group. *Anim. Behav.* 11:455–460.

Bernstein, I. S. and Sharpe, I. G. 1966. Social roles in a rhesus monkey group. *Behav.* 26:(1–2):91–104.

Biegert, J. 1957. Der Formwandel des Primatenschädels und seine Beziehungen zur ontogenetischen Entwicklung und den phylogenetischen Spezialisationen der Kopforgane. *Morph. Jb.* 98:77–199.

————1963. The evaluation of characteristics of the skull, hands, and feet for primate taxonomy. In *Classification and human evolution,* pp. 116–145. Washburn, S. L., ed. Chicago: Aldine Publ. Co.

Bishop, A. 1962. Control of the hand in lower primates. *Ann. N.Y. Acad. Sci.* 102(2):316–337.

————1964. Use of the hand in lower primates. In *Evolutionary and genetic biology of primates* vol. 2, pp. 133–225. Buettner-Janusch, J., ed. New York: Academic Press.

Bishop, W. W., Miller, J. A., and Fitch, F. J. 1969. New potassium-argon age determinations relevant to the Miocene fossil mammal sequence in East Africa. *Amer. Jour. Sci.* 267:669.

Bishop, Y. M. M. 1971. Effects of collapsing multidimensional contingency tables. *Biometrics.* 27:545–562.

Bishop, Y. M. M. and Fienberg, S. E. 1969. Incomplete two-dimensional contingency tables. *Biometrics* 22:119–128.

Blankenhorn, M. L. P. 1921. Aegypten. In *Handbook of regional geology* 7(9):1. Heidelberg: C. Winter.

Blackith, R. E. and Kevan, D. K. McE. 1967. A study of the genus *Chrotogonus* (Orthoptera), 8: Patterns of variation in external morphology. *Evolution* 21:76–84.

Blinkov, S. M. and Glezer, I. I. 1968. *The human brain in figures and tables.* New York: Plenum Press, Basic Books, Inc.

Bock, R. D. and Haggard, E. A. 1968. *The use of multivariate analysis of variance in behavioral sciences.* Whitla, D. K., ed. Reading, Mass.: Addison-Wesley Publ. Co., Inc.

Bollinger, A. and Hardy, M. H. 1945. The sternal integument of *Trichosurus vulpecula. Jour. Roy. Soc. N.S.W.* 78:122–133.

Bonin, G. von 1963. *The evolution of the human brain.* Chicago: University of Chicago Press.

Boorman, S. A. 1970. *Metric spaces of complex objects.* Senior honor thesis. Cambridge, Mass.: Division of Engineering and Applied Physics, Harvard College.

————In preparation. *Asymptotic models of hierarchy and dominance.*

Booth, C. 1962. Some observations on behavior of Cercopithecus monkeys. *Ann. N.Y. Acad. Sci.* 102:477–488.

Boyce, A. J. 1964. The value of some methods of numerical taxonomy with reference to hominoid classification. In *Phenetic and phylogenetic classification*, pp. 47–65. Heywood, V. H. and McNeill, J., eds. The Systematics Association, publ. no. 6.

————1969. Mapping diversity: a comparative study of some numerical methods. In *Numerical taxonomy*, pp. 1–31. Cole, A. J., ed. London: Academic Press.

Brace, C. L. 1969. The australopithecine range of variation. In *Proc. 38th Ann. Meeting of AAPA.* Abstr.: *Amer. Jour. Phys. Anthrop.* 31:255.

Brace, C. L. and Montagu, M. F. A. 1965. *Man's evolution: an introduction to Physical Anthropology.* New York: Macmillan.

Brain, C. K. 1967a. The Transvaal Museum's fossil project at Swartkans. *S. Afr. Jour. Sci.* 63:368–384.

————1967b. Bone weathering and the problem of bone pseudo-tools. *S. Afr. Jour. Sci.* 63:97–99.

————1967c. Hottentot food remains and their bearing on the interpretation of fossil bone assemblages. *Scient. Pap. Namib Desert Res. Stn.* no. 32:1–11.

————1968. Who killed the Swartkrans ape men? *S. Afr. Mus. Assoc. Bull. (SAMAB)* 9:127–139.

————1969. The probable role of leopards as predators of the Swartkrans australopithecines. *S. Afr. Archaeol. Bull.* 24:170–171.

Bray, R. S. 1968. Zoonotic potential of blood parasites. In *Infectious blood diseases of man and animals* 1. Weinman, D. and Ristic, M., eds. New York: Academic Press.

Bridgman, P. W. 1927. *The logic of modern physics.* New York: Macmillan.

Broca, P. 1873. Quelques resultats de la détermination trigonometrique de l'angle alvéolo-condylien et de l'angle biorbitaire. *Bull. Soc. Anthrop. Paris,* ser. 2, 8:150–179.

Brown, T. 1967. *Skull of the Australian aboriginal: a multivariate analysis of craniofacial associations.* Adelaide: Department of Dental Science, University of Adelaide.

Bunak, V. V. 1969. Sur l'évolution de la forme du crâne humain. *Symp. Biol. Hungarica* 9:51–63.

Burton, M. 1962. *Systematic dictionary of mammals of the world.* London: Museum Press Limited.

Butler, J. W. 1968. Automatic analysis of bone autoradiographs. In *Pictorial pattern recognition,* pp. 75–85. Cheng, G. C., ed. Washington: Thompson.

Campanella, S. J. and Robinson, G. S. 1970. Digital sequency decomposition of voice signals. In *Proceedings of symposium on Applications of Walsh Functions,* pp. 230–237. Bass, C. A., ed. Naval Research Laboratory.

Campbell, B. G. 1966. *Human evolution: an introduction to man's adaptations.* Chicago: Aldine Publ. Co.

———1967. *Human evolution.* London: Heinemann.

Carlsöö, S. 1964. Influence of frontal and dorsal loads on muscle activity and on the weight distribution in the feet. *Acta orthop. Scandinav.* 34:299–309.

Carpenter, C. R. 1934. A field study of the behavior and social relations of howling monkeys. *Comp. Psych. Monog.* 10(2).

———1939. Behavior and social relations of free-ranging primates. *Sci. Mon. N.Y.* 48:319–325.

———1940. A field study in Siam of the behavior and social relations of the gibbon *(Hylobates lar). Comp. Psych. Monog.* 16(5).

———1942a. Societies of monkeys and apes. *Biol. Symp.* 8:177–204.

———1942b. Characteristics of social behavior in nonhuman primates. *Trans. N.Y. Acad. Sci.* 4:248–258.

———1953. Grouping behavior of howling monkeys. *Extrait des Arch. Neerlandaises de Zoologie* 10:45–50.

———1954. Tentative generalizations on the grouping behavior of nonhuman primates. *Human Biol.* 26(3):269–276.

———1964a. A field study of the behavior and social relations of howling monkeys *(Alouatta palliata).* In Carpenter, C. R., *Naturalistic behavior of nonhuman primates,* pp. 3–92. University Park: Pennsylvania State University Press.

———1964b. Behavior of red spider monkeys in Panama. In Carpenter, C. R., *ibid.,* pp. 93–105.

———1964c. A field study in Siam of the behavior and social relations of the gibbon *(Hylobates lar).* In Carpenter, C. R., *ibid.,* pp. 145–271.

Cartmill, M. 1970. *The orbits of arboreal mammals: a reassessment of the arboreal theory of primate evolution.* Ph.D. dissertation, The University of Chicago.

Cartmill, M. and Tuttle, R. 1966. Mammalian social patterns in a savanna environment. *Amer. Jour. Phys. Anthrop.* 25:202.

Caspari, E. 1961. Some genetic implications of human evolution. In *Social life of early man,* pp. 267–277. Washburn, S. L., ed. Viking Fund Publ. in Anthrop. no. 31.

Cave, A. J. E. 1967. Observations on the platyrrhine nasal fossa. *Amer. Jour. Phys. Anthrop.* 26:277–288.

Chalmers, N. R. 1968. Group composition, ecology, and daily activities of free-living mangabeys in Uganda. The social behavior of free-living mangabeys in Uganda. *Folia primat.* 8:247–262, 263–287.

Charles-Dominique, P. and Martin, R. D. 1970. Evolution of lorises and lemurs. *Nature* 227:257–260.

Chivers, D. J. 1969. On the daily behaviour and spacing of howling monkey groups. *Folia. primat.* 10:48–102.

Clark, J. D., Haynes, C. V., and Mawby, J. E. 1971. Interim report on palaeoanthropological investigations in the Lake Malawi Rift. *Proc. 6th Pan-African Congress on prehistory and quaternary studies,* Dakar, 1967.

Clark, J. D., Stephens, E. A., and Coryndon, S. C. 1966. Pleistocene fossiliferous lake beds of the Malawi (Nyasa) Rift: a preliminary report. *Amer. Anthrop. Spec. Publ.: Recent studies in palaeo-anthropology* 68(2):48–87. Clark, J. D. and Howell, F. C., eds.

Clark, W. E. LeGros 1924. Notes on the living tarsier (*Tarsius spectrum*). *Proc. Zool. Soc. Lond.* 1924:217–233.

————1945. Note on the paleontology of the lemuroid brain. *Jour. Anat.* 79:123–126.

————1947. Observations on the anatomy of the fossil australopithecineae. *Jour. Anat.* 81:300.

————1957. *History of the primates.* Chicago: University of Chicago Press.

————1959. *The antecedents of man: an introduction to the evolution of the primates.* Chicago: Quadrangle Books.

————1962. *The antecedents of man.* 2nd ed. Edinburgh: University Press.

————1964. *The fossil evidence for human evolution.* 2nd ed. Chicago: University of Chicago Press.

Clark, W. E. LeGros, Cooper, D. M., and Zuckerman, S. 1936. The endocranial cast of the chimpanzee. *Jour. Roy. Anthrop. Inst. Gr. Brit.* 66:249–268.

Clark, W. E. LeGros and Leakey, L. S. B. 1951. The Miocene Hominoidea of East Africa. *Foss. Mamm. Af., Brit. Mus.* (*Nat. Hist.*) 1:1.

Clark, W. E. LeGros and Thomas, D. P. 1951. Associated jaws and limb bones of Limnopithecus macinnesi. *Foss. Mamm. Af., Brit. Mus.* (*Nat. Hist.*) 3:1.

Coker, E. G. and Filon, L. N. G. 1957. *A treatise on photo-elasticity.* Cambridge: University Press.

Cockburn, A. 1963. *The evolution and eradication of infectious diseases.* Baltimore: Johns Hopkins.

Cohen, J. E. 1970. Men, models, and mathematics. *Harvard Bulletin* 72(11):19–23.

————1971. *Casual groups of monkeys and men: stochastic models of elemental social systems.* Cambridge: Harvard University Press.

————In press. Social grouping and troop size in yellow baboons. *Proc. 3rd Inter. cong. of primat.,* Zurich 1970. Basel: S. Karger.

Coleman, J. S. 1964a. Mathematical models and computer simulation. In *Handbook of modern sociology.* Faris, R. E. L., ed. Chicago: Rand McNally.

————1964b. *Introduction to mathematical sociology.* New York: Free Press.

Collias, N. and Southwick, C. 1952. A field study of population density and social organization in howling monkeys. *Proc. Amer. Phil. Soc.* 96:143–156.

Collins, E. T. 1921. Changes in the visual organs correlated with the adoption of arboreal life and with the assumption of the erect posture. *Trans. Ophthalm. Soc. U.K.* 41:10–90.

Conaway, C. H. and Sade, D. S. 1965. The seasonal spermatogenic cycle in free-ranging rhesus monkeys. *Folia primat.* 3:1–12.

Conel, J. L. 1939–63. *The postnatal development of the human cerebral cortex* vols. 1–7. Cambridge: Harvard University Press.

Connolly, C. J. 1950. *External morphology of the primate brain.* Springfield, Ill.: C. C. Thomas.

Contacos, P. G. and Collins, W. E. 1969. *Plasmodium malariae:* transmission from monkey to man by mosquito bite. *Science* 165:918–919.

Cooke, H. B. S. 1963. Pleistocene mammal faunas of Africa, with particular reference to Southern Africa. In *African ecology and human evolution,* pp. 65–116. Howell, F. C. and Bourliere, F., eds. Viking Fund Publ. in Anthrop.

————1967. The Pleistocene sequence in South Africa and problems of correlation. In *Background to evolution in Africa*, pp. 175–184. Bishop, W. W. and Clark, J. D., eds. Chicago: University of Chicago Press.

Cooley, W. W. and Lohnes, P. R. 1962. *Multivariate procedures for the behavioral sciences*. New York: John Wiley & Sons.

Coon, C. S. 1962. *The origin of races*. New York: Knopf.

———1963. *The origin of races*. London: Jonathan Cape.

Coon, C. S., Garn, S. M. and Birdsell, J. B. 1950. *Races: a study of the problems of race formation in man*. Springfield, Ill.: C. C. Thomas.

Count, E. W. 1958–59. Eine biologische Entwiklungsgeschichte der menschliche Sozialitat. *Homo* 9:129–146; 10:1–35, 65–92.

Court-Brown, W. M. 1967. *Human population genetics*. New York: John Wiley & Sons.

Crompton, A. W. and Hiiemae, K. 1970. Molar occlusion and mandibular movements during occlusion in the American opossum, *Didelphis marsupialis* L. *Zool. Jour. Lin. Soc.* 49:21–47.

Crook, J. H. 1966. Gelada baboon herd structure and movement; a comparative report. *Symp. Zool. Soc. Lond.* 18:237–258.

———1967. Evolutionary change in primate societies. *Science Jour.* 3(6):66–70.

———1970. Social organization and the environment: aspects of contemporary social ethology. *Anim. Behav.* 18:197–209.

Crook, J. H. and Aldrich-Blake, P. 1968. Ecological and behavioural contrasts between sympatric grounddwelling primates in Ethiopia. *Folia primat.* 8:192–227.

Crook, J. H. and Gartlan, J. S. 1966. Evolution of primate societies. *Nature* 210 (5042):1200–1203.

Crusafont-Pairo, M. 1965. El desarrollo de los caninos en algunos driopitecidos del Vallesiense en Catuluña. *Not. y. Com. Inst. Geol. y. Min de España* 80:179.

Currey, J. D. 1968. The adaptation of bones to stress. *Jour. Theoret. Biol.* 20:91–106.

Dabelow, A. 1929. Über Korrelationen in der phylogenetischen Entwicklung der Schädelform. 1. Die Beziehungen zwischen Rumpf und Schädelform. *Morph. Jb.* 63:1–49.

Daitz, H. 1953. Note on the fibre content of the fornix system in man. *Brain* 76:509–512.

Dart, R. A. 1926. Taungs and its significance. *Nat. Hist.* 26:315–327.

———1964. The ecology of the South African man-apes. In *Ecological studies in Southern Africa*, pp. 49–66. Davis, D. H. S., ed. The Hague: Junk.

Davis, D. D. 1955. Primate evolution from the viewpoint of comparative anatomy. In *The nonhuman primates and human evolution*, pp. 33–41. Gavan, J. A., ed. Detroit: Wayne State University Press.

———1962. Mammals of the lowland rain-forest of North Borneo. *Bull. Natn. Mus. St. Singapore* 31:1–129.

Davis, J. C. 1970. Optical processing of microporous fabrics. In *Data Processing in Biology and Geology*, pp. 69–87. Cutbill, J. L., ed. Systematics Association Special Volume 3.

Davis, R. T., Leary, R. W., Smith, C., Dell, M., and Thompson, R. F. 1968. Species differences in the gross behavior of nonhuman primates. *Behav.* 31(3–4):326–338.

Day, M. H. 1967. Olduvai hominid 10: a multivariate analysis. *Nature* 215:323–324.

Day, M. H. and Napier, J. R. 1966. A hominid toe bone from Bed I, Olduvai Gorge, Tanzania. *Nature* 211:929–930.

Day, M. H. and Wood, B. A. 1969. Hominoid tali from East Africa. *Nature* 222:591.

Delattre, M. A. 1958. La formation du crâne humain. In *Le processus de l'hominisation*, pp. 37–57. Paris: Colloques Internationaux du Centre National de la Recherche Scientifique.

Delgado, J. M. R. 1967. Social rank and radio-stimulated aggressiveness in monkeys. *Jour. Nerv. and Mental Disease* 144(5):383–390.

Delmas, A. and Pineau, H. 1959. Les poids des vertèbres comme élément significatif de leur interdépendence fonctionelle. *Compt. Rend. Assoc. Anat.* XLVIe:214–223.

Delmas, A., Pineau, H. and Kyncl, H. 1958. Signification fonctionelle du poids du rachis. *Compt. Rend. Assoc. Anat.* XLVe:810–820.

Dempster, A. P. 1969. *Elements of continuous multivariate analysis.* Reading, Mass.: Addison-Wesley.

Deutsch, J. 1966. *Analysis of positional behavior in the rhesus monkey (Macaca mulatta).* Technical Report no. 66–1. Tulane University, Delta Regional Primate Research Center.

DeVilliers, H. 1968. *The skull of the South African Negro: a biometrical and morphological study.* Johannesburg: Witwatersrand University Press.

DeVore, I. 1963. A comparison of the ecology and behavior of monkeys and apes. In *Classification and human evolution*, pp. 301–319. Washburn, S. L., ed. Viking Fund Publ. in Anthrop. no. 37. New York: Wenner-Gren Foundation.

DeVore, I. and Hall, K. R. L. 1965. Baboon ecology. In *Primate behavior. Field studies of monkeys and apes*, pp. 20–52. DeVore, I., ed. New York: Holt, Rinehart and Winston.

DeVore, I. and Washburn, S. L. 1963. Baboon ecology and evolution. In *African ecology and human evolution*, pp. 355–367. Howell, F. C. and Bourliere, F., eds. Chicago: Aldine Publ. Co.

Diamond, I. T. and Hall, W. C. 1969. Evolution of neocortex. *Science* 164:251–262.

Dobzhansky, Th. 1944. On species and races of living and fossil man. *Amer. Jour. Phys. Anthrop.* 2:251–265.

Dorr, J. A. 1952. Early Cenozoic stratigraphy and vertebrate paleontology of the Hoback Basin, Wyoming. *Bull. Geol. Soc. Amer.* 63:59–94.

Dubois, E. 1897. Über die Abhängigkeit des Hirngewichts von der Körpergrösse bei den Säugetieren. *Arch. Anthrop.* 25:1–28.

DuMond, F. V. and Hutchinson, T. 1967. Squirrel monkey reproduction: the "fatted" male phenonenon and seasonal spermatogenesis. *Science* 158:1467–1470.

Edinger, T. 1938. Mitteilungen über Wirbeltierreste aus dem Mittelpliocän des Natrontales (Aegypten). 9. Das Gehirn des *Libypithecus*. *Zentralblatt f. Min., Jahrg. Abt. B*, no. 4:122–128.

——1961. Antropocentric misconceptions in paleoneurology. *Proc. Rud. Virchow Med. Soc. N.Y.* 19:56–107.

Eggeling, H. 1896. Zur morphologie der dammuskulatur. (Schluss.) *Gegenbaur's Morph. Jahrb.* 24:511–631. 768–774.

Ehrenberg, A. S. C. 1962. Some questions about factor analysis. *The Statistician* 12:191–208.

Eisenberg, J. 1969. Field observations on the behavior of the Ceylon leopard, *Panthera pardus fusca. 11th Inter. Etholog. Conf.*, Rennes, France.

Eisenberg, J. F. and Kuehn, R. E. 1966. The behavior of *Ateles geoffroyi* and related species. *Smithson. misc. coll.* 151(8):1–63.

Elftman, H. O. 1929. Functional adaptations of the pelvis in marsupials. *Bull. Amer. Mus. Nat. Hist.* 58:189–232.

————1932. The evolution of the pelvic floor of primates. *Amer. Jour. Anat.* 51:307–346.

Ellefson, J. O. 1968. Territorial behavior in the common white-handed gibbon, *Hylobates lar Linn.* In *Primates,* pp. 180–199. Jay, P. C., ed. New York: Holt, Rinehart and Winston.

Emlen, J. T. and Schaller, G. 1960. In the home of the mountain gorilla. In *Animal kingdom: Bull. N.Y. Zool. Soc.* 63:98–108.

Enders, R. K. 1935. Mammalian life histories from Barro Colorado Island, Panama. *Bull. Mus. Comp. Zool. (Harvard)* 78(4):383–502.

Engen, C. W. 1970. *Faunal behavior at Diaz Tinaja, Organ Pipe Cactus National Monument, Arizona.* Unpublished senior thesis, Thomas Jefferson College, Grand Valley State College.

Epple, G. and Lorenz, R. 1967. Vorkommen, Morphologie und Function der Sternaldrüse bei den Platyrrhini. *Folia primat.* 7:98–126.

Erikson, G. E. 1963. Brachiation in New World monkeys and in anthropoid apes. *Symp. Zool. Soc. Lond.* 10:135–164.

Etkin, W. 1954. Social behavior and the evolution of man's mental faculties. *Amer. Natural.* 88:129–143.

————1963. Social behavioral factors in the emergence of man. *Human Biol.* 35:299–310.

Evans, F. G. 1957. *Stress and strain in bones.* Springfield, Ill.: C. C. Thomas.

Eyre, S. R. 1963. *Vegetation and soils.* Chicago: Aldine Publ. Co.

Farmer, B. 1957. Ceylon. In *India and Pakistan,* pp. 745–786. Spate, O., ed. London: Methuen.

Feinstein, B., Lindegard, B., Nyman, E. and Wohlfart, G. 1955. Morphological studies of motor units in normal human muscles. *Acta anat.* 23:127–142.

Fick, R. 1895. Vergleichend anatomische studien an einem erwaschsenen orangutan. *Arch. Anat. und Physiol., Anat. Abth.,* 1–96.

Fienberg, S. E. 1969. Preliminary graphical analysis and quasi-independence for two-way contingency tables. *Jour. Roy. Stat. Soc. C* (Applied Stats.) 18:153–168.

————1970. The analysis of multidimensional contingency tables. *Ecology* 51(2):419–433.

Fiennes, R. 1967. *Zoonoses of primates.* Ithaca: Cornell University Press.

Floyd, W. F. and Silver, P. H. S. 1955. The function of the erectores spinae muscles in certain movements and postures in man. *Jour. Physiol.* 129:184–203.

Flynn, P. D., Feder, J. C., Gilbert, J. T. and Roll, A. A. 1962. Some new techniques for dynamic photoelasticity. *Proc. Soc. Exp. Stress Anal.* 19:159–160.

Fourtau, R. 1918. *Contribution à l'étude des vertebrés de l'Égypt.* Cairo: Government Press.

Friedlaender, J. S., Sgaramella-Zonta, L. A., Kidd, K. K., Lai, L. Y. C., Clark, P., and Walsh, R. J. 1971. Biological divergences in south-central Bougainville: an analysis of blood polymorphism gene frequencies and anthropometric measurements utilizing tree models, and a comparison of these variables with linguistic, geographic, and migrational "distances." *Am. Jour. Human Genet.* 23:253–270.

Fukada, E. and Yasuda, I. 1957. On the piezo-electric effect of bone. *Jour. Phys. Soc. Japan* 12:1158–1162.

Fulton, J. F. and Dusser de Barenne, J. G. 1933. The representation of the tail in the motor cortex of primates, with special reference to spider monkeys. *Jour. Cellular and Comp. Physiol.* 2:399–426.

Furuya, Y. 1957. Grooming behavior in the wild Japanese monkeys. *Primates* 1:47–68.

———1961–62. The social life of silvered leaf monkeys, *Trachypithecus cristatus. Primates* 3:41–60.

Gans, C. and Bock, W. J. 1965. The functional significance of muscle architecture: a theoretical analysis. *Ergebn. anat. Entw. Gesch.* 38:115–142.

Gartlan, J. S. 1970. Preliminary notes on the ecology and behavior of the drill, *Mandrillus leucophaeus* Ritgen 1824. In *The Old World monkeys,* pp. 445–480. Napier, J. A. and Napier, P. H., eds. New York: Academic Press.

Gartlan, J. S. and Brain, C. K. 1968. Ecology and social variability in *Cercopithecus aethiops* and *C. mitis.* In *Primates,* pp. 253–292. Jay, Phyllis C., ed. New York: Holt, Rinehart and Winston.

Gazin, C. L. 1968. A new primate from the Torrejon middle Paleocene of the San Juan Basin, New Mexico. *Proc. Biol. Soc. of Wash.* 81:629–634.

Gelfand, P. 1970. *Faunal behavior at Agua Dulce Spring, Cabeza Prieta Game Range, Arizona.* Unpublished senior thesis, Thomas Jefferson College, Grand Valley State College.

Gellhorn, E. 1947. Patterns of muscular activity in man. *Arch. Phys. Med.* 28:568–574.

Geschwind, N. 1964. *Development of brain and evolution of language,* pp. 155–170. Georgetown University Monog. Series on Lang. and Ling. 17.

———1965. Disconnexion syndromes in animals and man. *Brain* 88:237–294, 585–644.

Geschwind, N. and Levitcky, W. 1968. Human Brain: Left-Right Asymmetries in Temporal Speech Region. *Science* 161:186–187.

Gjukic, M. 1955. Ein Beitrag zum Problem der Korrelation zwischen Hirngewicht und Körpergewicht. *Z. Morphol. Anthrop.* 47:43–57.

Goffman, E. 1963. *Behavior in public places.* New York: Free Press.

Goffman, W. and Warren, K. S. 1970. An application of the Kermack-McKendrick theory to the epidemiology of schisotosomiasis. *Amer. Jour. Trop. Med. and Hyg.* 19(2):278–283.

Goodman, G., Snyer, F. N., Stimson, G. W. and Rankin, J. J. 1969. Phylogenetic changes in the proportion of two kinds of lactate dehydrogenase in primate brain regions. *Brain Research* 14:447–459.

Goodman, J. W. 1968. *Introduction to Fourier optics.* San Francisco: McGraw-Hill.

Goodman, L. A. 1968. The analysis of cross-classified data. *Jour. Amer. Stat. Assoc.* 63:1091–1131.

Gottlieb, H. H. 1914. Die Anticlinie der Wirbelsäule der Säugetiere. *Morph. Jahrb.* 49:179–220.

Gower, J. C. 1967. A comparison of some methods of cluster analysis. *Biometrics* 23:623–637.

Gower, J. C. and Ross, G. J. S. 1966. Minimum spanning trees and single linkage cluster analysis. *Applied Statistics* 18:54–64.

Grand, T. I. 1967. The functional anatomy of the ankle and foot of the slow loris (*Nycticebus coucang*). *Amer. Jour. Phys. Anthrop.* 26:207–218.

———1968a. The functional anatomy of the lower limb of the howler monkey (*Alouatta* caraya). *Amer. Jour. Phys. Anthrop.* 28:168–182.

———1968b. The functional anatomy of the upper limb of the howler monkey

(*Alouatta caraya*). In *The biology of the howler monkey. Biblio. primat.* 7:104–125. Malinow, M. R., ed. Basel and New York: Karger.

Grand, T. I. and Lorenz, R. 1968. Functional analysis of the hip joint in *Tarsius bancanus* (Horsfield 1821) and *Tarsius syrichta* (Linnaeus 1758). *Folia primat.* 9:161–181.

Gray, E. G. and Basmajian, J. V. 1968. Electromyography and cinematography of leg and foot (normal and flat) during walking. *Anat. Record* 161:1–15.

Gregory, W. K. 1915. On the classification and phylogeny of the Lemuroidea. *Bull. Geol. Soc. Amer.* 26:426–446.

———1920. On the structure and relation of *Notharctus,* an American Eocene primate. *Mem. Amer. Mus. Nat. Hist. n.s.* 3:49–243.

Gregory, W. K., Hellman, M. and Lewis, G. E. 1938. *Fossil anthropoids of the Yale-Cambridge Indian Expedition of 1935.* Carnegie Inst. Wash. Publ. 495:1.

Grossman, S. P. 1967. *A textbook of physiological psychology.* New York: John Wiley & Sons.

Guhl, A. M. 1956. Social order of chickens. *Scient. Amer.* 194:42–46.

Haines, R. W. 1950. The interorbital septum in mammals. *J. Linn. Soc. Lond.* (*Zool*). 41:585–607.

———1958. Arboreal or terrestrial ancestry of placental mammals. *Quart. Rev. Biol.* 33:1–23.

Hall, E. R. and Dalquest, W. W. 1963. The mammals of Veracruz. *Univ. Kans. Publ. Mus. Nat. Hist.* 14:165–362.

Hall, K. R. L. 1963a. Variations in the ecology of the chacma baboon. In *The Primates.* Napier, J. A. and Barnicot, N. A., eds. Symp. Zool. Soc. Lond., no. 10.

———1963b. Tool-using performances as indicators of behavioral adaptability. *Curr. Anthrop.* 4:479–494.

———1965. Behaviour and ecology of the wild patas monkey, *Erythrocebus patas,* in Uganda. *Jour. Zool.* 148:15–87.

———1967. Social interactions of the adult males and adult females of a patas monkey group. In *Social communication among primates,* pp. 261–280. Altmann, S. A., ed. Chicago and London: University of Chicago Press.

Hall, K. R. L. and DeVore, I. 1965. Baboon social behavior. In *Primate behavior. Field studies of monkeys and apes,* pp. 53–110. DeVore, I., ed. New York: Holt, Rinehart and Winston.

Hall, M. C. 1965. The locomotor system. In *Functional anatomy,* vol. 1, p. 562. Springfield, Ill.: C. C. Thomas.

Hall-Craggs, E. C. B. 1966. Rotational movements in foot of *Galago senegalensis. Anat. Rec.* 154:287–294.

Hallowell, A. I. 1961. The protocultural foundations of human adaptation. In *Social life of early man,* pp. 236–255. Washburn, S. L., ed. Viking Fund Publ. in Anthrop., no. 31.

Hamilton, W. J., Jr. 1930. The food of the Soricidae. *Jour. Mammal.* 11:26–39.

Harris, H. 1970. *Principles of human biochemical genetics.* Amsterdam: North-Holland.

Harrisson, B. 1962. Getting to know *Tarsius. Malayan Nat. Jour.* 16:197–204.

Haughton, S. 1864. Notes on animal mechanics. 2: On the muscles of some of the smaller monkeys of the genera *Cercopithecus* and *Macacus. Proc. Roy. Irish Acad.* 8:467–471.

———1865. Notes on animal mechanics. 7: On the muscular anatomy of the *Macacus nemestrinus. Proc. Roy. Irish Acad.* 9:277–294.

————1873. *Principles of animal mechanics*. London: Longmans Green.

Hendey, Q. B. 1969. Quaternary vertebrate fossil sites in the southwestern Cape Province. *S. Afr. Archaeol. Bull.* 24:96–105.

Herrnstein, R. J. 1970. On the law of effect. *Jour. Exp. Analy. Behav.* 13:243–266.

————In press. Quantitative hedonism. *Scient. Amer.*

Hershkovitz, P. 1970. Cerebral fissural patterns in platyrrhine monkeys. *Folia primat.* 13:213–240.

Hickman, V. V. and Hickman, J. L. 1960. Notes on the habits of the Tasmanian dormouse phalangers *Cercaertus nanus* (Desmarest) and *Eudromicia lepida* (Thomas). *Proc. Zool. Soc. Lond.* 135:365–374.

Hicks, J. H. 1954. The mechanics of the foot. 2. The plantar aponeurosis and the arch. *Jour. Anat.* 88:25–31.

Hiernaux, J. 1956. Analyse de la variation des caractères physiques humains en une région de l'Afrique centrale: Ruanda-Urundi et Kivu. *Ann. de Mus. Roy. du Congo Belge, serie in 8, Anthropologie* vol. 3.

Hiiemae, K. and Jenkins, F. 1969. The anatomy and internal architecture of the muscles of mastication in *Didelphis marsupialis*. *Postilla* 140:1–49.

Hill, W. C. O. 1966. *Primates. Comparative anatomy and taxonomy* vol. 6, Cercopithecoidea. Edinburgh: University Press.

————1967. Taxonomy of the baboon. In *The baboon in medical research* vol. 2, pp. 3–11. Vagtborg, H., ed. Austin: University of Texas Press.

Hinde, R. A. and Spencer-Booth, Y. 1968. The study of mother-infant interaction in captive group-living rhesus monkeys. *Proc. Roy. Soc. Brit.* 169:177–207.

Hinde, R. G. and Rowell, T. E. 1962. Communication by postures and facial expression in the rhesus monkey (*Macaca mulatta*). *Proc. Zool Soc. Lond.* 138:1–21.

Hirsch, J. F. and Coxe, W. S. 1958. Representation of cutaneous tactile sensibility in cerebral cortex of *Cebus*. *Jour. Neurophys.* 21:481–498.

Hockett, C. F., and Ascher, R. 1964. The human revolution. *Current Anthrop.* 5:135–152.

Hofer, H. 1957. Über die Bewegungen des Potto. *Natur und Volk* 87:409–418.

Hofer, H. and Tigges, J. 1964. Makromorphologie des Zentralnervensystems. *Handb. der Zoologie* vol. 8, p. 34.

Holister, A. S. 1961. Recent developments in photoelastic coating techniques. *Jour. Roy. Aero. Soc.* 65:661–669.

Holloway, R. L. 1964. *Some aspects of quantitative relations in the primate brain.* Unpublished Ph.D. dissertation. Berkeley: University of California.

————1966a. Cranial capacity, neural reorganization, and hominid evolution: a search for more suitable parameters. *Amer. Anthrop.* 68:103–121.

————1966b. Cranial capacity and neuron number: critique and proposal. *Amer. Jour. Phys. Anthrop.* 25:305–314.

————1967. The evolution of the human brain: some notes toward a synthesis between neural structure and the evolution of complex behavior. *General Systems* 12:3–19.

————1968. The evolution of the primate brain: some aspects of quantitative relations. *Brain Research* 7:121–172.

————1969a. Some questions on parameters of neural evolution in primates. *Ann. N.Y. Acad. Sci.* 167:332–340.

————1969b. Culture: a human domain. *Curr. Anthrop.* 10:395–412.

————1970. Australopithecine endocast (Taung specimen, 1924); a new volume determination. *Science* 168:966–968.

Hooijer, D. A. 1951. Questions relating to a new large anthropoid ape from the Mio-Pliocene of the Siwaliks. *Amer. Jour. Phys. Anthrop.* 9:79.

————1963. Miocene mammalia of Congo. *Ann. Mus. Roy. de l'Afrique Centrale. Tervaren, Belgique (Sci. Geol.)* 46:1.

Hooton, E. A. 1930. Doubts and suspicions concerning certain functional theories of primate evolution. *Human Biol.* 2:223–249.

————1942. *Man's poor relations.* Garden City, N.Y.: Doubleday and Doran.

Howell, A. B. and Straus, W. L., Jr. 1933. The muscular system. In *The anatomy of the rhesus monkey (Macaca mulatta)*, pp. 89–175. Hartman, Carl G. and Straus, William L., Jr., eds. New York: Hafner Publ. Co.

Howell, F. C. 1967. Recent advances in human evolutionary studies. *Quart. Rev. Biol.* 42:471–513.

Howell, F. C. and Clark, J. 1963. Acheulian hunter-gatherers of sub-Saharan Africa. In *African ecology and human evolution*, pp. 458–533. Howell, F. C. and Bourliere, F., eds. Chicago: Aldine Publ. Co.

Howells, W. W. 1947. *Mankind so far.* Garden City, N.Y.: Doubleday.

————1951. Factors of human physique. *Amer. Jour. Phys. Anthrop.* 9:159–192.

————1957. The cranial vault: factors of size and shape. *Amer. Jour. Phys. Anthrop.* 15:19–48.

————1966. *Craniometry and multivariate analysis. 1: The Jomon population of Japan, a study of discriminant analysis of Japanese and Ainu crania.* Papers, Peabody Mus. Arch. and Etholo., Harvard 57:1–68.

————1969a. The use of multivariate techniques in the study of skeletal populations. *Amer. Jour. Phys. Anthrop.* 31:311–314.

————1969b. Comments by W. W. Howells. *Human Biol.* 41:295–297.

————1970a. Multivariate analysis of human crania. *Proc. 8th Inter. Cong. Anthrop. and Ethnolog. Sci.* 1968, vol. 1(A-1):1–3.

————1970b. Mount Carmel man: morphological relationships, *ibid*, vol. 1(S-1):269–271.

————1970c. Multivariate analysis for the identification of race from crania. In *Personal identification in mass disasters*, pp. 111–121. Stewart, T. D., ed. Washington, D.C.: Smithsonian Institution.

————. In press. Cranial variation in man. Peabody Museum Papers.

Hubel, D. H. and Wiesel, T. N. 1970. Cells sensitive to binocular depth in area 18 of the macaque monkey cortex. *Nature* 225:41–42.

Hunt, E. E., Jr. 1960. The continuing evolution of modern man. *Cold Spring Harbor Symp. on Quantitative Biol.* 24:245–254.

Hürzeler, J. 1954. Contribution à l'odontologie et à la phylogenie du genre *Pliopithecus* Gervais. *Ann. Paleontol.* 40:5.

Huxley, T. H. 1863. *Man's place in nature.* 1959 reprint. Ann Arbor: University of Michigan Press.

Itani, J. et al. 1963. The social construction of natural troops of Japanese monkeys in Takasakiyama. *Primates* 4(3).

Itani, J. and Suzuki, A. 1967. The social unit of chimpanzees. *Primates* 8:335–381.

Jacobson, A. 1968. *A morphological and metrical study of the teeth, the jaws, and the bony palate of several large groups of South African Bantu-speaking negroids.* Ph.D. thesis, Johannesburg: University of the Witwatersrand, Department of Anatomy.

Jay, P. 1963a. Mother-infant relations in langurs. In *Maternal behavior in mammals*, pp. 282–304. Rheingold, H., ed. New York: John Wiley & Sons.

————1963b. *The social behavior of the langur monkey.* Unpublished doctoral dissertation, Chicago: The University of Chicago.

————1965. The common langur of North India. In *Primate behavior,* pp. 197–249. DeVore, I., ed. New York: Holt, Rinehart and Winston.

Jenkins, F. A., Jr. 1969. The evolution and development of the dens of the mammalian axis. *Anat. Rec.* 164(2): 173–184.

————1970. Anatomy and function of expanded ribs in certain edentates and primates. *Jour. Mammal.* 51(2): 288–301.

Jepsen, G. L. 1930. Stratigraphy and paleontology of northeastern Park County, Wyoming. *Proc. Amer. Phil. Soc.* 69(7): 463–528.

Jerison, H. J. 1963. Interpreting the evolution of the brain. *Human Biol.* 35:263–291.

Johnson, G. L. 1901. Contributions to the comparative anatomy of the mammalian eye, chiefly based on ophthalmoscopic examination. *Philos. Trans. Roy. Soc. Lond.* 194:1–82.

Jolly, A. 1966. *Lemur behavior.* Chicago: University of Chicago Press.

Jolly, C. J. 1965. The evolution of the baboons. In *The baboon in medical research,* pp. 23–50. Vagtbord, H., ed. Austin: University of Texas Press.

————1970. The seed-eaters: a new model of hominid behavioral differentiation based on a baboon analogy. *Man* 5:5–26.

Jones, F. W. 1917. *Arboreal man.* Reprint 1964. New York: Hafner.

Jones, K. J. In press. *Multivariate methodology.* New York: McGraw-Hill.

Jones, R. L. 1941. The human foot: an experimental study of its mechanics and the role of its muscle and ligaments in the support of the arch. *Amer. Jour. of Anat.* 68:1–39.

————1945. The functional significance of the declination of the axis of the subtalar joint. *Anat. Rec.* 93:151–159.

Jouffroy, F. K. and Lessertisseur, J. 1960. Les specializations anatomiques de la main chez les singes à progression suspendue. *Mammalia* 24:93–151.

Kanda, S. and Kurisu, K. 1967, 1968. Factor analytic studies on the Japanese skulls; and Factor analysis of Japanese skulls, 2. *Med. J. Osaka Univ.:* 18:1–9, 315–318.

Karlin, S. 1969. *Equilibrium behavior of population genetic models with non-random mating.* London: Gordon and Breach.

Kaufmann, J. H. 1967. Social relations of adult males in a free-ranging band of rhesus monkeys. In *Social communication among primates.* Altmann, S. A., ed. Chicago: University of Chicago Press.

Kaufman, J. C. and Rosenblum, L. A. 1969. The waning of the mother-infant bond in two species of macaque. In *Determinants of infant behavior* 4:41–59.

Kawai, M. 1958. On the system of social ranks in a natural troop of Japanese monkeys. *Primates* 1(2): 84–98, 111–130. English trans. in *Japanese monkeys: a collection of translations.* Imanishi, K. and Altmann, S. A., eds. Atlanta, 1965.

Kawamura, S. 1958. Matriarchal social ranks in the Minoo-B troop: a study of the rank system of Japanese monkeys. *Primates* 1–2:149–156.

Keith, A. 1902. The extent to which the posterior segments of the body have been transmuted and suppressed in the evolution of man and allied primates. *Jour. Anat. Lond.* 37:18–40.

————1931. *New discoveries relating to the antiquity of man.* London: Williams and Norgate.

————1940. "Fifty years ago." *Amer. Jour. Phys. Anthrop.* 26:251–277.

Kellogg, R. 1944. A new macaque from an island off the east coast of Borneo. *Proc. Biol. Soc. Wash.* 57:75–76.

Kendall, M. G. and Moran, P. A. P. 1963. *Geometrical probability.* New York: Hafner.

Klaauw, C. J. van der 1929. On the development of the tympanic region of the skull in the Macroscelididae. *Proc. Zool. Soc. Lond.* 1929:491–560.

Klausen, K. 1965. The form and function of the loaded human spine. *Acta physiol. scandinav.* 65:176–190.

Klopfer, P. H., Adams, D. K. and Klopfer, M. S. 1964. Maternal "imprinting" in goats. *Proc. Nat. Acad. Sci. Wash.* 52:912–974.

Knussman, R. 1967. Das proximale Ende der Ulna von *Oreopithecus bambolii* und seine Aussage über dessen systematische Stellung. *Zeitschr. fur Morph. u. Anthrop.* 59:57–76.

Koch, J. C. 1917. The laws of bone architecture. *Amer. Jour. Anat.* 21:177–298.

Koenigswald, G. H. R. von 1949. Bemerkungen zu *Dryopithecus giganteus* Pilgrim. *Ecl. geol. helv.* 42:515.

————1967. Evolutionary trends in the deciduous molars of the Hominidae. *Jour. dent. Res., Supp: Inter. Symp. on Dental Morph.* 46:779–786.

————1968. The phylogenetical position of the Hylobatinae. In *Taxonomy and phylogony of Old World primates with references to the origin of man.* Chiarelli, B., ed. Torino: Rosenberg and Sellier.

————1969. Miocene Cercopithecoidea and Oreopithecoidea of East Africa. In *Fossil vertebrates of Africa* vol. 1. Leakey, L. S. B., ed. New York: Academic Press.

Koford, C. B. 1963a. Group relations in an island colony of rhesus monkeys. In *Primate social behavior.* Southwick, C. H., ed. Princeton: von Nostrand.

————1963b. Rank of mothers and sons in bands of rhesus monkeys. *Science* 147:356–357.

————1966. Population changes in rhesus monkeys: Cayo Santiago, 1960–1964. *Tulane Studies in Zool.* 13:1–7.

Kollman, M. 1925. Études sur les lémuriens. La fosse orbito-temporale et l'os planum. *Mém. Soc. Linn. Normandie* (Caen), n.s., sec. Zool. 1:1–20.

Kollman, M. and Papin, L. 1925. Études sur les lémuriens. Anatomie comparée des fosses nasales et de leurs annexes. *Arch. Morph.* 22:1–60.

Koyama, N. 1967. On dominance rank and kinship of a wild Japanese monkey troop in Arashiyama. *Primates* 8:189–216.

Krantz, G. 1968. Brain size and earliest hunting in man. *Current Anthrop.* 9:450–451.

Krishnamurti, A. 1966. The external morphology of the brain of the slow loris *Nycticebus coucang coucang. Folia primat.* 4:361–380.

Krishnamurti, A. and Welker, W. I. 1965. Somatic sensory area in the cerebral neocortex of slow loris (*Nycticebus coucang coucang*). Abst.: *Fed. Proc.* 24(2):140.

Kummer, B. 1959. *Bauprinzipien des Säugerskeletes.* Stuttgart: Thieme.

————1965. Die Biomechanik der aufrechten Haltung. *Mitt. naturf. Ges. Bern* n.f. 22:239–259.

Kummer, H. 1968. *Social organization of Hamadryas baboons: a field study.* Chicago: University of Chicago Press.

————1970. Behavioral characters in primate taxonomy. In *The Old World monkeys,* pp. 25–36. Napier, J. A. and Napier, P. H. eds. New York: Academic Press.

Kummer, H. and Kurt, F. 1965. A comparison of social behavior in captive and wild hamadryas baboons. In *The baboon in medical research,* vol. 1, pp. 65–80. Vagtborg, H., ed. Austin: University of Texas Press.

Lancaster, J. 1968. Primate communication systems and the emergence of human language. In *Primates: studies in adaptation and variability,* pp. 439–457. Jay, P., ed. New York: Holt, Rinehart and Winston.

Landau, H. G. 1968. Models of social structure. *Bull. Math. Biophysics* 30:215–224.

Landauer, C. A. 1962. A factor analysis of the facial skeleton. *Human Biol.* 34:239–253.

Langer, C. 1879. Die musculatur der extremitaten des orang als grundlage einer vergleichend-myologischen untersuchung. *S. B. Math. Naturwiss. Cl. Akad. Wiss. zu Wien* 79:177–222.

Lasker, G. 1969. Human biological adaptability. *Science* 166:1480–1486.

Lawick-Goodall, J. van 1967. Mother-offspring relationship in free-ranging chimpanzees. In *Primate ethology,* pp. 287–346. Morris, D., ed. London: Weidenfeld and Nicolson.

———1968. The behavior of free-living chimpanzees in the Gombe Stream Reserve. *Anim. Behav. Monog.* 1(3):161–311.

Leakey, L. S. B. 1962. A new Lower Pliocene fossil primate from Kenya. *Ann. Mag. Nat. Hist.* 4:689.

———1967. An Early Miocene member of Hominidae. *Nature* 213:155.

———1968. Upper Miocene primates from Kenya. *Nature* 218:527.

Leakey, L. S. B., Tobias, P. V., and Napier, J. R. 1964. A new species of the genus *Homo* from Olduvai Gorge. *Nature* 202:7–9.

Lenneberg, E. H. 1964. A biological perspective of language. In *New directions in the study of language,* pp. 65–88. Lenneberg, E. H., ed. Cambridge: M.I.T. Press.

———1967. *Biological foundations of language.* New York: John Wiley & Sons, Inc.

Leutenegger, W. 1970a. Beziehungen zwischen Neugeborenengrösse und dem Sexualdimorphismus am Becken bei simischen Primaten. *Folia primat.* 12:224–235.

———1970b. Das Becken der rezenten Primaten. *Morph. Jahrb.* 115:1–100.

Leven, M. M. 1955. Quantitative three-dimensional photoelasticity. *Proc. Soc. Exp. Stress Anal.* 12:157–171.

Lewis, O. J. 1964a. The evolution of the long flexor muscles of the leg and foot. *Int. Rev. gen. expl. Zool.* 1:165–185.

———1964b. The homologies of the mammalian tarsal bones. *Jour. Anat. Lond.* 98:195–208.

———1965. Evolutionary change in the primate wrist and inferior radio-ulnar joints. *Anat. Rec.* 151:275–286.

———1969. The hominoid wrist joint. *Amer. Jour. Phys. Anthrop.* 30:251–268.

———1970. The development of the human wrist joint during the fetal period. *Anat. Rec.* 166:499–516.

———1971. The contrasting morphology found in the wrist joints of semi-brachiating monkeys and brachiating apes. *Folia primat.* In press.

Lewis, O. J., Hamshere, R. J., and Bucknill, T. M. 1970. The anatomy of the wrist joint. *Jour. Anat. Lond.* 106:539–552.

Liberson, W. G. 1965. Biomechanics of gait: a method of study. *Arch. Phys. Med.* 46:37–48.

Lorenz, K. Z. 1958. The evolution of behavior. *Scient. Amer.* 199:67–74.

Lorrain, F. and White, H. C. 1971. Structural equivalence of individuals in social networks. *Jour. Math. Sociol.* 1(1):49–80.

Lovejoy, C. O. and Heiple, K. G. 1970. A reconstruction of the femur of *Australopithecus africanus, Amer. Jour. Phys. Anthrop.* 32:33–40.

Loy, J. D. In press. Estrus behavior of free-ranging rhesus monkeys (*Macaca mulatta*). *Primates*.

Lyon, M. W., Jr. 1908. Mammals collected in Western Borneo by Dr. W. L. Abbott. *Proc. U.S. Natn. Mus.* 33:547–572.

MacConnaill, M. A. 1946. Some anatomical factors affecting the stabilizing functions of muscles. *Irish Jour. Med. Sci.* 6:160–164.

———1949. The movements of bones and joints. 2. Function of the musculature. *Jour. Bone and Joint Surg.* 31-B:100–104.

MacConnaill, M. A. and Basmajian, J. V. 1969. *Muscles and movements: a basis for human kinesiology.* Baltimore: Williams and Wilkins.

Macdonald, G. 1957. *The epidemiology and control of malaria.* London: Oxford University Press.

MacKinnon, I. L., Kennedy, J. A. and Davies, T. V. 1956. The estimation of skull capacity from Roentgenologic measurements. *Amer. Jour. Roentgenology, Radiation Therapy, and Nuclear Med.* 76:303–310.

Mann, A. 1968. *A palaeo-demography of Australopithecus.* Ph.D. thesis. Berkeley: University of California, Department of Anthropology.

Marler, P. 1969. *Colobus guereza:* territoriality and group composition. *Science* 163:93–95.

Marshall, D. S. 1969. Book review: Craniometry and multivariate analysis, Parts 1, 2, by W. W. Howells and J. M. Crichton. *Human Biol.* 41:291–295.

Martin, R. 1928. *Lehrbuch der Anthropologie.* 2nd ed. 3 vols. Jena: Fischer.

Martin, R. D. 1967. *Behavior and taxonomy of tree shrews (Tupaiidae).* Ph.D. thesis. Oxford University.

———1968. Towards a new definition of primates. *Man* 3(3):376–401.

Maruyama, M. 1963. The second cybernetics: deviation-amplifying mutual-causal processes. *Amer. Sci.* 51:164–179.

Masali, M. 1968. The vertebral column as indication of taxonomic and postural distinction among Old World primates. In Chiarelli, *Taxonomy and phylogeny of Old World primates,* pp. 87–94.

Mason, W. A. 1966. Social organization of the South American monkey, *Callicebus moloch:* a preliminary report. *Tulane Studies in Zool.* 13:23–28.

———1968. Use of space by *Callicebus* groups. In *Primate studies in adaptation and variability.* Jay, P., ed. New York: Holt, Rinehart and Winston.

Matthew, W. D. 1904. The arboreal ancestry of the Mammalia. *Amer. Nat.* 38:811–818.

Mayr, E. 1959. Darwin and the evolutionary theory in biology. In *Evolution and anthropology: a centennial appraisal,* pp. 3–12. Anthrop. Soc. of Wash.

———1970. *Populations, species, and evolution.* Cambridge: Harvard University Press.

McCann, C. 1933. Observations on some of the Indian langurs. *Jour. Bombay Nat. Hist. Soc.* 36:618–628.

McClure, H. E. 1964. Some observations of primates in climax diptocarp forest near Kuala Lumpur, Malaya. *Primates* 5:39–58.

McDowell, S. B. 1958. The Greater Antillean insectivores. *Bull. Amer. Mus. Nat. Hist.* 115(3):113–214.

McKenna, M. C. 1960. Fossil mammalia from the early Wasatchian Four Mile fauna, Eocene of northwest Colorado. *Univ. Calif. Publ., Geol. Sci.* 37(1):1–130.

———1961. A note on the origin of rodents. *Amer. Mus. Novitates* 2037:1–5.

———1966. Paleontology and the origin of primates. *Folia primat.* 4(1):1–25.

Meienberg, G. P. 1962. Beitrag zur Morphogeneses des oberen Endes der Halswirbelsäule. *Z. Morph. Anthrop.* 52:76–89.

Meltzer, B., Searle, N. H. and Brown, R. 1967. Numerical specification of biological form. *Nature* 216:32–36.

Meulders, M., Gybels, J., Bergmans, J., Gerebtzoff, M. A. and Goffart, M. 1966. Sensory projections of somatic, auditory, and visual origin to the cerebral cortex of the sloth (*Choloepus hoffmanni* Peters). *Jour. Comp. Neur.* 126:535–546.

Michael, R. P. and Keverne, E. B. 1968. Pheromones in the communication of sexual status in primates. *Nature* 218:746–749.

Michaelis, P. 1903. Beitrage zur vergleichenden myologie des *Cynocephalus babuin, Simia satyrus, Troglodytes niger. Arch. Anat. Phys. Anat. Abth.,* 205–256.

Miller, G. S., Jr. 1907a. Mammals collected by Dr. W. L. Abbott in the Karimata Islands, Dutch East Indies. *Proc. U.S. Natn. Mus.* 31:55–66.

————1907b. The mammals collected by Dr. W. L. Abbott in the Rhio-Linga Archipelago. *Proc. U.S. Natn. Mus.* 31:247–286.

Miller, J. H. 1960. *Papio doguera,* dog face baboon, a primate reservoir host of S. mansoni in East Africa. *Trans. Roy. Soc. of Tropical Medicine and Hygiene* 54(1):44–46.

Miller, R. A. 1932. Evolution of the pectoral girdle and forelimb in the primates. *Amer. Jour. Phys. Anthrop.* 17:1–56.

Miller, R. A. 1945. The ischial callosities of primates. *Amer. Jour. Anat.* 76:67–91.

Mizuhara, H. 1964. Social changes of Japanese monkey troops in the Takasakiyama. *Primates* 5(1–2):27–52.

Morton, D. J. 1935. *The human foot: its evolution, physiology, and functional disorders.* 1964 reprint. New York: Hafner.

Moynihan, M. 1967. Comparative aspects of communication in New World primates. In *Primate ethology,* pp. 236–266. Morris, D., ed. Chicago: Aldine Publ. Co.

Mukherjee, R., Rao, C. R. and Trevor, J. C. 1955. *The ancient inhabitants of the Jebel Moya.* Cambridge: University Press.

Murray, P. D. F. 1936. *Bones: a study in the development and structure of the vertebrate skeleton.* Cambridge: University Press.

Nachemson, A. 1966. Electromyographic studies of the vertebral portion of the psoas muscle. *Acta orthop. scandin.* 37:177–190.

Napier, J. R. 1955. The form and function of the carpo-metacarpal joint of the thumb. *Jour. Anat.* 89:362–369.

————1956. The prehensile movements of the human hand. *Jour. Bone and Joint Surg.* 38-B:902–913.

————1957. The foot and the shoe. *Physiotherapy* 43:65–74.

————1963. Brachiation and brachiators. In *The primates,* pp. 183–195. Napier, J. and Barnicot, N. A., eds. *Symp. Zool. Soc. Lond.* no. 10.

————1967. Evolutionary aspects of locomotion. *Amer. Jour. Phys. Anthrop.* 27(3):333–342.

Napier, J. R. and Davis, P. R. 1959. The forelimb skeleton and associated remains of *Proconsul africanus. Foss. mamm. Af., Brit. Mus. (Nat. Hist.)* 16:1.

Napier, J. R. and Napier, P. H. 1967. *A Handbook of living primates.* London, New York: Academic Press.

Napier, J. R. and Walker, A. C. 1967a. Vertical clinging and leaping: a newly recognized category of locomotor behavior of primates. *Folia primat.* 6:204–219.

————1967b. Vertical clinging and leaping in living and fossil primates. In

Progress in primatology, pp. 64–69. Starck, D., Schneider, R., and Kuhn, H. J., eds. Stuttgart: Gustav Fischer Verlag.

Nathan, R. and Selzer, R. H. 1968. Digital video data handling: Mars, the moon, and men. In *Image processing in biological science,* pp. 177–210. Ramsey, D. M., ed. Los Angeles: University of California Press.

Neville, M. K. 1968. A free-ranging rhesus monkey troop lacking adult males. *Jour. Mamm.* 49:771–773.

Osborn, H. F. 1907. The Fayûm expedition of the American Museum. *Science, New York* n.s. 25:513.

———1908. New fossil mammals from the Fayûm Oligocene of Egypt. *Bull. Amer. Mus. Nat. Hist.* 24:265.

———1910. *The age of mammals in Europe, Asia, and North America.* New York: Macmillan Co.

Oxnard, C. E. 1963. Locomotor adaptations in the primate forelimb. *Symp. Zool. Soc. Lond.* 10:165–182.

———1966. Some functional osteometric features of the primate innominate bone. *Proc. Inter. Primat. Soc. Frankfurt:* 59–60.

———1967. The functional morphology of the primate shoulder as revealed by comparative anatomical, osteometric, and discriminant function techniques. *Amer. Jour. Phys. Anthrop.* 26:219–240.

———1968a. Primate evolution: a method of investigation. *Amer. Jour. Phys. Anthrop.* 28:289–302.

———1968b. The architecture of the shoulder in some mammals. *Jour. Morph.* 126:249–290.

———1969a. Evolution of the human shoulder: some possible pathways. *Amer. Jour. Phys. Anthrop.* 30:319–331.

———1969b. The combined use of multivariate and clustering analyses in functional morphology. *Jour. Biomech.* 2:73–88.

———1969c. Mathematics, shape and function: a study in primate anatomy. *Amer. Sci.* 57(1):75–96.

Oxnard, C. E. and Neely, P. M. 1969. The descriptive use of neighborhood limited classification in functional morphology: an analysis of the shoulder in primates. *Jour. Morph.* 129:127–148.

Oxnard, C. E. and Tuttle, R. H. 1969. An analysis of the mechanical efficiency of the digital ray in the chimpanzee in relation to knuckle-walking postures and movement. *Amer. Jour. Phys. Anthrop.* 31:265.

Patterson, B. and Howells, W. W. 1967. Hominid humeral fragment from early Pleistocene of northwestern Kenya. *Science* 156:64–66.

Paulian, R. 1946. La voûte de la forêt tropicale, milieu biologique. *Revue Scientifique* 83:281–286.

———1947. Observations écologiques en forêt de Basse Côte d'Ivoire. *Encycl. Biogeog. Ecol., Paris,* vol. 2.

Pauwels, F. 1965. *Gesammelte Abhandlungen zur funktionellen Anatomie des Bewegungsapparates.* Berlin, Heidelberg, New York: Springer.

Pearson, K. and Davin, A. G. 1924. On the biometric constants of the human skull. *Biometrika* 16:328–363.

Petter, J.-J. 1962a. Recherches sur l'écologie et l'éthologie des Lémuriens Malgaches. *Mém. Mus. nat. d'Hist. nat.* (Paris) 27:1–146.

———1962b. Ecological and behavioral studies of Madagascar lemurs in the field. *Ann. N.Y. Acad. Sci.* 102(2):267–281.

Pfuhl, W. 1926. Zur mechanik der Zwerchfellbewegung. *Z. Konstit. Lehre* 12:158–177.

———1937. Die gefiederten Muskeln, ihre Form und ihre Wirkungsweise. *Z. Anat. Entwickl. Gesch.* 106:749–769.

Phillips, W. W. A. 1935. *Manual of the mammals of Ceylon*. Ceylon: Colombo Museum.

Pielou, E. C. 1969. *An introduction to mathematical ecology*. New York: John Wiley & Sons.

Pilbeam, D. R. 1968. The earliest hominids. *Nature* 219:1335.

———1969a. Tertiary Pongidae of East Africa: evolutionary relationships and taxonomy. *Bull. Peabody Mus.* no. 31. New Haven: Yale University.

———1969b. Possible identity of Miocene tali from Kenya. *Nature* 223:648.

———1969c. Early Hominidae and cranial capacity. *Nature* 224:386.

———1970. *Gigantopithecus* and the origins of Hominidae. *Nature* 225:516.

Pilbeam, D. R. and Simons, E. L. 1965. Some problems of hominid classification. *Amer. Sci.* 53:237–257.

———1971. Humerus of *Dryopithecus* from St. Gaudens, France. *Nature* 229:406–407.

Pilgrim, G. E. 1927. A Sivapithecus palate and other primate fossils from India. *Paleont. Ind.* n.s. 14:1.

Pincus, H. J. 1969. Sensitivity of optical data processing to changes in rock fabric. *Inter. Jour. Rock Mech. Min. Sci.* 6:259–272.

Pinto-Hamuy, T., Bromiley, R. B. and Woolsey, C. M. 1956. Somatic afferent areas I and II of dog's cerebral cortex. *Jour. Neurophysiol.* 19:485–499.

Pirlot, P. and Stephan, H. 1970. Encephalization in Chiroptera. *Can. Jour. Zool.* 48:433–444.

Piveteau, J. 1950. Recherches sur l'encephale de lemuriens disparus. *Ann. Paleont.* 36:85–103.

———1957. Histoire paléontologique des Primates. In *Traité de paléontologie* vol. 7. Piveteau, J., ed. Paris: Masson.

Plutchik, R. 1964. The study of social behavior in primates. *Folia primat.* 2:67–92.

Pocock, R. I. 1915. On the feet and glands and other external characters of the paradoxurine genera *Paradoxurus, Arctictis, Arctogalidea,* and *Nandinia*. *Proc. Zool. Soc. Lond.* 1915:387–412.

———1922. On the external characters of the beaver (Castoridae) and of some squirrels (Sciuridae). *Proc. Zool. Soc. Lond.* 1922(2):1171–1212.

———1925. The external characters of the catarrhine monkeys and apes. *Proc. Zool. Soc. Lond.* 2:1479–1579.

Poglayen-Neuwall, I. 1966. On the marking behavior of the kinkajou (*Potos flavus* Schreber). *Zoologica* 51:137–141.

Poirier, F. E. 1967. *The ecology and social behavior of the Nilgiri langur (Presbytis johnii) of south India*. Unpublished Ph.D. dissertation. University of Oregon.

———1968a. Analysis of a Nilgiri langur (*Presbytis johnii*) home range change. *Primates* 9:29–43.

———1968b. The Nilgiri langur (*Presbytis johnii*)mother-infant dyad. *Primates* 9:45–68.

———1968c. Nilgiri langur (*Presbytis johnii*) territorial behavior. *Primates* 9:351–364.

———1969. The Nilgiri langur (*Presbytis johnii*) troop: its composition, structure, function and change. *Folia primat.* 10:20–47.

———1970. Dominance structure of the Nilgiri langur (*Presbytis johnii*) of south India. *Folia primat.* 12:161–186.

Polyak, S. 1958. *The vertebrate visual system*. Chicago: University of Chicago Press.

Powell, T. P. S., Guillery, R. W. and Cowan, W. M. 1957. A quantitative study of the fornix-mammillo-thalamic system. *Jour. Anat.* 91:419–437.

Pratt, J. W., Raiffa, H. and Schlaifer, R. 1965. *Introduction to statistical decision theory.* New York: McGraw-Hill.

Preston, F. W., Green, D. W. and Davis, J. C. 1969. Numerical characterization of reservoir rock pore structure. *2nd Ann. Rep. to the Amer. Petroleum Inst.,* pp. 1–84.

Preuschoft, H. 1961. Muskeln und gelenke der hinterextremitat des gorillas (*Gorilla gorilla* Savage et Wymann 1847). *Morph. Jahrb.* 101:432–540.

————1963. Muskelgewichte bei gorilla, orangutan, und mensch. *Anthrop. Anz.* 26:308–317.

————1965. Muskeln und gelenke der vorderextremitat des gorillas (*Gorilla gorilla* Savage et Wymann 1847). *Morph. Jahrb.* 107:99–183.

Prost, J. H. 1965. A definitional system for the classification of primate locomotion. *Amer. Anthrop.* 67:1198–1214.

Radcliffe, C. W. 1962. The biomechanics of below-knee prostheses in normal level bipedal walking. *Artif. Limbs* 6:16–24.

Radinsky, L. 1967a. Relative brain size: a new measure. *Science* 155:836–837.

————1967b. The oldest primate endocast. *Amer. Jour. Phys. Anthrop.* 27:385–388.

————1968a. A new approach to mammalian cranial analysis, illustrated by examples of prosimian primates. *Jour. Morph.* 124:167–180.

————1968b. Evolution of somatic sensory specialization in otter brains. *Jour. Comp. Neurol.* 134:495–506.

————1970. The fossil evidence of prosimian brain evolution. In *Advances in primatology* vol. 1, *The primate brain.* Noback, C., ed. New York: Appleton-Century-Crofts.

Rahaman, H. and Parthasarathy, M. D. 1967. A population survey of the bonnet monkey (*Macaca radiata Goeffroy*) in Bangalore, South India. *Jour. Bombay Nat. Hist.* 64:251–255.

Randall, F. E. 1943. The skeletal and dental development and variability of the gorilla. *Human Biol.* 15:236–254, 307–337.

Remane, A. 1936. Die Wirbelsäule und ihre Abkömmlinge. *Handb. vergl. Anat. Wirbelt.* vol. 4, pp. 1–206. Berlin: Urban and Schwarzenberg.

Rensch, B. 1956. Increase of learning capability by increase of brain size. *Amer. Nat.* 90:81–95.

————1959. Trends toward progress of brains and sense organs. *Cold Spring Harbor Symp. on Quant. Biol.* 24:291–303.

Reynolds, V. 1963. An outline of the behaviour and social organization of forest living chimpanzees. *Folia primat.* 1:95–102.

Ride, W. D. 1964. A review of Australian fossil marsupials. *Jour. Roy. Soc. W. Australia* 47:97–131 .

Ripley, S. 1965. *The ecology and social behavior of the Ceylon gray langur, Presbytis entellus theristes.* Unpublished Ph.D. dissertation. Berkeley: University of California.

————1967a. Intertroop encounters among Ceylon gray langurs (*Presbytis entellus*). In *Social communication among primates,* pp. 237–253. Altmann, S., ed. Chicago: University of Chicago Press.

————1967b. The leaping of langurs: a problem in the study of locomotor adaptation. *Amer. Jour. Phys. Anthrop.* 26:149–170.

————1970. Leaves and leaf-monkeys. In *The Old World monkeys,* pp. 481–509. Napier, J. R. and Napier, P. H., eds. New York: Academic Press.

Roberts, F. S. 1969. *On nontransitive indifference.* Santa Monica, Calif.: RAND Corp. RM-5782-PR.

Robinson, J. T. 1956. *The dentition of the Australopithecinae.* Transvaal Mus. Mem. no. 9:1–179.

Robinson, J. T. 1962. The origin and adaptive radiation of the Australopithecines. G. Kurth (ed.), *Evolution and Hominisation.* Stuttgart, Gustav Fischer Verlag.

———1965. *Homo "habilis"* and the australopithecines. *Nature, Lond.* 205:121–124.

———1966. The distinctiveness of *Homo habilis. Nature* 209:957–960.

———1967. Variation and the taxonomy of the early hominids. In *Evolutionary biology* vol. 1, pp. 69–100. Dobzhansky, Th., Hecht, M. K. and Steere, W. C., eds. New York: Appleton-Century-Crofts.

———1970. Two new early hominid vertebrae from Swartkrans. *Nature* 225 (5238):8–10.

Robinson, P. 1967. The paleontology and geology of the Badwater Creek Area, central Wyoming. 4. Late Eocene primates from Badwater, Wyoming, with a discussion of material from Utah. *Ann. Carn. Mus.* 39(19):307–326.

Rockwell, H. F., Evans, G. and Pheasant, H. C. 1938. The comparative morphology of the vertebral spinal column: its form as related to function. *Jour. Morph.* 63:87–117.

Rohen, J. W. and Castenholz, A. 1967. Über die Zentralisation der Retina bei Primaten. *Folia primat.* 5:92–147.

Rosenblum, L. A. and Kaufman, I. C. 1967. Laboratory observations of early mother-infant relations in pigtail and bonnet macaques. In *Social communication among primates,* pp. 33–41. Altmann, S. A., ed. Chicago: University of Chicago Press.

Rosenblum, L. A., Kaufman, I. C., and Stynes, A. J. 1964. Individual distance in two species of macaque. *Anim. Behav.* 12:338–342.

Rosenfeld, A. 1969. *Picture processing by computer.* New York: Academic Press.

Roux, W. 1885. Beiträge zur Morphologie der funktionellen Anpassung. *Arch. Anat., Physiol., Anat. Abt.* 9:120–158.

Rowell, T. E. 1966. Forest living baboons in Uganda. *Jour. Zool.* 149:344–364.

———1967. A quantitative comparison of the behavior of a wild and a caged baboon group. *Anim. Behav.* 15:499–509.

Russell, D. E. 1964. Les mammifères paléocènes d'Europe. *Mém. Mus. nat. d'Hist. nat. (Paris) n.s.* 13:1–324.

———1967. Sur *Menatotherium* et l'âge du gisement de Menat. *Probls. Act. de Paleont.,* Coll. Inter. Cent., Nat. Recherche Sci., no. 163, pp. 483–490.

Russell, D. E., Louis, P. and Savage, D. E. 1967. Primates of the French early Eocene. University of Calif. Publ., *Geol. Sci.* 73:1–46.

Saban, R. 1963. Contribution à l'étude de l'os temporal des Primates. *Mém. Mus. nat. d'Hist. nat. (Paris)* n.s. 4, sér. A. (Zool.) 29:1–378.

Sacher, G. 1959. *Relation of lifespan to brain weight and body weight in mammals,* pp. 115–133. In Ciba Foundation symp. on lifespan of animals.

Sade, D. S. 1965. Some aspects of parent-offspring relations in a group of rhesus monkeys, with a discussion of grooming. *Amer. Jour. Phys. Anthrop.* (n.s.) 23:1–17.

———1966. *Ontogeny of social relations in a group of free-ranging rhesus monkeys (Macaca mulatta* Zimmerman). Unpublished Ph.D. dissertation. Berkeley: University California.

———1967. Determinants of dominance in a group of free-ranging rhesus monkeys. In *Social communication among primates.* Altmann, S. A., ed. Chicago: University of Chicago Press.

———1968. Inhibition of son-mother mating among free-ranging rhesus monkeys. *Sci. and Psychoanalysis* 12:18–38.

———1969a. *An algorithm for dominance relations among rhesus monkeys:*

rules for adult females and sisters. Paper presented at annual meeting, American Assoc. of Physical Anthropologists, Mexico City, April 1969.

————1969b. *Short- and long-term views of organization in a monkey society.* Paper presented at meetings of American Anthropological Assoc., New Orleans, November 1969.

————1970. *Interobserver consistency in a longitudinal study of social behavior of free-ranging rhesus monkeys.* Paper presented at meetings of American Assoc. of Physical Anthropologists, Washington, D.C., March 1970.

Salmons, S. 1969. 8th Inter. Conf. on Med. and Biological Engineering. *BioMed Eng.*: 467–474.

Sanderson, I. T. 1957. *The Monkey Kingdom.* Garden City, New York: Hanover House.

Sanides, F. and Krishnamurti, A. 1967. Cytoarchitectonic subdivisions of sensorimotor and prefrontal regions and of bordering insular and limbic fields in slow loris (*Nycticebus coucang coucang*). *Jour. Hirnforschung* 9:225–252.

Sarbin, R. T. 1954. Role theory. In *Handbook of social psychology* vol. 1. Lindzey, G., ed. Cambridge, Mass.: Addison-Wesley.

Sarich, V. M. 1968. The origins of hominids: an immunological approach. In *Perspectives on human evolution* vol. 1, p. 94. New York: Holt, Rinehart and Winston.

Schade, J. P., Meeter, K. and Groeningen, W. B. van 1962. Maturational aspects of the dendrites in the human cerebral cortex. *Acta Morph. neerscandin.* 5:37–48.

Schaeffer, B. 1948. The origin of a mammalian ordinal character. *Evolution* 2:164–175.

Schaller, G. B. 1963. *The mountain gorilla: ecology and behavior*, p. 431. Chicago: University of Chicago Press.

————1967. *The deer and the tiger.* Chicago: University of Chicago Press.

Schaller, G. and Lowther, G. 1969. The relevance of carnivore behavior to the study of early hominids. *Southw. Jour. Anthrop.* 25:307–341.

Schepers, G. W. H. 1946. *The South African fossil ape-men*, part 2. Broom, R. and Schepers, G. W. H., eds. Transvaal Mus. Mem. no. 2.

————1950. *Sterkfontein ape-man Plesianthropus*, part 2. Broom, R., Robinson, J. T. and Schepers, G. W. H., eds. Transvaal Mus. Mem. no. 4.

Schlosser, M. 1911. Beiträge zur Kenntnis der Oligozänen Landsäugetiere aus dem Fayûm: Aegypten. *Beitr. Paläont. Geol. Öst.-Ung.*, Wien 24:51.

Schmidt, E. 1886. Über die Wirbelsäule der Primaten. *Korr. Bl. Ges. Anthrop. Urgesch.* 17:5–6.

Schmidt-Nielson, K. 1964. *Desert animals.* Oxford: Clarendon.

Schoener, T. W. 1970. Nonsynchronous spatial overlap of lizards in patchy habitats. *Ecology* 51(2):408–418.

Schreiber, H. 1936. Die extrembewegungen der schimpansehand. *Gegenb. Morph. Jahrb.* 77:22–60.

Schultz, A. H. 1929. The technique of measuring the outer body of human fetuses and of primates in general. Carn. Inst., Wash., Publ. no. 394. *Contri. Embryo* 117:213–257.

————1930. The skeleton of the trunk and limbs of higher primates. *Human Biol.* 23:303–438.

————1933a. Growth and development. In *The Anatomy of the rhesus monkey* (*Macaca mulatta*). pp. 10–28. Hartman, Carl G. and Straus, William L., Jr., eds. New York: Hafner.

————1933b. Observations on the growth, classification, and evolutionary specialization of gibbons and siamangs. *Human Biol.* 5:212–255, 385–428.

————1936. Characters common to higher primates and characters specific to man. *Quart. Rev. Biol.* 11:259–283, 425–455.

————1938. The relative length of the regions of the spinal column in Old World primates. *Amer. Jour. Phys. Anthrop.* 24:1–22.

————1949. Sex differences in the pelves of primates. *Amer. Jour. Phys. Anthrop.* n.s. 7:401–424.

————1956. Postembryonic age changes. In *Primatologia* vol. 1, pp. 887–964. Basel, New York: Karger.

————1960. Einige Beobachtungen und Masse am Skelett von *Oreopithecus* im Vergleich mit anderen catarrhinen Primaten. *Z. Morph. Anthrop.* 50:136–149.

————1961. Vertebral column and thorax. In *Primatologia* vol. 4(5), pp. 1–66.

————1963. Relations between the lengths of the main parts of the foot skeleton in primates. *Folia primat.* 1:150–171.

————1965. The cranial capacity and the orbital volume of hominoids according to age and sex. *Homenaje a Juan Comas en su 65 Anniversario,* part 2, pp. 337–357. Editorial libros de Mexico, 1965.

————1969. Observations on the acetabulum of primates. *Folia primat.* 11:181–199.

Schultze-Westrum, T. 1964. Nachweis differenzierter Duftstoffe beim Gleitbeulter *Petaurus breviceps papuanus* Thomas. *Naturwiss.* 51:226–227.

Seal, H. 1964. *Multivariate statistical analysis for biologists.* London: Methuen.

Selzer, R. H. 1968. *Improving biomedical image quality with computers, pp. 1–22.* National Aeronautics and Space Administration, tech. rep. no. 32–1336. Pasadena: Jet Propulsion Laboratory, California Institute of Technology.

Shelman, C. B. and Hodges, D. 1970. A general purpose program for the extraction of physical features from a black and white picture. In *IEEE Conf. Rec. of the symp. on feature recognition and selection in pattern recognition,* pp. 135–144. Yau, S. S. and Garnett, J. M., eds. New York: Inst. Elec. and Elect. Eng.

Sigmon, B. A. and Robinson, J. T. 1967. On the function of gluteus maximus in apes and in man. Abst., *Amer. Jour. Phys. Anthrop.* 27:245–246.

Simmons, N. 1969. *The social organization, behavior, and environment of the desert bighorn sheep on the Cabeza Prieta Game Range, Arizona.* Unpublished Ph.D. dissertation. Tucson: University of Arizona.

Simonds, P. E. 1965. The bonnet macaque in South India. In *Primate behavior. Field studies of monkeys and apes,* pp. 175–196. Devore, I., ed. New York: Holt, Rinehart and Winston.

Simons, E. L. 1959. An anthropoid frontal bone from the Fayum Oligocene of Egypt: the oldest skull fragment of a higher primate. *Amer. Mus. Nat. Hist. Novitates* 1976:1–16.

————1961a. Notes on Eocene tarsioids and a revision of some Necrolemurinae. *Bull. Brit. Mus. (Nat. Hist.), Geol.* 5:45–69.

————1961b. The phyletic position of *Ramapithecus. Postilla, Peabody Mus.,* no. 57. New Haven: Yale University.

————1962a. Two new primate species from the African Oligocene. *Postilla, Peabody Mus.,* no. 64(1). New Haven: Yale University.

————1962b. Fossil evidence relating to the early evolution of primate behavior. *Ann. N.Y. Acad. Sci.* 102:282–294.

————1963. A critical reappraisal of Tertiary primates. In *Evolutionary and genetic biology of primates* vol. 1, pp. 65–129. Buettner-Janusch, J., ed. London, New York: Academic Press.

————1964a. On the mandible of *Ramapithecus. Proc. Nat. Acad. Sci.* 51:528.

————1964b. The early relatives of man. *Scient. Amer.* 211:50–62.

————1965. New fossil apes from Egypt and the initial differentiation of Hominoidea. *Nature* 205:135.

————1967a. The significance of primate paleontology for anthropological studies. *Amer. Jour. Phys. Anthrop.* 27:307.

————1967b. The earliest apes. *Scient. Amer.* 217(6):28.

————1967c. Fossil primates and the evolution of some primate locomotor systems. *Amer. Jour. Phys. Anthrop.* 26:241–253.

————1968. A source for dental comparison of *Ramapithecus* with *Australopithecus* and *Homo*. *S. Afr. Jour. Sci.* 64:92.

————1969a. Late Miocene hominid from Fort Ternan, Kenya. *Nature* 221(5179):448.

————1969b. Miocene monkey (Prohylobates) from Northern Egypt. *Nature* 223(5207):687.

————1969c. The origin and radiation of the primates. *Ann. N.Y. Acad. Sci.* 167(1):319.

————1970. The deployment and history of Old World monkeys (Cercopithecidae, Primates). In *Classification of Old World monkeys*. New York: Academic Press.

Simons, E. L. and Chopra, S. R. K. 1969. *Gigantopithecus* (Pongidae, Hominoidea): a new species from North India. *Postilla, Peabody Mus.,* no. 138. New Haven: Yale University.

Simons, E. L. and Ettel, P. C. 1970. *Gigantopithecus. Scient. Amer.* 222:76.

Simons, E. L. and Pilbeam, D. R. 1965. Preliminary revision of the Dryopithecinae (Pongidae, Anthropoidea). *Folia primat.* 3:81–152.

————1971. A gorilla-sized ape from the Miocene of India. *Science* 173:23–27.

Simons, E. L. and Tattersall, I. M. 1969. Notes on some little-known primate fossils from India. *Folia primat.* 10:146.

Simons, E. L. and Wood, A. E. 1968. Early Cenozoic mammalian faunas: Fayûm province, Egypt. Part 1, Simons; Part 2, Wood. *Bull. Peabody Mus.* no. 28. New Haven: Yale University.

Simpson, G. G. 1928. A new mammalian fauna from the Fort Union of southern Montana. *Amer. Mus. Novitates* 279:1–15.

————1933. The "plagiaulacoid" type of mammalian dentition: a study of convergence. *Jour. Mammal.* 14(2):97–107.

————1935. The Tiffany fauna, upper Paleocene. 2. Structure and relationships of Plesiadapis. *Amer. Mus. Novitates* 816:1–30.

————1937. The Fort Union of the Crazy Mountain Field and its mammalian faunas. *Bull. U.S. Natl. Mus.* 169:1–287.

————1940. Studies on the earliest primates. *Bull. Amer. Mus. Nat. Hist.* 77:185–212.

————1944. Tempo and mode in evolution. New York: Columbia University Press.

————1955. The Phenacolemuridae, new family of early primates. *Bull. Amer. Mus. Nat. Hist.* 105:411–442.

————1961. *Principles of animal taxonomy*. New York: Columbia University Press.

Singh, S. D. 1968. Social interactions between the rural and urban monkeys, *Macaca mulatta. Primates* 9:69–74.

Slijper, E. J. 1946. Comp. biol.-anat. investigation on the vertebral column and spinal musculature of mammals. *Kon. ned. Akad. Ver.* (Tweede sec.) 42:1–128.

Slobodkin, L. B. 1967. *Growth and regulation of animal populations*. New York: Holt, Rinehart and Winston.

Smith, G. E. 1924. *The evolution of man*. London: Oxford University Press.

Sneath, P. H. A. 1967. Trend surface analysis of transformation grids. *Jour. Zool. Lond.* 151:65–122.

Snell, O. 1892. Die Abhängigkeit des Hirngewichts von dem Körpergewicht und den geistigen Fähigkeiten. *Arch. Psychiatr. Nervenkrankh.* 23:436–446.

Sody, H. J. V. 1949. Notes on some primates, Carnivora and the Barbirusa from the Indo-Malayan and Indo-Australian regions. *Treubia* 20:121–190.

Sokal, R. R. and Sneath, P. H. A. 1963. *Principles of numerical taxonomy.* San Francisco: Freeman.

Southwick, C. H., Beg, G. and Siddiqi, M. 1965. Rhesus monkeys in North India. In *Primate behavior,* pp. 111–159. DeVore, I., ed. New York: Holt, Rinehart and Winston.

————1967. A population survey of rhesus monkeys in villages, towns, and temples of northern India. *Ecology* 42:538–547.

Southwick, C. H. and Siddiqi, M. 1967. The role of social tradition in the maintenance of dominance in a wild rhesus group. *Primates* 8:341–353.

Spatz, W. B. 1968. Die Bedeutung der Augen für die Sagittale Gestaltung des Schädels von *Tarsius* (Promisiae, Tarsiiformes). *Folia primat.* 9:22–40.

Spinage, C. 1969. Territoriality and social organization of the Uganda defassa waterbuck. *Jour. Zool.* 159:329–361.

Starck, D. 1954. Morphologische Untersuchungen am Kopf der Säugetiere, besonders der Prosimier: ein Beitrag zum Problem des Formwandels des Säugerschädels. *Z. wiss. Zool.* 157:169–219.

Steegman, A. T., Jr. 1970. Cold adaptation and the human face. *Amer. Jour. Phys. Anthrop.* 32:243–250.

Steininger, F. 1967. Ein weiterer Zahn von *Dryopithecus* (Mammalia, Pongidae) aus dem Miozän des Wiener Beckens. *Folia primat.* 7:243.

Stephan, H. 1959. Vergleichend-anatomische Untersuchungen an Insektivorengehirnen. 3. Hirn-Körpergewichtsbeziehungen. *Morph. Jahrb.* 99:853–880.

————1961. Vergleichend-anatomische Untersuchungen an Insektivorengehirnen. 5. Die quantitative Zusammensetzung der Oberflächen des Allocortex. *Acta anat.* 44:12–59.

————1967. Zur Entwicklungshöhe der Insektivoren nach Merkmalen des Gehirns und die Definition der "Basalen Insektivoren." *Zool. Anz.* 179:177–199.

————1969a. Quantitative investigations on visual structures in Primate brains. *Proc. 2nd Inter. Cong. Primat., Atlanta 1968* vol. 3, pp. 34–42. Basel, New York: Karger.

————1969b. Vergleichende metrische und morphologische Untersuchungen an den circumventriculären Organen bei Insektivoren und Primaten. In *Zirkumventrikuläre Organe und Liquor,* pp. 139–145. Sterba, G., Hrsg. Jena: Fischer.

Stephan, H. and Andy, O. J. 1969. Quantitative comparative neuro-anatomy of primates: an attempt at a phylogenetic interpretation. In *Comparative and evolutionary aspects of the vertebrate central nervous system. Ann. N.Y. Acad. Sci.* 167:370–387.

————In preparation. *Quantitative comparison of the amygdala in insectivores and primates.*

Stephan, H. and Bauchot, R. 1965. Hirn-Körpergewichtsbeziehungen bei den Halbaffen (Prosimii). *Acta Zool.* 46:209–231.

Stephan, H., Bauchot, R. and Andy, O. J. 1970. Data on size of the brain and of various brain parts in insectivores and primates. In *The Primate brain: advances in primatology* vol. 1, pp. 289–297. Noback, Ch. and Montagna, W., eds. New York: Appleton-Century-Crofts.

Stephan, H. and Pirlot, P. 1970. Volumetric comparisons of brain structures in bats. *Z. Zool. Syst. Evolutionsforsch.* 8:200–236.

Stewart, T. D. 1962. Neanderthal cervical vertebrae. *Bibl. primat.* vol. 1, pp. 130–153.

Story, H. E. 1945. The external genitalia and perfume gland in *Arctictis binturong. Jour. Mamm.* 26:64–66.

Straus, W. L., Jr. 1963. The classification of *Oreopithecus. Viking Fund Publ. in Anthrop.* 37:146–177.

————1964. The classification of *Oreopithecus.* In *Classification and human evolution,* pp. 146–177. Washburn, S. L., ed. London: Methuen.

Stromer, E. 1913. Mitteilungen über die Wirbeltiere aus dem Mittelpliozän des Natrontales (Aegypten). *Z. dtsch. geol. Ges.* (Berlin) 65:350.

————1920. Mitteilungen über die Wirbeltierreste aus dem Mittelpliocän des Natrontales (Aegypten). 5. Nachtrag zu 1. Affen. 6. Nachtrag zu 2. Raubtiere. *S. B. Bayer. Akad. Wiss.* Munich.

Struhsaker, T. T. 1967a. Ecology of vervet monkeys (*Cercopithecus aethiops*) in the Masai-Amboseli Game Reserve, Kenya. *Ecology* 48:891–904.

————1967b. Social structure among vervet monkeys (*Cercopithecus aethiops*). *Behav.* 29:83–121.

————1969. Correlates of ecology and social organization among African Cercopithecines. *Folia primat.* 11:80–118.

————1970. Phylogenetic implications of some vocalizations of *Cercopithecus* monkeys. In *The Old World monkeys,* pp. 365–444. Napier, J. A. and Napier, P. H., eds. New York: Academic Press.

Struhsaker, T. T. and Gartlan, J. S. 1970. Observations on the behavior and ecology of the Patas monkey (*Erythrocebus patas*) in the Waza Reserve, Cameroon. *Jour. Zool. Lond.* 161:49–63.

Subramoniam, S. 1957. Some observations on the habits of the slender loris, *Loris tardigradus* L. *Jour. Bombay Nat. Hist. Soc.* 54:387–398.

Sugiyama, Y. 1964. Group composition, population density, and some sociological observations of Hanuman langurs (*Presbytis entellus*). *Primates* 5:7–37.

————1965. Behavioral development and social structure in two troops of Hanuman langurs (*Presbytis entellus*). *Primates* 6:213–247.

————1967. Social organization of Hanuman langurs. In *Social communication among primates,* pp. 221–236. Altmann, S., ed. Chicago: University of Chicago Press.

Sugiyama, Y., Yoshiba, K. and Parthasarathy, M. D. 1965. Home range, breeding season, male group, and intertroop relations in Hanuman langurs (*Presbytis entellus*). *Primates* 6:73–106.

Sutherland, D. H. 1966. An electromyographic study of the plantar flexors of the ankle in normal walking on the level. *Jour. Bone and Joint Surg.* 48-A:66–71.

Sutton, D. L. 1962. Surface and needle electrodes in electromyography. *Dental Progress,* 2: 127–131.

Symington, J. 1916. Endocranial casts and brain form: a criticism of some recent speculations. *Jour. Anat. Physiol.* 50:111–130.

Szalay, F. S. 1968a. The beginnings of primates. *Evolution* 22(1):19–36.

————1968b. The Picrodontidae, a family of early primates. *Amer. Mus. Novitates* 2329:1–55.

————1968c. Origins of the Apatemyidae (Mammalia, Insectivora). *Amer. Mus. Novitates* 2352:1–11.

————1969a. Mixodectidae, Microsyopidae, and the insectivore-primate transition. *Bull. Amer. Mus. Nat. Hist.* 140(4):193–330.

————1969b. Uintasoricinae, a new subfamily of early Tertiary mammals (?Primates). *Amer. Mus. Novitates* 2363:1–36.

————1969c. Origin and evolution of function of the mesonychid condylarth feeding mechanism. *Evolution* 23(4):703–720.

Tanaka, J. 1965. Social structure of Nilgiri langurs. *Primates* 6:107–128.

Tappen, N. C. 1955. Relative weights of some functionally important muscles of the thigh, hip, and leg in a gibbon and in man. *Amer. Jour. Phys. Anthrop.* 13:415–420.

————1961. A method for analyzing muscles function in locomotion. *Surgical Forum* 12:440–442.

Tattersall, I. 1969. Ecology of North Indian *Ramapithecus*. *Nature* 221:451.

Thenius, E. 1969. Stammesgeschichte der Säugetiere (einschliesslich der Hominiden). *Handb. d. Zool., Bd. 8 (Mammalia)*, 47. Lieferung, 1–368.

Thomas, F. D. and Sellers, E. E. 1969. *Biomedical sciences instrumentation* vol. 6. Pittsburgh: Instrument Soc. of Amer.

Thorington, R. W., Jr. 1967. Feeding and activity of Cebus and Saimiri in a Columbian forest. In *Progress in primatology*, pp. 180–184. Starck, D., Schneider, R. and Kuhn, H. J., eds. Stuttgart: Gustav Fischer Verlag.

Thuma, B. D. 1928. Studies on the diencephalon of the cat. 1. The cytoarchitecture of the corpus geniculatum laterale. *Jour. Comp. Neur.* 46:173–198.

Tilney, F. 1928. *The brain from ape to man.* New York: Hoeber.

Tobias, P. V. 1963. The cranial capacity of *Zinjanthropus* and other australopithecines. *Nature* 197:743–746.

————1964. The Olduvai Bed I hominine with special reference to its cranial capacity. *Nature* 202:3–4.

————1966. A member of the genus *Homo* from Ubeidiya. *Israel Acad. of Sci. and Human. (Jerusalem)* 1966:1–12.

————1967. *The cranium and maxillary dentition of Australopithecus* (Zinjanthropus) *boisei. Olduvai Gorge*, vol. 2. Cambridge: University Press.

————1968a. *New African evidence on human evolution.* Address to Supper Conference of Wenner-Gren Foundation for Anthropological Research, New York, April 5, 1968.

————1968b. Cranial capacity in anthropoid apes, *Australopithecus* and *Homo habilis*, with comments on skewed samples. *S. Afr. Jour. Sci.* 64:81–91.

————1968c. The age of death among the australopithecines. In *The Anthropologist*, special vol. 1968:23–28.

————1968d. Cultural hominisation among the earlier African Pleistocene hominids. *Proc. Prehist. Soc.* n.s. 33:367–376.

————1968e. The taxonomy and phylogeny of the australopithecines. In *Taxonomy and Phylogeny of Old World Primates*, B. Chiarelli, ed. pp. 277–315. Turin: Rosenberg and Sellier.

————1969. Commentary on new discoveries and interpretations of early African fossil hominids. *Yearbook of Phys. Anthrop., 1967*, 15:24–30.

————1970. Brain size, grey matter and race—fact or fiction? *Amer. Jour. Phys. Anthrop.* 32:3–26.

————1971a. The men who came before Malawian history. In *Early history of Malawi*. Pachai, B., ed. London: Messrs. Longmans.

————1971b. *The brain in hominid evolution.* New York and London: Columbia University Press.

————1971c. The distribution of cranial capacity values among living hominoids. *Proc. 3rd Inter. Cong. of Primat.*, Zurich, August 2–5, 1970.

Tobias, P. V. and Hughes, A. R. 1969. The new Witwatersrand University excavation at Sterkfontein. *S. Afr. Archaeol. Bull.* 24:158–169.

Tobias, P. V. and Koenigswald, G. H. R. von 1964. A comparison between the

Olduvai hominines and those of Java, and some implications for hominid phylogeny. *Nature* 204:515–518.

Tobien, H. 1936. Mitteilungen über Wirbeltierreste aus dem Mittelpleistocän des Natrontales (Aegypten). *2 Dtsch. Geol. Ges.* 88:41.

Travill, A. and Basmajian, J. V. 1961. Electromyography of the supinators of the forearm. *Anat. Rec.* 139:557–560.

Trotter, M. 1964. Accessory sacroiliac articulations in East African skeletons. *Amer. Jour. Phys. Anthrop.* 22:137–142.

Tugby, D. J. 1953. A new internal casting method. *Amer. Jour. Phys. Anthrop.* 11:437–440.

Tuttle, R. H. 1965. *The anatomy of the chimpanzee hand, with comments on hominoid evolution.* Ph.D. thesis. Berkeley: University of California.

———1967. Knuckle-walking and the evolution of hominoid hands. *Amer. Jour. Phys. Anthrop.* 26:171–206.

———1968. Some swamp forest adaptations of the orangutan. *Bull. Amer. Anthrop. Assoc.* 1:141.

———1969a. Quantitative and functional studies on the hands of the Anthropoidea. 1. The Hominoidea. *Jour. Morph.* 128:309–364.

———1969b. Knuckle-walking and the problem of human origins. *Science* 166:953–961.

———1969c. *Terrestrial trends in the hands of the Anthropoidea: a preliminary report.* vol. 2, pp. 192–200. Proc. 2nd Inter. Cong. Primat., Atlanta, Georgia. Basel: Karger.

———1970. Postural, propulsive, and prehensile capabilities in the cheiridia of chimpanzees and other great apes. In *The Chimpanzee* vol. 2, pp. 167–253. Bourne, G., ed. Basel, New York: Karger.

Ullrich, W. 1961. Zur biologie und soziologie der Colobusaffen (*Colobus guereza caudatus* Thomas 1885). *Der Zoologische Garten* 25:305–368.

United Nations 1967. *Manual IV: Methods of estimating basic demographic measures from incomplete data.* Department of Economic and Social Affairs Population Study 42. New York: United Nations.

Uyeno, E. T. 1967. Lysergic acid diethylamide and dominance behavior of the squirrel monkey. *Arch. Inter. de Pharmacodynamie et de Thérapie* 169(1): 66–69.

Vallois, H. V. 1954. La capacité cranienne chez les Primates supérieurs et le "Rubicon cerebral." *C.R. Acad. Sci.* 238:1349–1351.

Vandenbergh, J. G. 1967. The development of social structure in free-ranging rhesus monkeys. *Behav.* 29(2–4):179–194.

Van Valen, L. 1965. Tree shrews, primates, and fossils. *Evolution* 19:137–151.

———1966. Deltatheridia, a new order of mammals. *Bull. Amer. Mus. Nat. Hist.* 132(1):1.

———1969. A classification of the primates. *Amer. Jour. Phys. Anthrop.* 30: 295–296.

Van Valen, L., and Sloan, R. E. 1965. The earliest primates. *Science* 150: 743–745.

Vogel, C. 1970. *Verhaltensunterschiede der indischen Langurenart Presbytis entellus in den Kumaon-Hills (S. Himalaya) und in Rajasthan.* Paper read at the Third International Congress of Primatology, Zurich.

Vogt, C. and Vogt, O. 1907. Zur Kenntnis der elektrisch erregbaren Hirnrindengebiete bei den Säugetieren. *Jour. Psychol. Neurol.* 8:277–456.

Wagner, S. S. 1970. The maximum-likelihood estimate for contingency tables with zero diagonal. *Jour. Amer. Stat. Assoc.* 65(331):1362–1383.

Wagstaffe, W. W. 1874. *On the mechanical structure of the cancellous tissue of bone.* St. Thomas Hosp., Rep. 5 (n.s.):192–214.

Walker, A. C. 1967a. Patterns of extinction among the subfossil Madagascan lemuroids. In *Pleistocene extinctions,* pp. 425–432. Martin, P. S. and Wright, H. E., Jr., eds. New Haven: Yale University Press.

———1967b. *Locomotor adaptations in Recent and fossil Madagascan lemurs.* Ph.D. thesis. London: University of London.

———1969. The locomotion of the lorises, with special reference to the potto. *E. Afr. Wildlife Jour.* 7:1–5.

———1970. Nuchal adaptations in *Perodicticus potto. Primates* 11:135–144.

Walker, A. C. and Rose, M. D. 1968. Fossil hominoïd vertebra from the Miocene of Uganda. *Nature* 217:980.

Walls, G. L. 1942. *The vertebrate eye and its adaptive radiation.* 1965 reprint. New York: Hafner.

Washburn, S. L. 1947. The relation of the temporal muscle to the form of the skull. *Anat. Rec.* 99:239–248.

———1951a. The analysis of primate evolution with particular reference to the origin of man. *Cold Spring Harbor Symp. on Quant. Biol.* 15:67–78.

———1951b. The new physical anthropology. *Trans. N.Y. Acad. Sci.* ser. 2, 13:198–304.

———1957. Ischial callosities as sleeping adaptations. *Amer. Jour. Phys. Anthrop.* 15:269–276.

———1960. Tools and human evolution. *Scient. Amer.* 203:63.

Washburn, S. L. and DeVore, I. 1961a. Social behavior of baboons and early man. In *Social life of early man,* pp. 91–105. Washburn, S. L., ed. Chicago: Aldine Publ. Co.

———1961b. The social life of baboons. *Scient. Amer.* 204:62–71.

Washburn, S. L. and Jay, P. C. 1968. *Perspectives on human evolution* vol. 1. New York: Holt, Rinehart and Winston.

Watt, K. E. F. 1968. *Ecology and resource management.* New York: McGraw-Hill.

Weidenreich, F. 1945. The brachycephalization of recent mankind. *Southw. Jour. Anthrop.* 1:45–98.

———1946. *Apes, giants, and man.* Chicago: University of Chicago Press.

Weiner, J. S. 1954. Nose shape and climate. *Amer. Jour. Phys. Anthrop.* 12: 615–618.

Weir, J. S. 1960. A possible course of evolution of animal drinking holes (pans) and reflected changes in their biology. *Proc. 1st Fed. Sci. Cong.,* Salisbury.

Welker, W. I. and Campos, G. B. 1963. Physiological significance of sulci in somatic sensory cerebral cortex in mammals of the family Procyonidae. *Jour. Comp. Neur.* 120:19–36.

Welker, W. I. and Seidenstein, S. 1959. Somatic sensory representation in the cerebral cortex of the raccoon (*Procyon lotor*). *Jour. Comp. Neur.* 11:469–501.

Wells, L. H. 1969. Faunal subdivision of the Quaternary in Southern Africa. *S. Afr. Archaeol. Bull.* 24:93–95.

Welt, C. 1962. *Topographical organization of somatic sensory and motor areas of the cerebral cortex of the gibbon (Hylobates) and chimpanzee (Pan).* Ph.D. thesis. Chicago: University of Chicago.

Whillis, J. 1940. The development of synovial joints. *Jour. Anat. Lond.* 74:277–283.

Whitla, D. K., ed. 1968. *Handbook of measurement and assessment in behavioral sciences.* Reading, Mass.: Addison-Wesley.

Wickler, W. 1967. Socio-sexual signals and their intraspecific imitation among primates. In *Primate ethology,* pp. 69–147. Morris, D., ed. Chicago: Aldine Publ. Co.

Willis, J. C. and Willis, M. 1911. *A revised catalogue of the flowering plants and ferns of Ceylon.* Colombo: H. C. Cottle.

Wilson, E. G. and Worcester, J. 1939. Note on factor analysis. *Psychometrika* 4:133–148.

Wimer, R. E., Wimer, C. E., and Roderick, T. H. 1969. Genetic variability and forebrain structures between inbred straints of mice. *Brain Research* 16: 257–264.

Wishart, D. 1969. Mode analysis: a generalisation of nearest neighbour which reduces chaining effects. In *Numerical taxonomy,* pp. 282–308. Cole, A. J., ed. London: Academic Press.

Wolff, J. 1870. Über die innere Architectur der Knochen und ihre Bedeutung für die Frage vom Knochenwachstum. *Virchow's arch. path. anat.* 50:389–453.

Wolffson, D. M. 1950. Scapula shape and muscle function, with special reference to the vertebral border. *Amer. Jour. Phys. Anthrop.* 8:331–343.

Wolpoff, M. R. 1968. Climatic influence on the skeletal nasal aperture. *Amer. Jour. Phys. Anthrop.* 29:405–424.

———1969. Cranial capacity and taxonomy of Olduvai Hominid 7. *Nature* 223:182–183.

Woo, J. K. 1962. The mandibles and dentition of *Gigantopithecus. Palaeont. Sinica* n.s. D, 11:1–94.

Wood, A. E. 1962. The early Tertiary rodents of the family Paramyidae. *Trans. Amer. Phil. Soc.,* new ser., vol. 52(1):1–261.

Woolsey, C. N. 1958. Organization of somatic sensory and motor areas of the cerebral cortex. In *Biological and biochemical bases of behavior.* Harlow, H. F. and Woolsey, C. N., eds. Madison: University of Wisconsin Press.

Wortman, J. L. 1903–04. Studies of Eocene mammalia in the Marsh Collection, Peabody Mus. 2. Primates. *Amer. Jour. Sci.,* vols. 15–17.

Wray, S. H. 1969. Innervation ratios for large and small muscles in the baboon. *Jour. Comp. Neur.* 137:227–250.

Wrobel, K. H. 1966. Untersuchungen an den Blutgefässen des Primaten Greifschwanzes. *Z. Anat. Entwicklungsgesch.* 125:177–188.

Yamada, M. 1963. A study of blood-relationship in the natural society of the Japanese macaque. *Primates.* 4:43–65.

Yau, S. S. and Garnett, J. M. 1970. *Conference record of the symposium on feature extraction and selection in pattern recognition.* New York: Institute of Electrical and Electronic Engineering.

Yoshiba, K. 1968. Local and intertroop variability in ecology and social behavior of common Indian langurs. In *Primates,* pp. 217–242. Jay, P., ed. New York: Holt, Rinehart and Winston.

Zapfe, H. 1958. The skeleton of *Pliopithecus* (*Epipliopithecus*) *vindobonensis* Zapfe and Hürzeler. *Amer. Jour. Phys. Anthrop.* 16:441.

Ziegler, A. C. 1964. Brachiating adaptations of chimpanzee upper limb musculature. *Amer. Jour. Phys. Anthrop.* 22:15–32.

Zuckerman, S. 1933. *Functional affinities of man, monkeys, and apes.* New York: Harcourt, Brace.

———1966. Myths and methods in anatomy. *Jour. Roy. Coll. Surg.* (*Edinburgh*) 11:87–114.

———1970. *Beyond the ivory tower.* London: Weidenfeld and Nicolson.

Zuckerman, S., Ashton, E. H., Oxnard, C. E. and Spence, T. F. 1967. The functional significance of certain features of the innominate bone in living and fossil primates. *Jour. Anat. Lond.* 101:608-609.

Zuckerman, S. and Fulton, J. F. 1941. The motor cortex in *Galago* and *Perodicticus. Jour. Anat. Lond.* 75:447–456.

Index

Aardvark, 97
Abelson, R. P., 435, 437
Accessory olfactory bulb: *see* Bulbus olfactorius accessorius
Acheulian, 65, 375
Activity profiles, 410, 411
Adams, D. K., 452
Adapids, 6, 7, 32
Adapis, 119
Adaptive grades of primates, 421
Adaptive shifts, 97, 98; to diurnal, herbivorous mode, 119; of prey seizing functions of hand, 117
Aegyptopithecus zeuxis, 41, 43, 44, 45, 47, 181
Aeolopithecus chirobates, 33, 44, 46, 56
Aethechinus algirus, 167
Aggression: chasing, 368–69; control of, 414; intra-group, 197, 201, 369, 388, 407; reduction of, 197, 201; in rhesus, 384, 385, 421; at waterholes, 363, 364, 376
Aging, x, xx
Agriotherine bear, 66
Alarm calls, 118
Aldrich-Blake, P., 258, 357, 444
Allee, W. C., 104, 437
Allen, G. M., 246, 257, 437
Allen's rule, 257
Allometry, xiv, 306; dental arcade, 117; formula, 156, 173; method, 156; orbital frontation, 111; orbital-margin convergence, 111; tarsal elongation, 118
Alouatta, 165, 172, 179, 182, 183, 184, 219, 233, 265, 266, 319, 356, 368, 380, 412, 413
Alouatta palliata, 356, 368
Alouatta seniculus, 167
Alouattinae, 54, 230, 232, 234, 235, 318
Alouattines: *see* Alouattinae
Altmann, J., 421, 424, 432, 437
Altmann, S. A., 368, 378, 394, 410, 421, 423, 424, 426, 432, 437
Amphipithecus, 33
Amygdaloid, 158, 168
Anaptomorphidae, 9, 29, 31
Anaptomorphids: *see* Anaptomorphidae
Ancepsoides, 17
Andrew, R. J., 114, 437
Andrews, C. W., 39, 437
Andrews, H. C., 338, 347, 437
Andrews, H. N., 103, 437
Andrews, P., 41, 55, 438
Andy, O. J., 156, 158, 163, 172, 188, 438, 463
Angiosperms, 116
Ankel, F., xv, 55, 224, 225, 226, 229, 231, 232, 233, 234, 235, 236, 239, 240, 438
Anteater, 103, 107
Antelope, 41
Anthony, J., 179, 182, 184, 438
Anthracocerus coronatus, 372
Anthracothere, 40, 42, 43
Anthropoidea, 112, 178, 180, 266, 318
Anthropoids: *see* Anthropoidea
Anticlinal vertebra, 225, 230, 239

Aotus trivirgatus, 167, 171, 179, 182, 184
Apatemids, 6
Apes, 99, 198, 211, 212, 215, 221, 228, 250, 262, 269, 271, 280, 289, 315, 325, 330, 346, 420; *see also* Pongidae, Hylobatidae, Hominoidea
Apidium, 43, 47
Apterodon, 40
Aquino, C. F., 346, 438
Arambourg, C., xii, 39, 41, 438
Arao, T., 114, 438
Arboreal adaptations: grasping branches of small diameter, 108; in Hominoidea, 222; in mammals, 105; primate, 100–108; primate-like, 121; varieties of, 103–7
Arboreal habitation: beginning of, 155; in Mesozoic primate ancestors, 100; vicissitudes of, xii
Arboreal pathways, xvii; movement on, 108
Arboreal theory of primate evolution, 100–102, 105, 107, 121
Archaeolemur, 177, 178, 181
Arctocebus, 179, 229, 318
Arctogalidia, 105
Armstrong, 40
Arrangement of chapters, x
Arsinotheres, 43
Arctictis, 105, 107
Artiodactyla, 215, 395
Ascending primate scales, 162, 163, 168, 169, 171, 173, 174
Ascher, R., 200, 449
Ashby, R., 380, 438
Ashton, E. H., 90, 194, 259, 265, 266, 310, 311, 312, 313, 327, 330, 438, 468
Asmussen, E., 302, 438
Astronomy, 307
Asymmetry, xx
Ateles, 55, 56, 167, 179, 180, 181, 182, 229, 233, 235, 266, 357, 413
Atelinae, 54, 177, 230, 234, 255, 318
Atelines: *see* Atelinae
Auditory bulla, 5, 6; early Paleocene primate, 16; lorisoid, 7; in rodents, 16; tupaiid, 6
Auditory sense, x
Auditory tube, 5; microchoerid, 7; in *Phenacolemur,* 7; in *Plesiadapis,* 7; tarsiid, 7
Aulaxinus, 41
Australopithecinae, 81, 191, 195, 200, 203
Australopithecine: *see* Australopithecinae
Australopithecus, 38, 39, 57, 59, 62, 70, 81, 87, 88, 125, 162, 195, 346
Australopithecus africanus, 77, 80, 81, 82, 83, 86, 90, 91, 185, 237
Australopithecus boisei, 67, 87, 91, 162
Australopithecus habilis, 39
Australopithecus robustus, 68, 76, 77, 80, 81, 82, 83, 87, 88, 90, 91, 237
Avahi, 111, 118
Avahi laniger, 167
Avis, V., 210, 211, 215, 438
Axis axis, 367, 371, 372, 373, 374, 375

469

Axis deer: *see Axis axis*
Aye-aye: *see Daubentonia madagascariensis*
Azuma, S., 413, 438

Baboon: *see Papio*
Baldwin, J. D., 382, 397, 413, 438
Bales, R. F., 421, 426, 439
Banks, E., 104, 439
Barnicot, N. A., 455
Bartholomew, D. J., 435, 439
Bartlett, M. S., 435, 439
Basal insectivore, 156, 158, 159, 161, 168, 169,
   171, 172, 173
Basenji hound, 185
Basicranial anatomy: in early primates, 6–7; of
   early rodents, 16; in Paromomyidae, 8; of
   Plesiadapis, 16, 34
Basmajian, J. V., xvi, 267, 268, 292, 296, 298,
   299, 300, 301, 302, 303, 323, 439, 448, 454,
   465
Bassett, C. A. L., 337, 439
Bat: *see Chiroptera*
Batadon, 6
Battye, C. K., 300, 304, 439
Bauchot, R., 156, 157, 158, 161, 162, 177, 439,
   463
Bazant, F. J., 298, 439
Beadnell, H. J. L., 39, 40, 42, 439
Bear macaque: *see Macaca arctoides*
Beck, B., xvii
Becker, H. C., 439
Becker, R. O., 337, 338, 439
Beg, G., 354, 463
Behavior, vii; activity cycles, 353; air biting,
   368; aunt, 406; categories of, 385–86; displacement, 368; evolution of hominid, 196–
   97, 375; field, 62; frequency of, 415, 426,
   432–34; inferential analyses of, 310; intraspecific, 356; interspecific, 352, 353, 356; naturalistic, ix, 416; patterns, 421; presenting,
   368; qualification of, xviii, 415; relation to
   morphology, 310; spacing, 368, 394, 405,
   414; species-specific, 185, 186, 411; subsistence, xvii, xix; symbolizing, 197; terrestrial, xvii, 358; time units of, 379, 381–83;
   type of measurement of, 380–81; units of,
   185, 380
Behavioral observations, viii; bias, 380, 384;
   episodic, xvii; laboratory, ix; mobile, xvii,
   351, 352; naturalistic, ix; night, 370, 377;
   short term, xvii; stationary, xvii, 351, 352,
   377; from tree blind, 355
Behavioral studies: on apes, 269; on Cercopithecoidea, 269; cross-sectional, 421; focus of
   field, 379, 380; laboratory, 387; longitudinal,
   x, xvii, 381, 382, 384, 387, 391, 394–95,
   422; primate, xvii, 352; selection of sites for,
   402–4; short-term, 387, 391
Benjamin, R. M., 182, 439
Bennett, K. A., 440
Bensley, 101
Bergmann's rule, 257
Bergmans, J., 455
Biegert, J., 33, 118, 440
Bernstein, I. S., xvii, xviii, 259, 359, 363, 394,
   400, 401, 402, 406, 407, 409, 410, 411, 412,
   413, 434, 440
Berruvius, 10, 37, 121
Biegert, J., 32, 118, 440
Biogram, 396
Biomechanics, 292, 304; forces, 307; of living
   forms, 330; of movement, 322; studies, 307;
   theoretic, 292–94
Biomechanical principles, ix, xvi
Biomolecular primatology, viii

Bipedalism: apprenticeship for, 216; on
   branches, 118, 289; in early hominids, 76, 81,
   197, 201, 375; evolutionary origin of human,
   295, 304; human, xvi, 202, 237; langur, 375,
   376; selection for, 200
Bird, 175, 371
Birdsell, J. B., 124, 444
Bishop, A., 104, 117, 440
Bishop, W. W., 48, 440
Bishop, Y. M. M., 426, 427, 429, 440
Blankenhorn, M. L. P., 41, 441
Blackith, R. E., 311, 441
Blinkov, S. M., 186, 441
Bock, R. D., 311, 441
Bock, W. J., 267, 268, 447
Body proportions, 262, 311
Body size: of *Aegyptopithecus*, 43; affects on
   dietary requirements and preferences, 27; of
   African dryopithecines, 54; of australopithe-
   cines, 81, 196; brain volume related to, 158,
   202; of Carpolestidae, 21; of fossil hominids,
   195–96; of *Gigantopithecus blacki*, 61; increase in, 197, 202; increase during prosimian
   evolution, 120; increase from *Australopithecus*
   to *Homo erectus*, 196; of Paromomyidae, 8;
   of Plesiadapidae, 17; of *Pronothodectes*, 17;
   of *Propliopithecus*, 46
Body weight, xx; cranial capacity to estimate,
   191, 196; of fossils, 177; in *Macaca*, 257;
   relationship of brain weight to, xiv, 156, 157,
   159, 173, 177, 188; support of, 309; of *Tarsius*, 266
Bollihger, A., 114, 441
Bond: early interactional, 387; mother-daughter,
   390; mother-infant, 387
Bone-joint-muscle relationships, xvi, 322
Bones, viii; adaptation of, 310; compact, 306,
   309; piezo-electric phenomena in, 337; semi-
   conducting phenomena in, 337; shapes of, 331;
   stresses, 337; trabecular, 306, 309; wrist, xv
Bonin, G. von, 81, 441
Bonnet macaque, *see Macaca radiata*
Boorman, S. A., 434, 441
Booth, C., 404, 441
Boyce, A. J., 332, 441
Brace, C. L., 111, 194, 441
Brachial index: in earliest primates, 35; of
   Notharctus, 34; of *Plesiadapis*, 34; in vertical
   clingers and leapers, 34
Brachiation: *see* Suspensory behaviors
Brachyodus, 43
Brachyteles, 179, 182, 229
Brain, ix; of *Aegyptopithecus*, 43; australopithecine, 199; behavioral qualities from morphology of, 185; comparative studies on, ix, 175,
   186–89; correlation between size and longevity, 177; enlargement, 99, 200, 202,
   203; evolution, ix, 186; evolution in Hominoidea, xiv, 174, 186; external morphology and endocasts of, 176–77; fossil: 155;
   function of parts in, 155; functional significance of external morphology of, 176; intelligence, 177; internal structure of, xiii; Jerison's
   coefficient of efficiency, 177; localization of
   function in, 189; of primitive insectivores,
   156; reorganization, 188, 196; sexual dimorphism of, 187; size, xx, 155, 177, 188; volume, 177, 188, 191; volumes of components
   in, 158–59, 163–72; weight, xiii, xiv, 156–
   58, 159–63, 188
Brain, C. K., 66, 81, 83, 402, 404, 414, 441,
   447
Braincase: *see* Neurocranium
Bray, R. S., 434, 441
Bridgman, P. W., 436, 441

Broca, P., 109, 199, 441
Bromiley, R. B., 457
Broom, R., 90
Brow development, 124
Brown, R., 347, 455
Brown, T., 129, 130, 131, 132, 133, 140, 441
Bucknill, T. M., 209, 213, 453
Bulbus olfactorius, 158, 173, 177; in Chiroptera, 173; regression of, 168, 171; in tree-shrews, 173
Bulbus olfactorius accessorius, 158, 171, 173
Bunak, V. V., 125, 441
Burg Wartenstein, x; symposium, viii
Burramys, 26, 104
Burton, M, 246, 442
Butler, J. W., 338, 442

Cacajao, 179
Caenolestes, 122
Caenolestid, 25
Calcaneum: elongation in Eocene primates, 118; elongation in *Galago*, 207; elongation in *Microsyops*, 33; elongation in *Tarsius*, 207; tuber calcanei of, 115
Callicebus, 104, 179, 368, 412
Callicebus moloch, 167, 368
Callithricidae, 107, 411
Callithricids: see Callithricidae
Callithrix, 105, 235
Callithrix jacchus, 167
Callosciurus, 105
Callosciurus prevosti, 112
Caluromys, 104, 114
Camel, 41
Campanella, S. J., 339, 347, 442
Campbell, B. G., xii, 80, 81, 87, 257, 442
Campos, G. B., 180, 467
Canine: in Carpolestidae, 24; in *Gigantopithecus*, 61; of gorilla, 57; in Picrodontidae, 14, 16; in *Pronothodectes*, 17; of pygmy chimpanzee, 57; reduction in hominids, 60; reduction in Paleocene primates, 29; reduction and weapon use, 60; in *Tarsius*, 28
Canine-premolar hone-complex: in *Aegyptopithecus*, 44; in platyrrhine monkeys, 44
Canis aureus, 363, 371
Canonical analysis, 123, 124, 310, 311, 312, 313, 316, 317, 319, 320, 326, 327, 328, 330, 334, 419
Capuchin: see *Cebus*
Caribbean Primate Research Center, xviii
Carlsöö, S., 301, 302, 303, 442
Carnivora, 40, 214, 377
Carnivore: see Carnivora
Carpal bones: early hominid, xv, 77 (see also Centrale, Hamate, Pisiform, Scaphoid, Trapezium, and Triquetral)
Carpenter, C. R., 357, 368, 380, 394, 412, 413, 442
Carpodaptes, 23, 24, 31
Carpodaptes aulacodon, 22
Carpodaptes hazelae, 22
Carpodaptes hobackensis, 22, 24
Carpolestes, 23, 24
Carpolestes aquilae, 22, 25
Carpolestes cygneus, 22
Carpolestes dubius, 22, 23
Carpolestes nigridens, 23
Carpolestidae, 8, 20, 21, 23, 24, 29, 30, 32, 121
Carpolestids: see Carpolestidae
Carpus: see wrist
Cartesian coordinates, 312–13, 418
Cartmill, M., xii, xiii, xvii, 103, 105, 111, 112, 117, 119, 352, 442
Caspari, E., 200, 442

Castenholz, A., 109, 459
Cat, 113, 180
Catfish, 42
Caudal region: see Tail
Cave, A. J. E., 115, 442
Cayo Santiago Island, xvii, xviii, 383, 403
Cebidae, 179
Cebids: see Cebidea
Ceboidea, 44, 101, 102, 107, 162, 171, 172, 177, 185, 214, 216, 226, 230
Ceboids, see Ceboidea
Cebuella, 235
Cebus, 104, 167, 172, 177, 179, 182, 232, 233, 234, 413
Cebus albifrons, 167
Cebus nigrivittatus, 218
center of gravity: position in hominoids and monkeys, 237; relation to joints, 330
Centrale: fusion to scaphoid, 211
Cercartetus, 104, 108, 117
Cercocebus, 54, 233, 269, 276, 281, 285, 286, 287, 288, 290, 381, 413
Cercocebus albigna, 167
Cercocebus atys, 276, 278, 280
Cercopithecidae, 398
Cercopithecinae, 269, 280, 281, 282, 286, 290, 291
Cercopithecoidea, 54, 55, 162, 171, 172, 177, 184, 216, 226, 230, 249, 250, 260, 269, 270, 271, 274, 276, 277, 279, 283, 284, 285, 286, 287
Cercopithecoids: see Cercopithecoidea
Cercopithecus, 44, 225, 233, 249, 269, 280, 290, 412
Cercopithecus aethiops, 43, 269, 277, 278, 280, 282, 286, 288, 403, 404, 412, 414, 423, 425
Cercopithecus ascanius, 167
Cercopithecus (Miopithecus) talapoin, 162, 165, 167
Cercopithecus mitis, 167, 356
Cercopithecus nictitans, 104, 276, 277, 278, 279, 280, 282, 283, 286, 288
Cerebellum, 158, 177, 198; in gibbons, 178; progression indices of, 168–69, 170
Cerebral cortex: Broca's area, 199; infraparietal, 199; inferior parietal, 198; parietal, 198, 199, 201; primitive patterns of, 156; superior parietal, 198; temporal, 198, 199, 201; Wernicke's area, 199
Cervical vertebrae, 225, 226; atlas, 224, 227, 228; axis, 224, 227, 228, 236; bodies of, 228–29; early hominid, 78; elongated dorsal spines in *Perodicticus*, 229, 236; elongated dorsal spines in Pongidae, 229, 236; function of, 235; in Neanderthal men, 236; specializations of, 227–29; spinal processes of, 229
Cetaceans, 176
Ceylonese gray langur: see *Presbytis entellus thersistes*
Chadronia, 40
Chalmers, N. R., 380, 381, 404, 442
Chameleon, 116
Charles-Dominique, P., 120, 442
Chasing, 427, 428
Cheek teeth: see Molars
Cheiridia: development of clasping in Cretaceous insectivorans, 116; exaggerated prehensility of, 120, 121
Cheirogaleine: see Cheirogaleinae
Cheirogaleinae, 108, 113, 114, 119, 164
Cheirogaleus, 105, 120
Cheirogaleus major, 167
Cheirogaleus medius, 167
Cherry, R., 85
Chimpanzee: see *Pan troglodytes*

Chiromyoides, 16, 17, 18, 28, 31, 32, 37
Chiromyoides campanisus, 19
Chiroptera, 27, 97, 103, 156, 157, 165, 173
Chivers, D. J., 356, 368, 442
Chlorotalpa stuhlmanni, 167
Chopra, S. R. K., 61, 462
Chronology, viii
Civet: *see Genetta*
Clark, J. D., 65, 375
Clark, W. E. Le Gros, xiv, 48, 49, 50, 54, 87, 88, 98, 104, 117, 175, 177, 178, 180, 197, 250
Clark, P., 446
Clavicle: in *Daubentonia*, 327; early hominid, 77; from Maboko Island, 51; Olduvai, 330; pliopithecine, 55; shape of, 311, 327
Claws, 108; of *Daubentonia*, 108; of *Plesiadapis*, 34, 99
Climatic adaptations: in *Macaca*, 257
Cockburn, A., 434, 443
Cocker spaniel, 186
Cognitive reorganization, 199
Cohen, J. E., xi, xviii, 147, 149, 150, 416, 422, 424, 443
Coker, E. G., 323, 443
Coleman, J. S., 422, 423, 435, 443
Collagen, 306
Collias, N., 413, 443
Collins, E. T., 100, 101, 443
Collins, W. E., 435, 443
Colobinae, 269, 278, 279, 280, 281, 290, 318, 359, 363, 375, 411
Colobines: *see* Colobinae
Colobus, 55, 229, 266, 269, 286, 368
Colobus badius, 167
Colobus guerza, 368
Colobus monkey: *see Colobus*
Colobus polykomos, 275, 276, 278, 280, 281, 284, 286, 287, 288, 290
Communication: human, 422; signals of, 394, 421
Comparative morphologist, x
Comparative morphology: postcranial problems in, xv; studies needed on living hominoids, 89,
Competition between males, 397
Conaway, C. H., 397, 443
Condylarth, 29
Conel, J. L., 187, 443
Connolly, C. J., 177, 182, 443
Contacos, P. G., 435, 443
Cooke, H. B. S., 86, 443
Cooley, W. W., 142, 444
Coon, C. S., 76, 124, 444
Cooper, D. M., 443
Coppens, Y., xii, 80, 92, 93
Copulation: frequency of, 381; inhibitions of mother-son, xvii, 392–93; rhesus, 384; ventral-ventral, 197, 201
Coronoid process: in carpolestids, 30; in early condylarths, 20; in early leptictids, 29; in paromomyids, 29; in picrodontids, 29; in plesiadapids, 29
Corpus geniculatum laterale, 158, 172, 175
Coryndon, S. C., 443
Coryphodon, 61
Costal morphology, xv
Count, E. W., 396, 444
Court-Brown, W. M., 434, 444
Coxe, W. S., 179, 181, 182, 449
Cowan, W. M., 457
Cranial capacity, xv, 187, 189–91, 197; australopithecine, 191, 195, 198, of *Australopithecus*, 80–81, 88; and cultural complexity, 196; early hominid, 80–81, 187, 191–96; formula, 193; hominoid, 89, 90; of Hylobatidae,

89; of Javanese *Homo erectus*, 195; of Olduvai Hominid #7, 195; of Olduvai Hominid #9, 195; sexual dimorphism in, 89, 187, 191, 194; variability of, 191
Cranial index, 125, 151
Cranial morphology, vii, viii, xii; anthropoid trends in, xii; arboreal adaptations in, xii; distinctions between human populations, xiii; human intrapopulation variation, xiii; tarsian trends in, xii
Cranium: *see* Skull
Cretaceous primates, 3, 8
Crocidura occidentalis, 167
Crocidura russula, 167
Crocodile, 42, 109, 116, 371
Crompton, A. W., 26, 444
Crook, J. H., 27, 250, 258, 357, 412, 421, 444
Crown-rump length in *Macaca*, 243
Crural index: of earliest primates, 35; of *Notharctus*, 34; of *Plesiadapis*, 34
Crusafont-Pairo, M., 37, 444
Cryptoprocta, 103
Culture, 191; adaptability, 196; learning, 196; skills, 200; symbolic behavior, ix
Currey, J. D., 338, 444
Curtis, 39, 40
Cynocephalus babuin, 265
Cynomys, 112

Dabelow, A., 109, 444
Dactylonax, 104, 108, 112
Dactylopsila, 104, 108, 112
Dahlquest, 104
Daitz, H., 187, 444
Dalquest, W. W., 448
Dart, R. A., 81, 197, 444
Data: methods of quantifying, x; methods of recording, x; methods of transforming, x
Daubentonia madagascariensis, 104, 107, 108, 111, 114, 162, 165, 167, 170, 171, 172, 180, 183, 184, 328, 329
Daubentonia robusta, 162
Davies, P. R., 79
Davies, T. V., 454
Davin, A. G., 126, 456
Davis, D. D., 98, 246, 444
Davis, J. C., 322, 335, 339, 340, 342, 444
Davis, P. R., 43, 48, 81, 220, 221
Davis, R. T., 258, 444, 455
Day, M. H., 49, 79, 80, 81, 150, 444
De Lattre, M. A., 139, 445
Delgado, J. M. R., 434, 445
Dell, M., 444
Delmas, A., 240, 445
Deme, 397
Demography, xx; australopithecine, 81–83; variables of, 191
Dempster, A. P., 148, 445
Dens: *see* Odontoid process
Dental arcade: relative length of, 117; in *Tarsius*, 119
Dental formulae: of Carpolestidae, 21–24; of Paromomyidae, 10; of Plesiadapidae, 17; reduction in Paleocene primates, 30–31
Dental mechanisms, 62
Dental specializations: of Eocene anaptomorphids and microchoerids, 29; in Paleocene primates, 27–28; in Paromomyidae, 8, 10–11; in *Plesiadapis*, 34; in rodents, 97
Dental taxa, 4
Dentition: *see* Teeth
Dermopterans, 103
Desmana moschata, 167
Deutsch, J., 248, 445

DeVilliers, H., 89, 445
De Vore, I., 259, 352, 371, 378, 380, 402, 413, 414, 445, 448, 467
Diamond, I. T., 113, 445
Didelphids, 103, 107, 109, 115
Didelphis, 103, 105, 108, 110, 112
Diencephalon, 158; nuclei of, 158; progression indices of, 168–69, 170
Diet, 421; of Carpolestidae, 25; gelada, 60; of Picrodontidae, 16; of *Ramapithecus*, 60
Diffraction pattern, 339
Diprotodonts: *see* marsupials
Display, 385, 394
Diseases, 434–35
Dissection, xvi, 124
Diurnal habits: of early prosimians, 120
Division of labor, 400
Dobzhansky, T., 123, 445
Dog, 180, 184, 371
Dolichopithecus ruscinensis, 41
Dolphin, 42
Dominance: determination of rank, 387–91, 406; hierarchies, xvii, 378, 379, 388, 400, 407; intramale, 398; langur male, 368, 375; ordering of females, xvii; prestige factors, 407; rank (status), 381, 384, 386, 406, 415; reversals in, 388, 406; rhesus female, 388, 390, 395, 396; rhesus male, 388, 390, 392, 393; study of, 434
Door, J. A., 445
Dorr, J. A., 22, 23, 24, 445
Draper, W. A., 434, 440
Dremomys, 104
Drinking, 356; langur, 359–63, 365, 366, 367, 370, 372, 375, 377; toque macaque, 376
Dryopithecinae, 47, 52, 54, 220
Dryopithecine: *see* Dryopithecinae
Dryopithecus, 37, 42, 44, 48, 51, 52, 53, 57, 62
Dryopithecus (Proconsul) africanus, xv, xx, 43, 48, 51, 54, 218, 220, 221
Dryopithecus fontani, 37, 50, 51, 52, 54
Dryopithecus indicus, 53, 61
Dryopithecus laietanus, 52
Dryopithecus major, 49, 50, 51, 54
Dryopithecus nyanzae, 50, 51, 52
Dryopithecus sivalensis, 50, 53, 54, 57
Dubois, E., 156, 445
Du Mond, F. V., 382, 397, 445
Dusser de Baranne, J. G., 181, 447
Dwarf galago: *see Galago demidovii*
Dwarfing in *Miopithecus*, 165
Dwarf lemur: *see Cheirogaleus*

Echinops telfairi, 167
Ecology, vii, ix; cercopithecoid, xvii; early hominid, 375; geographic variations in, x; new approach to study of primates, 352; relation of tail to, 256; relation to social structure, 399, 414; resource utilization, 401, 405; seasonal variations in, x; study of primate, xvii, 416
Ectotympanic, 5; bone in placentals, 98; cartilaginous, 6; loss of, 7; ossification center, 98; of *Plesiadapis*, 16; ring-like, 7; in rodents, 16
Edinger, T., 177, 239, 445
Eggeling, H., 249, 445
Ehrenberg, A. S. C., 147, 445
Eisenberg, J. F., 357, 371, 374, 402, 413, 445, 446
Elbow of *Dryopithecus africanus*, 48
Electromyography, xvi, 62, 267, 269, 292–304, 307, 323; apparatus, 296–97; fine-wire electrodes, 294, 296; inserted electrodes, 296; multichannel, 294; surface electrodes, 296; telemetering, 310

Elephantid, 28
Elephantulus fuscipes, 167
Elftman, H. O., 235, 249, 260, 446
Ellefson, J. O., 56, 368, 412, 446
Elphidotarsius, 21, 22, 23, 24, 30
Elphidotarsius florencae, 22, 23
Emigration, 405, 407
Emlen, J. T., 413, 446
Encephalization, xiv, 156, 159–63, 173–74; coefficients of, 188; indices of, 162
Enders, R. K., 108, 446
Endocasts, vii, 175–84, 191–200; of *Aegyptopithecus*, 43, 181; australopithecine, xiv, 80–81, 191, 203; comparative studies of, ix, 176, 203; details on, 176–77; early hominid, xiii, 68, 76, 78, 80–81, 186; evolutionary inferences from, xiii; fossil, ix, 175, 183, 188; functional inferences from, xiii, 180; of *Homo erectus*, 191; of large-brained mammals, xiv; latex, xiv; MLD 37/38, 193; Neanderthal, 191; of Olduvai Hominid #5, 192, 198; otter, 180; pongid, 198; preparation of, 176; SK 1585, 192, 198; of small mammals, xiv; STS 5, 192; STS 19, 192, 193; STS 71, 193; studies of, xiv, 184; of Taung child (STS 60), 192, 197, 198; of *Tetonius homunculus*, 178
Endocranial volume: *see* Cranial capacity
Engen, C. W., 446
Enhydriodon, 66
Entepicondylar foramen, 62
Entomolestes, 61
Environment: responses to perturbations of, x
Epigamic features, 194; dimorphism in, 197; increase in, 201
Epiphysis, 158, 172
Epple, G., 114, 446
Erectness: *see* Truncal uprightness
Erikson, G. E., 230, 239, 446
Erinaceidae, 156
Erinaceotans, 6
Erinaceus, 156, 239
Erinaceus europaeus, 167
Erythrocebus patas, 269, 276, 278, 280, 282, 283, 285, 286, 287, 288, 290, 291, 352, 357, 363, 412
Esthonyx, 61
Ethology, 377
Etkin, W., 200, 446
Ettel, P. C., 60, 462
Eudromicia, 104, 108, 111, 112
Eutamias, 104
Evans, F. G., 230, 337, 446, 459
Evolution: arboreal theory of primate, 100–102; convergent, 331; hominid, xv, 91, 189, 191; hylobatid, xv–xvi; inferences on, 425; later prosimian, 119–21; mosaic, 91, 191, 203; parallel, 215, 221, 331; quantum, 98; of social systems, 425
Evolutionary anthropology, xiv
Evolutionary trends, 102; in arboreal mammals, 102–103; onset of primate, 115; primate, 99, 102, 107, 120; toward recession of snout, 116–17
Experiments, viii, xix; ablation, 175, 189; on biological bases of behavior, 395; electrophysiological mapping, 180, 188; field, 405; muscle removal, 124; neurological stimulation, 189; photoelastic, 307, 336; stress analysis, 307, 308, 313, 322–26, 331, 346
Eyeball: enlargement of, 155; hypertrophy in *Tarsius*, 114; size, 111
Eye-hand coordination, 200, 203; importance to primates, 116; selection for, xii
Eyre, S. R., 103, 446

Face: cold adaptation in, 124; width, 124

Factor analysis, xiii, 123, 127–29, 149, 310; applications of, 151, 315; of human crania, 129-40, 143, 145; limitations of, 147–48

Factors: base breadth, 130, 151; brain size, 129; breadth of interorbital region, 135–36; compared with functions, 143–47; cranial-skeletal size, 129; eigenvector, 128, 140; face height, 130, 132–133, 144, 146, 151; forward extensions of facial skeleton, 131–32, 144, 145, 146, 151; frontal bone flatness, 137–38, 145, 146; frontal bone length and size, 137, 145; general midfacial size, 133, 151; general sagittal lengthening, 134; general upper facial breadth, 133, 145, 146, 151; generalized rotated, 131–34, 139; horizontal interorbital profile, 135, 145, 146, 151; horizontal profile of orbits, 134, 144, 145, 146, 147, 151; "length," 129, 131, 151; loading, 327; local rotated, 134–39; lower face, 130; masticatory or ruggedness, 129; malar size, 137, 145, 146; occipital curvature, 138–39, 151; occipital size and height, 139, 147; parietal size and curvature, 138; prognathism, 136–137, 145, 146; prominence of nasalia, 135, 145, 156; subnasal flatness, 136–37, 145, 147; testing of, 149–50; vault breadth, 132, 144, 146, 151

False gavial, 42

Families of Paleocene primates, 3

Farmer, B., 446

Fatigue, 297–99

Faunal correlations: between North Africa and Europe, 40; between North America and Eurasia, 61

Fayum hominoids, 43–47

Feder, J. C., 446

Feeding adaptations: of Paleocene primates, 27–32; picrodontid, 27, 29; prey-catching, 28; in terminal branches, 55

Feeding in langurs, 365, 366, 369–70, 374

Feinstein, B., 268, 446

Felis, 103

Felis bengalensis, 112

Femur: cylindrical head of, 118; early hominid, 77, 78; extension with media rotation of, 293; flexion with lateral rotation of, 293; from Maboko Island, 51

Fibula for early hominid, 78

Fick, R., 265, 446

Fienberg, S. E., 426, 427, 429, 431, 440, 446

Fiennes, R., 434, 446

Filon, L. N. G., 323, 443

Fisher, 151, 432

Fist-walking, 212, 420; adducted, 212

Fitch, F. J., 48, 440

Fleeing, 427, 428

Flinn, R. M., 311, 312, 327, 438

Floyd W. F., 302, 446

Flynn, P. D., 324, 446

Food sources, xvii

Foot: arch-support in human, 298–99, 300, 301; clawed, 108; dorsiflexion of, 300; of Dryopithecus africanus, 48; grasping branches, 107–108, 290; grasping digits II–V in Pongo, 289; heel-strike, 300; inversion of, 300; manipulatory functions of, 269; muscles of, 270–76, 282–88; prehensile functions of, 269; propulsive functions of, 269; reduction of digits II–V in gorilla, 274, 289; specialization for grasping, 99, 107; toe-off, 300, 301, 303

Foramen magnum: central placement of, 32–118; as indicator of brain-size, 177

Forces: compressive, 227; gravitational, 302, 303; mechanical, 320; shearing, 227; tensile, 227, 311, 312, 319

Ford, D. C., 85

Forelimb: bones of Dryopithecus africanus, 220, 221; elongation, 271; elongation as criterion for brachiation, 215; elongation in man, 332; see also Upper limb

Foreplay, 201

Forest-floor predators, 103

Fornix, 187

Forrest, W. J., 296, 439

Fossils: analysis of, viii, 308; ape, 420; casts, 308; equatorial australopithecine, 64, 65, 69; evolutionary inferences from, xv, xvii; functional analysis of, xi; functional inferences from, xv, xvii; hominid, 64, 125, 198; mode of accumulation at Makapansgat, 83–84; mode of accumulation at Swartkrans, 83–84; Mount Carmel man, 419; photographs of, 308; recovery of: viii, 36; search for superlatives, xi; subtropical australopithecine, 64, 65, 69; taxonomy, xi

Fossil men: racial affinities of, xiii; skulls of, xiii

Fossil sites: Abbey Wood, Kent, 37; Bolt's Farm, 66; in Burma, 40; Can Llobateres, 37; Can Ponish, 37; Cernay-lès-Reims, 37; Chelmer, Rhodesia, 65; Chemeron, 74, 85, 87; Chesowanja, 68, 74, 77; Chinji, 52, 56; Chiwando, Lake Malawi Rift, 65; Congo, 38; dating, vii; Dhok Pathan, 52, 61; Djetis, 39; early Eocene, 16; in East Africa, 64–66, 93; in Ethiopia, xii, 52, 61; European, 37, 51, 54; Fayum, xi, 38, 40, 42, 43; Fort Ternan, 47, 59; Garusi, 65, 74; Gladysvale, 66; Haritalyangar, India, 39, 53, 54, 56, 57; Hopefield (Elandsfontein), 66; Hordwell, Hampshire, 37; Ileret, 92; in Indonesia, 39; in Iran, 38; in Iraq, 38; Isle of Wight, 37; Isimila, Tanzania, 65; Jebel Coquin, 38; Jebel el Quatrani, 40, 41, 43, 44, 47; Kalambo Falls, Zambia, 65; Kamlial, 52; Kanapoi, 74, 85; Kavirondo Gulf, Lake Victoria, 38, 47; Kenya, xi, 36, 38, 42, 43, 47, 48, 54; Koobi Fora, 85, 92; Koru, 48; Kromdraai, 65, 68, 74, 75, 76, 77, 78, 82, 84, 86, 88; Lake Rudolf, 38, 65, 67, 68, 74, 75, 77, 80, 85, 92, 93, 191, 195; Langebaanweg, 66; late Cretaceous, 8; Losodok, 48; Lothagam, 67, 74, 85, 87; Maboko Island, 48, 50; Makapansgat, 65, 67, 68, 74, 75, 76, 77, 82, 83, 84, 85, 86, 90, 93; Moroto, 47, 48; Mount Elgon, 47; Nagri, 39, 52, 54, 56; Napak, 47, 48; Nigeria, 38; in North Africa, 38; in northern Spain, 37; Olduvai Gorge, 36, 65, 67, 68, 73, 74, 75, 76, 77, 79, 80, 84, 85, 86, 93, 191; Oman, 38; Omo River Valley, 41, 67, 68, 74, 75, 76, 77, 80, 84, 85, 86, 92, 93, 191, 195; Oran, Algeria, 39; Paleocene, 8, 37; Peninj, 74, 77, 84, 85; post-deposition perturbations, viii; Purgatory Hill, 10, 16; Quercy, 37; reopening, xi; in Republic of South Africa, 64-66, 86, 93; Rusinga Island, 38, 48; Saint Gaudens, 37, 51; Salt Range, 52; Saudi Arabia, 38; Siwalik Hills, 38, 52; Songhor, 48; in Southern China, 60; Sterkfontein, 65, 66–67, 68, 73, 74, 76, 77, 78, 82, 84, 85, 86, 90, 93; in Sudan, 38; Swartkrans, 65, 66, 68, 73, 74, 76, 77, 78, 82, 83, 84, 86, 88, 90, 93; in Tanzania, xii, 38; Taung, 65, 73, 74, 75, 76, 82, 86, 90; Trinil, 39; Twin Rivers, Zambia, 65; Ubeidiya, Israel, 65; in Uganda, 38, 43, 47, 48, 54; Vallés Panedés, 52; Wadi Moghara,

Egypt, 39, 41, 42; Wadi Natrum, Egypt, 39, 41; Yale Quarry I (Fayum), 40
Fourtau, R., 39, 41, 446
Fox, R., 150
Friction skin on tail, 232
Freidlander, J. S., 148, 446
Frontal bone from Tabon Cave, Phillipines, 150
Frontal cortex of *Loris,* 114
Frontal lobes, 117; on endocasts, 192
Fukada, E., 337, 446
Fulton, J. F., 180, 181, 447, 468
Furuya, V., 387, 447

Gait: evolution of man's, 292; human, 297, 303–4
Galaginae, 34, 111, 313
Galagine: *see* Galaginae
Galago, 104, 113, 117, 118, 207, 233
Galago crassicaudautus, 104, 118, 167
Galago demidovii, 167, 171
Galago elegantulus, 104
Galagos: *see Galago*
Galago senegalensis, 104, 111, 167
Galapagos finch, 104
Galemys pyrenaicus, 167
Game management, xx
Ganglia, 163
Gans, C., 267, 268, 447
Garn, S. M., 124, 144
Garnett, J. M., 339, 468
Gartlan, J. S., 27, 258, 352, 357, 363, 402, 404, 412, 413, 421, 444, 447, 464
Gazin, C. L., 447
Gelada: *see Theropithecus gelada*
Gelfand, P., 447
Gellhorn, E., 299, 447
Gene flow, 397
Geneological relationships, xviii, 387–91, 398, 412, 434; influence on social interactions, 405–6
Generalized distance, 306, 310, 315, 317, 320, 328
Generation, 381
Genetta, 103
Geochronology, xii; of Djetis beds, 39; of earliest Oligocene in North America and Europe, 40, 41; of East African sites, 84–85; of East Rudolf, 67; of Fayum, 40; fission track, 85; K/A, 191; of Koru, 48; of Lothagam, 67; of Moroto, 48; of Napak, 48; of Omo, 67; paleomagnetism, 85; radiocarbon, 331; relative, 86; of Rusinga Island, 48; of Songhor, 48; of south African sites, 84–86; of Sterkfontein, 67; thermoluminescence, 85; uranium/thorium, 85; X-ray diffraction, 85
Geology, 307, 343
Gerebtzoff, M. A., 455
Geschwind, N., 199, 447
Giant otter, 66
Gibbon, 89, 177, 178, 209, 211, 214, 215, 236, 250, 260, 269, 271, 274, 289, 317, 368, 412, 417
Gidley, 22
Gigantopithecus, 60, 61, 87
Gigantopithecus bilaspurensis, 61
Gigantopithecus blacki, 60
Gilbert, J. T., 446
Gill, E., 86
Gjukic, M., 161, 447
Glezer, I. I., 186, 441
Goffart, M., 455
Goffman, E., 422, 447
Goffman, W., 435, 447
Goodman, G., 187, 447
Goodman, J. W., 321, 447

Goodman, L. A., 423, 426, 447
Gorilla: *see Gorilla gorilla*
Gorilla gorilla, 27, 49, 51, 57, 162, 165, 167, 168, 170, 172, 186, 193, 194, 196, 198, 210, 211, 212, 213, 215, 217, 219, 221, 228, 229, 233, 237, 269, 270, 271, 274, 276, 279, 283, 284, 285, 286, 287, 289, 295, 325
Gorilla gorilla beringei, 210
Gottlieb, H. H., 230, 239, 447
Gower, J. C., 316, 320, 447
Gracile australopithecine: *see Australopithecus africanus*
Grand, T. I., 119, 219, 240, 265, 447, 448
Granger, 22
Gray, E. G., 300, 301, 448
Gray langur: *see Presbytis entellus*
Green, D. W., 339, 458
Greenlaw, R. K., 301, 302, 303, 439
Green monkey: *see Cercopithecus aethiops*
Gregory, W. K., 53, 54, 98, 119, 120, 448
Groeningen, W. B. van, 460
Grooming: langur, 368; rhesus, 384
Grossman, S. P., 356, 448
Growth, xx, 189, 201; rates of, 196
Guenon: *see Cercopithecus*
Guhl, A. M., 426, 448
Guillery, R. W., 457
Gybels, J., 455
Gyrus: parietal (post central), 179

Haar, 347
Habiline: *see Homo habilis*
Habitats: arboreal, 308; baboon, 434; captive compound, 403; Chinji, 52; choice of, 415; control of, 409; definition of natural, 401; Dhok Pathan, 52; of *Dryopithecus africanus,* 48; of *Dryopithecus major,* 49; European Middle and Late Miocene, 51; of *Gigantopithecus blacki,* 60–61; Kamlial, 52; Koru, 48; of *Limnopithecus,* 55; of long-tailed *Macaca,* 242, 259–60; of medium-tailed *Macaca,* 242–243, 259–60; modifications by man, 401, 404; Moroto, 48; Nagri, 52; Napak, 48; natural, xviii, 402, 404, 409; outdoor enclosure, 403; Rusinga Island, 48; savannah, 200, 201, 404; of short-tailed *Macaca,* 242, 259–60; Songhor, 48; studies of, 62; swamp forest, 289, 404; temperate forest, 404; terrestrial, 308; tropical rain forest, 404, 405; zoo island, 403
Habituation, 402, 404, 405
Habitus: of ancestral primates, 103; features, 4; stalking and capturing insects, 116; terrestrial-arboreal of hanuman langurs, 290
Hadamard, 339, 347
Haemapophyses, 226, 232, 235; forming chevron bones, 232
Haggard, E. A., 311, 441
Haines, R. W., 114, 115, 116, 448
Hair length in *Macaca,* 257
Hall, E. R., 104, 448
Hall, K. R. L., 357, 379, 402, 412, 413, 414, 448
Hall, M. C., 240, 448
Hall, W. C., 113, 445
Hall-Craggs, E. C. B., 207, 448
Hallowell, A. I., 200, 448
Hallux (first or great toe): capacity for grasping in gorilla, 289; divergent, 207; nail-bearing, 121; opposable, 7; prehensile in Hylobatidae, 289; prominence in Hylobatidae, 273; of *Ratufa,* 105; reduction in *Erythrocebus patas,* 276; reduction in *Pongo,* 284–85, 289; reduction in *Theropithecus gelada,* 276; webbing in gorilla, 285

Hamadryas baboon: *see Papio hamadryas*
Hamate, 219; hook of, 219
Hamilton, W. J., 117, 448
Hamshere, R. J., 209, 213, 453
Hand: biomechanics of middle phalanges in, 279, 291; clawed, 108; distal phalanges in terrestrial cercopithecoids, 279; grasping branches, 107–8; grasping of digits II–V in *Pongo*, 289; great ape, 420; as gripping hook, 309; hook-like grasp in *Colobus polykomos*, 290; human, 309; manipulatory functions, 269; muscles of, 270–82; Olduvai, 331; prehensile functions, 269; prey-seizing functions, 117; propulsive functions, 269; sensitivity in racoon, 180; specialization for grasping, 99, 107; studies of, 311; study of digital ray, 325
Hanuman langur: *see Presbytis entellus*
Hapalemur, 164
Hapalemur simus, 167
Haplorhines, 111
Hardy, M. H., 114, 441
Hare, 41
Harris, H., 434, 448
Harrisson, B., 448
Harrison, 104
Haughton, S., 264, 448
Hawk, 113
Haynes, C. V., 65, 442
Hays, R., 67
Healy, M. J. R., 310, 311, 438
Hedgehog: *see Erinaceus*
Heiple, K. G., 196, 453
Heliosciurus, 105
Hellman, M., 53, 54, 448
Hemiacodon, 33, 97
Hemicentetes semispinosis, 167
Hendey, Q. B., 66, 449
Herbivores: canopy, 104; ground squirrels, 104; low canopy and ground, 104; vertically ranging, 105, 108
Heritage features, 4
Herrnstein, R. J., 432, 433, 449
Hershkovitz, P., 184, 449
Hickman, J. L., 104, 449
Hickman, V. V., 104, 449
Hicks, J. H., 300, 449
Hiernaux, J., 148, 449
Hiiemae, K., 26, 109, 444, 449
Hill, W. C. O., 249, 250, 449
Hinde, R. A., 387, 449
Hinde, R. G., 256, 449
Hindlimb: functional importance in Hylobatidae, 289; *see also* Lower limb
Hipbone (os coxa, innominate): early hominid, 78; from Sterkfontein, 328–30; studies on, 313
Hippocampus, 158, 173, 187; progressions in, 168, 170–71
Hippopotamus: *see Hippopotamus*
Hippopotamus, 41, 42
Hippopotamus protamphibius, 41
Hirsch, J. F., 179, 181, 182, 449
Hobbs, J., 85
Hockett, C. F., 200, 449
Hodges, D., 339, 347, 461
Hofer, H., 229, 235, 449
Holister, A. S., 324, 449
Holloway, R. L., xiii, xix, 81, 155, 177, 185, 186, 189, 192, 198, 199, 200, 202, 449
Home range, 375, 405, 414; of *Presbytis entellus,* 356
Hominid: *see* Hominidae
Hominidae, xi, xii, xiv, 56, 60, 61, 63, 189, 198, 215, 221, 375; divergence from Miocene

Pongidae, 60; early, xi; sympatry and synchrony of robust and gracile, 67–68
Hominization, 87, 91, 375
Hominoidea, xiv, xv, 43, 62, 162, 168, 172, 198, 207, 215, 220, 222, 226, 229, 232, 237, 269, 276, 277, 279, 280, 281, 285, 286
Homo, 44, 49, 65, 70, 76, 88, 99, 159, 162, 165, 170, 171, 172, 174, 187, 198, 212, 219, 220, 228, 230, 232, 233, 236, 237, 276, 277, 280, 281, 291, 317, 325, 330, 332, 414, 423, 432
Homo erectus, 39, 90, 161, 162, 191, 195, 196, 200
Homo habilis, 67, 87, 90, 91, 195, 196, 200
Homo sapiens, xiii, xiv, xv, 61, 161, 162, 167, 174, 220, 279, 281, 399, 401, 413
Hooijer, D. A., 38, 53, 459
Hooton, E. A., 101, 102, 111, 459
Hornbill, 372
Horwich, 368
Howell, A. B., 249, 450
Howell, F. C., xii, 80, 88, 93, 191, 195, 375, 450
Howells, W. W., xi, xii, xiii, 81, 101, 131, 148, 150, 310, 311, 332, 419, 450, 456
Howler monkey: *see Alouatta*
Hubel, D. H., 113, 450
Hughes, A. R., 67, 465
Huia, 104
Human populations: Africans, 144, 145, 146; American Indians, 145, 146; Atayals, 150; Australians, 144, 145; Australo-Melanesians, 146; Buriats, 146; Chinese, 150; Europeans, 145, 146; Filipinos, 150; Greenland Eskimos, 145, 149; Japanese, 150; Mokapu Hawaiians, 147, 149; Oceanic Negroids, 144; Southwest Pacific, 144
Humerus: of *Dryopithecus,* 51; early hominid, 78; Maboko Island, 51; Kanapoi, 81, 150; pliopithecine, 55
Hunt, E. E., 124, 450
Hunting, 201; dependence on, 196, 202
Hürzeler, J., 54, 450
Hutchinson, T., 382, 397, 445
Huxley, T. H., 98, 450
Hydroxyapatite, 306
Hylobates, 89, 211, 213, 215, 219, 220, 221, 225, 233, 270, 276, 277, 279, 280, 283, 285, 287, 418
Hylobates agilis, 89
Hylobates cinereus, 89
Hylobates concolor, 89
Hylobates hoolock, 89
Hylobates klossii, 89
Hylobates lar, 89, 209, 217, 286, 368
Hylobates pileatus, 89
Hylobatid: *see* Hylobatidae
Hylobatidae, 47, 55, 56, 226, 228, 237, 269, 271, 274, 276, 277, 280, 281, 283, 285, 286, 289, 421
Hypophysis, 158
Hypsiprymnodon, 26
Hyracotherium, 61
Hyrax, 42

Immigration, 405, 407
Imprinting, 395
Incisors: of *Australopithecus africanus,* 77; of *Australopithecus robustus,* 77; of *Carpolestes aquilae,* 25; of Carpolestidae, 24; of *Dryopithecus africanus,* 48; enlargement in Paleocene primates, 27, 29; of *Gigantopithecus,* 61; of Paromomyidae, 8, 11; of Picrodontidae, 14, 16; of *Platychoerops,* 19; of *Pronothodec-*

tes, 17; of *Ramapithecus,* 60; of *Saxonella,* 19, 20; of *Tarsius,* 28
Indopithecus giganteus, 53
Indri, 104, 171, 181, 184, 232, 318
Indri indri, 167
Indriidae, 6, 111, 114, 117, 119, 178
Indriids: *see* Indriidae
Insectivora, xiii, 121, 122, 157; adaptations for stalking and capturing insects, 117; canopy, 104; shrub-layer, 103–4, 108; *see also* Basal insectivore and progressive insectivore
Insectivoran-primate transition, 98, 99
Insectivore: *see* Insectivora.
Intelligence, 191
Intermembral index: in earliest primates, 35; of *Notharctus,* 34; of *Plesiadapis,* 18, 34
International Society of Electromyographic Kinesiology, 294
Interorbital breadth, 109
Interorbital septum: apical, 114–15, 119, 120; *Tarsius,* 114, 115
Interspecies relationships at waterholes, xvii
Interstitial wear: in *Australopithecus,* 59; chimpanzee, 59; in *Ramapithecus,* 60
Intervertebral discs, 224, 228; reduction in sacra, 232; in tail, 232
Ischial callosity: adaptive advantage of, 260; area of, 243, 244, 251, 255; in *Macaca,* 250–51; shape of, 242, 250–51, 258; as sitting pad, 250; size of, 242, 254–55, 258, 261
Ischial tuberosity: area of, 255; shape of, 242, 251; size of, 242, 251, 254–55, 261
Itani, J., 413, 434, 450

Jackel: *see Canis aureus*
Jacobson, A., 89, 450
James, J., 422, 424, 425
Japanese macaque: *see Macaca fuscata*
Jay, P. C., 354, 357, 361, 362, 368, 369, 371, 412, 450, 467
Jenkins, F. A., 109, 227, 229, 240, 449, 451
Jepsen, G. L., 22, 451
Jerison, H. J., 162, 177, 451
Johnson, G. L., 113, 451
Joints: biomechanics of shoulder (glenohumeral), 298; carpometacarpal, 294; closepacked position of, 293, 294; composite movements of, 293; habitual motions at, 293–94; hinge, 299; hip, 293, 303; interphalangeal, 294; locking movements of, 293; midcarpal, 211; sacroiliac, 232, 236; shoulder, 313, 315, 317, 319, 320, 330; shoulder as suspensory mechanism, 311; spin in, 293; stabilizing the intervertebral, 303; subtalar, 207; swing in, 293; two-joint, 303; unlocking movements of, 293
Jolly, A., 117, 118, 451
Jolly, C., 60, 200, 259, 451
Jones, F. W., xiii, 100, 101, 102, 107, 113, 115, 117, 451
Jones, K. J., 142, 451
Jones, R. L., 301, 451
Joseph, J., 300, 304, 439
Jouffroy, F. K., 211, 240, 451
Jump: *see* Leaping

Kanda, S., 129, 451
Kangaroo: *see* Macropodidae
Karlin, S., 434, 451
Kaufman, I. C., 406, 412, 451
Kaufman, J. C., 387, 459
Kaufmann, J. H., 378, 451
Kawai, M., 382, 388, 434, 451
Kawamura, S., 388, 392, 451
Keith, A., 100, 125, 246, 263, 451

Kellogg, R., 246, 452
Kendall, M. G., 435, 452
Kennedy, J. A., 454
Kenyapithecus africanus, 50
Kenyapithecus wickeri, 59
Kevan, D. K., 311, 441
Keverne, E. B., 114, 455
Kidd, K. K., 446
Klaatsch, H., xiii, 101
Klaauw, C. J. van der, 98, 452
Klausen, K., 302, 452
Klopfer, M. S., 395, 452
Klopfer, P., 452
Knuckle-walkers, 49–50, 212, 237; *Dryopithecus major* as, 54
Knuckle-walking: 291, 331; of African apes, 212, 274, 420; of chimpanzees, 309; in *Dryopithecus africanus,* 49; in *Dryopithecus major,* 50; evolution of, 423; mechanisms of, xvi, 212
Knussman, R., 150, 452
Koala: *see Phascolarctos*
Kobus defassa, 377
Koch, J. C., 337, 452
Koenigswald, G. H. R. von, 42, 53, 54, 55, 79, 452, 465
Koford, C. B., 378, 388, 397, 452
Kollman, M., 114, 452
Koyama, N., 388, 452
Krantz, G., 197, 452
Krishnamurti, A., 180, 181, 452, 460
Kuehn, R. E., 402, 413, 446
Kummer, B., 227, 240, 452
Kummer, H., 380, 402, 412, 421, 452
Kurisu, K., 129, 451
Kurt, F., 402, 452
Kyncl, H., 445
Kyphosis: rump, 227; thoracic, 227

Lactic dehydrogenase: 187–88
Lagothrix, 55, 179, 182, 230, 233, 235, 319
Lagothrix lagotricha, 167
Lagomorph, 5
Lai, L. Y. C., 446
Lancaster, J., 199, 453
Landau, H. G., 434, 453
Landauer, C. A., 129, 130, 133, 453
Langer, C., 265, 453
Language: 189–202; development of, 201; expansion of parietal cortex and, 199–200; laterality of capabilities for, 199–200; primate, 421; relation to handedness, 199; symbols, 199
Langur: *see Presbytis*
Lapouge, 125
Lariscus, 104
Lasker, G., 375, 453
Lawick-Goodall, J. van., 351, 413, 453
Leaf-monkey: *see* Colobinae
Leakey, L. S. B., xii, 36, 38, 48, 49, 50, 54, 55, 59, 79, 88, 443, 453
Leakey, M. D., 67, 69
Leakey, R. E., 67, 92, 191, 195
Leaping: by *Colobus polykomos,* 290; in early primate ancestors, 100; indriid, 119; langur display, 362, 369; langur progression by, 372; by *Lepilemur,* 119; by *Macaca,* 257; by *Notharctus,* 34; pliopithecine, 55; related to visual field overlap, 112–13; by *Tarsius,* 119
Learning, 201, 202, 413
Leary, R. W., 444
Lemur, 11, 62, 118, 165, 170, 175, 181, 184, 214, 233
Lemur: *see Lemur*
Lemur catta, 17, 104, 118

Lemur fulvus, 104, 162, 167
Lemuridae, 6, 114
Lemurids: *see* Lemuridae
Lemuriformes, 101, 108
Lemurinae, 114
Lemurines: *see* Lemurinae
Lemur rufiventer, 159
Lemur variegatus, 104, 167
Lemuroidea, 6, 8, 34, 178, 313
Lemuroids: *see* Lemuroidea
Lenneberg, E. H., 199, 453
Leontocebus oedipus, 167
Leopard: *see Panthera pardus*
Lepilemur, 117, 118, 119, 159, 164, 165
Lepilemurini, 163, 165
Lepilemur leucopus, 17
Lepilemur ruficaudatus, 167
Leptacodon, 6
Leptacodon ladae, 9
Leptictidae, 6, 7, 29
Leptictids: *see* Leptictidae
Lessertisseur, J., 211, 451
Leutenegger, W., 232, 239, 453
Leven, M. M., 323, 453
Levitcky, W., 447
Lewis, G. E., 53, 54, 448
Lewis, O. J., xv, 48, 207, 209, 210, 213, 216, 453
Liberson, W. G., 303, 453
Libypithecus, 41, 177
Libytherium, 41
Ligament: 298; coracohumeral, 298; dorsal radiocarpal, 213; ligamentum longitudinale ventrale, 130; ligamentum transversum, 227; palmar intracapsular of wrist, 210; palmar radiocarpal, 210, 211, 212, 213; palmar ulnocarpal, 210, 213; pisohamate, 219; plantar, 299, 301; radial collateral, 210, 212; radiotriquetral, 210; ulnar collateral, 210
Limbic system, 171
Limnogale mergulus, 167
Limnopithecus, 43, 44, 54, 55
Limnopithecus legetet, 55
Limnopithecus macinnesi, 55
Lindegard, B., 446
Line of gravity, 301
Lipton, S., 310, 438
Lizard, 427
Locomotion, 188, 239; of *Aegyptopithecus,* 44; acrobatic in squirrels, 112; arboreal climbing, 271, 274, 289; arboreal quadrupedal, 118, 256–57, 261, 365, 366, 370; categories of, 62, 237, 266, 267; changes in, 200; climbing tree trunks, 289, 291, 311; hauling, 290; human, xvi, 300–303; modified palmigrade walking, 420; ontogeny, x; patterns of, 311; of *Plesiadapis,* 34, 35; progression in langurs, 359, 360; quadrupedal beneath branches, 118; stance phase of, 300, 301, 302; on steeply sloping and vertical supports, 108; studies on, 269, 294; swing phase of, 300, 302; tail use in, 242; techniques for study of, ix; terristrial, 271; terrestrial crossing in langurs, 363–67, 372; terrestrial plantigrade, 289; terrestrial quadrupedal, 261, 291; walking, 297, 303; *see also* Bipedalism and sussory behaviors
Locomotive system, ix; evolutionary biology of, xvi; functional biology of, xvi
Locomotor adaptations: of Paleogene primates, 32–35; in vertebral column, 235–36
Lohnes, P. R., 142, 444
Lordosis: lumbar, 277, 303; neck, 227
Lorenz, K. Z., 420, 453
Lorenz, R., 114, 119, 265, 446, 448

Loris, 104, 108, 112, 113, 114, 170, 179, 180, 184, 318
Loris: *see Loris*
Lorisidae, 226, 229
Loris gracilis, 167
Lorisiformes, 114
Lorisinae, 112, 120, 184, 313
Lorisines: *see* Lorisinae
Loris tardigradus, 112, 120
Lorrain, F., 434, 453
Louis, P., 459
Lovejoy, C. O., 196, 453
Lowther, G., 377, 460
Loy, J. D., 381, 392, 453
Lumbar vertebrae: 225, 226, 230; bodies in Hominoidea, 230, 236; of *Dryopithecus major,* 49; early hominid, 78; function of, 235, 239; human, 230, 303, 343; longitudinal ventral keeling of, 237; processes accessorii of, 230; specializations of, 230; transverse processes of, 230
Lunula, 209, 214
Lutong: *see Presbytis cristatus*
Lutra, 41
Lutra libyca, 41
Lyon, M. W., 246, 454

Macaca, xvi, 41, 55, 175, 180, 182, 241, 242, 246, 249, 250, 252, 257, 258, 261, 269, 276, 278, 280, 281, 282, 283, 284, 285, 286, 287, 288, 290, 291, 387, 388, 396, 397, 412
Macaca arctoides, 242, 246, 247, 248, 249, 250, 251, 252, 254, 255, 256, 258, 259, 260, 271, 276, 278, 280
Macaca fascicularis, 233, 242, 244, 246, 247, 248, 249, 250, 251, 252, 254, 255, 256, 258, 259, 260, 276, 278, 280
Macaca fuscata, 259, 382, 388, 392, 398
Macaca mulatta, xvii, 167, 232, 233, 235, 242, 244, 246, 247, 248, 249, 250, 251, 252, 254, 255, 256, 258, 259, 260, 278, 354, 378, 380, 381, 383, 388, 397, 398, 421
Macaca nemestrina, 242, 246, 247, 248, 249, 250, 251, 252, 253, 254, 255, 256, 258, 259, 260, 278, 280, 285, 288, 405, 406, 407, 412
Macaca radiata, 259, 260, 412, 413
Macaca silenus, 259
Macaca sinica, 260, 352, 356, 360, 375
Macaca speciosa: see Macaca arctoides
Macaca sylvana, 259
Macaque: *see Macaca*
MacConaill, M. A., 268, 292, 298, 454
Macdonald, G., 435, 454
Machaerodus, 41
MacKinnon, I. L., 193, 454
Macropodidae, 25, 27
Macropodids: *see* Macropodidae
Macroscelididae, 103, 170
Macroscelidids: *see* Macroscelididae
Maguire, B., 67
Mahalanobis, 306
Malagasy lemur: *see* Lemuriformes
Mammals: *see* Mammalia
Mammalia, 175
Mammillo-thalamic system, 187
Mammuthus, 66
Man: *see Homo*
Mandible: of *Aegyptopithecus zeuxis,* 43, 44; of *Aeolopithecus chirobates,* 47; of *Carpodaptes hobackensis,* 24; of Carpolestidae, 29; of Chiromyoides, 19; condyles in Paleocene primates, 30; of *Dryopithecus africanus,* 48; of *Dryopithecus fontani,* 51; of *Dryopithecus indicus,* 52; of *Dryopithecus sivalensis,* 57;

early hominid, 68, 70, 92, 93; of *Elphidotarsius*, 22, 23; evolution in Hylobatidae, 417–18, 421; extreme changes in Paleocene primates, 32; of *Gigantopithecus*, 61; Lothagam, 87; of modern man, 124; of *"Moeripithecus" markgrafi*, 46; of *Oligopithecus savagei*, 47; of Omo *Australopithecus*, 57; of Paromomyidae, 8, 10–11, 29; of Picrodontidae, 29; of *Picrodus*, 12–13; of Plesiadapidae, 29; of *Propliopithecus haeckeli*, 44; of *Ramapithecus punjabicus*, 56, 60; of *Saxonella*, 19–20; symphysis, 28; of vertical clingers and leapers, 33, 118

Mandrillus, 412; *see also Papio sphinx*

Mangabey: *see Cercocebus*

Manipulation, 188; abilities for, 200; of arboreal foods, 107; grip in early hominids, 78; mechanisms, xvi; preadaptation of hands for, 216; primate forelimb, 100

Mann, A., 81, 454

Marler, P., 368, 454

Marmosa, 104, 108, 112

Marmoset: *see* Callithricidae

Marmota, 104

Marshall, D. S., 332, 454

Marsupials, 25, 29, 103, 107, 114, 116, 214

Marsupicarnivora, 122

Martin, R., 262, 454

Martin, R. D., 7, 98, 114, 120, 442, 454

Maruyama, M., 200, 454

Masali, M., 239, 454

Mason, W. A., 368, 409, 412, 440, 454

Masritherium, 41, 42

Mastadon, 41

Masticatory apparatus (complex): viii; australopithecine, 79; in Hominidae, 91; marsupial, 110

Mating: *see* Copulation

Matrix: analyses, 411; communication, 385; correlations, 127; in discriminant analysis, 140; of expectation, 430; interaction, 386; interpoint distance, 315, 316; of numbers, 127, 141; socio, 384, 385, 386; sociometric grooming, 384; transaction flow, 426, 427, 428

Matthew, W. D., 11, 22, 101, 115, 454

Mawby, J. E., 65, 442

Maxillae: of *Dryopithecus fontani*, 51; early hominid, 68, 70, 92, 93; of *Ramapithecus punjabicus*, 56; Taung, 76

Mayr, E., 127, 454

McCann, C., 361, 454

McClure, H. E., 368, 454

McDowell, S. B., 6, 7, 454

McIver, J., 85

McKenna, M. C., 6, 7, 9, 16, 98, 454

Mckennatherium libitum, 9

Measurement (mensuration), xvi; of central tendency, 380; correlation, 127; covariation, 127; cranial, 151; genotypic, 419–20; of mechanical factors, 124; methods of, 306; of nearest neighbors, 435; phenotypic, 419, 420; of recent human skulls, 123; of shape, 305, 307; of social interaction, 395, 409, 415, 426; of social patterning, 421; social response, 410; transformation of, 127, 129, 420

Mechano-electric properties of tissues, 307

Medulla oblongata, 158, 171–72; progressions in, 168

Meeter, K., 460

Megalapadis, xiv, 183

Megalapadis edwardsi, 177

Meganthropus paleojavanicus, 39

Meienberg, G. P., 227, 454

Meltzer, B., 347, 455

Memory, 203

Menatotherium, 17

Meniscus: in ankle joint of marsupials, 207, in wrist of *Dryopithecus africanus*, 221; in wrist of *Gorilla*, 219; in wrist of *Homo sapiens*, 213, 214, 219, 220; in wrist of *Hylobates lar*, 209; in wrist of *Pan troglodytes*, 209, 212, 219; in wrist of *Pongo pygmaeus*, 210, 220

Mesencephalon, 158, 171–72; progressions in, 168

Mesopithecus, 177

Metasinopa, 40

Metatarsal bones: early hominid, 78

Meulders, M., 180, 455

Meyers, P. H., 338, 439

Michael, R. P., 114, 455

Michaelis, P., 455

Microcebus, 104, 108, 118, 120, 164

Microcebus murinus, 27, 120, 167

Microcephalic human, 198

Microchoeridae, 7, 29

Microchoerids: *see* Microchoeridae

Microchoerus, 28, 33

Microparamys, 61

Microsyopidae, xiii, 6, 8, 31, 33

Microsyopids: *see* Microsyopidae

Microsyopoidea, xiii, 8, 99, 121

Microsyopoids: *see* Microsyopoidea

Microsyops, 33, 99

Middle ear, 5, 98; adapid, 17; of *Phenacolemur*, 16–17; of *Plesiadapis*, 17; of prosimians, xi

Miller, G. S., 246, 455

Miller, J. A., 48, 440

Miller, J. H., 435, 455

Miller, R. A., 124, 250, 260, 455

Miopithecus, 165

Mizuhara, H., 434, 455

Model: building, 150, 261, 307, 418; for conjoint action, 430; of early hominid behavior, xiii; of early hominid evolution, xv, 186, 200–203; evolutionary, viii; genetic, 319; of hominid brain evolution, xiii; of hominid phylogeny, xvi; of human child-rearing, 411; of independence, 430; of interindividual relations, xviii; linear one-step transistion (LOST), 423–25, 427, 431; mathematical, xi, xviii, 313, 415, 416–17, 421–36; nonisomorphic, 386; nonmathematical, xviii, 417, 420; plastic, 323; of population dynamics, xviii, 434; probability, 426; of quasi-independence, 429, 430, 431; of roles in human society, 434; "Rubicon" of language, 191, 199; of social organization, 398, 401, 426–32; stochastic, 423, 435; testing, 150, 346

Moeripithecus markgrafi, 43, 46

Molars: of *Australopithecus*, 59; of Carpolestidae, 21, 23; of *Dryopithecus africanus*, 48; of *Dryopithecus major*, 49; of *Dryopithecus sivalensis*, 57; of earliest primates, 5; early hominid, 70; of *Elphidotarsius florencae*, 23; of *Gigantopithecus*, 61; hypertrophy of M3 in Paleocene radiation, 31; Paromomyidae, 8, 10–11; patterns in Paleocene primates, 28, 32; Picrodontidae, 11, 15–16; of *Pronothodectes*, 17; of *Propliopithecus haeckeli*, 46; pygmy chimpanzee, 57; of *Ramapithecus punjabicus*, 57, 59, 60; relationship between *P. haeckeli* and *A. zeuxis*, 45; *Saxonella*, 19 monkey, 62, 99, 105, 175, 176, 178, 187, 210, 214, 216, 221, 226, 228, 230, 232, 234, 236, 237, 315, 319, 424

Monodelphis, 103, 112

Montagu, M. F. A., 111, 441

Moran, P. A., 435, 452

480   INDEX

Mortality: differential, 397; infant, 407; rates, xx
Morton, D. J., 115, 455
Motor unit, 268, 295; potentials, 295–96
Mounting: in langurs, 367–68; rhesus, 384–93
Mouse, 101, 187
Mouse lemur: see Microcebus
Moynihan, M., 114, 455
Mukherjee, R., 310, 332, 455
Multidimensional contingency tables, 415, 426–31
Multidisciplinary studies, xix; paleoanthropological, xii
Multiple discriminant function analysis, xiii, 123, 140–42, 148, 149; applications of, 151; of human skulls, 142–43
Multituberculates, 29
Multivariate analysis, 123, 124, 125–27; advantages of, 148–49; applications of, 149–51; limitations of, 147–48; of social organization, 381
Mungai, 80
Murray, P. D. F., 337, 455
Muscles, 307; abdominal, 302, 303; abductor hallucis, 300; absence of transverse head of adductor hallucis in E. patas, 286; adductor hallucis, 286–87, 290, 300; adductors of hip, 301, 302; adductor pollicis, 280–81, 291; antagonist, 299; antigravity, 297; back, 240; brachialis, 299; cheiridial, xvi, 262, 265, 266; coordination of, 299; deltoid, 298; extrinsic (long) manual digital flexor, 211, 212, 268, 276–79, 289, 290; facial, viii; flexor carpi radialis, 299; flexor digitorum brevis, 300; flexor digitorum profundus, 276, 278, 279, 290, 291; flexor digitorum superficialis, 276, 278, 279, 289, 290, 291; flexor hallucis longus, 300, 301; forearm, 271; forelimb, 265; gastrocnemius, 268; glutei, 301; gluteus maximus, 265, 295, 303, 304; hamstring: see ischiocrural; hindlimb, 265; hip, 301–2; iliocaudalis, 249–50, 254, 261; iliopsoas, 301, 302, 303; intrinsic of back, 302, 303; intrinsic foot, 285; intrinsic hallucal, 274, 275, 276, 285–86, 290; intrinsic hallucal flexor and abductor, 287–88; intrinsic hand, 271, 279–80; intrinsic pollical, 274, 275, 276, 279–80, 290, 291; ischiocaudalis, 249–50, 254; ischiocrural, 301–3; law of minimal shunt action, 298; law of minimal spurt action, 298; leg, 282–84, 300–301; manual interosseous, 271, 274; manual lumbrical, 271, 274; palm, 271, 289; pedal interosseous, 271, 274, 289; pedal long digital flexor muscles, 282–84, 289, 290; pedal lumbrical, 271, 274, 289; pelvo-caudal, 242, 243, 249–50, 258, 260; peronius brevis, 284; peronius longus, 284–85, 289, 290, 291, 300, 301; physiological factors of, 267–69; plantar flexors, 301, 303; pretibial, 303; prime movers, 299; pronator quadratus, 299; psoas major, 303; pubocaudalis, 249–50, 254, 261; quadruceps femoris, 301, 303; reciprocal inhibition of, 299; relative mass of cheiridial, 262–91; relative mass of pelvocaudal, 243, 254, 255; sacrospinalis, 302, 303; shoulder, 265, 266; sparing, 297–99; spasticity, 299; supraspinatus, 298; synergy in, 299; tail, 234; temporalis in Paleocene primates, 30; tibialis anterior, 300; tibialis posterior, 300, 301; thenar eminence, 281–82; thigh, 301–2; triceps brachii, 299; triceps surae: see plantar flexors; weights, 244
Myosciurus, 112

Nachemson, A., 303, 455
Nandinia, 105

Napier, J. R., xi, xii, 32, 33, 34, 43, 48, 79, 81, 88, 102, 117, 118, 119, 215, 220, 221, 294, 301, 445, 453, 455
Napier, P. H., 102, 117, 455
Nasal index, 125, 151
Nasua, 105
Nathan, R., 338, 456
Natural selection, 98, 102, 175, 188, 189, 191, 196; for grasping branches, 290; for locomotor patterns, 319; for nose and face shape, 124; pressures, 400, 402; for tree-climbing, 100
Navajovius, 9, 10, 11
Navajovius kohlhaasae, 10
Navajovius mckennai, 10
Navicular: elongation in Eocene primates, 118; elongation in Galago, 207; elongation in Microsyops, 33; elongation in Tarsius, 207
Necrolemur, 7, 28, 33
Necrolemurinae, 9
Neely, P. M., 316, 335, 456
Neighborhood limited classification, 148, 316, 317–19, 331, 333
Neocortex, xiv, 159, 163, 174; area striata of, 159; 172, 175; differentiation of, 188; face area in Archaeolemur, 181; progression indices for, 166–67; volumetric comparisons of 168
Neocorticalization: in forerunners of Hominoidea, 168; indices of, 164; phylogenetic interpretations of, 165–68
Neomys fodiens, 167
Nerve cells: see Neurons
Neural arch, 224, 226, 233, 234, 235
Neural development in fossil hominids, xv
Neural reorganization, xiv, 188, 191, 199, 202, 203
Neural structures, 186; hyperplasia of, 196; hypertrophy of, 196; variability in, 186
Neuroanatomy: see Neurobiology
Neurobiologist, ix
Neurobiology, vii, xiv; comparative, 155; comparative studies on extant hominoids, xv, 186
Neurochemistry, 186
Neurocranium, 114, 120; enlargement of, 155
Neurological disorders, 187
Neurological genetics, 186
Neurons, 163; dendritic branching of, 187; density of, 187, 188; nuclear volume of, 187, 188; neural/glial ratios, 187, 188; size of, 187
Neurophysiology, ix, 184, 186
Neuronal differentiation, 155
Neurotransmitter substances, 187, 188
Neville, M. K., 392, 456
New World monkeys: see Ceboidea
Nesogale talazaci, 167
Nice, C. M., 338, 439
Niche, 175, 197, 399; of Dryopithecus nyanzae, 50; of earliest primates, xi; flexible branch, 211, 215; of gray langur, 359; of hominids in Southern and Eastern Africa, 91; insect, 116; protohominid, 375; terrestrial, 257; woodpecker, 104
Nilgiri langur: see Presbytis johnii
Nose: shape related to climate, 124; width, 124
Notharctus, 119
Notharctus tenebrosus, 34
Nucleus tractus olfactorii lateralis, 158
Nycticebus, 104, 112, 118, 179, 180, 181, 184, 266, 318
Nycticebus coucang 167, 266
Nyman, E., 446

Object-naming, 199

Occipital lobes, 177
Odontoceti, 162
Odontoid process, 227–28
Old World monkeys: *see* Cercopithecoidea
Olfactory apparatus (system), 171, 174; affects of orbital approximation on, 114; atrophy of, 99; constriction of, 114; displacement in *Microcebus*, 120; of mammals, 102; reduction of, xiv; regression of, 113–15, 121; reversions in, xiv
Olfactory lobes: atrophy of, 113; in *Loris*, 114
Olfactory sense in *Tetonius*, 178
Oligopithecus savagei, 47
Omomyidae, 119
Omomyids: *see* Omomyidae
Operant conditioning, 374, 415
Opossum: *see* Didelphis
Optic adnexa, 109
Optic axis: in higher primates, 109; shift of, 155
Orangutan: *see* Pongo pygmaeus
Orbit: dimensions of, viii; neurocranial, viii
Orbital axis, 109
Orbital convergence, 99, 101, 107, 108–13, 114, 116, 121; biometrical studies on, xii; definition of 109; in *Microcebus*, 120; in non-primate predators, 113; related to leaping, 112
Orbital frontation: definition of, 109; *Didelphis*, 110; pneumatization of frontal bone affects, 112; relation to frontal pole of brain, 112
Oreopithecus bambolii, 4, 62, 150, 237
Orlosky, H., 89
Osborn, H. F., 40, 42, 43, 456
Os Daubentonii: in gibbons, 209, 211, 212, 213, 216; in man, 212
Os planum, 7
Ossification of orbital wall (periorbital ossification), 99, 121
Otter: *see* Lutra
Owl, 113
Oxnard, C. E., xi, xvi, xvii, 124, 125, 148, 149, 150, 259, 264, 265, 266, 305, 310, 311, 312, 313, 316, 325, 327, 328, 330, 331, 332, 335, 418, 419, 420, 438, 456, 468

Palaechthon, 9, 10, 11, 12, 30, 99, 121
Palaechthonines: *see* Palaechthonini
Palaechthonini, 9, 28
Palaeoryctids, 6
Palaeopropithecus, 183
Palate: early hominid, 68, 70; *Platychoerops*, 19; *Zanycteris*, 12
Palenochtha, 9, 10, 12, 30
Palenochtha minor, 10
Paleocene primates, 3, 35
Paleocenus: 11
Paleocortex, 158, 168
Paleohistology, ix
Paleoneurobiologist, ix
Paleoneurology, 186
Paleopithecus, 57
Paleoprimatologist, viii, x
Paleoprimatology, vii, viii
Paleosinopa, 61
Pan: *see* Waterhole
Pan, 49, 175, 211, 212, 213, 215, 219, 220, 221, 226, 228, 233, 237
Pangolins, 102, 122
Pan gorilla: see Gorilla gorilla
Pan paniscus, 48, 51, 57, 60
Panthera pardus, 83, 364, 371, 372
Pan troglodytes, 48, 51, 57, 59, 60, 126, 162, 167, 168, 186, 193, 195, 198, 209, 212, 217, 218, 265, 267, 269, 270, 271, 274, 276, 279, 283, 284, 285, 286, 287, 289, 295, 309, 325, 331, 368, 413

Papin, L., 114, 452
Papio, 49, 177, 230, 258, 269, 276, 281, 284, 286, 317, 359, 371, 374, 380, 413, 414, 421, 424, 425, 432, 433, 434
Papio anubis: see Papio doguera
Papio doguera, 236, 276, 278, 282, 283, 285, 286, 287, 288, 290, 291, 398
Papio hamadryas, 233, 286, 290, 380, 398, 412
Papio papio, 233
Papio sphinx, 232, 233
Paradoxurus, 105
Parallax, 113
Paramys, 61
Paranthropus: see Australopithecus robustus
Parapithecus, 33, 47, 417, 418
Parapithecus fraasi, 43
Paromomyidae, xiii, 6, 8, 10, 29, 30, 121
Paromomyids: *see* Paromomyidae
Paromomyinae, 9
Paromomyini, 9
Paromomys, 9, 10, 12, 23, 30
Paromomys maturus, 10
Parthasarathy, M. D., 413, 458, 464
Partridge, T., 85
Patas monkey: *see* Erythrocebus patas
Patterson, B., 42, 81, 150, 456
Paulian, R., 104, 456
Pauwels, F., 240, 456
Pearson, K., 126, 456
Pelvis, 231, 239; anatomy of, 241; dimensions of, 313; inlet of, 231–32; studies of, 311, 319, 320, 326; support for viscera of, 250, 258
Pelycodus, 17, 61
Penrose, 306
People: *see* Homo
Peratherium, 61
Periventricular organs, 172
Perkins, E., 114, 438
Perodicticus, 104, 179, 229, 230, 233, 236, 266, 318
Perodicticus potto, 167, 230
Personality, 400, 408
Petrosal bulla, 5; diagnostic feature of primates, 7, 98, 121; of *Plesiadapis*, 16
Petter, J. J., 108, 456
Pfuhl, W., 267, 456
Phalangeridae, 25, 103, 111
Phalangerids: *see* Phalangeridae
Phalanger maculatus, 112
Phalanges of foot: distal hallucal phalanx from Olduvai, 150; early hominid, 78
Phalanges of hand in early hominids, 77
Phaner, 104, 118
Phascolarctidae, 103
Phascolarctids: *see* Phascolarctidae
Phascolarctos, 103, 104, 112
Pheasant, H. C., 230, 459
Phenacodus, 61
Phenacolemur, 6, 7, 8, 9, 10, 11, 16, 23, 29, 31, 33, 61, 99
Phenacolemur jepseni, 11
Phenacolemuridae, 9
Phillips, W. W. A., 104, 457
Phylogeny, xi
Physical techniques, xvi
Picidae, 104
Picrodontidae, xiii, 8, 11, 27, 29, 30, 31, 32, 121
Picrodontids: *see* Picrodontidae
Picrodus, 12, 30
Picrodus silberlingi, 11
Pielou, E. C., 435, 457
Pig, 41; *see also Sus scrofa*
Pigeon, 432

Pigtail monkey: see Macaca nemestrina
Pilbeam, D. R., xi, 48, 49, 50, 51, 52, 53, 56, 60, 195, 200, 222, 457, 462
Pilgrim, G. E., 57, 457
Pincus, H. J., 339, 457
Pineau, H., 240, 445
Pinto-Hamuy, T., 180, 457
Pirlot, P., 156, 173, 457, 463
Pisiform, 207, 209, 210, 211, 214, 216, 219, 220, 221; migration of, 220
Pithecia, 179
Pithecia monacha, 167
Piveteau, J., 177, 178, 457
Plantarflexion, 115
Platychoerops, 16, 19, 31
Platychoerops richardsoni, 19
Platyrrhines: see Ceboidea
Play, 203, 427, 428, 430–32; in geladas, 408; groups, 422, 428; in humans, 422, 424; in langurs, 359, 367–68, 370, 375; partnership, 415; rhesus, 384, 421
Plesiadapidae, xiii, 8, 16, 17, 29, 30, 121
Plesiadapids: see Plesiadapidae
Plesiadapinae, 16
Plesiadapis, 7, 16, 17, 18, 19, 23, 25, 31, 34, 35, 37, 61, 99
Plesiadapis gidleyi, 31
Plesiadapis tricuspidens, 17
Plesiadapis walbackensis, 17, 18
Plesiadapoidae, 8
Plesiadapoids: see Plesiadapoidea
Plesiadapoidea, 29
Plesiolestes, 9, 10, 30
Pliopithecines: see Pliopithecinae
Pliopithecinae, 55
Pliopithecus, xv, 44, 54, 55, 233, 417, 418
Pliopithecus vindobonensis, xv, 236
Ploydolopids, 25
Plutchik, R., 378, 457
Pocock, R. I., 108, 114, 250, 457
Poglayen-Neuwall, I., 114, 457
Poirier, F. E., 357, 358, 359, 361, 363, 366, 367, 368, 371, 374, 412, 457
Pollex: in Colobinae, 275, 280, 289; divergence of, 115; grasping, 207; metacarpal of, 294; nail-bearing, 121; opposable, 7, 115; opposition of, 293; in Pongo, 280, 289; prehension in Hylobatidae, 289; Ratufa, 105; reposition of, 293
Polyak, S., 109, 113, 457
Polyandry, 413
Polygenic characters, 419
Polyprotodont: see Marsupials
Pongidae, xiv, 53, 54, 56, 162, 174, 187, 198, 207, 210, 211, 212, 213, 221, 226, 228, 229, 230, 232, 236, 269, 276, 277, 280, 281, 285, 291
Pongids: see Pongidae
Pongo pygmaeus, 54, 210, 212, 217, 219, 220, 228, 229, 233, 237, 265, 267, 270, 271, 274, 276, 279, 280, 283, 284, 286, 287, 289, 295, 325
Population dynamics, x, 383
Postcranial skeleton, xiii, 123; of Aegyptopithecus, 43, 44; differentiation between pongids and hominids, 56; of Dryopithecus africanus, 48; of Dryopithecus fontani, 51, 52; of Dryopithecus major, 49; early hominid, 68, 69, 70, 72, 73, 75, 77, 84, 92, 93; early Tertiary mammals, 34; Eocene prosimian, 118; of Fayum primates, 43; hominid remains from Maboko Island, 50; identification of fossil primate, 4, 5; of Limnopithecus, 55; occurrence, 62; of Plesiadapis, 18, 34; of Pliopithe-

cus, 55; Sterkfontein, 81; studies of fossil hominid, 81
Postorbital bar, 116, 121; lack in Palaechthon, 121
Postorbital ligament, 116; ossification of, 116
Posture: digitigrade, 260; evolution of upright, 330; human, xvi, 292, 300–303; palmigrade, 211, 260; pronated quadrupedal hand, 216; relation of callosity to, 258; relation of tail to sitting, 256; sitting and sleeping, 242; study of, 294
Potamogale velox, 167
Potos, 104, 105, 107, 108
Powell, T. P. S., 187, 402, 457
Power spectrum, 321, 322, 339, 341, 343, 344, 345
Pratt, J. W., 458
Pratt, W. K., 338, 437
Predation: see Predatory habits
Predators: forest-floor, 116; langur defense against, 370–74; on Madagascar, 117–18; mutual protection against, 371, 373, 374; pressure from, 357; protection against, 201, 375; visually-directed, 113
Predatory habits: lack in plesiadapoids, 29; manual, 117
Prehensile functions: correlated with relative muscle mass, 269–91; power grisp, 294, 399; prehension grip, 294
Prehistory, viii
Premolars: of Aegyptopithecus zeuxis, 46; of Carpolestidae, 21, 23, 24, 25–26; convergent adaptations to common vegetable diet, 25–26; of Elipthidotarsius florencae, 23; of Navajovius, 11; human, xvi, 292, 300–303; of Paromomyidae, 8; of Picrodontidae, 14; of Plagiaulacoid, 26, 29; of Propliopithecus haeckeli, 46; of Saxonella, 19; of Tarsius, 28
Presbytis, xvii, 284, 357, 358, 359, 361, 362, 363, 365, 366, 367, 368, 369, 371, 373, 374, 375, 412, 414
Presbytis cristatus, 275, 276, 278, 280, 284, 285, 286, 287, 288, 290, 412
Presbytis entellus, 236, 260, 269, 275, 276, 278, 279, 280, 286, 287, 288, 290, 353, 356, 357, 359, 361, 366, 368, 369, 371, 375
Presbytis entellus entellus, 354
Presbytis entellus thersistes, 352, 354, 357
Presbytis johnii, 357, 358, 359, 363, 366, 367, 368, 371, 374
Preston, F. W., 339, 458
Preuschoft, H., 265, 458
Primates, 3, 155, 157, 159, 165, 170, 171, 187, 209, 378, 414, 415, 416, 421, 423, 434; diagnosis of Order, 98; basal adaptations of, 121; earliest, xi; evolutionary trends in, 99; initial attribute of, 155; Neogene, xi; ordinal differentiation of, 121–22; origins from palaeoryctoid–erinaceotan stock, 6; Paleogene, xi; predecessors of Eocene, xiii
Primate research centers, xx; see also Yerkes Regional Primates Research Center
Principal components, 310, 315; analysis, 127
Pristophoca aff. occitana, 41
Proboscidean, 42, 176
Proconsul, see also Dryopithecus [Proconsul] 51, 52, 53
Procyon, 103
Procyonidae, 103, 112, 114, 180
Procyonids: see Procyonidae
Progressive insectivores, 156, 159, 170, 172
Prohylobates, 42
Pronation, 210, 299
Pronothodectes, 16, 17, 18, 21, 30
Pronycticebus, 119

Propithecus, 118, 181, 184, 235, 266, 318
Propithecus verrauxi, 104, 118, 167
Propliopithecus, 46, 417, 418
Propliopithecus haeckeli, 43, 44, 45, 46, 47
Propliopithecus markgrafi, 47
Prosimians: *see* Prosimii
Prosimii, 99, 105, 111, 116, 118, 120, 158, 159, 161, 162, 163, 164, 165, 168, 169, 170, 171, 172, 173, 180, 184, 226, 228, 230, 313, 318, 319
Prosimii-Anthropoidea subordinal arrangement, 8
Prost, J. H., 239, 458
Provisioning, 351, 402, 403, 404
Pseudocheirus, 104
Pseudocheirus lemuroides, 112
Pseudoloris, 119
Psychologist, 381, 434
Pterodon, 40
Ptilocercus, 104, 120
Ptolemaia, 40
Pulvinar, 198
Purgatoriinae, 9
Purgatorius, 9, 10

Quadrupedalism: *see* Locomotion
Quantitative comparisons, viii; of cheiridial muscles, xvi
Quantum evolution, 98

Race, 191
Rachigrams, 239–40
Racial difference, 124
Racoon: *see* Procyon
Radcliffe, C. W., 303, 458
Radiation: adaptive, 98; evolutionary, 97; of Madagascan primates, 28, 101, 120; mammalian, 98; manual predation as pre-adaptive foundation for later primate, 117; Paleocene primates, 3, 8–26, 31
Radinsky, L., xiii, xiv, 43, 114, 155, 176, 177, 178, 181, 183, 184, 458
Radius: early hominid, 78
Rage, 189
Rahaman, H., 413, 458
Raiffa, H., 458
Ramapithecus, 50, 52, 53, 54, 55, 56, 57, 59, 197, 200
Ramapithecus punjabicus, 56, 59, 60, 222
Ramapithecus wickeri, 59, 60
Randall, F. E., 194, 458
Rankin, J. J., 447
Rao, C. R., 310, 332, 455
Rat, 186, 187, 432
Ratufa, 104, 105, 108
Ratufa bicolor, 112
Ray, 42
Recession of snout: *see* Rostral reduction
Reinforcement, 433
Reithroparamys, 61
Remane, A., 226, 228, 458
Rensch, B., 177, 200, 458
Retinal summation, 111
Reynolds, V., 368, 413, 458
Rhesus Monkey: *see* Macaca mulatta
Rhinosciurus, 103
Rhynchocyon stuhlmanni, 167
Ribs, 226, 235; broadened in *Arctocebus,* 229; capitulum of, 230; early hominid, 78; hominoid, 229; in springing and brachiating primates, 239; tuberculum costae, 230
Ride, W. D., 122, 458
Ripley, S., 239, 266, 354, 357, 358, 359, 360, 361, 362, 367, 368, 369, 371, 458
Roberts, F. S., 434, 458
Robinson, G. S., 339, 347, 442

Robinson, J. T., 79, 80, 81, 87, 88, 90, 237, 295, 458
Robinson, P., 10, 459
Robust australopithecine: *see Australopithecus robustus*
Rockwell, H. F., 230, 459
Rodent: *see* Rodentia
Rodentia, 16, 31, 33, 97, 103, 176; ancestry of, 16
Roderick, T. H., 467
Rohen, J. W., 109, 459
Roles: bachelor male in geladas, 408; components of society, 400, 401; differentiation of, 400; key, 408; male intragroup, xvii; paternal, 400; patterns, 409
Roll, A. A., 446
Rose, M. D., 466
Rosenblum, L. A., 387, 406, 412, 451, 459
Rosenfeld, A., 338, 459
Ross, G. J. S., 320, 447
Rostral reduction, 99, 117, 118, 121, 155; biometrical studies on, xii
Roux, W., 337, 459
Rowell, T. E., 256, 387, 402, 413, 414, 449, 459
Rusa unicolor, 371
Russell, D. E., 10, 17, 18, 19, 37, 99, 459
Russell, L., 22

Saban, R., 120, 459
Sabre tooth tiger, 41
Sacher, G., 177, 459
Sacral index, 234–35; of *Pliopithecus vindobonensis,* 236
Sacral vertebrae, *see* Sacrum
Sacrum, 224, 226, 232,; of *Australopithecus africanus,* 238; distal (caudal) opening of, 234; early hominid, 78; foramina in tersacralia, 226; human, 226, 231; neural canal of, 234, 235, 238; number of vertebrae correlated with caudal number, 232; promontory of, 231; proximal (cranial) opening of, 234
Sade, D. S., xvii, 379, 380, 383, 384, 385, 387, 388, 390, 397, 398, 434, 443, 459
Sagittal keeling, 124
Saguinus, 105
Saguinus tamarin, 167
Saimiri, 179, 182, 183, 233, 234, 236, 382, 397, 398, 413
Saimiri sciureus, 167
Salmons, S., 323, 460
Sanderson, 257
Sanides, F., 180, 181, 460
Sarbin, R. T., 400, 460
Sarich, V. M., 460
Savage, D. E., 459
Sawfish, 42
Saxonella, 17, 19, 20, 25, 29, 31
Saxonellinae, 16
Scaphoid tubercle, 212
Scapula: of *Daubentonia,* 149, 327; early hominid, 77; human, 298; pliopithecine, 55; shape of, 311; Sterkfontein, 330; studies on, 320
Scent glands, 113–14
Scent-marking behavior, 113–14
Schade, J. P., 187, 460
Schaeffer, B., 215, 460
Schaller, G., 371, 377, 413, 446, 460
Schepers, G. W. H., 80, 197, 460
Schizocortex, 158, 170; progression in, 168
Schlaifer, R., 430, 458
Schlosser, M., 40, 42, 43, 44, 46, 460
Schmidt, E., 232, 356, 460
Schmidt-Nielson, K., 460

Schoener, T. W., 427, 460
Schreiber, H., 265, 460
Schultz, A. H., vii, 88, 89, 90, 118, 226, 229, 231, 232, 237, 239, 244, 246, 262, 332, 460
Schultze-Westrum, T., 114, 461
Sciuridae, 104, 107, 108, 112
Sciurids: see Sciuridae
Sciurillus, 112
Sciurus, 105, 113
Sciurus carolinensis, 112
Sea cow, 41
Seal, 41
Seal, H., 310, 461
Searle, N. H., 347, 455
Seidenstein, S., 180, 467
Sellers, E. E., 339, 465
Selzer, R. H., 338, 344, 456, 461
Semibrachiators, 266
Septum, 158, 170; nuclei of, 158–59; phylogentic development of, 172; progression in, 168
Setifer setosus, 167
Sexual dimorphism, xx, 421; in Aegyptopithecus and Propliopithecus, 46; australopithecine, xii, 87–89; in brain, 187, 191, 194; in cranial capacities, 89; of Dryopithecus major, 49; of early hominids in Africa, 78; in hominoids, xii, 89, 197; in hominoids at Haritalyangar, 54; hylobatid, 55; in langurs, 370; pliopithecines, 55; reduction of, 201; in tail mobility, 248; of teeth, 80
Sexual receptivity of females, 197, 200
Sgaramella-Zonta, L. A., 446
Shark, 42
Sharpe, L. G., 393, 400, 440
Shelman, C. B., 339, 347, 461
Sherrington, 299
Shine, G., 296, 439
Shoulder girdle: canonical analysis of, 316; of Daubentonia, 327–28; human, 312; neighborhood limited classification of, 316; in nonprimate mammals, vxi, 328; in primates, xvi; studies on, 311, 325
Shrew, 27, 156
Siamang: see Symphalangus syndactylus
Siddiqi, M., 354, 394, 463
Sigmon, B. A., 295, 461
Silver, P. H. S., 302, 446
Simians, 158, 159, 161, 162, 163, 165, 168, 170, 171, 172, 173, 174
Simmons, N., 352, 461
Simonds, P. E., 413, 461
Simons, E. L., xi, 17, 18, 34, 39, 40, 41, 42, 43, 45, 46, 47, 50, 51, 52, 53, 55, 56, 57, 60, 61, 99, 102, 115, 119, 121, 200, 215, 222, 457, 461, 462
Simpson, G. G., 8, 9, 11, 16, 18, 22, 23, 25, 33, 34, 98, 119, 462
Singh, S. D., 354, 462
Sivachoerus, 41
Sivapithecus, 53
Skinner box, 432
Skulls, viii; of Aegyptopithecus, 41, 43; Broken Hill, 150; of Dryopithecus africanus, 48; early hominid, 68, 69, 70, 73, 75, 84, 92, 93; Fish Hoek, 311; historical (recent) human, xiii, 123, 129–51; Keilor, 311; Lake Rudolf, 191, 195; MLD 1 occipital fragment, 193; metrical study of, 125; modern human, 124, 126; Mount Carmel, 419; Neanderthal, xiii, 125, 147, 148, 150; Ngandong, 150; Olduvai, 81, 191, 195; Omo, 191, 195; paromomyid, 11; picrodontid, 15; of Plesiadapis tricuspidens, 17–18; prehistorical human, xiii; primate, 98; Saldanha, 66; STS 71, 193; Trinil, 125

Sleeping, 427, 428; groups of vervets, 423–24; in langurs, 370; sites, xvii
Slijper, E. J., 240, 462
Sloan, 9, 10
Slobodkin, L. B., 383, 462
Sloth, 103, 180, 312
Smilodectes, 33
Smith, C., 444
Smith, E., xiii, 100, 101, 102, 116
Smith, G. E., 462
Sneath, P. H. A., 316, 320, 332, 462, 463
Snell, O., 156, 462
Snyder, F. N., 447
Social behavior, x, 188, 200; adaptation, 200; human, 421; maternal care, 411, 412; methodology for study of primate, 378; models for, 426–32; paternal care, 411; quantitative recording of, 426; socializing, 424
Social group, 200, 415; activeness of langur, 358; age/sex composition of, 357, 377, 381, 407, 413; all-male, 357; basic social unit of, 412; captive, 405; casual, 422, 423, 425, 426; cohesiveness of langur: 358, 364; 366; dispersal of males from rhesus, 396, 397; fractionization of langur, 364–67, 368; human, 414, 422, 423; integration of newcomers, 401, 414; intertaxonal comparisons in primates, xviii, 399; nuclear family, 412; one-male heterosexual, 357; population genetics of, 434; size, 421, 423, 427, 430; size in howling monkeys, 380; size in langurs, 357, 365; stability, 381, 391–92; temporary associations, 412; unit of, 413; withdrawal of male from, 397
Socialization, 407, 414
Social organization, xvii, 318; baboon, 380; binary socialization, 427; coalition, 400; cross-sectional approaches to, 405, 407; differences in Cercopithecoidea, 398; during progression, 370; effects of provisioning on, 402; evolution of, 398; human, 399, 413; inferences about, 379; longitudinal studies of, 383, 395, 405, 406, 407, 408, 409, 410, 412; patterns of, 405; patterns of stability and change, 384; role of male in rhesus, 393–94; short-term studies of, 395; specification of, 410; social units, 401; subgrouping (cliques), 386, 387; types of, 400
Sociobiology, 416; rhesus, xviii
Sody, H. J. V., 246, 462
Sokal, R. R., 316, 320, 463
Solenodon paradoxus, 167
Solitary males, 357 rhesus, 390–91
Sonoran Desert, 352, 356
Sorex araneus, 167
Sorex minutus, 167
Soricidae, 103, 156
Soricids: see Soricidae
Sound recording, x
Southwick, C. H., 354, 380, 394, 413, 443, 463
Space: contiguous, 422; Euclidean, 316, 334; neighborhood, 316
Spatz, W. B., 156, 463
Speech, ix
Spence, T. F., 194, 311, 312, 313, 327, 330, 438, 468
Spencer-Booth, V., 387, 449
Sperber, G., 79, 89
Spermophilus, 104
Spider monkey: see Ateles
Spinage, C., 377, 463
Splanchnocranium, 114
Squirrel, 103, 104, 105, 107, 108, 109, 312
Squirrel monkey: see Saimiri
Starck, D., 114, 463
Statistics: bivariate, 306, 386; clustering anal-

yses, 316; coefficient of variation, 244; correlation, 306; fiducial limits, 265, 266, 267; Gaussian distributions, 315; grouping methods, 315–20; intertaxonal comparisons, 265, 266, 268; linear analysis, 420; mean, 126, 150, 244, 265, 266, 267; multivarate, xi, xiii, xvi, 127, 148, 306, 310–20, 419; multivariate transformations, 311; Poisson, 411, 422; ratios, 268–69, 269–88; skewed samples, 187, 191; standard deviation, 126, 244; standard error of mean, 265; studies on body proportions, 262; studies on bones, 262; studies on muscles, 262, 264; tests of significance, 142, 265; truncated Poisson, 423, 424, 425, 427, 430, 431; truncated negative binomial, 423; 425, 427, 430, 431; univariate, 126, 148, 266, 306, 313; use of indices, 306; variance, 265, 380

Stecko, G., 296, 299, 439

Steegman, A. T., 124, 463

Steininger, F., 52, 463

Stephan, H., xiii, xiv, 156, 157, 158, 161, 162, 163, 171, 172, 173, 177, 188, 438, 439, 457, 463

Stephens, E. A., 443

Stereoscopic vision: adaptation for predation, 113; primate evolutionary trend, 100; related to short face, 32

Sternum, 229

Stewart, T. D., 236, 463

Stimson, G. W., 447

Story, H. E., 114, 463

Straus, W. L., 215, 237, 249, 450, 463

Striatum, 159; progession indices of, 168

Stromer, E., 39, 41, 464

Struhsaker, T. T., 357, 363, 402, 404, 412, 413, 423, 424, 425, 426, 464

Stynes, A. J., 406, 459

Subcommissural organ (or body), 158, 172

Subfornical body, 158, 172, 173

Subramomiam, S., 108, 464

Substrates for locomotion, 269; branches compared to ground, 107; leafy, 290

Sugiyama, Y., 354, 357, 358, 361, 363, 412, 464

Sulcus, 176, 182; central, 81; cortical localization, 180, 184; intraparietal, 179, 181, 182, 183, 184; longitudinal orientation in lemurs, 180; lunate, 197, 198; sylvian, 179, 181, 182, 183, 184; transverse orientation in anthropoids, 180

Suncus murinus, 167

Sundasciurus, 104

Supination, 210, 211

Suspensory behaviors, 274, 289, 325, 331; in African dryopithecines (Proconsul), 54; arm-swinging, xv, xvi, 210, 215, 271, 291; bimanual walking beneath branches, 118; brachiation, xv, 56, 182, 210, 211, 216, 222, 237; in *Dryopithecus africanus,* 48, 220, 221; in *Dryopithecus fontani,* 52; hang and feed, 118, 289; hindlimb, 101, 107; for locomotion, 207, 211, 215, 216; in pliopithecines, 55; ricochetal arm-swinging, 289

Sus scrofa, 363

Sutherland, D. H., 301, 464

Sutton, 296

Suzuki, A., 413, 450

Symbolic behaviors: see Culture

Symington, J., 177, 464

Symphalangus syndactylus, 89, 233, 250, 274, 276, 277, 279, 280, 287, 289, 368

Syntheses, xix

Szalay, F., xi, 5, 6, 8, 9, 10, 11, 15, 28, 30, 33, 99, 117, 121, 464

Tactile perceptions, x

Tail: distal segments of, 224, 232, 234; drinking posture in langurs, 359; function, 235, 242, 256; length in *Macaca,* 243, 246–47, 252; mobility at base (root), 233, 234, 247, 252, 253; mobility in *Macaca,* 242, 243, 247–248, 252, 255, 256; mobility of proximal region, 233, 248, 252; mobility of tip, 234, 248, 252; number of vertebrae in *Macaca,* 243, 244–47, 252; of *Pliopithecus vindobonensis,* xv, 55, 236; prehensile, 107, 108, 232, 234, 235; projection areas on brain, 180, 181–83; proximal, 232, 233, 234; reduction of, 232, 234; reduction in apes, 236; reduction in *Macaca,* xvi, 241–61, 252, 255, 260; reduction in man, 236; reduction in *Oreopithecus,* 237; size in primates, xv; specialization in *Saimiri,* 234, 236; vertebrae, 226; volume in *Macaca,* 243, 247, 252

Talapoin monkey: see *Cercopithecus (Miopithecus) talapoin*

Talpa, 122

Talpa europaea, 167

Talus: double-pulley, 97; early hominid, 78; Songhor, 49

Tamias, 104, 112

Tanaka, J., 357, 359, 374, 464

Tank: see Waterhole

Tappen, N. C., 265, 267, 464

Tarsal Bones: early hominid, 78; elongation among quadrupeds, 118; elongation in saltatory mammals, 119; see also Calcaneum, Navicular, and Talus

Tarsier: see *Tarsius*

Tarsiidae, 7, 114, 226

Tarsiids: see Tarsiidae

Tarsius, 28, 34, 104, 107, 111, 113, 114, 115, 117, 118, 119, 165, 172, 207, 228, 229, 233, 235, 265, 266

Tarsius syrichta, 167

Tattersall, I. M., 56, 462, 465

Taxonomy, viii; alpha, 4; placement of taxonomic boundaries, 98; supraspecific, 4

Techniques and approaches, xix; cinematography, 294, 302, 310; computer, 126, 264, 297, 306, 307, 321, 333, 338, 386; cross-sectional, 405; dating fossil sites in South Africa, 85–86; dioptographic tracings of endocasts, 194; dissection, 243; electron microscopy, 187; EMG, see Electromyography; field, 404; films, x, 240; Fourier analysis, 307, 321, 335, 340, 346; frozen, 323; laser, 333, 338; mathematical, xii, xvi, 308, 434; optical data analysis, 307, 320–22, 335, 337–47; photography, 306; radiography, 243, 306, 337, 338; sampling, 380, 381; sociometric, 411; sound spectrographs, 347, 369; split-line, 240; standardization of, 90; telemetering, 294, 310, 323, test retest, 410; time sampling, 410; videotaping, 294; weights of muscles, 267; X-rays: see radiography; X-ray cinematography, 62, 240

Teeth, viii; deciduous of early hominids, 70, 71, 73, 75, 76, 77, 79–80, 92, 93; early hominid, 68, 69, 84; erinaceoid, 6; eruption, xx; mammal, 4; permanent of early hominids, 70, 71, 73, 75, 77, 79–80, 92, 93; taxonomic employment of, 4, 91; wear on, xx

Telanthropus capensis, 90

Temporal bone from Chemeron site, 87

Tenrec, 156

Tenrec ecaudatus, 167

Tenrecidae, 156

Tenrecoidea, 7

Tenrecoids: see Tenrecoidea

Territorality, 405, 414; in man, 413–14

Testes regression, 397
Testosterone production, 396
Tetonius, 33, 119
Tetonius homunculus, 178
Thenius, E., 155, 161, 162, 465
Theropithecus gelada, 60, 269, 271, 276, 278, 279, 281, 282, 284, 285, 286, 287, 288, 290, 291, 357, 408, 412
Thomas, D. P., 54, 443
Thomas, F. D., 339, 465
Thompson, R. F., 418, 444
Thoracic vertebrae, 225, 226; bodies in Hominoidea, 229; early hominid, 78; elongated dorsal spines in *Perodicticus,* 229, 236; elongated dorsal spines in Pongidae, 229, 236; foramina in lorises, 229, 236; function of, 235
Thorax, 226, 229, 235; hominoid, 229, 230, 236, 237; in springing and brachiating primates, 239
Thorington, R. W., 413, 465
Thuma, B. D., 113, 465
Thumb: *see* Pollex
Thylacosmiles, 122
Tibia: early hominid, 77; of Mycrosyopidae, 33
Tigges, J., 235, 449
Tilney, F., 186, 465
Titi: *see Callicebus*
Tobias, P. V., xi, 65, 67, 68, 70, 81, 82, 83, 87, 88, 89, 90 162, 187, 192, 195, 453, 465
Tobien, H., 465
Tools, 196, 197, 199, 200, 202
Tool-making, 202; capacities, ix; by *Ramapithecus,* 60; by *Pan* troglodytes
Tool-use: by Galapagos finch, 104; by huia, 104; by langur, 375; by *Pan troglodytes,* 60; preadaptation of hands for, 216; by *Ramapithecus,* 60
Toque macaque: *see Macaca sinica*
Torrejonia, Wilsoni, 9
Toyoshima, A., 413, 438
Trabeculae, 320, 321, 322, 337–39, 344–346
Transition periods of maturation, 395
Trapezium, 293
Travill, A., 299, 439, 465
Tree shrew: *see* Tupaiidae
Trevor, J. C., 310, 332, 455
Triangular articular disc of wrist, 207, 210, 213; in *Gorilla,* 210; in *Homo sapiens,* 213; in *Hylobates lar,* 209; in *Pan troglodytes,* 209; in *Pongo pygmaeus,* 212
Trichosurus, 104
Triquetral, 207, 209, 210, 214, 216, 219, 220, 221
Trogolemur, 9
Troop: *see* Social group
Trotter, M., 232, 465
Truncal uprightness, 226, 231, 237; in early hominids, 78; in rodents, 34; in vertical clingers and leapers, 33
Tugby, D. J., 176, 466
Tupaia, 113, 175
Tupaia glis, 112, 167
Tupaia minor, 112
Tupaia tana, 103
Tupaiidae, 6, 7, 8, 103, 105, 107, 109, 114, 120, 163, 164, 171, 173
Tupaiids *see* Tupaiidae
Turbinal apparatus, 114; atrophy of, 113; *Ratufa,* 105
Tuttle, R. H., 211, 212, 237, 243, 259, 260, 264, 265, 266, 267, 269, 289, 295, 310, 325, 352, 420, 423, 442, 456, 466
Tympanic membrane, 5

Ullrich, W., 368, 466
Ulna, 207; early hominid, 77; exclusion from wrist, 209; from Omo, 84; *Oreopithecus,* 150; retreat from wrist, 210, 212; styloid process of, 209, 210, 211, 212, 213, 214, 216, 219, 220, 221
Ungulates 5, 176, 371, 374
Uniformitarian arguments, ix
United Nations, 435, 466
Urogale, 103, 112
Urogale everetti, 167
Uyeno, E. T., 434, 466

Vallois, H. V., 89, 466
Vandenbergh, J. G., 434, 465
Van Valen, L., 6, 8, 9, 10, 16, 19, 40, 98, 99, 121, 466
Variation: in child rearing 414; increase with tail reduction in *Macaca,* 252; intrapopulation, xii, xvi, 78, 130, 140, 144, 146, 315, 327; interpopulation, xii, xvi, 130, 140, 142, 144, 146; in maternal care patterns, 412, 414; patterns of cranial, 123; of population revealed by multivariate statistics, 126; racial, 124; in social organization, 399; in shape, 124
Vector, 126, 334, resultant of forces, 308–9
Vertebral morphology, xv, 223–40; general, 223–26; of *Paranthropus,* 237; specializations of, 227–36, 238–39; studies of, 239–40
Vertical clinging and leaping: in *Aeolopithecus,* 33, in *Amphipithecus,* 33; an artificial category, 119; in Eocene primates, xi, 33; galagine, 34; lemuroid, 34; in *Microchoerus,* 33; morphological correlates, 32, 35; Napier-Walker hypothesis, xi, xii, 32–35, 117–19; in *Necrolemur,* 33; in *Parapithecus,* 33; in *Tarsius,* 34; in *Tetonius,* 33
Vervet monkey: *see Cercopithecus aethiops*
Victoriapithecus, 42
Vision in *Tetonius,* 178
Visual acuity, 109
Visual axis: in higher primates, 109; of nocturnal animals, 109
Visual cortex: diminution of, 198; expansion in *Aegyptopithecus,* 181
Visual field overlap, 111, 112; adaptation for predation: 113, 116; related to leaping: 112–13
Viverridae, 103, 105, 107, 112, 114
Viverrids: *see* Viverridae
Vocalization: alarm, 371; growling in langurs, 369; langur alarm, 371; whooping in langurs, 362, 365, 368–69
Vogel, C., 354, 357, 466
Vogel, J., 86
Vogt, O., 180, 466
Vogt, D., 180, 466
Vondra, 40

Wagner, S. S., 410, 426, 429, 437, 466
Wagstaffe, W. W., 338, 466
Walker, A. C., xi, xii, 32, 33, 49, 80, 117, 118, 119, 455, 466
Walls, G. L., 111, 467
Wallace, J. W., 79
Walsh, R. J., 339, 347, 446
Wankie National Park, Rhodesia, 352
Warren, K. S., 435, 447
Washburn, S. L., 60, 124, 235, 250, 260, 265, 295, 352, 371, 380, 402, 413, 414, 445, 467
Waterbuck: *see Kobus defassa*
Waterholes, xvii, 375; aprons of, 352; in Cameroon, 352; in Ceylon, 352; intergroup encounters at, 362–63; langur presence at 358; Nelun Wila, 354, 363; in Rhodesia, 352;

study strategy, xvii, 351–53, 377; structure of, 353–54, 358
Watt, K. E. F., 435, 467
Weidenreich, F., 124, 125, 467
Weidenreich-Coon hypothesis, 124
Weiner, J. S., 467
Weir, J. S., 353, 467
Welker, W. I., 180, 182, 439, 452, 467
Wells, L. H., 86, 467
Welt, C., 182 186, 467
Wernicke, 199
Whale, 28
Whillis, J., 216, 467
White, H. C., 422, 423, 434, 453
Whitla, D. K., 435, 467
Whitman, C. O., 420
Wickler, W., 250, 467
Wiesel, T. N., 113, 450
Wildlife management, 377, 381
Willis, J. C., 370, 467
Willis, M., 370, 467
Wilpattu National Park, 354
Wilson, D., xvi
Wilson, E. G., 147, 467
Wilson, R. W., 11
Wimer, C. E., 467
Wimer, R. E., 187, 467
Wishart, D., 316, 467
Wohlfart, G., 446
Wolf, 41
Wolff, J., 337, 468
Wolffson, D. M., 124, 468
Wolpoff, M. R., 124, 195, 468
Woo, J. K., 468
Wood, A. E., 16, 40, 42, 462, 468
Wood, B. A., 445
Woodpecker: *see* Picidae
Woodpecker avatars, 104, 108

Woolsey, C. M., 180, 457, 468
Woolly monkey: *see Lagothrix*
Worcester, J., 147, 467
Wortman, J. L., 33, 468
Wray, S. H., 268, 468
Wright, I. H., 85
Wright, 151
Wrist: bones: *see* Carpal; of chimpanzee, 212; of *Dryopithecus africanus,* xv, 48, 220, 222; evolution of hominid, 215–16; of gibbon, 209, 211; hominoid, 207, 222; human, 212–15, 299; limitation of extension in, 211; organization in *Homo sapiens,* xv; organization in Pongidae, xv; of *Pliopithecus,* xv; pongid, xv
Wrobel, K. H., 235, 468

Xerus, 104
Xerus inauris, 112

Yamada, M., 387, 468
Yasuda, I., 337, 446
Yau, S. S., 339, 468
Yerkes Regional Primate Research Center, xviii, 295, 406, 407
Yoshiba, K., 354, 357, 363, 402, 414, 464, 468

Zalambdodont insectivorans, 103
Zanycteris, 11, 12
Zanycteris paleocenus, 15
Zapfe, H., 55, 468
Ziegler, A. C., 265, 468
Zihlman, A., 81
Zinjanthropus: see Australopithecus boisei
Zuckerman, S., 90, 111, 180, 313, 328, 330, 443, 468
Zygapophyses, 224; postzygapophyses, 224, 226, 230; prezygapophyses, 224, 225, 230; of tail, 232